D0966468

Lecture Notes in Statistics

Edited by P. Bickel, P. Diggle, S. Fienberg, K. Krickeberg, I. Olkin, N. Wermuth, S. Zeger

114

Springer
New York
Berlin
Heidelberg
Barcelona
Budapest
Hong Kong
London
Milan
Paris
Santa Clara
Singapore
Tokyo

C.C. Heyde
Yu V. Prohorov
R. Pyke
S.T. Rachev
(Editors)

Athens Conference on Applied Probability and Time Series Analysis

Volume I: Applied Probability
In Honor of J.M. Gani

 Springer

236 25

C.C. Heyde
Stochastic Analysis Program, SMS
Australian National University
Canberra ACT 0200
Australia

Yu V. Prohorov
Academy of Sciences
Mathematical Institute
Vavilov Street 42
Moscow 333
Russia

Ronald Pyke
University of Washington
Department of Mathematics
Seattle, WA 98195

S.T. Rachev
University of California, Santa Barbara
Department of Statistics and Applied Probability
Santa Barbara, CA 93106

CIP data available.
Printed on acid-free paper.

© 1996 Springer-Verlag New York, Inc.
All rights reserved. This work may not be translated or copied in whole or in part without the written
permission of the publisher (Springer-Verlag New York, Inc., 175 Fifth Avenue, New York, NY 10010,
USA), except for brief excerpts in connection with reviews or scholarly analysis. Use in connection with
any form of information storage and retrieval, electronic adaptation, computer software, or by similar or
dissimilar methodology now known or hereafter developed is forbidden.
The use of general descriptive names, trade names, trademarks, etc., in this publication, even if the former
are not especially identified, is not to be taken as a sign that such names, as understood by the Trade
Marks and Merchandise Marks Act, may accordingly be used freely by anyone.

Camera ready copy provided by the author.
Printed and bound by Braun-Brumfield, Ann Arbor, MI.
Printed in the United States of America.

9 8 7 6 5 4 3 2 1

ISBN 0-387-94788-4 Springer-Verlag New York Berlin Heidelberg   SPIN 10538550

Editorial

The Athens Conference on Applied Probability and Time Series Analysis was held at the Titania Hotel in Athens over the period March 22-26, 1995. It was jointly sponsored by the University of California, Santa Barbara, the University of Pireus, Greece, and the Australian National University, Canberra. It was designed to bring together researchers in applied probability and time series analysis from across the world. However, there was the specific intention to honour J. Gani and E.J. Hannan for their pioneering work in these fields. Unfortunately, Professor Hannan passed away before the conference was held, so its time series component became a memorial to him.

The published proceedings appear in two volumes: Volume 1, Applied Probability. In Honor of J.M.Gani, and Volume 2, Time Series. In Memory of E.J.Hannan.

Biographical information concerning J.M.Gani can be found in *Austral. J.Statist.* 30A (1988), Preface and pp. 1-13 and *Statistical Science* 10 (1995), 214-230. For a detailed Obituary of E.J.Hannan see *Historical Records of Australian Science* 10 (1994), 173-185.

The Editors
March 1996

Preface

It is over 30 years since three enthusiastic Australians and the London Mathematical Society (LMS) founded the Applied Probability Trust (APT) in 1964, and launched its first publication, the *Journal of Applied Probability* (JAP). By the early 1960s, it had become obvious that a journal dedicated to the applications of probability theory to the biological, physical, social and technological sciences was very much needed. It remained only to create it.

I recognized this need only very gradually, and began in 1962-3 by convincing two Australian colleagues at the Australian National University, where I was then working, to help me satisfy it. After some discussion, they agreed to assist me in raising finance for the journal. Their names were Norma McArthur of the Department of Demography and Ted Hannan of the Department of Statistics. Eventually, after a meeting between the Council of the LMS and me in 1963, we raised enough financial support to create the APT and start JAP. The founding trustees of the APT were Sir Edward Collingwood for the LMS, Ted Hannan, Norma McArthur and myself; sadly, I find myself the sole survivor of this group. But the APT lives on; its current trustees, apart from me, are Daryl Daley, Chris Heyde and Sir John Kingman for the LMS.

The first volume of JAP published in 1964 consisted of two issues totalling just under 400 pages; the last complete volume of 4 issues in 1995 numbered 1149 pages. Its companion volume,*Advances in Applied Probability* begun in 1969, also had 4 issues of 1191 pages in 1995. Thus, after 31 years, the number of pages in applied probability published in the APT's journals alone had increased by a factor of 6.

But this hardly represents the true measure of growth of the subject in the past three decades. There are now another 30 or more journals, many of them published only within the last decade, which contain some material in applied probability, including time series analysis. Among the better known of these are

Annales de l'Institut Henri Poincaré, Section B. Probabilités et Statistique
Annals of Applied Probability
Applied Stochastic Models and Data Analysis
Communications in Statistics, Stochastic Models
International Journal of Forecasting
Journal of Forecasting
Journal of the Royal Statistical Society, Series B
Journal of Time Series Analysis
Mathematics of Operations Research
Operations Research
Probability in the Engineering and Informational Sciences
Probability Theory and Related Fields
Queueing Systems, Theory and Applications
Scandinavian Actuarial Journal
Sequential Analysis
Stochastic Analysis and Applications
Stochastic Processes and their Applications
Stochastics and Stochastics Reports
Theory of Probability and its Applications.

This list is by no means exhaustive. One is led to conclude that applied probability and time series analysis remain very active fields of research, whose growth seems destined to continue well into the future.

Randomness is an integral part of many natural and social phenomena, so that it is not surprising to find probabilistic methods applied to a wide range of problems. As probability theory itself becomes deeper and more sophisticated, it is used increasingly to model such problems. Two recent examples are the use of the Black-Scholes equation in probabilistic financial models, and of martingale methods in estimating the infection parameters of epidemics.

The contributions to this Conference attest to the breadth of applied probability and time series analysis. I shall leave it to Peter Robinson and Murray Rosenblatt to comment in detail on the time series volume, and will restrict myself to a brief summary of the applied probability proceedings. These include sections on probability and probabilistic methods in recursive algorithms and stochastic models, Markov and other stochastic models such as Markov chains, branching processes and semi-Markov systems, biomathematical and genetic models, epidemiological models including S-I-R (Susceptible-Infective-Removal), household and AIDS epidemics, financial models for option pricing and optimization problems, random walks, queues and their waiting times, spatial models for earthquakes and inference on spatial models. There would hardly seem to be any area of real life or scientific research that is resistant to probabilistic analysis.

I should like to think of Laplace as one of the early godfathers of applied probability. In his *Théorie Analytique de Probabilités* (1812), he applied probabilistic reasoning to various demographic models. In Chapter 6 of his book, on the probability of causes of future events based on observed events (pp. 377-401), for example, Laplace considered three topical problems. First he analyzed the different ratios of boys to girls born in London, Naples and Paris; he went on to discuss mortality tables, probabilities of survival and a method for estimating the "population of a large empire". He concluded his Chapter by considering the ratios of christenings of boys and girls in Paris, which had been 25/24 between 1745 and 1784, and predicted that the probability that it would be greater than 1 for 100 years thereafter was $P = 0.782$. I do not know if this result has ever been verified against subsequent data, but I can attest to the fascination of Laplace's arguments.

Let me end on a note of gratitude. I personally enjoyed the Athens Conference greatly, not only because of the quality of the participants' papers, and the liveliness of their discussions, but also because of the warm welcome we received from our Greek colleagues. Ted Hannan would have loved the occasion, had he only been with us.

To all the participants who helped to celebrate the vigour and inventiveness of applied probability and time series analysis, to our generous Greek hosts, and to Bessy Athanasopoulos for her help in the organization of the Conference, I offer my warmest thanks.

Reference

Laplace, P.-S. (1812) *Théorie Analytique des Probabilités*. Vol. 7 of the *Oeuvres Complétes de Laplace*, Paris.

Joe Gani
Stochastic Analysis Group, SMS
Australian National University
March 1996

VOLUME 1. APPLIED PROBABILITY

CONTENTS

x

HALF-PROPHETS AND ROBBINS' PROBLEM
OF MINIMIZING THE EXPECTED RANK

F. Thomas Bruss*, *Université Libre de Bruxelles*
Thomas S. Ferguson**, *University of California, Los Angeles*

Summary: Let $X_1, X_2, \ldots, X_n$ be i.i.d. random variables with a known continuous distribution function. Robbins' problem is to find a sequential stopping rule without recall which minimizes the expected rank of the selected observation. An upper bound (obtained by memoryless threshold rules) and a procedure to obtain lower bounds of the value are known, but the difficulty is that the optimal strategy depends for all $n > 2$ in an intractable way on the whole history of preceding observations. The goal of this article is to understand better the structure of both optimal memoryless threshold rules and the (overall) optimal rule. We prove that the optimal rule is a "stepwise" monotone increasing threshold-function rule and then study its property of, what we call, full history-dependence. For each n, we describe a tractable statistic of preceding observations which is sufficient for optimal decisions of decision makers with half-prophetical abilities who can do generally better than we. It is shown that their advice can always be used to improve strictly on memoryless rules, and we determine such an improved rule for all n sufficiently large. We do not know, however, whether one can construct, as $n \to \infty$, *asymptotically relevant* improvements.

Keywords: Sequential selection - Full information - Memoryless threshold rules - "Stepwise" monotonicity - Prophets - Order statistics.

AMS 1991 Subject Classification: Primary 60G40, Secondary 62L15.

§1. Introduction.

1.1 The Problem. Let $X_1, X_2, \ldots, X_n$ be i.i.d. r.v.'s with c.d.f. F. We assume F to be continuous so that the X_k's are uniquely rankable (a.s.). Since the payoffs we consider depend only on the ranks of the X_i, we may and do assume, w.l.o.g., that F is uniform on $[0, 1]$. The relative rank of observation X_k is defined as

$$r_k = \sum_{j=1}^{k} I(X_j \le X_k), \quad 1 \le k \le n, \tag{1.1}$$

---

* Address: Université Libre de Bruxelles, Département de Mathématique et Institut de Statistique, CP 210, B-1050 Brussels, Belgium, e-mail: tbruss@ulb.ac.be

** Address: Department of Mathematics, University of California, Los Angeles, CA 90095, USA, e-mail: tom@math.ucla.edu

1

where $I(A)$ denotes the indicator of the event A. Note that we define the smaller observations to be the better ones, which proves here to be more convenient. Let

$$R_k^* = r_k + \sum_{j=k+1}^{n} I(X_j \leq X_k) \qquad (1.2)$$

denote the absolute rank of X_k, and let T denote the set of all stopping rules,

$$T = \{\tau : \{\tau = k\} \in \mathcal{F}_k, \forall k = 1, 2, \ldots, n\}, \qquad (1.3)$$

where $\mathcal{F}_k$ denotes the σ-field generated by $X_1, X_2, \ldots, X_k$. Since R_k^* is not $\mathcal{F}_k$-measurable, we replace it by $R_k = E(R_k^* | \mathcal{F}_k) = r_k + (n-k)X_k$ (see also Assaf & Samuel-Cahn (A&S-C, 1992)). Robbins' problem is then to find

$$V(n) = \inf_{\tau \in T} E(R_\tau) \qquad (1.4)$$

and the stopping rule τ which attains this value. This is the full-information version of the problem studied by Chow et al. (1964).

1.2 Motivation. The motivation to study Robbins' problem goes beyond the ambition to solve the $4th$ secretary problem (see Bruss and Ferguson (B&F),1993). Once we think about it, we see that this problem stands indeed for a whole class of problems, about which little is known. The central question is what to do if the optimal rule is seemingly dependent on the whole history (in a sense which we will make precise) and if we do *not* have an idea about the value. Replacing at each stage the history by some adequately looking summary of the available information, on which we would base the decisions, is of little help, because if we cannot assess the error, how can we evaluate the trial? In Robbins' problem, it turns out, that we have reasonable bounds to the value, and with a lower bound of 1.908 and an upper bound of 2.3232, we have indeed little to worry about. But what to do if the bounds are not close at all? - The problem would stay the same, and any improvement would be desirable.

1.3 Organization of the Paper. The paper is organized as follows: In Section 2 we review memoryless threshold rules, prove the uniqueness of the optimal rule in this class and briefly discuss computational methods. Section 3, which prepares the study of the overall optimal rule, introduces the notion of a half-prophet. This is a decision maker who turns into a prophet *provided* he decides not to choose the present observation and to go on. Thus a half-prophet has more power than we have, and it will be shown that he still needs all preceding information to realize the optimal value. Section 4 contains the main results of the paper. We show in Lemma 4.1 that the overall optimal rule is a "stepwise" monotone threshold function rule which contrasts the lack of monotonicity with respect to preceding observations. We then discuss the notion of *full history-dependence* of a stopping rule and show that the overall optimal rule for Robbins' problem has this property.

2

In Section 5 we describe a rule which strictly improves on the optimal memoryless rule for finite n.

§2. Memoryless Threshold Strategies.

By a threshold strategy or rule we mean a stopping rule, where the decision whether or not to select the observation X_k (i.e. stop at k) depends only on whether or not X_k is smaller (respectively larger, depending on the problem) than some real (threshold) value p_k. The p_k's are often thought of as being constants (see also Kennedy & Kertz (1990)). We must classify here threshold rules more precisely and prefer to call these *memoryless* (in short m^0-) threshold rules to distinguish them from those threshold rules where each p_k may itself be a function of some or all preceding observations. Thus a m^0-threshold rule is defined by

$$N(\mathbf{p}) = \inf\{k \geq 1 : X_k \leq p_k\}, \tag{2.1}$$

where $\mathbf{p}$ is a pre-determined sequence $(p_1, p_2, \ldots, p_n)$ with $0 \leq p_k \leq 1, k = 1, 2, \ldots, n - 1$ and $p_n = 1$. In the case of Robbins' problem we know (see also A&S-C, (1992)) that if we use m^0-threshold rules we can confine our interest to the class of monotone rules $N(\mathbf{p})$, with $\mathbf{p}$ satisfying

$$p_1 \leq p_2 \leq \ldots \leq p_n = 1. \tag{2.2}$$

The expected rank obtained by the $N(\mathbf{p})$-rule is given by formula (2.1) of B&F (1993). The optimal p_k's , denoted by $p_1^*, p_2^*, \ldots, p_{n-1}^*$ depend themselves on n, but we use the additional indexation by n only where necessary to prevent confusion. For small n the optimal p_k^*'s can be obtained from by solving the corresponding system of partial derivatives equations. We have

Lemma 2.1: The system of partial differential equations

$$\left\{\frac{\partial(E(R_{N(\mathbf{p})}))}{\partial p_k} = 0\right\}, \quad k = 1, 2, \ldots, n - 1, \tag{2.3}$$

has for $n \geq 2$ a unique solution $\mathbf{p}^* = (p_1^*, p_2^*, \ldots, p_{n-1}^*, 1)$ minimizing $E(R_{N(\mathbf{p})})$.

Proof: Since $E(R_{N(\mathbf{p})})$ is a continuous function of the p_k's, it must take its minimum somewhere on $[0, 1]^{n-1}$. Let $V^{m^0}(n)$ denote this minimum value and let

$$\mathbf{p}^* = (p_1^*(n), p_2^*(n), \cdots, p_{n-1}^*(n), 1) \tag{2.4}$$

be a sequence of m^0-thresholds which achieves it. Note that we do not imply so far uniqueness of these thresholds. Clearly

$$\forall n = 2, 3, \cdots : V^{m^0}(n) > 1, \tag{2.5}$$

3

because, for $n > 1$, no sequential rule selects rank 1 with probability one.

We first show that $p_1^*(n)$ must be greater than 0 for all n. This is intuitively clear, since we feel it cannot be optimal to refuse almost surely *any* first observation. Formally: The statement is true for $n = 1$, since $p_1^*(1) = 1$. Suppose now that $p_1^*(n + 1) = 0$ for some $n \geq 1$. Then the observation X_1 will be almost surely refused, so that memoryless optimal play on the $X_2, ..., X_{n+1}$ yields an expected total loss

$$e_{n+1}(0) = V_{p_1(n+1)=0}^{m^0}(n + 1) = W^{m^0}(n) + P(X_{\tau_n} \geq X_1), \qquad (2.6)$$

where τ_n denotes the stopping time induced by $(p_2^*(n+1), p_3^*(n+1), \cdots, p_n^*(n+1), 1)$ and $W^{m^0}(n)$ the corresponding expected loss for the remaining n observations only, i.e. disregarding the contribution of X_1. Clearly $W^{m^0}(n) \geq V^{m^0}(n) > 1$. Now, if we use instead of 0 the threshold p_1, say, with $1 > p_1 > 0$, then we obtain similarly by conditioning on X_1

$$e_{n+1}(p_1) = p_1(1 + np_1/2) + (1 - p_1)[W^{m^0}(n) + P(X_{\tau_n} \geq X_1 | X_1 > p_1)]. \qquad (2.7)$$

Clearly, the conditional probability term in (2.7) is non-increasing in p_1 on $[0, 1]$ so that using (2.6) in (2.7) implies

$$e_{n+1}(p_1) \leq p_1(1 + np_1/2) + (1 - p_1)e_{n+1}(0). \qquad (2.8)$$

Note that the last RHS of (2.8) can be made smaller than $e_{n+1}(0)$ by choosing for p_1 any value such that $1 < 1 + np_1/2 < e_{n+1}(0)$ or equivalently $0 < p_1 < 2(e_{n+1}(0) - 1)/n$. This is possible as $e_{n+1}(0) \geq W^{m^0} > 1$ for all $n > 1$. Thus the choice $p_1^*(n + 1) = 0$ would be sub-optimal for any $n \geq 1$, i.e. $p_1^*(n + 1) > 0$ for all $n \geq 1$.

Secondly, since $p_n^*(n) \equiv 1$, it is easy to see that we must have $p_{n-1}^*(n) < 1$. Indeed, if we used the threshold value $p_{n-1} = 1$, then we would select $\min\{X_{n-1}, X_n\}$ with probability 1/2. Any threshold value p_{n-1} with $0 < p_{n-1} < 1$ however would do this with a higher probability and therefore yield a smaller rank, because this probability $q_{n-1}(p_{n-1})$, say, equals

$$q_{n-1}(p_{n-1}) = \int_0^1 \{(1 - x)I(x \leq p_{n-1}) + xI(x > p_{n-1})\} \, dx$$
$$= 1/2 + p_{n-1}(1 - p_{n-1}) > 1/2. \qquad (2.9)$$

Thus, in particular, $p_{n-1}^*(n) < 1$. By the monotonicity of the optimal p_k^*'s with $1 \leq k \leq n - 1$ we can thus assure that no local or global minimum of $E(R_{N(\mathbf{p})})$ can lie on the boundary of the set $[0, 1]^{n-1}$. Since all partial derivatives of $E(R_{N(\mathbf{p})})$ exist on $(0, 1)^{n-1}$ and the minimum (or minima) must lie in this open set $(0, 1)^{n-1}$, (2.3) must have at least one solution.

4

Finally, looking at the version (3.1) of B&F (1993) of the formula for $E(R_{N(\mathbf{p})})$ we see that it is a multiquadratic function of the p_k's, i.e. it is quadratic in each p_k (with positive sign) by holding the other p_j's constant. Each partial differential equation of the system (2.3) has thus at most one solution. Therefore the system (2.3) itself can have at most one solution. Since it has at least one, it must have a unique solution, which must be, accordingly, the minimum. ∎

Remark 2.2. It is tempting to maximize q_{n-1} in (2.9) with respect to p_{n-1}, i.e. to choose $p_{n-1} = 1/2$ with $q_{n-1}(1/2) = 3/4$, but this does not give the optimal threshold $p_{n-1}^*(n)$ (see for instance Table 1), because this simple probability argument neglects relative ranks and the fact that, given X_{n-2} has been passed over, $X_1, X_2, ..., X_{n-2}$ are no longer i.i.d $U[0,1]$ random variables. We shall see however later that $p_{n-1}^*(n) \to 1/2$ as $n \to \infty$.

Computational aspects. Due to the multiquadratic property, the computation of the p_k^*'s is numerically still tractable for larger n. Table 1a displays the sequences of thresholds for $n = 1, 2, \ldots, 12$, which we will use for examples in Section 5. The corresponding values are denoted by $V_n^{m^0}$ and displayed in Table 1b.

n	p_1	p_2	p_3	p_4	p_5	p_6	p_7	p_8	p_9	p_{10}	p_{11}
1	1										
2	.5	1									
3	.3603	.5285	1								
4	.2858	.3825	.5351	1							
5	.2385	.3021	.3918	.5359	1						
6	.2052	.2509	.3110	.3954	.5349	1					
7	.1809	.2148	.2582	.3156	.3977	.5333	1				
8	.1619	.1881	.2211	.2628	.3183	.3983	.5315	1			
9	.1465	.1679	.1936	.2254	.2659	.3200	.3982	.5298	1		
10	.1341	.1516	.1724	.1974	.2284	.2679	.3208	.3978	.5283	1	
11	.1244	.1383	.1547	.1758	.2002	.2304	.2691	.3213	.3973	.5 265	1
12	.1125	.1288	.1423	.1582	.1779	.2040	.2322	.2702	.3142	.3 967	.5256

Table 1a

m^0-threshold values

n	1	2	3	4	5	6
$V_n^{m^0}$	1	1.25	1.4009	1.5606	1.5861	1.6490

n	7	8	9	10	11	12
$V_n^{m^0}$	1.7002	1.7430	1.7794	1.8109	1.8384	1.8627

Table 1b

Expected loss for the m^0-rules

5

The following lemma shows that "most" of the $p_j^*(n)$'s tend to 0 as $n \to \infty$. This will prove to be an essential tool in Section 5.

Lemma 2.3: Let $\mathbf{p}^*(n) = (p_1^*(n), p_2^*(n), \ldots, p_{n-1}^*(n), 1)$ be the sequence of m^o-optimal thresholds for Robbins' problem with n observations. Further let α be a constant with $0 < \alpha < 1$ and $\alpha_n = [\alpha n]$, where $[x]$ denotes the floor of x. Then, as $n \to \infty$,

$$\forall 0 < \alpha < 1, \forall k = k(n) \le \alpha_n : p_{k(n)}(n) \to 0.$$

Proof: Suppose the contrary, i.e. there exists $\epsilon > 0$ such that for all n and some $0 < \alpha < 1$ there would exist an integer $k = k(n) < \alpha_n$ with $p_k^*(n) > \epsilon$. Since $p_k^*(n)$ is m^o-optimal this would then imply that the observation $X_k = \epsilon$, say, must be accepted. However, we shall see that we can construct another m^0-rule yielding a strictly smaller expected loss. We first note that,

$$E(R_k | X_1, X_2, \cdots, X_{k-1}, X_k = \epsilon) = r_k(\epsilon) + (n - k)\epsilon, \qquad (2.10)$$

where $r_k(\epsilon)$ denotes the relative rank of X_k for $X_k = \epsilon$. Let δ be fixed but arbitrarily chosen with $0 < \delta < \epsilon$, and let τ_δ be the stopping time defined by

$$\tau_\delta = \min\{n, \inf\{j > k : X_j < \delta\}\}. \qquad (2.11)$$

Then

$$E(R_{\tau_\delta} | X_1, X_2, \cdots, X_{k-1}, X_k = \epsilon) \qquad (2.12)$$

$$= E(r_{\tau_\delta} | X_1, X_2, \cdots, X_{k-1}, X_k = \epsilon) + E((n - \tau_\delta) X_{\tau_\delta} | X_1, X_2, \cdots, X_{k-1}, X_k = \epsilon)$$

$$< E(r_{\tau_\delta} | X_1, X_2, \cdots, X_{k-1}, X_k = \epsilon) + (n - k) E(X_{\tau_\delta} | X_1, X_2, \cdots, X_{k-1}, X_k = \epsilon).$$

Conditioning on the outcome of X_{τ_δ} we obtain

$$E(R_{\tau_\delta} | X_1, X_2, \cdots, X_{k-1}, X_k = \epsilon) \qquad (2.13)$$

$$= E[E(R_{\tau_\delta} | X_1, X_2, \cdots, X_{k-1}, X_k = \epsilon, X_{\tau_\delta}) | X_1, X_2, \cdots, X_{k-1}, X_k = \epsilon]$$

$$< \left(1 - (1 - \delta)^{n-k-1}\right) E(R_{\tau_\delta} | X_1, X_2, \cdots, X_{k-1}, X_k = \epsilon, X_{\tau_\delta} < \delta) + n(1 - \delta)^{n-k-1},$$

where we used for $\tau_\delta = n$ the worst-case bound n as an upper bound for $R_n \le n$.

As $n \to \infty$, the first factor of the first term in the last line of (2.13) clearly tends to 1. The second summand in this line tends to 0 since $n - k - 1 \ge n - \alpha_k - 1 \to \infty$ more quickly than some $(1 - \alpha)n$. Therefore, the RHS of (2.13) is close to $E(R_{\tau_\delta} | X_{\tau_\delta} < \delta)$ as $n \to \infty$. Since $E(R_{\tau_\delta} | X_{\tau_\delta} < \delta)$ is clearly a strictly increasing function of δ on $(0, \epsilon]$, the inequality $E(R_{\tau_\delta} | X_k = \epsilon) < E(R_{\tau_\epsilon} | X_{\tau_\epsilon} < \epsilon)$ a.s. must hold for all n sufficiently large. Now, since $E(R_{\tau_\epsilon} | X_{\tau_\epsilon} < \epsilon) < E(R_k | X_k = \epsilon)$ a.s. it follows in particular that, for all n sufficiently large, $E(R_{\tau_\delta} | X_k = \epsilon) < E(R_k | X_k = \epsilon)$ a.s.. This means that, for all n sufficiently large, the rule τ_δ defined by (2.11) yields a smaller expected loss than accepting some $X_k \ge \epsilon$. However, this contradicts the hypothesis of $p_k^*(n) > \epsilon$ for all n, and thus proves the lemma. $\blacksquare$

6

Now let $V(n)$ denote the overall optimal value of Robbins' problem for n variables. Clearly $V(n) \leq V^{m^0}(n)$ for all n so that

$$V = \lim_{n \to \infty} V(n) \leq \lim_{n \to \infty} V^{m^0}(n) = V^{m^0}. \tag{2.14}$$

We know (see B&F (1993)) that $V^{m^0} = 2.326....$ and that $V > 1.908$, where the latter lower bound was obtained by computing the optimal strategy for a truncated modification of the expected loss. We could also show that this procedure would converge to the correct limiting value V, but the computation times increase exponentially. The optimal strategy is very "sensitive" to the past, and no statistic is known, which summarizes enough of the information of the past to allow for a ϵ-optimal strategy while still being sufficiently tractable. We will introduce the notion of *full history dependence* to describe the mentioned sensitivity.

Apart from the above bounds we know little about V. We even do not know yet whether $V > 2$, although one may, in principle, be able to settle this question numerically by the truncation method proposed in B&F (1993). Gnedin (1995, INFORM Conference on Applied Probability) reported that the strict inequality, $V < V^{m^0}$, can be obtained by an adequate embedding of the $n = \infty$ case into a Poisson process. We focus on finite n. The $V(n)$ are only known for n up to 5, and computation times are prohibitive to try to do much better. This raises the question how to find tractable improvements on optimal threshold strategies, with which we deal in Section 5 and where half-prophets prove their usefulness again.

§3. Half-prophets.

The motivation to introduce the notion of a *half-prophet* (h-prophet) is to overcome one side of the deadlock resulting from the apparent history dependence, namely the future side. An h-prophet, called in at time k say, can (as we can) see the observations $X_1, X_2, \ldots, X_k$ and nothing else. However, given that he decides not to select X_k and to go on, then he foresees at time $k+1$ the whole future $X_{k+1}, X_{k+2}, \ldots, X_n$ (whereas we see of course only X_{k+1}). The value of the game for an h-prophet depends, apart from n clearly also on k, and as we shall see, on the whole past $X_1, X_2, \ldots, X_k$. We have to define more precisely:

Definition 3.1: An *h(k)-prophet* for a (discrete) sequential decision problem is a decision maker who is able to foresee the complete future if and only if he decides to enter stage $k+1$. An *h-prophet* is a decision maker, which can be elected to be an h(k)-prophet for exactly one $k \geq 1$.

Note that an h-prophet in Robbins' problem faces a much simpler decision problem. He must only decide whether to select the present observation, or alternatively, to wait for the best later on. This "binary" feature implies that his value at time k, denoted by $h(k, n)$, say, can be computed in a straightforward manner.

7

Lemma 3.2: The value of an $h(k)$-prophet in Robbins' problem with n observations conditional on $X_1, \ldots, X_k$ is equal to

$$h(k,n) = \min\Big\{r_k + (n-k)X_k, 1 + \sum_{j=1}^{k}(1-X_j)^{n-k}\Big\}, \; k = 1, 2, \cdots, n.$$

Proof: We first note that if the $h(k)$-prophet selects X_k then his expected loss (denoted by L_k) is equal to

$$L_k = E(R_k|X_1, X_2, \ldots, X_k) = r_k + E\Big(\sum_{j=k+1}^{n} I(X_j \le X_k | X_1, X_2, \ldots, X_k)\Big). \quad (3.1)$$

By independence of X_j with $j > k$ of the past, the second term simplifies to $(n-k)X_k$ so that

$$L_k = r_k + (n-k)X_k. \quad (3.2)$$

Suppose now that the $h(k)$-prophet refuses X_k. Then he enters time $k+1$ and optimal behavior forces him to select $\iota(k) := \inf_{k<j\le n}\{X_j\}$. The expected absolute rank of $\iota(k)$ given $X_1, X_2, \ldots, X_k$ depends on $X_1, X_2, \ldots, X_k$, but the distribution function $F_{\iota(k)}(\cdot)$ of $\iota(k)$ itself is simply the distribution function of the smallest order statistic of $(n-k)$ i.i.d. uniform r.v.'s on $[0,1]$, independent of $X_1, X_2, \ldots, X_k$. Denoting the absolute rank of $\iota(k)$ by $R^{\iota(k)}$ we obtain by conditioning on the value of $\iota(k)$ and by using its independence of $X_1, X_2, \ldots, X_k$,

$$L_k^{(+)} := E(R^{\iota(k)} | X_1, X_2, \ldots, X_k) = E\big(E(R^{\iota(k)} | X_1, X_2, \ldots, X_k; \iota(k))\big).$$

Now note that $\iota(k)$ can (by definition) only be preceded by the $X_1, X_2, \ldots, X_k$, and by none of the future values. If $\iota(k)$ is smaller than all $X_1, X_2, \ldots, X_k$, then its rank equals 1, and it moves up by 1 with each X_j it surpasses. Therefore $L_k^{(+)}$ can be written as

$$L_k^{(+)} = \int_0^1 \Big(1 + \sum_{m=1}^{k} I(X_m \le s)\Big) dF_{\iota(k)}(s) \quad (3.3)$$

where $dF_{\iota(k)}(s) = d(1 - (1-s)^{n-k}) = (n-k)(1-s)^{n-k-1}ds$. Interchanging the order of integration and summation and then adjusting the boundaries of integration according to the indicators yields by straightforward calculation

$$L_k^{(+)} = 1 + \sum_{j=1}^{k}(1-X_j)^{n-k}, \quad k = 1, 2, \ldots, n. \quad (3.4)$$

By the optimality principle,

$$h(k,n) = \min\{L_k, L_k^{(+)}\}, \quad (3.5)$$

8

and thus the statement of the Lemma follows from (3.2), (3.4) and (3.5). ∎

Remark 3.3: The most powerful h-prophet — apart from the h(0)-prophet, who is equivalent to a *prophet* — is, by definition, the h(1)-prophet. He can simply ignore the decision problem any later h-prophet will face. Since his probability to select rank 1 tends to one as $n \to \infty$ one feels that he is "asymptotically" as good as a prophet, and that indeed the same should be true for any h(k)-prophet for fixed k. This is true for the expected rank, but a worst case analysis shows, that all h-prophets are distinctly inferior to a prophet in the sense that $\sup_x \inf_\tau E(R_\tau | X_1 = x) > 1$. To see this, note that according to (3.2), (3.4) and (3.5)

$$\sup_x h(1,n) := \sup_x h(1,n)_{|X_1=x} = 1+\sup_x\{\min((n-1)x,(1-x)^{n-1})\} = 1+c_n, \quad (3.6)$$

say. Since $(n-1)x$ is increasing and $(1-x)^{n-1}$ decreasing on $[0,1]$, and both intersect there, c_n must be the unique solution of $(n-1)x = (1-x)^{n-1}$. Putting $x = \xi/(n-1)$ in (3.6) yields the limiting equation $\xi = e^{-\xi}$ with solution $\xi = \lim_{n\to\infty} c_n = .5671\cdots$. Thus,

$$\forall 1 \le k \le n : \sup h(k,n) \ge \sup h(1,n) \to 1.5671 \cdots \quad \text{as } n \to \infty, \quad (3.7)$$

whereas a prophet can always achieve the value 1.

§4. Properties of the optimal rule.

We turn now to the main results of this paper. The first is Lemma (4.1) which shows a monotonic feature which contrasts many non-monotonic features of Robbins' Problem. This will then be used to show that the optimal rule depends in a certain sense on the full history of the process.

4.1 Stepwise Monotonicity of the Optimal Threshold Functions. We denote by $\tau^* = \tau^*(n)$ the (overall) optimal rule for Robbins' problem. An observation X_k, which can be selected (respectively, must be refused) under the rule $\tau^*(n)$) will be called shortly *acceptable*, (respectively, *unacceptable*).

Lemma 4.1: $\tau^* = \tau^*(n)$ is a *stepwise* monotone increasing threshold function rule for all n, i.e. it is of the form

$$\tau^*(n) = \inf\{1 \le k \le n : X_k \le p_k(X_1, X_2, \ldots, X_{k-1})\}, \quad (i)$$

where the functions $p_k(...)$ satisfy

$$p_k(X_1, X_2, \ldots, X_{k-1}) \le p_{k+1}(X_1, X_2, \ldots, X_{k-1}, X_k) \text{ a.s.} \quad (ii)$$

and

$$p_{n-1}(X_1, X_2, \cdots, X_{n-2}) < 1 \text{ a.s.} \quad (iii)$$

9

Proof: (*i*) We show first, that $\tau^*(n)$ is of the described form. For this we have to show that, at any stage k, the following is true: Whenever it is optimal to stop at X_k it would be optimal to stop for any $X'_k \leq X_k$, and whenever it is optimal to refuse X_k it is optimal to refuse any $X'_k \geq X_k$. To see this let $L_k(x)$ be defined as in (3.2) with X_k being replaced by x. Thus

$$L_k(x) = \sum_{j=1}^{k-1} I(X_j < x) + 1 + (n-k)x \qquad (4.1)$$

describes our expected loss by accepting x at stage k. Clearly, $L_k(x)$ is strictly increasing in x for all $1 \leq k < n$. On the other hand, the expected loss by refusing x at stage k and continuing optimally thereafter equals

$$L_k^*(x) = \text{ess inf}_{\{\tau > k\}} E(R_\tau | X_1, X_2, \ldots, X_{k-1}, X_k = x), \qquad (4.2)$$

where it is understood that n is arbitrarily chosen but fixed, and $1 \leq k < n$. Let τ_x^* now be that rule which achieves $L_k^*(x)$ given in (4.2). The latter can then be written in the form

$$L_k^*(x) = \sum_{j=1}^{k-1} P(X_{\tau_x^*} > X_j) + 1 + P(X_{\tau_x^*} \geq x) + E\Big(\sum_{j=k+1, j \neq \tau_x^*}^{n} I(X_j \leq X_{\tau_x^*}) \Big). \quad (4.3)$$

Further, let $\tilde{L}_k(x, y)$ be that modification of (4.3) which replaces x (only) by y without replacing τ_x^* by τ_y^*, i.e. formally

$$\tilde{L}_k(x, y) = L_k^*(x) - P(X_{\tau_x^*} \geq x) + P(X_{\tau_x^*} \geq y). \qquad (4.4)$$

$\tilde{L}_k(x, y)$ is then the conditional expected loss, given $X_1, X_2, \ldots, X_{k-1}, X_k = y$, of the value obtained by the suboptimal continuation τ_x^* (which is optimal for $x = y$). Since $P(X_{\tau_x^*} \geq y)$ is non-increasing in y, it follows that $\tilde{L}_k(x, y)$ is non-increasing in y. Thus

$$\forall y \in [x, 1] : \tilde{L}_k(x, y) \leq \tilde{L}_k(x, x) = L_k^*(x) \text{ a.s.} \qquad (4.5)$$

On the other hand, suboptimality of τ_x^* for the history $X_1, X_2, \ldots X_{k-1}, X_k = y$ implies

$$\tilde{L}_k(x, y) \geq \tilde{L}_k(y, y) = L_k^*(y) \text{ a.s..} \qquad (4.6)$$

Combining inequalities (4.5) and (4.6) yields

$$\forall 0 \leq x \leq y \leq 1 : L_k^*(x) \geq L_k^*(y), \text{ a.s..} \qquad (4.7)$$

An increasing $L_k(x)$ faces therefore a non-increasing $L_k^*(x)$ as a function of x. (4.1) tells us further that for all $k = 1, 2, \ldots, n$, $L_k(0) = 1$ and $L_k(1) = n$ a.s., so that $L_k^*(x)$ defined in (4.2) must lie in this range for any $x \in [0, 1]$. This proves for each

10

stage k and $1 \le j \le k$ the existence of optimal thresholds $p_j(\cdots)$ such that any $X_j > p_j(\cdots)$ must be refused whereas the first $X_k \le p_k(\cdots)$ must be accepted.

(ii) The proof of the second part (optimal thresholds are monotone increasing) becomes shorter when we note that, by definition, the $p_k(...)$'s need only be defined for those $X_1, X_2 \ldots, X_{k-1}$ which are unacceptable. Therefore it suffices to show that if we replace an acceptable $X_k = x$ by an unacceptable $X_k' > x$ then any $X_{k+1} \le x$ is again acceptable. Let $X_k = x \le p_k(...)$. This X_k is thus, by definition, acceptable. Now, if X_k is replaced by an unacceptable X_k' then, by definition, $X_k' > x$. It follows then from (4.1) that for a fixed history $X_1, X_2, ..., X_{k-1}$ we have $L_{k+1}(x) = L_k(x) - x$, because $X_k' > x$ leaves the relative rank of x at stage $k+1$ unchanged. Therefore $L_{k+1}(x) < L_k(x)$ a.s. It suffices now to show that $L_k^*(x)$ (see (4.2)) is non-decreasing in k, because then we have

$$\forall k = 1, \ldots, n-1 : L_k(x) \le L_k^*(x) \text{ a.s.} \Rightarrow L_{k+1}(x) \le L_{k+1}^*(x) \text{ a.s,} \qquad (4.8)$$

i.e. $X_{k+1} = x$ is then acceptable, so that $p_{k+1}(...) \ge p_k(...)$ a.s. for any k and any history $X_1, X_2, \cdots, X_{k-1}$.

To see that $L_k^*(x)$ is non-decreasing in k a.s., we first note that $L_k^*(x)$ is invariant a.s. with respect to any permutation of the history. Indeed, if the stopping rule τ_x^* achieves $L_k^*(x)$ in (4.3) then τ_x^* achieves the same value for the history $(X_{i_1}, X_{i_2}, \cdots, X_{i_{k-1}}, X_{i_k} = x)$, because such a permutation only changes the order of summation of the first k terms on the RHS and since $(X_{k+1}, \cdots, X_n)$ is independent of the past. Now

$$L_{k+1}^*(x) = \text{ ess inf}_{\{k+1 < \tau \le n\}} E(R_\tau | X_1, \ldots, X_{k-1}, X_k, X_{k+1} = x)$$

$$= \text{ess inf}_{\{k+1 < \tau \le n\}} E(R_\tau | X_1, \ldots, X_{k-1}, X_k = x, X_{k+1}) =: \tilde{L}_{k+1}^*(x, X_{k+1}), \quad (4.9)$$

say, since $L_{k+1}^*(x)$ has the above described invariance property, and since X_k and X_{k+1} are (unconditioned) i.i.d. r.v.'s. But then

$$L_{k+1}^*(x, X_{k+1}) \ge \text{ess inf}_{\{k < \tau \le n\}} E(R_\tau | X_1, \cdots, X_{k-1}, X_k = x), \qquad (4.10)$$

because optimal behavior on the set $\{k < \tau \le n\}$ allows for stopping at stage $k+1$, or to continue optimally otherwise. Thus, from (4.9) and (4.10), $L_k^*(x) \le L_{k+1}^*(x)$ a.s. so that (ii) is proved.

(iii) Finally, the proof of $p_{n-1}(X_1, X_2, \cdots, X_{n-2}) < 1$ a.s. follows immediately from $p_n(X_1, X_2, \cdots, X_{n-1}) \equiv 1$ and the argument we used in (2.9), because otherwise *any* memoryless threshold value p_{n-1} would do strictly better in expectation. ∎

4.2 Full History Dependence of the Optimal Rule. In this section we will see that the optimal rule has the property of being fully history dependent, that is at no time may one discard past information. At each stage $k \le n-1$, the exact values of all past observations may play a role in future optimal decisions. Clearly X_k could not be accepted if X_j for some $1 \leq j < k$ would have been accepted so that the events $\{\tau = k\}$ and $\{\tau > k\}$ always depend on all preceding observations in this sense. However, this does not imply that all preceding observations are relevant for future optimal decisions. A fully history dependent rule allows for no fading of memory at all. Varying some X_j (in certain cases even *any* X_j) by an arbitrary little amount may keep $X_j, \ldots, X_{k-1}$ unacceptable but be decisive for X_k to be acceptable or not.

To see that τ^* is fully history-dependent, we show that for each stage $s \leq n-1$, τ^* leads with positive probability to a situation where all preceding information $X_1, X_2, \cdots, X_{s-1}$ is needed to decide whether to stop and accept or, alternatively, to go on. Note that no information at all is needed at stage $s = n$, since $p_n(X_1, X_2, \cdots, X_{n-1}) \equiv 1$. Therefore we first look at $s = n - 1$. The probability to reach stage $n - 1$ equals

$$P(\tau^* > n - 2) = E(\prod_{j=1}^{n-2} (1 - p_j(X_1, X_2, \cdots, X_{j-1}))) \text{ a.s.}, \qquad (4.11)$$

where we used the definition of optimal threshold functions as given in Lemma 4.1 and the definition (convention) $p_1(\cdot) := p_1$. According to Lemma 4.1 we have $p_j(X_1, X_2, \cdots, X_{j-1}) \leq p_{j+1}(X_1, X_2, \cdots, X_j) < 1$ a.s. for all $1 \leq j \leq n-2$, so that $P(\tau^* > n - 2) > 0$. Similarly we obtain $P(\tau^* > n - 2 | \tau^* \geq k) > 0$ for all $1 \leq k \leq n - 2$.

Suppose now that $\tau^* I(\tau^* > n - 2)$ depends on all preceding information, i.e. that τ^* is fully history-dependent on the set $\{\tau^* > n - 2\}$. At stage k, the decision to go on depends by the principle of optimality on the optimal (future) expected loss. This expected loss involves, with $P(\tau^* > n - 2 | \tau^* \geq k)$ being a.s. strictly positive, by definition the expected loss on the set $\{\tau^* > n - 2\}$. In return, the latter depends according to our hypothesis on $X_1, X_2, \cdots, X_{n-2}$, i.e. in particular on $X_1, X_2, \cdots, X_{k-1}$. Therefore, if all preceding information is needed at stage $s = n - 1$, then all preceding information is needed at each stage $2 \leq k \leq n - 2$, and thus it suffices to show that τ^* is fully history-dependent on $\{\tau^* > n - 2\}$.

To see the latter we recall first that, at stage $s = n - 1$, we have the same capacity as a half-prophet. According to Lemma 3.2 we therefore accept X_{n-1} if

$$r_{n-1} + X_{n-1} \leq n - \sum_{j=1}^{n-1} X_j \qquad (4.12)$$

and accept X_n otherwise. If we write $r_{n-1}(x)$ for the rank of X_{n-1} given $X_{n-1} = x$, then the decision criterion becomes

$$r_{n-1}(x) + 2x \leq n - \sum_{j=1}^{n-2} X_j, \qquad (4.13)$$

12

where the RHS does not depend on x. We shall now distinguish between two cases.

Case 1: Suppose first that there exists x^*, say, such that in (4.13) equality holds for both sides. Then, by definition, $X_{n-1} = x^*$ is acceptable. But since all $X_1, X_2, \cdots, X_{n-2}$ are a.s. different, their nearest neighbor distance ϵ, say, is a.s. positive.

Now move an arbitrarily chosen X_j on the RHS to $X_j + \epsilon/2$, say. This leaves by construction the relative ranks of $X_1, X_2, \cdots, X_{n-2}$ unchanged and the chosen jth observation stays unacceptable (becomes less desirable) by Lemma 4.1. Now $X_1, X_2, \cdots X_{j-1}$ clearly stay unacceptable because they precede X_j chronologically. On the other hand $X_{j+1}, X_{j+2}, \cdots, X_{n-2}$ are unacceptable as well, because they *were* unacceptable before the change and face with the change from X_j to $X_j+\epsilon/2$ a reduced optimal expected loss by being passed over. Therefore the modified history $X_1, X_2, \cdots, X_{j-1}, X_j + \epsilon/2, X_{j+1}, \ldots, X_{n-2}$ stays within the set $\{\tau^* > n - 2\}$. However, the RHS of the criterion (4.13) decreases by $\epsilon/2$, so that now $X_{n-1} = x^*$ is unacceptable.

Case 2: Suppose now that (for a given history) there is no x^* to yield equality of both sides of (4.13). Then the critical value must coincide with one of $X_1, X_2, \cdots, X_{n-2}$. Suppose it coincides with X_k. For $\epsilon > 0$ chosen as the distance to the nearest neighbor let $X_{n-1} := x \in (X_k, X_k + \epsilon/2)$. Then X_{n-1} lies to the right of the critical value and is therefore unacceptable. But now move X_k into the interval $(x, x+\delta)$ with $0 < \delta < \epsilon/2$. This leaves the relative ranks of the new X_k and all other X_j's with $1 \leq j \leq n-2$ unchanged whereas the relative rank of X_{n-1} drops by one. The LHS of the criterion is therefore reduced by $1 - 2\delta$, whereas the RHS decreases by δ. But since $r_{n-1}(X_k)$ differs from $n - (X_1 + X_2, + \cdots + X_{n-2})$ by less than one, there must exist a sufficiently small δ such that this move renders $X_{n-1} = x$ acceptable.

However, as in case 1, for such a δ the observations $X_k + \delta, X_{k+1}, \cdots, X_{n-2}$ stay unacceptable. This means that the modified history stays also within the set $\{\tau^* > n-2\}$, and that X_k determines whether $\tau^* = n-1$, or, alternatively $\tau^* = n$.

Remark 4.3: Note that in case 2 (when a jump of the LHS occurs in X_k) the influence of X_k on the decision to accept or to refuse happens to be particularly strong, but that the inequalities in the criterion may be reversible by keeping X_k and X_{n-1} fixed and only varying the other preceding observations.

§5. Improving on Optimal Memoryless Rules.

In this section we shall describe a strategy which is superior to the optimal m^0-rule and (still) tractable. The construction depends on the following property of m^0-optimal thresholds $p^*_{n-1}(n)$ (see Section 2):

Lemma 5.1: The m^0-optimal threshold value $p^*_{n-1}(n)$ satisfies: $p^*_{n-1}(n) \to 1/2$ as $n \to \infty$.

Proof: Suppose we have observed $X_1, \ldots, X_{n-1}$ and are wondering whether to stop with $X_{n-1} = x$. The memoryless rule may not use any of the past information involving $X_1, \ldots, X_{n-2}$, except that we know $X_j > p_j(n)$ for $j = 1, \ldots, n-2$. If we stop with $X_{n-1} = x$ we expect to pay

$$E\left(\sum_{j=1}^{n-2} I(X_j < x)|X_1 > p_1(n), \ldots, X_{n-2} > p_{n-2}(n)\right) + 1 + x$$

$$= 1 + x + \sum_{j=1}^{n-2} \frac{(x - p_j(n))^+}{1 - p_j(n)}. \tag{5.1}$$

If we continue with $X_{n-1} = x$, we must stop at the next observation and so expect to pay

$$E\left(\sum_{j=1}^{n-2} I(X_j < X_n)|X_1 > p_1(n), \ldots, X_{n-2} > p_{n-2}(n)\right) + (1 - x) + 1$$

$$= 2 - x + \sum_{j=1}^{n-2} \frac{(1 - p_j(n))^2}{2}, \tag{5.2}$$

since under the conditioning, $X_1, \ldots, X_{n-2}, X_n$ are independent with X_j uniform on $(p_j(n), 1)$ for $j = 1, \ldots, n-2$ and X_n uniform on $(0, 1)$. Therefore the critical value for X_{n-1} is that value of $x = p_{n-1}(n)$ at which (5.1) is equal to (5.2), namely,

$$2p_{n-1}(n) + \sum_{j=1}^{n-2} \frac{(p_{n-1}(n) - p_j(n))^+}{1 - p_j(n)} = 1 + \sum_{j=1}^{n-2} \frac{(1 - p_j(n))^2}{2}. \tag{5.3}$$

Recall now Lemma 2.3 which states that for every $\alpha \in (0, 1)$ and every sequence $k(n) \le n\alpha$, we have $p_{k(n)}(n) \to 0$ as $n \to \infty$. This implies that the average of the $p_j(n)$'s must converge to zero, because

$$\frac{1}{n-2} \sum_{j=1}^{n-2} p_j(n) \le \frac{1}{n-2} \sum_{j=1}^{\lceil n\alpha \rceil} p_j(n) + 1 - \alpha$$

$$\le \frac{1}{n-2} p_{\lceil n\alpha \rceil}(n) + 1 - \alpha \to 1 - \alpha. \tag{5.4}$$

Since this is true for all $\alpha < 1$, we have $\sum_{j=1}^{n-2} p_j(n)/(n-2) \to 0$. This implies that the right side of (5.3), when divided by $n - 2$, converges to $1/2$.

A similar argument shows that the left side of (5.3), when divided by $n - 2$, must eventually for large n be within a preassigned $\epsilon > 0$ of $p_{n-1}(n)$. Now divide equality (5.3) by $n - 2$ and then subtract $p_{n-1}(n)$ on both sides. This preserves equality for the new LHS $\lambda(n)$ and the new RHS $\rho(n)$. Since $\lambda(n) \to 0$ by the

14

preceding argument and $\rho(n) = 1/2 + \delta_n - p_{n-1}(n)$ for some δ_n, which converges, according to (5.4), to zero, $\lambda(n) - \rho(n) = 0$ cannot hold unless $p_{n-1}(n) \to 1/2$ as $n \to \infty$. $\blacksquare$

Combining m^0-rules with half-prophet rules. Let $\tau(n)$ be the optimal m^0-rule defined by $p_1^*(n), p_2^*(n), \ldots, p_{n-1}^*(n), 1$, i.e.

$$\tau(n) = \inf\{1 \le k \le n : X_k \le p_k^*(n)\}. \tag{5.5}$$

We know from Lemma 2.1 that this rule is unique. Further let

$$\tau_h(n) := \min\{1 \le k \le n : L_k = h(k, n)\}, \tag{5.6}$$

i.e. $\tau_h(n)$ denotes the earliest time k, at which a $h(k)$-prophet would stop. Finally, let

$$\sigma^*(n) := \min\{\tau(n), \tau_h(n)\}. \tag{5.7}$$

Theorem 5.2 For all n sufficiently large: $E(R_{\sigma^*(n)}) < E(R_{\tau(n)})$.

Proof: We first note that $\tau_h(n) \le n$ since $h(n, n) \le n$. Also, if we stop at stage k with $\tau_h(n) = k$, whereas $\tau(n) > k$, then we act optimally since, according to (3.3), (3.4) and (3.5),

$$L_k^{(+)} = \inf_{\tau > k} \left(\int_0^1 (1 + \sum_{m=1}^n I(X_m \le x)) dF_\tau(x) \right) \le L_k^*, \tag{5.8}$$

and therefore $L_k \le L_k^*$. Thus $E(R_{\sigma^*(n)}) \le E(R_{\tau(n)} | \tau(n) > t_h(n))$ for all n. Moreover, the preceding inequality becomes strict for $n > 2$, because if there are at least two more observations to come, any strategy will miss the smallest of these with a positive probability. It is therefore sufficient to show that for all sufficiently large n,

$$P(\sigma^*(n) < \tau(n)) = P(\tau_h(n) < \tau(n)) > 0. \tag{5.9}$$

Now let $a_n = p_{n-1}^*(n)$, $b_n = (1 + a_n)/2$ and $A_n = \{X_k \in (a_n, b_n), 1 \le k \le n\}$. Since $a_n < b_n < 1$ for all $n \ge 2$ we have $P(A_n) > 0$ for all $n \ge 2$. Moreover, since on A_n no X_k with $1 \le k \le n-1$ is acceptable under the optimal m^0-rule, we obtain $P(\tau(n) = n) \ge P(A_n) > 0$. Now $P(\tau_h(n) < \tau(n)) \ge P(\tau_h(n) \le n - 1 | A_n) P(A_n)$, so that it suffices to show that $P(\tau_h(n) \le n - 1 | A_n) > 0$. Note that

$$P(\tau_h(n) \le n - 1 | A_n) \ge P(\tau_h(n) \le n - 1 | A_n, r_{n-1} = 1) P(r_{n-1} = 1 | A_n).$$

By Rényi's record value theorem (Rényi (1962)) for i.i.d. r.v's we have $P(r_{n-1} = 1 | A_n) = 1/(n - 1) > 0$, so that it suffices again to show that $P(\tau_h(n) \le n - 1 | A_n, r_{n-1} = 1) > 0$, for all n sufficiently large. This latter term equals indeed one for almost all n, since

$$P(\tau_h(n) \le n - 1 | A_n, r_{n-1} = 1) = P(L_{n-1} \le L_{n-1}^{(+)} | A_n, r_{n-1} = 1)$$

15

$$= P(1 + X_{n-1} \le 1 + \sum_{j=1}^{n-1}(1 - X_j)|A_n) \ge P(2X_{n-1} \le 1 + (n-2)(1 - b_n)),$$

where we used $X_1, X_2, \cdots, X_{n-2} < b_n$ on A_n. Using it again for X_{n-1} yields

$$P(2X_{n-1} \le 1 + (n-2)(1 - b_n)) \ge P(0 \le n - 1 - nb_n) = 1 \text{ for almost all } n,$$

because, by Lemma 5.1, $b_n \to 3/4$ as $n \to \infty$, and this completes the proof. ∎

Remark 5.3: Table 1a of Section 2 strongly suggests that $p_{n-1}(n)$ increases for $n = 2$ up to $n = 5$ with value $.5359 \cdots$ and decreases thereafter, converging (as we know) to $1/2$, but the proof of this does not seem to be straightforward. One can easily verify that it would imply strict improvement of the rule $\sigma^*(n)$ for all $n \ge 3$. More important than this observation however is the fact (see e.g. Example 5.4) that the strict improvement is not only due to what the proof is based on, namely to detecting late *record* values.

Example 5.4: Let $n = 12$, $X_1 = 0.12, X_2 = 0.13$. Further let $X_3, X_4, \cdots, X_{10}$ all be approximately $.6$, and finally let $X_{11} = 0.55$. Then all X_k for $1 \le k \le 11$ are unacceptable under the optimal m^0-rule (see Table 1a), but (5.10) and Lemma 3.2 show that $\sigma^*(12) = \tau_h(12) = 11$ and stops with a 3-record only. Nevertheless, the expected loss by stopping at stage 11 equals 3.55, whereas stopping at stage 12 yields a much higher expected loss of $12 - 0.12 - 0.13 - 0.55 - (X_3 + \cdots + X_{10}) \approx 6.4$, so that even for non-record values the difference can be quite large. Clearly it would go up, if we moved X_1 and/or X_2 to the right of 0.6. ∎

We could also give examples to show that the improvement can become effective earlier than at stage $n - 1$. $\sigma^*(n)$ is so far the "uniformly" best strategy we know.

The half-prophet rule $\tau_h(n)$ *alone* would not do well for larger n; it succeeds more often than the optimal m^0-rule to select the smallest or second smallest observation but allows for worse outliers than the optimal m^0-rule. Lemma 3.2 shows the influence of the powers $n - k$, which let $L_k^{(+)}$ grow only slowly as k increases.

Acknowledgement: This paper was begun in summer 1993 while the first author was visiting UCLA and finished in spring 1995 while the second author was visiting ULB in Brussels. We would like to thank our respective host Departments for their warm hospitality. We are also grateful to the National Science Foundation of Belgium for financial support.

References:

Assaf D. & E. Samuel-Cahn. (1992) *The secretary problem: minimizing the expected rank with full information.* Preprint, The Hebrew University of Jerusalem.

Bruss, F. Thomas & Thomas S. Ferguson (1993) *Minimizing the expected rank with full information.* J. Appl. Prob., Vol. 30 , pp 616-626.

Chow, Y.S., S. Moriguti, H. Robbins & S.M. Samuels (1964) *Optimal selection based on relative ranks*, Israel Journal of Mathematics, Vol. 2, pp 81-90.

Gnedin, A.V. (1995) personal communication.

Kennedy, D.P. & R.P. Kertz (1990) *Limit Theorems for threshold-stopped random variables with applications to optimal stopping.* Adv. Appl. Prob., Vol. 22, pp 396-411.

Rényi, A. (1962) *Théorie des éléments saillants d'une suite d'observations.* Proc. Colloq. Comb. Meth. in Prob. Theory (Aarhus Univ.), pp 104-115.

Robbins, H. (1990) personal communication.

Analysis of recursive algorithms by the contraction method

M. Cramer
L. Rüschendorf
University of Freiburg

ABSTRACT Several examples of the asymptotic analysis of recursive algorithms are investigated by the contraction method. The examples concern random permutations and binary search trees. In these examples it is demonstrated that the contraction method can be applied successfully to problems with contraction constants converging to one and with nonregular normalizations as logarithmic normalizations, which are typical in search type algorithms. An advantage of this approach is its generality and the possibility to obtain quantitative approximation results.

1   Introduction

The contraction method is a general method for the study of recursive algorithms. It is based essentially on the application of suitable probability metrics which reflect the structure of the algorithm. The contraction principle was introduced in its general form in Rachev and Rüschendorf [RR91]. A wellknown problem of this method and the application of probability metrics to limit theorems are nonregular normalizations of logarithmic type (which are quite typical in a wide area of algorithms) since this behaviour is not reflected in the regularity structure of probability metrics, while power normalizations n^a can be captured more easily. A second difficulty are contraction factors converging asymptotically to one.

In this paper we deal with several examples which show this problematic behaviour by the use of a modified version of the contraction method. In sections 2 and 3 we consider the number of inversions and the 'MAX'-algorithm for random permutations and in sections 4 and 5 we consider successful and unsuccessful searching in binary random trees. Each of these examples needs some special arguments in order to achieve approximation by a limit distribution; so in general the contraction method can not be considered as an 'automatic' method.

The advantage of the contraction method are its generality which allows e.g. to consider recursions in very general spaces and the fact that it often allows to obtain quantitative approximations.

The examples of this paper which are based on parts of the dissertation of Cramer [Cr95] show that the method can also deal with more difficult situations concerning the contraction factors and the normalizations.

Some of the probability metrics used in this paper are the Kolmogorov-metric ρ for real random variables X, Y defined by

$$\rho(X,Y) := \sup\left\{|F_X(x) - F_Y(x)| : x \in \mathbb{R}^1\right\}, \tag{1.1}$$

F_X, F_Y the corresponding distribution functions, and the Zolotarev-metric

$$\zeta_r(X,Y) := \sup\left\{|\mathbb{E}[f(X) - f(Y)]| : \left|f^{(m)}(x) - f^{(m)}(y)\right| \le |x - y|^\alpha\right\} \tag{1.2}$$

where $r = m + \alpha$, $m \in \mathbb{N}_0$, $0 < \alpha \le 1$. Convergence w.r.t. ζ_r implies weak convergence.

For $\mathbb{E}\,X^j = \mathbb{E}\,Y^j$, $j = 1, \ldots, m$ and $r > 1$

$$\zeta_r(X,Y) \le \frac{\Gamma(1 + 1/p)}{\Gamma(1 + r)}\,\kappa_r(X,Y). \tag{1.3}$$

where

$$\kappa_r(X,Y) := r \int |x|^{r-1} \cdot |F_X(x) - F_Y(x)|\,dx \tag{1.4}$$

is the r-difference pseudomoment. Observe that

$$\kappa_r(X,Y) \le \mathbb{E}\,|X|^r + \mathbb{E}\,|Y|^r. \tag{1.5}$$

For the general properties of these and related metrics we refer to Rachev [Ra91].

2 The number of inversions of a random permutation

In a permutation $\sigma = (a_1, \ldots, a_n)$ a pair (a_i, a_j), $i < j$ is called inversion if $a_i > a_j$. Denote by I_n the number of inversions in a random permutation of size n. Then one obtains the following recursion

$$I_n \overset{d}{=} I_{n-1} + X_n, \quad I_1 = 0 \tag{2.1}$$

where $X_n \sim U(\{0, \ldots, n-1\})$ is uniformly distributed on $0, \ldots, n-1$ and I_{n-1}, X_n are independent. This leads to the moment generating function and first moments

$$G_n(z) = \mathbb{E}z^{I_n} = \frac{1}{n!} \cdot \frac{(1 - z^2) \cdot \ldots \cdot (1 - z^n)}{(1 - z)^{n-1}} \tag{2.2}$$

$$\mathbb{E}\,I_n = \frac{n(n-1)}{4}, \quad \text{Var}\,I_n = \frac{(n-1)\,n\,(2n+5)}{72} \tag{2.3}$$

(cf. Hofri [Ho87, pg. 122-124]). For the normalized version

$$\widehat{I}_n := \frac{I_n - \mathbb{E} I_n}{\sqrt{\text{Var } I_n}} \tag{2.4}$$

one obtains the following Berry-Esseen type result (Note that we assume that all the occuring random variables are defined on the same probability space).

Theorem 2.1 *For $n \geq 7$ holds*

$$\rho\left(\widehat{I}_n, \mathcal{N}(0,1)\right) \leq C \cdot n^{-\frac{1}{2}}, \tag{2.5}$$

with $C = 2.75 \cdot \frac{8^4}{6 \cdot 128} \sqrt{\frac{7}{6}}$.

Proof: W.l.g. we assume that $I_n = \sum_{i=1}^{n} X_i$ where X_i are independent, $X_i \sim U(\{0, \ldots, i-1\})$. By the Berry-Esseen theorem (cf. Bhattacharya and Ranga Rao [BR76, Th. 12.4])

$$\rho\left(\widehat{I}_n, \mathcal{N}(0,1)\right) \leq 2.75 \frac{S_{n,3}}{(S_{n,2})^{3/2}} \tag{2.6}$$

where

$$S_{n,m} := \sum_{k=1}^{n} \mathbb{E}|X_k - \mathbb{E} X_k|^m. \tag{2.7}$$

We have $X_1 \equiv 0$ and for $k \geq 2$ (by some calculations)

$$\mathbb{E}|X_k - \mathbb{E} X_k|^3 \leq \frac{k^3}{32}, \quad \text{Var } X_k = \frac{k^2 - 1}{12}.$$

This implies $\sum_{k=1}^{n} \text{Var } X_k \geq \frac{(n-1)^3}{36}$ and $\sum_{k=1}^{n} \mathbb{E}|X_k - \mathbb{E} X_k|^3 \leq \frac{(n+1)^4}{128}$. So from (2.6) we obtain for $n \geq 7$

$$\rho\left(\widehat{I}_n, \mathcal{N}(0,1)\right) \leq 2.75 \frac{6^3}{128} \left(\frac{n+1}{n-1}\right)^4 \sqrt{\frac{n}{n-1}} \, n^{-\frac{1}{2}} \leq C \, n^{-\frac{1}{2}}.$$

$\square$

Recursion (2.1) leads to a sum of independent variables and, therefore, allows the application of the classical tools of the central limit theorem. On the other hand it is an interesting test example for the contraction method, since the contraction factors of the normalized recursion converge to one. Also the approximation result in terms of the ζ_3-metric is of independent interest. It gives the same convergence rate as in Theorem 2.1 uniformly for integrals of functions $f(I_n)$ with $\|f^{(3)}\|_\infty \leq 1$, where $\widetilde{I}_n$ is the normalization defined by

$$\widetilde{I}_n := \frac{I_n - \mathbb{E} I_n}{n^{3/2}}. \tag{2.8}$$

Theorem 2.2 Let $\sigma_n^2 := \text{Var}(\widetilde{I}_n)$ and $Z_n \sim \mathcal{N}(0, \sigma_n^2)$, then for some $C > 0$ and all $n \in \mathbb{N}$

$$\zeta_3(\widetilde{I}_n, Z_n) \leq C\, n^{-\frac{1}{2}}. \tag{2.9}$$

Proof: $\widetilde{I}_n$ satisfies the modified recursion

$$\widetilde{I}_n \stackrel{d}{=} \left(\frac{n-1}{n}\right)^{3/2} \widetilde{I_{n-1}} + \widetilde{X}_n \tag{2.10}$$

where $\widetilde{X}_n := \frac{X_n - \mathbb{E} X_n}{n^{3/2}}$. Let (Z_n) be independent $Z_n \sim \mathcal{N}(0, \sigma_n^2)$ and define the 'adapted' normal copy of (2.10)

$$Z_n^\star := \left(\frac{n-1}{n}\right)^{3/2} Z_{n-1} + \widetilde{X}_n. \tag{2.11}$$

Let $Y_i \sim \mathcal{N}(0, \tau_i^2)$ be independent of $(\widetilde{X}_i, Z_{i-1})$ where $\tau_i^2 := \sigma_i^2 - \left(\frac{i-1}{i}\right)^3 \sigma_{i-1}^2 = \frac{\text{Var}\, X_i}{i^3} \geq 0$, then

$$Z_i \stackrel{d}{=} \left(\frac{i-1}{i}\right)^{3/2} Z_{i-1} + Y_i. \tag{2.12}$$

Using the regularity of order three of ζ_3 we obtain

$$\zeta_3(\widetilde{I}_n, Z_n) \leq \zeta_3\left(\left(\frac{n-1}{n}\right)^{3/2}\widetilde{I_{n-1}} + \widetilde{X}_n, \left(\frac{n-1}{n}\right)^{3/2}Z_{n-1} + \widetilde{X}_n\right)$$
$$+ \zeta_3(Z_n^\star, Z_n)$$
$$\leq \left(\frac{n-1}{n}\right)^{9/2} \zeta_3(\widetilde{I_{n-1}}, Z_{n-1}) + \zeta_3(Z_n^\star, Z_n).$$

and by iteration using $Z_1 = \widetilde{I}_1 = 0$ one obtains the 'basic estimate':

$$\zeta_3(\widetilde{I}_n, Z_n) \leq \sum_{i=2}^{n} \left(\frac{i}{n}\right)^{9/2} \zeta_3(Z_i^\star, Z_i). \tag{2.13}$$

Note that $\mathbb{E}\, Z_i = \mathbb{E}\, Z_i^\star = 0$ and $\mathbb{E}\, Z_i^2 = \mathbb{E}(Z_i^\star)^2$. Therefore, by (1.3), (1.5) and (2.12) and some calculations (cf. Cramer [Cr95])

$$\zeta_3(Z_i^\star, Z_i) = \zeta_3\left(\widetilde{X}_i + \left(\frac{i-1}{i}\right)^{3/2}Z_{i-1}, Y_i + \left(\frac{i-1}{i}\right)^{3/2}Z_{i-1}\right)$$
$$\leq \zeta_3\left(\widetilde{X}_i, Y_i\right) \leq \frac{\Gamma(2)}{\Gamma(4)}\kappa_3\left(\widetilde{X}_i, Y_i\right)$$
$$= \int_{-\infty}^{0} x^2\, |F_{\widetilde{X}_i}(x) - F_{Y_i}(x)|\, dx$$
$$\leq \frac{7}{2^6 \cdot 3^2} i^{-3/2} + \frac{1}{2^5} i^{-5/2}.$$

Therefore, by some simple calculations

$$
\begin{aligned}
\zeta_3(\widetilde{I}_n, Z_n) &\leq \sum_{i=2}^{n} \left(\frac{i}{n}\right)^{9/2} \zeta_3(Z_i^\star, Z_i) \\
&\leq \sum_{i=2}^{n} \left(\frac{i}{n}\right)^{9/2} \left(\frac{1}{2^5} i^{-5/2} + \frac{7}{2^6 \cdot 3^2} i^{-3/2}\right) \\
&\leq \frac{1}{n^{9/2}} \left[\frac{1}{2^5}\left(n^2 + \frac{1}{3}n^3\right) + \frac{7}{2^6 \cdot 3^2}\left(n^3 + \frac{1}{4}n^4\right)\right] \\
&= \frac{7}{2^8 \cdot 3^2} \cdot \frac{1}{\sqrt{n}} + \mathcal{O}\left(n^{-\frac{3}{2}}\right).
\end{aligned}
$$

$\square$

Note that the contraction factor in this example is only of order $\left(\frac{n-1}{n}\right)^{\frac{3}{2}}$ and so we can not obtain a uniform bound but have to estimate carefully the individual terms.

3   The number of records

The number of records in a sequence $x_1, \ldots, x_n$ is relevant for the behaviour of the 'MAX'-algorithm which determines the maximum element of the sequence (cf. Hofri [Ho87, pg. 112/113]). Let M_n denote the number of maxima of a random permutation read from the left to the right. Then one obtains the recursion

$$
M_n \overset{d}{=} M_{n-1} + X_n, \tag{3.1}
$$

where $X_n \sim \mathfrak{B}\left(1, \frac{1}{n}\right)$ and X_n, M_{n-1} are independent. Define $M_1 = 0$, then

$$
M_n \overset{d}{=} \sum_{i=2}^{n} X_i \tag{3.2}
$$

(X_i) independent, and

$$
\mathbb{E}\, M_n = \mathrm{H}_n - 1, \quad \operatorname{Var} M_n = \mathrm{H}_n - \mathrm{H}_n^{(2)} \tag{3.3}
$$

where $\mathrm{H}_n^{(k)} = \sum_{j=1}^{n} \frac{1}{j^k}$, $\mathrm{H}_n = \mathrm{H}_n^{(1)} = \ln n + \gamma + \mathcal{O}\left(n^{-1}\right)$,
$\mathrm{H}_n^{(2)} \underset{n \to \infty}{\longrightarrow} \zeta(2) = \frac{\pi^2}{6}$ (cf. Hofri [Ho87]).
Define the normalized version

$$
\widehat{M_n} := \frac{M_n - \mathbb{E}\, M_n}{\sqrt{\operatorname{Var} M_n}}, \tag{3.4}
$$

then as in section 2 we obtain the normal approximation but with logarithmic rate only.

Theorem 3.1 *For all* $n \in \mathbb{N}$ *and some constant* $C > 0$ *holds*

$$\rho\left(\widehat{M_n}, \mathcal{N}(0,1)\right) \le \frac{C}{\sqrt{\ln n}}. \tag{3.5}$$

Proof: Follows from the Berry-Esseen bound (2.6), where $\mathbb{E}\, X_k = \frac{1}{k}$, $\operatorname{Var} X_k = \frac{k-1}{k^2}$, $\mathbb{E}|X_k - \mathbb{E}\, X_k|^3 = \frac{k^3 - 3k^2 + 4k - 2}{k^4}$.

Therefore, $\sum_{k=2}^{n} \mathbb{E}|X_k - \mathbb{E} X_k|^3 \sim \ln n$ and $\sum_{k=2}^{n} \operatorname{Var} X_k \sim \ln n$ and so we obtain (3.5) (C can easily be calculated in explicit form). $\qquad \square$

The normalization of M_n is logarithmic in n. For a convergence result as in (3.5) by the contraction method we shall make use of the ζ_3-metric. It turns out that in this example one obtains contraction factors of order $\sqrt{\frac{\ln(n-1)}{\ln n}}$ which converge fast to one. Nevertheless by careful estimation of the individual terms the method as expounded in the proof of Theorem 2.2 also can be applied to this case. Define

$$\widetilde{M_n} := \frac{M_n - \mathbb{E}\, M_n}{\sqrt{\ln n}}. \tag{3.6}$$

Theorem 3.2 *For* $\sigma_n^2 := \operatorname{Var} \widetilde{M_n}$, $Z_n \sim \mathcal{N}(0, \sigma_n^2)$ *holds*

$$\zeta_3(\widetilde{M_n}, Z) = \mathcal{O}\left(\frac{1}{\sqrt{\ln n}}\right). \tag{3.7}$$

Proof: $\widetilde{M_n}$ satisfies the recursion

$$\widetilde{M_n} \stackrel{d}{=} \sqrt{\frac{\ln(n-1)}{\ln n}}\, \widetilde{M_{n-1}} + \widetilde{X_n} \tag{3.8}$$

where $\widetilde{X_n} := \frac{X_n - \mathbb{E}\, X_n}{\sqrt{\ln n}}$. Let (Z_n) be independent, $Z_n \sim \mathcal{N}(0, \sigma_n^2)$,

$$Z_n^\star := \sqrt{\frac{\ln(n-1)}{\ln n}}\, Z_{n-1} + \widetilde{X_n} \tag{3.9}$$

the normal 'copy' and

$$Y_n \sim \mathcal{N}(0, \tau_n^2), \quad \tau_n^2 := \sigma_n^2 - \frac{\ln(n-1)}{\ln n}\sigma_{n-1}^2 = \operatorname{Var} \widetilde{X_n} \tag{3.10}$$

then

$$Z_n \stackrel{d}{=} \sqrt{\frac{\ln(n-1)}{\ln n}}\, Z_{n-1} + Y_n. \tag{3.11}$$

and as in section 2

$$\zeta_3(\widetilde{M_n}, Z_n) \le \zeta_3(\widetilde{M_n}, Z_n^\star) + \zeta_3(Z_n^\star, Z_n)$$

$$\le \left(\frac{\ln(n-1)}{\ln n}\right)^{3/2} \zeta_3(\widetilde{M_{n-1}}, Z_{n-1}) + \zeta_3(Y_n, \widetilde{X_n})$$

which yields by iteration

$$\zeta_3(\widetilde{M_n}, Z_n) \leq \left(\frac{\ln 2}{\ln n}\right)^{3/2} \zeta_3(\widetilde{M_2}, Z_2) + \sum_{i=3}^{n} \left(\frac{\ln i}{\ln n}\right)^{3/2} \zeta_3(Y_i, \widetilde{X_i}). \quad (3.12)$$

By the moment estimate $\zeta_3(\widetilde{M_2}, Z_2) < \infty$
and since $\operatorname{Var} \widetilde{X_i} = \operatorname{Var} Y_i = \tau_i^2 = \frac{1}{\ln i} \cdot \frac{i-1}{i^2}$

$$\zeta_3(Y_i, \widetilde{X_i}) \leq \frac{1}{6} \left(\mathbb{E}|Y_i|^3 + \mathbb{E}|\widetilde{X_i}|^3 \right)$$

$$\leq \frac{1}{(\ln i)^{3/2}} \cdot \frac{1}{6} \left(\frac{1}{i} + \frac{\sqrt{8}}{\sqrt{\pi}} \cdot \frac{1}{i\sqrt{i}} \right) \quad (3.13)$$

using that $\mathbb{E}\left|X_i - \frac{1}{i}\right|^3 \leq \frac{1}{i}$.
Therefore, we finally obtain from (3.12)

$$\zeta_3(\widetilde{M_n}, Z_n) \leq \frac{1}{6} \cdot \left(\frac{\ln 2}{\ln n}\right)^{3/2}$$

$$+ \sum_{i=3}^{n} \left(\frac{\ln i}{\ln n}\right)^{3/2} \cdot \frac{1}{(\ln i)^{3/2}} \cdot \frac{1}{6} \left(\frac{1}{i} + \frac{\sqrt{8}}{\sqrt{\pi}} \cdot \frac{1}{i\sqrt{i}} \right)$$

$$= \frac{1}{6(\ln n)^{3/2}} \left[(\ln 2)^{3/2} + \sum_{i=3}^{n} \frac{1}{i} + \sum_{i=3}^{n} \frac{1}{i\sqrt{i}} \cdot \frac{\sqrt{8}}{\sqrt{\pi}} \right]$$

$$\leq \frac{1}{6(\ln n)^{3/2}} \left[(\ln 2)^{3/2} + 2\ln n \right]$$

$$= \frac{1}{3} \cdot \frac{1}{\sqrt{\ln n}} + \mathcal{O}\left((\ln n)^{-3/2} \right).$$

$\square$

4   Unsuccessful searching in binary search trees

An introduction to random search trees is given in Mahmoud [Ma92]. In this and the following section we deal with the analysis of inserting and retrieving randomly ordered data in binary search trees by the contraction method.

Let U_n denote the number of comparisons necessary to insert a new random element in a random search tree. A search tree is called random if it arises from a random permutation. An element to be inserted in a tree is called random if any of the $n+1$ free leaves of the tree has probability $\frac{1}{n+1}$ to be attained.

U_n satisfies the recursion

$$U_n \stackrel{d}{=} U_{n-1} + Y_n, \quad U_0 = 0, \quad (4.1)$$

where U_{n-1}, Y_n are independent, $Y_n \sim \mathfrak{B}(1, \frac{2}{n+1})$. For $n = 1$ one comparison with the root is necessary. For $n \geq 2$ insertion of the $(n+1)$-th element needs as many comparisons in the n-tree as in the $(n-1)$-tree except in the case that one comparison with the n-th element is necessary. The probability that no comparison with this element is necessary is $\frac{n-1}{n+1}$.

From (4.1) one obtains

$$\mathbb{E}\,U_n = 2\,(H_{n+1} - 1), \quad \mathrm{Var}\,U_n = 2\,H_{n+1} - 4\,H_{n+1}^{(2)} + 2. \tag{4.2}$$

Brown and Shubert (1984) (cf. Mahmoud [Ma92, pg. 76]) proved a central limit theorem for U_n based on Lyapunov's theorem and generating functions. Since by (4.1)

$$U_n \stackrel{d}{=} \sum_{i=1}^{n} Y_i, \quad Y_i \sim \mathfrak{B}(1, \frac{2}{i+1}), \quad (Y_i) \text{ independent,} \tag{4.3}$$

this argument can be simplified to yield:

Theorem 4.1 *Define* $\widehat{U_n} := \frac{U_n - \mathbb{E}\,U_n}{\sqrt{\mathrm{Var}\,U_n}}$, *then for some constant* $C > 0$ *and all* n

$$\rho\left(\widehat{U_n}, \mathcal{N}(0,1)\right) \leq \frac{C}{\sqrt{\ln n}}. \tag{4.4}$$

Proof: Observe that $\frac{S_{n,3}}{S_{n,2}^{3/2}} \sim \frac{1}{\sqrt{2\ln n}}$ (cf. Mahmoud [Ma92, pg. 77]). Therefore, (4.4) is a consequence of (2.6). $\qquad\square$

The contraction method can be applied in much the same way as in section 3. We, therefore, only give a sketch of this application. For more details we refer to Cramer [Cr95].

Theorem 4.2 *Define* $\widetilde{U_n} := \frac{U_n - \mathbb{E}\,U_n}{\sqrt{\ln n}}$, $\sigma_n^2 := \mathrm{Var}\,\widetilde{U_n}$, $Z_n \sim \mathcal{N}(0, \sigma_n^2)$, *then for some* $C > 0$ *and all* $n \in \mathbb{N}$

$$\zeta_3\left(\widetilde{U_n}, Z_n\right) \leq \frac{C}{\sqrt{\ln n}}. \tag{4.5}$$

Proof: $\widetilde{U_n}$ satifies the recursion

$$\widetilde{U_n} \stackrel{d}{=} \sqrt{\frac{\ln(n-1)}{\ln n}}\,\widetilde{U_{n-1}} + \widetilde{Y_n}, \quad \widetilde{Y_n} := \frac{Y_n - \mathbb{E}\,Y_n}{\sqrt{\ln n}}. \tag{4.6}$$

Define

$$Z_n^\star := \sqrt{\frac{\ln(n-1)}{\ln n}}\,Z_{n-1} + \widetilde{Y_n}, \tag{4.7}$$

$$\tau_n^2 := \sigma_n^2 - \frac{\ln(n-1)}{\ln n}\,\sigma_{n-1}^2 = \mathrm{Var}\,\widetilde{Y_n}. \tag{4.8}$$

Then for $W_n \sim \mathcal{N}(0, \tau_n^2)$ independent of (Z_n), (Y_n)

$$Z_n \overset{d}{=} \sqrt{\frac{\ln(n-1)}{\ln n}} Z_{n-1} + W_n. \tag{4.9}$$

And so as in section 3

$$\zeta_3\left(\widetilde{U}_n, Z_n\right) \leq \left(\frac{\ln 2}{\ln n}\right)^{3/2} \zeta_3\left(\widetilde{U}_2, Z_2\right) + \sum_{i=3}^{n} \left(\frac{\ln i}{\ln n}\right)^{3/2} \zeta_3\left(W_i, \widetilde{Y}_i\right) \tag{4.10}$$

As $\mathbb{E}\,\widetilde{U}_2 = 0 = \mathbb{E}\,Z_2$, $\mathrm{Var}\,\widetilde{U}_2 = \sigma_2^2 = \mathrm{Var}\,Z_2$ one obtains

$$\zeta_3\left(\widetilde{U}_2, Z_2\right) \leq \tfrac{1}{6}\left(\mathbb{E}\left|\widetilde{U}_2\right|^3 + \mathbb{E}\,|Z_2|^3\right) < \infty. \text{ Also,}$$

$$\zeta_3\left(W_i, \widetilde{Y}_i\right) \leq \frac{1}{6}\left(\mathbb{E}|W_i|^3 + \mathbb{E}\left|\widetilde{Y}_i\right|^3\right)$$

$$= \frac{1}{6}\left[\frac{2\sqrt{2}}{\sqrt{\pi}}\tau_i^3 + \frac{1}{(\ln i)^{3/2}}\left(\frac{i-1}{i+1}\left(\frac{2}{i+1}\right)^3 + \frac{2}{i+1}\left(\frac{i-1}{i+1}\right)^3\right)\right]$$

$$\leq \frac{2}{6\,(\ln i)^{3/2}\,(i+1)}\left[1 + \frac{4}{\sqrt{\pi}\,(i+1)}\right].$$

Therefore,

$$\zeta_3\left(\widetilde{U}_n, Z_n\right) \leq \frac{1}{(\ln n)^{3/2}}\frac{1}{6}\left(\frac{10}{81} + \frac{8}{27\sqrt{\pi}}\right)$$

$$+ \frac{1}{(\ln n)^{3/2}}\sum_{i=3}^{n}\frac{1}{i+1}\cdot\frac{1}{3}\left(1 + \frac{4}{\sqrt{\pi(i+1)}}\right)$$

$$\leq \frac{1}{\sqrt{\ln n}} \qquad \text{for } n \geq n_0. \tag{4.11}$$

$$\square$$

Remark: Regarding recursion (4.6) one obtains convergence also under alternative distributional assumptions on $\widetilde{Y}_n$ (resp. Y_n). Let μ_r be any $(r, +)$-ideal, simple metric (cf. Rachev [Ra91]) then as in (4.10) one obtains for a recursion of this type the basic estimate

$$\mu_r\left(\widetilde{U}_n, Z_n\right) \leq \left(\frac{\ln 2}{\ln n}\right)^{r/2}\mu_r\left(\widetilde{U}_2, Z_2\right) + \sum_{i=3}^{n}\left(\frac{\ln i}{\ln n}\right)^{r/2}\mu_r\left(W_i, \widetilde{Y}_i\right) \tag{4.12}$$

Then

$$\mu_r\left(\widetilde{U}_n, Z_n\right) \xrightarrow[n\to\infty]{} 0 \tag{4.13}$$

if

$$\text{a) } \mu_r\left(\widetilde{U}_2, Z_2\right) < \infty, \quad \mu_r\left(W_i, \widetilde{Y}_i\right) < \infty, \quad i \geq 3$$

and

b) $\quad \mu_r\left(W_i, \widetilde{Y}_i\right) = o\left(\dfrac{1}{i \ln i}\right).$ \hfill (4.14)

For the proof of (4.13) let for $\varepsilon > 0$, $k_0 \in \mathbb{N}$ such that $\mu_r(W_k, \widetilde{Y_k}) \leq \dfrac{\varepsilon}{k \ln k}$,
then

$$
\begin{aligned}
\limsup_{n \to \infty} \mu_r\left(\widetilde{U_n}, Z_n\right) \;\leq\; & \limsup_{n \to \infty} \left(\frac{\ln 2}{\ln n}\right)^{r/2} \mu_r\left(\widetilde{U_2}, Z_2\right) \\
& + \limsup_{n \to \infty} \frac{1}{(\ln n)^{r/2}} \sum_{i=3}^{k_0 - 1} (\ln i)^{r/2} \mu_r\left(W_i, \widetilde{Y}_i\right) \\
& + \limsup_{n \to \infty} \frac{1}{(\ln n)^{r/2}} \sum_{i=k_0}^{n} (\ln i)^{r/2 - 1} \frac{1}{i}\, \varepsilon \\
\leq\; & 0 + 0 + \limsup_{n \to \infty} \varepsilon \frac{1}{\ln n} \sum_{i=k_0}^{n} \frac{1}{i} \;\leq\; \varepsilon
\end{aligned}
$$

$\square$

In the preceding example of unsuccessful searching the estimates of $\mu_r(W_i, \widetilde{Y}_i)$ of order $\frac{1}{i(\ln i)^{3/2}}$ even allow to obtain the convergence rate $\frac{1}{\sqrt{\ln n}}$.

5 Successful searching in binary search trees

Let for a random binary search tree as in section 4 S_n denote the number of comparisons to retrieve a randomly chosen element in the tree. Based on a formula for $\mathcal{P}(S_n = k)$ by Brown and Shubert (1984) a derivation of $\mathbb{E}\, S_n$, $\operatorname{Var} S_n$ and a central limit theorem for S_n due to Louchard (1987) is derived via generating functions in Mahmoud [Ma92, pg. 78-82]. We next derive a quantitative version of the central limit theorem by the contraction method and moment formulas based on the following (new) recursion for S_n:

$$
S_n \stackrel{d}{=} 1 + S_{I_n}, \quad S_0 = 0, \quad S_1 = 1, \tag{5.1}
$$

where I_n is independent of (S_i) and $\mathcal{P}(I_n = 0) = \frac{1}{n}$, $\mathcal{P}(I_n = j) = \frac{2j}{n^2}$, $1 \leq j \leq n - 1$.

It can be shown that this recursion does not allow the form of a sum of independent random variables as considered in the random search algorithm in Rachev and Rüschendorf [RR91]. Therefore, it does not allow the application of the Berry-Esseen type or Poisson type approximation result. (5.1) arises from the recursion

$$
\mathcal{P}(S_n = k) \;=\; \sum_{j=1}^{n} \mathcal{P}(S_n = k,\, j \text{ chosen})
$$

$$= \sum_{j=1}^{n} \sum_{i=1}^{n} \frac{1}{n^2} \left(\mathbb{1}_{\{i=j\}} \delta_{1k} + \mathbb{1}_{\{i<j\}} \mathcal{P}(S_{n-i} = k - 1) \right.$$

$$\left. + \mathbb{1}_{\{i>j\}} \mathcal{P}(S_{i-1} = k - 1) \right)$$

$$= \frac{\delta_{1k}}{n} + \sum_{i=1}^{n} \frac{n-i}{n^2} \mathcal{P}(S_{n-i} = k - 1)$$

$$+ \sum_{i=1}^{n} \frac{i-1}{n^2} \mathcal{P}(S_{i-1} = k - 1)$$

$$= \frac{\delta_{1k}}{n} + \sum_{j=1}^{n-1} \mathcal{P}(S_j = k - 1) \cdot \frac{2j}{n^2}. \tag{5.2}$$

An explicit formula for $\mathcal{P}(S_n = k)$ due to Brown and Shubert (1984) (cf. Mahmoud [Ma92, pg. 79]) is relatively involved. Based on this formula the first two moments are derived in Mahmoud [Ma92, pg. 80]. The recursion (5.1) allows a direct calculation.

Proposition 5.1

a) $\mathbb{E} S_n = 2 \left(1 + \frac{1}{n} \right) H_n - 3$ \hfill (5.3)

b) $\operatorname{Var} S_n = \left(2 + \frac{10}{n} \right) H_n - 4 \left(1 + \frac{1}{n} \right) \left[\frac{H_n^2}{n} + H_n^{(2)} \right] + 4$ \hfill (5.4)

Proof:

a) $\mathbb{E} S_n = 1 + \mathbb{E}\left(\mathbb{E}(S_{I_n} | I_n) \right) = 1 + \sum_{k=0}^{n-1} \mathcal{P}(I_n = k) \, \mathbb{E} S_k$

$$= 1 + \sum_{k=0}^{n-1} \frac{2k}{n^2} \mathbb{E} S_k. \tag{5.5}$$

With $Q_n := n \cdot \mathbb{E} S_n$, (5.5) leads to $Q_n = n + \frac{2}{n} \sum_{k=1}^{n-1} Q_k$, $Q_1 = 1$, which implies $Q_{n+1} = \frac{2n+1}{n+1} + \frac{n+2}{n+1} Q_n$. By iteration one obtains

$$Q_n = \frac{2n-1}{n} + \sum_{k=1}^{n-1} \frac{2k-1}{k} \cdot \frac{n+1}{k+1}$$

$$= (n+1) \left[\sum_{k=1}^{n} \frac{2}{k+1} - \sum_{k=1}^{n} \left(\frac{1}{k} - \frac{1}{k+1} \right) \right]$$

$$= (n+1) \left[2(H_{n+1} - 1) - 1 + \frac{1}{n+1} \right]$$

$$= 2(n+1) H_n - 3n.$$

b) $\mathbb{E} S_n^2 = 1 + 2 \mathbb{E} S_{I_n} + \mathbb{E} S_{I_n}^2$

$$= 1 + 2\,(\text{IE}\,S_n - 1) + \sum_{j=1}^{n-1} \frac{2\,j}{n^2}\,\text{IE}\,S_j^2.$$

With $P_n := n \cdot \text{IE}\,S_n^2$, $P_n = -n + 2\,Q_n + \frac{2}{n}\sum_{j=1}^{n-1} P_j$.

This yields $\frac{n+1}{2}\,P_{n+1} - \frac{n}{2}\,P_n = \frac{2n+1}{2} + 2\,Q_n + P_n$. By a), therefore, one obtains

$$P_{n+1} = 8\,H_n - \frac{10\,n - 1}{n+1} + \frac{n+2}{n+1}\,P_n$$

and by iteration

$$P_n = \sum_{j=1}^{n} \left(8\,H_j - \frac{10\,j - 3}{j} \right) \frac{n+1}{j+1}.$$

Using the relation

$$\sum_{j=1}^{n} \frac{H_j}{j} = \frac{H_n^{(2)} + H_n^2}{2}$$

one can explicitly calculate P_n and obtain (5.4).   □

Define the normalizations

$$\widetilde{S_n} := \frac{S_n - \text{IE}\,S_n}{\sqrt{2\,\ln n}}, \quad \widetilde{S_0} = \widetilde{S_1} = 0 \tag{5.6}$$

and

$$a(k,n) := 1 - \text{IE}\,S_n + \text{IE}\,S_k, \quad b(k) := \text{Var}\,S_k, \quad \sigma_n^2 := \text{Var}\,\widetilde{S_n}. \tag{5.7}$$

For our derivation we need the following (still unsettled) conjecture:

$$\text{C)} \quad \limsup_{n\to\infty} \int y^2 \left| \sum_{k=2}^{n-1} \frac{2\,k}{n^2} \left[\phi\left(\frac{y}{\sqrt{b(n)}} \right) - \phi\left(\frac{y - a(k,n)}{\sqrt{b(k)}} \right) \right] \right| dy < \infty \tag{5.8}$$

ϕ the distribution function of the standard normal distribution.

Let (Z_n) be independent of (S_n), $Z_n \sim \mathcal{N}(0, \sigma_n^2)$, then under C) we obtain the following quantitative result.

Theorem 5.2 *Under* C) *holds*

$$\zeta_3\left(\widetilde{S_n}, Z_n \right) = \mathcal{O}\left(\frac{1}{\sqrt{\ln n}} \right). \tag{5.9}$$

Proof: $(\widetilde{S_n})$ satisfies the recursion

$$\widetilde{S_n} \stackrel{d}{=} \sqrt{\frac{\ln I_n}{\ln n}}\,\widetilde{S_{I_n}} + c_n(I_n), \tag{5.10}$$

where $c_n(k) := \frac{1 - \text{IE}\,S_n + \text{IE}\,S_k}{\sqrt{2\,\ln n}}$.

Define the normal copy

$$Z_n^\star \overset{d}{=} \sqrt{\frac{\ln I_n}{\ln n}}\, Z_{I_n} + c_n(I_n). \tag{5.11}$$

Then we obtain from the wellknown properties of ζ_3, using the independence assumptions, the basic recursive estimate

$$\zeta_3\left(\widetilde{S}_n, Z_n\right) \leq \zeta_3\left(\widetilde{S}_n, Z_n^\star\right) + \zeta_3\left(Z_n^\star, Z_n\right)$$

$$\leq \sum_{k=0}^{n-1} \mathcal{P}(I_n = k)\,\zeta_3\left(\sqrt{\frac{\ln k}{\ln n}}\,\widetilde{S}_k + c_n(k), \sqrt{\frac{\ln k}{\ln n}}\, Z_k + c_n(k)\right)$$

$$+ \zeta_3(Z_n^\star, Z_n)$$

$$\leq \sum_{k=2}^{n-1} \frac{2\,k}{n^2}\left(\frac{\ln k}{\ln n}\right)^{3/2} \zeta_3\left(\widetilde{S}_k, Z_k\right) + \zeta_3\left(Z_n^\star, Z_n\right). \tag{5.12}$$

Since

$$\mathbb{E}\,Z_n^\star = \sum_{k=0}^{n-1} \mathcal{P}(I_n = k)\,\mathbb{E}\left(\sqrt{\frac{\ln k}{\ln n}}\, Z_k + c_n(k)\right)$$

$$= \sum_{k=0}^{n-1} \mathcal{P}(I_n = k)\,\frac{1 - \mathbb{E}\,S_n + \mathbb{E}\,S_k}{\sqrt{2\ln n}}$$

$$= (2\ln n)^{-1/2}\,[1 - \mathbb{E}\,S_n + \mathbb{E}\,S_{I_n}] = 0 = \mathbb{E}\,Z_n$$

and, similarly,

$$\mathbb{E}\left(Z_n^\star\right)^2 = \frac{1}{2\ln n}\,\mathrm{Var}\,S_n = \mathrm{Var}\,\widetilde{S}_n,$$

we obtain

$$\zeta_3\left(Z_n^\star, Z_n\right) \leq \frac{1}{6}\kappa_3\left(Z_n^\star, Z_n\right) = \frac{1}{2}\int x^2\left|F_{Z_n^\star}(x) - F_{Z_n}(x)\right|\,dx.$$

Furthermore,

$$F_{Z_n}(x) = \phi(x/\sigma_n) = \phi\left(x\sqrt{2\ln n}/\sqrt{b(n)}\right)$$

and

$$F_{Z_n^\star}(x) = \sum_{k=0}^{n-1} \mathcal{P}(Z_n^\star \leq x \mid I_n = k)\cdot\mathcal{P}(I_n = k)$$

$$= \sum_{k=2}^{n-1} \frac{2\,k}{n^2}\,\phi\left(\frac{x\sqrt{2\ln n} - a(k,n)}{\sqrt{b(k)}}\right) + \frac{1}{n}\,\mathbb{1}_{[1-\mathbb{E}\,S_n,\infty)}(x\sqrt{2\ln n})$$

$$+ \frac{2}{n^2}\,\mathbb{1}_{[2-\mathbb{E}\,S_n,\infty)}(x\sqrt{2\ln n}).$$

This implies using the substitution $y = x \cdot \sqrt{2 \ln n}$

$$\zeta_3 \left(Z_n^\star, Z_n \right) \leq \frac{1}{2 \cdot (2 \ln n)^{3/2}} \left[A_n + B_n + C_n \right] \tag{5.13}$$

where

$$A_n := \frac{1}{n} \int y^2 \left| 1\!\!1_{[1 - \mathbb{E} S_n, \infty)}(y) - \phi \left(\frac{y}{\sqrt{b(n)}} \right) \right| dy,$$

$$B_n := \frac{2}{n^2} \int y^2 \left| 1\!\!1_{[2 - \mathbb{E} S_n, \infty)}(y) - \phi \left(\frac{y}{\sqrt{b(n)}} \right) \right| dy,$$

$$C_n := \int y^2 \left| \sum_{k=2}^{n-1} \frac{2k}{n^2} \left[\phi \left(\frac{y - a(k,n)}{\sqrt{b(k)}} \right) - \phi \left(\frac{y}{\sqrt{b(n)}} \right) \right] \right| dy.$$

Under C) we obtain $C_n \leq M_C$, $\forall\, n \in \mathbb{N}$ and some constant M_C. Also for $n \geq n_0$, $\mathbb{E} S_n \geq 1$ and

$$A_n \leq \frac{1}{n} \int y^2 \left| \phi \left(\frac{y}{\sqrt{b(n)}} \right) - 1\!\!1_{[0,\infty)}(y) \right| dy + \frac{1}{n} \int y^2 \, 1\!\!1_{[1 - \mathbb{E} S_n, 0)}(y) \, dy$$

$$\leq \frac{1}{n} \cdot \frac{1}{3} 2 \frac{\sqrt{2}}{\sqrt{\pi}} \sqrt{b(n)}^3 + \frac{1}{n} \cdot \frac{1}{3} (\mathbb{E} S_n - 1)^3 \xrightarrow[n \to \infty]{} 0$$

as $b(n) = \operatorname{Var} S_n \sim 2 \ln n$ and $\mathbb{E} S_n \sim 2 \ln n$. Therefore, $A_n \leq M_A$, $\forall\, n$ and similarly $B_n \leq M_B$, $\forall n$ and, together, for a constant M we obtain

$$\zeta_3 \left(Z_n, Z_n^\star \right) \leq \frac{M}{(\ln n)^{3/2}}. \tag{5.14}$$

Application of Euler's summation formula (cf. Hofri [Ho87, pg. 19])

$$\sum_{j=1}^{n-1} f(j) = \int_1^n f(x) \, dx + \sum_{k=1}^m \frac{B_k}{k!} \left[f^{(k-1)}(n) - f^{(k-1)}(1) \right] + R_m, \tag{5.15}$$

where

$$R_m = \frac{(-1)^{m+1}}{m!} \int_1^n B_m(\{x\}) f^{(m)}(x) \, dx, \quad \{x\} = x - \lfloor x \rfloor,$$

and (B_k) the Bernoulli numbers, $B_m(x) = \sum_{k \geq 0} \binom{m}{k} B_k \, x^{m-k}$ the m-th Bernoulli-polynomial, to the function $f(x) = x \ln x$, $x \geq 1$ yields for $m = 2$ after some calculations

$$\sum_{j=2}^{n-1} j \ln j = \frac{1}{2} n^2 \ln n - \frac{1}{4} n^2 - \frac{1}{2} n \ln n + \mathcal{O}(\ln n). \tag{5.16}$$

Consider a sufficiently large n_0 such that for $n \geq n_0$, $\sum_{j=2}^{n-1} j \ln j \leq \frac{1}{2} n^2 \ln n - \frac{1}{4} n^2$. Choose $\widetilde{M}$ large enough that (5.9) holds for $n < n_0$ and define $K := \max\left(\widetilde{M}, 2\,M\right)$. Then we finally obtain from (5.12), (5.14) by induction, assuming (5.9) for all $k < n$, that

$$
\begin{aligned}
\zeta_3\left(\widetilde{S}_n, Z_n\right) &\le \sum_{k=2}^{n-1} \frac{2\,k}{n^2}\left(\frac{\ln k}{\ln n}\right)^{3/2} \frac{K}{\sqrt{\ln k}} + \frac{M}{(\ln n)^{3/2}} \\
&= \frac{1}{(\ln n)^{3/2}} \cdot \frac{2}{n^2} \cdot K \sum_{k=2}^{n-1} k\,\ln k + \frac{M}{(\ln n)^{3/2}} \\
&\le \frac{1}{(\ln n)^{3/2}}\left[\frac{2\,K}{n^2}\left(\frac{1}{2}\,n^2\,\ln n - \frac{1}{4}\,n^2\right) + M\right] \\
&\le \frac{1}{(\ln n)^{3/2}}\left[K\,\ln n - \frac{K}{2} + \frac{K}{2}\right] = \frac{K}{\sqrt{\ln n}}.
\end{aligned}
$$

$\square$

Remarks:

a) In the preceding example one can not give a direct proof for the convergence of $\widetilde{S}_n$ based on estimates for the terms in the 'basic estimate' but one obtains convergence by an induction argument which uses the Euler summation formula in a crucial way. This extension of the contraction technique seems to be potentially useful also for further examples.

b) To verify the conjecture C) used in our proof it seems not possible to put the absolute values under the sum. Numerical calculations with 'Maple' for $n \le 10\,000$ indicate that C) is correct. Denote the integral in (5.8) for $n \in \mathbb{N}$ by $f(n)$, then calculation on the range -25 to 25 with a Newton-Cote algorithm with exactness of order 10^{-5} one obtains the following curves of $f(n)$ against n resp. against $\ln(\ln(\ln n))$, which indicate boundedness of f.

Figure 1:

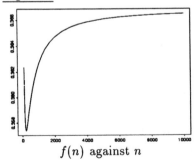
$f(n)$ against n

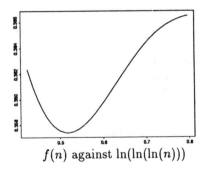
$f(n)$ against $\ln(\ln(\ln(n)))$

□

6 References

[BR76] R.N. Bhattacharya and R. Ranga Rao. *Normal approximation and asymptotic expansions.* Wiley, 1976

[Cr95] M. Cramer. Stochastische Analyse rekursiver Algorithmen mit idealen Metriken. *Dissertation, Universität Freiburg,* 1995

[Ho87] M. Hofri. *Probabilistic analysis of algorithms.* Springer, 1987

[Ma92] H.M. Mahmoud. *Evolution of random search trees.* Wiley, 1992

[Ra91] S.T. Rachev. *Probability metrics and the stability of stochastic models.* Wiley, 1991.

[RR91] S.T. Rachev and L. Rüschendorf. Probability metrics and recursive algorithms (1991). To appear in *Advances in Applied Probability*

M. Cramer und L. Rüschendorf
Institut für Mathematische Stochastik
Hebelstr. 27
79104 Freiburg

Comparison of completely positive maps
on a C^* algebra and a Lebesgue decomposition theorem

by

K.R. Parthasarathy
Indian Statistical Institute, Delhi Centre
7, S.J.S. Sansanwal Marg
New Delhi - 110016, India

e-mail : krp@isid.ernet.in

Key words : *completely positive map, Stinespring dilation, Radon-Nikodym density, Lebesgue decomposition*

AMS subject classification index 46L50

Acknowledgement : This work was begun at the university of Nottingham when the author was an INSA – Royal Society exchange visitor during 25 June – 24 July 1994. He is grateful to V.P. Belavkin for introducing him to the topic of the paper and J.M. Lindsay for several useful discussions.

34

1   Introduction

From the expositions in [M], [BP], [P2] it is now well understood that Status: R

unital completely positive maps on unital C^* algebras are the quantum probabilistic analogues of transition probability operators in Markov processes. In [BS], V.P. Belavkin and P.Staszewski introduced three different notions of absolute continuity of one completely positive (c.p.) map with respect to another on a C^* algebra and proved the existence of a Radon-Nikodym density under the condition of 'strong complete absolute continuity'. Here we combine their approach with the Hilbert space - theoretic proof of the classical Radon-Nikodym theorem (See the exercises in Section 31 of [H] or Section 47 of [P1] and obtain the Lebesgue decomposition of a unital c.p. map into its absolutely continuous and singular parts with respect to another such map. Analogues of chain rule and martingale properties of Radon-Nikodym derivatives and some examples are also included.

2   The Stinespring dilation of a unital c.p. map

We fix a unital C^* algebra $\mathcal{A}$. For any Hilbert space $\mathcal{H}$ denote by $\mathcal{B}(\mathcal{H})$ the C^* algebra of all bounded operators on $\mathcal{H}$. The identity element of any C^* algebra will be denoted by 1. By a c.p. map $S : \mathcal{A} \to \mathcal{B}(\mathcal{H})$ we mean a linear operator satisfying the inequality

$$\sum_{i,j}\langle u_i,\ S(X_i^* X_j)u_j\rangle \geq 0 \qquad (2.1)$$

for all $X_i \in \mathcal{A}$, $u_i \in \mathcal{H}$ and i varying over any finite set. If, in addition, $S(1) = 1$ we say that S is a unital c.p. (u.c.p.) map. Associate with any such u.c.p. map S the algebraic tensor product vector space $\mathcal{A} \otimes \mathcal{H}$ equipped with the nonnegative sesquilinear form B_S given by

$$B_S(\sum_i X_i \otimes u_i, \sum_j Y_j \otimes v_j) = \sum_{i,j} \langle u_i, S(X_i^* Y_j) v_j \rangle. \tag{2.2}$$

Following the standard GNS construction, say that $\xi, \eta \in \mathcal{A} \otimes \mathcal{H}$ are equivalent, $\xi \sim \eta$ in symbols, if $B_S(\xi - \eta, \ \xi - \eta) = 0$. Then $\sim$ is an equivalence relation. Denote by $[\xi]_S$ the equivalence class containing ξ. Such equivalence classes constitute a prehilbert space $\mathcal{D}_S$ with scalar product

$$\langle [\xi]_S, \ [\eta]_S \rangle = B_S(\xi, \eta).$$

Denote by $\mathcal{K}_S$ the Hilbert space obtained by its completion. For any $X \in \mathcal{A}$ there exists a unique bounded linear operator $\pi_S(X)$ on $\mathcal{K}_S$ satisfying

$$\pi_S(X) \left[\sum_i X_i \otimes u_i \right]_S = \left[\sum_i XX_i \otimes u_i \right]_S$$

for any $\sum_i X_i \otimes u_i$ in $\mathcal{A} \otimes \mathcal{H}$. The correspondence $\pi_S : X \to \pi_S(X)$ satisfies the following :

(i) $\|\pi_S(X)\| \leq \|X\|$;

(ii) π_S is a $*$ unital representation of $\mathcal{A}$ in $\mathcal{B}(\mathcal{K}_S)$;

(iii) The set $\{\pi_S(X)V_S u, \ X \in A, \ u \in \mathcal{H}\}$ where $V_S : \mathcal{H} \to \mathcal{K}_S$ denotes the isometry $V_S u = [1 \otimes u]_S$, is total in $\mathcal{K}_S$;

(iv) $S(X) = V_S^* \pi_S(X) V_S$;

(v) If $\mathcal{K}$ is another Hilbert space, $V : \mathcal{H} \to \mathcal{K}$ is an isometry, π is a $*$ unital representation of $\mathcal{A}$ in $\mathcal{B}(\mathcal{K})$, $S(X) = V^* \pi(X) V$ for all $X \in \mathcal{A}$ and $\{\pi(X)Vu, \ X \in A, \ u \in \mathcal{H}\}$ is total in $\mathcal{K}$ then there exists a unitary isomorphism $U : \mathcal{K} \to \mathcal{K}_S$ satisfying $UV = V_S$, $U\pi(X) = \pi_S(X)U$ for all $X \in \mathcal{A}$.

36

Any such triple $(\mathcal{K}, \pi, V)$ satisfying the conditions in (v) is called a *Stinespring dilation* of the unital c.p. map S. The triple $(\mathcal{K}_S, \pi_S, V_S)$ is one such realization of a Stinespring dilation of S.

3 Comparison of two unital c.p. maps

Let $\mathcal{A}, \mathcal{B}(\mathcal{H})$ be as in Section 2 and let S, T be two u.c.p. maps from $\mathcal{A}$ into $\mathcal{B}(\mathcal{H})$. We say that S is *uniformly dominated* by T, $S \prec T$ in symbols, if there exists a positive constant c such that $cT - S$ is a c.p. map. If $S \prec T$ and $T \prec S$ we say that S and T are *uniformly equivalent* and write $S \equiv_u T$. $\prec$ is a partial order and $\equiv_u$ is an equivalence relation. For any two u.c.p. maps S, T from $\mathcal{A}$ into $\mathcal{H}$, any two convex combinations $pS + qT$, $p'S + q'T (0 < p, p' < 1,\ p + q = p' + q' = 1)$ are uniformly equivalent. In the notations of Section 2 we have the following proposition.

Proposition 3.1 ([BS]) : Let $S, T : \mathcal{A} \to \mathcal{B}(\mathcal{H})$ be unital c.p. maps such that $S \prec T$. Then there exists a bounded linear operator $J(S, T) : \mathcal{K}_T \to \mathcal{K}_S$ satisfying the following:

(i) $J(S, T)[\xi]_T = [\xi]_S$ for all $\xi \in \mathcal{A} \otimes \mathcal{H}$;

(ii) $J(S, T)V_T = V_S$, $J(S, T)\pi_T(X) = \pi_S(X)J(S, T)$ for all $X \in \mathcal{A}$;

(iii) The positive bounded operator $\delta(S : T) = J(S, T)^* J(S, T)$ commutes with $\pi_T(X)$ for all X in $\mathcal{A}$ and

$$S(X) = V_T^* \delta(S : T)\pi_T(X)V_T;$$

(iv) If δ' is another (positive) bounded operator commuting with $\pi_T(X)$ for all X in $\mathcal{A}$ such that

$$S(X) = V_T^* \, \delta' \pi_T(X)V_T \text{ for all } X \in \mathcal{A}$$

37

then $\delta' = \delta(S : T)$;

(v) If R, S, T are unital c.p. maps from $\mathcal{A}$ into $\mathcal{B}(\mathcal{H})$ satisfying $R \prec S \prec T$ then

$$
\begin{aligned}
J(R,T) &= J(R,S)J(S,T), \\
\delta(R:T) &= J(S,T)^*\delta(R:S)J(S,T);
\end{aligned}
$$

(vi) If $S \equiv_u T$ then $J(S,T)J(T,S) = 1$, $J(T,S)J(S,T) = 1$ in $\mathcal{K}_S, \mathcal{K}_T$ respectively. In such a case $\delta(S : T)$ is a bounded operator with a bounded inverse and the representations π_S and π_T are unitarily equivalent;

(vii) If $R \prec T$, $S \prec T$ then, for $0 < p < 1$, $q = 1 - p$, $pR + qS \prec T$ and

$$
\delta(pR + qS : T) = p\delta(R : T) + q\delta(S : T).
$$

Proof: Let $c > 0$ be such that $cT - S$ is c.p. Define the map $J(S,T) : \mathcal{D}_T \to \mathcal{D}_S$ by

$$
J(S,T)[\xi]_T = [\xi]_S, \quad \xi \in \mathcal{A} \otimes \mathcal{H}.
$$

If $\xi = \sum_i X_i \otimes u_i$ then

$$
\begin{aligned}
\left\| \left[\sum_i X_i \otimes u_i \right]_S \right\|^2 &= \sum_{i,j} \langle u_i, S(X_i^* X_j)u_j \rangle \\
&\leq c \sum_{i,j} \langle u_i, T(X_i^* X_j)u_j \rangle \\
&= c \left\| \left[\sum_i X_i \otimes u_i \right]_T \right\|^2
\end{aligned}
$$

This shows that $J(S,T)$ is linear on $\mathcal{D}_T$ and extends uniquely to a bounded operator on $\mathcal{K}_T$ with norm not exceeding $\sqrt{c}$. This proves (i). Property (ii) is immediate from (i) and the boundedness of $J(S,T)$. The first part of (iii) is

38

immediate from (ii) and the second from the equalities

$$
\begin{aligned}
S(X) &= V_S^* \pi_S(X) V_S \\
&= V_T^* J(S,T)^* \pi_S(X) J(S,T) V_T \\
&= V_T^* \delta(S:T) \pi_T(X) V_T.
\end{aligned}
$$

To prove (iv) consider δ' satisfying the hypothesis in (iv) and observe that for $X_i, Y_j \in \mathcal{A}$, $u_i, v_j \in \mathcal{H}$

$$
\begin{aligned}
&\left\langle \left[\sum_i X_i \otimes u_i \right]_T, \ \delta' \left[\sum Y_j \otimes v_j \right]_T \right\rangle \\
&= \left\langle \sum_i \pi_T(X_i) V_T u_i, \ \delta' \sum_j \pi_T(Y_j) v_j \right\rangle \\
&= \sum_{i,j} \langle u_i, V_T^* \pi_T(X_i^*) \delta' \pi_T(Y_j) v_j \rangle \\
&= \sum_{i,j} \langle u_i, V_T^* \delta' \pi_T(X_i^* Y_j) v_j \rangle \\
&= \sum_{i,j} \langle u_i, \ S(X_i^* Y_j) v_j \rangle.
\end{aligned}
$$

On the other hand the same relations hold with δ' replaced by $\delta(S:T)$. Since $\mathcal{D}_T$ is dense in $\mathcal{K}_T$ it follows that $\delta' = \delta(S:T)$, proving (iv). Property (v) and the first two parts of (vi) are immediate from definitions. Since the invertibility of $J(S,T)$ implies that in its polar decomposition $J(S,T) = U(S,T)\sqrt{\delta(S:T)}$, the isometry $U(S,T)$ is, indeed, a unitary operator satisfying

$$
U(S,T)\pi_T(X)\delta(S:T)^{1/2} = \pi_S(X)U(S,T)\delta(S:T)^{1/2}
$$

and $\delta(S:T)$ is invertible it follows that π_T and π_S are intertwined by $U(S,T)$. This proves (vi). Property (vii) is immediate from (iii) and (iv). $\blacksquare$

Remark : If S and T are u.c.p. maps from $\mathcal{A}$ into $\mathcal{B}(\mathcal{H})$, $A = pS + qT$, $A' = p'S + q'T$, $0 < p, p' < 1$, $p + q = p' + q' = 1$ then π_A and $\pi_{A'}$ are unitarily equivalent. Indeed, this is immediate from the fact $A \equiv_u A'$.

39

Unital c.p. maps can be looked upon as quantum probabilistic analogues of conditional expectation maps or transition probability opeators [BP]. Thus it is not unreasonable to interpret $\delta(S : T)$ in (iii) of Proposition 3.1 as the 'Radon-Nikodym density' of S with respect to T whenever $S \prec T$. This is the interpretation suggested by Belavkin and Staszewski [BS]. Our next few propositions are based on this analogy with classical integration theory.

Proposition 3.2 : Let $\mathcal{A}_1, \mathcal{A}_2$ be unital C^* algebras and let $\alpha : \mathcal{A}_1 \to \mathcal{A}_2$ be a surjective C^* homomorphism. Suppose S and T are u.c.p. maps from $\mathcal{A}_2$ into $\mathcal{B}(\mathcal{H})$ and $S \prec T$. Then $S \circ \alpha$ and $T \circ \alpha$ are u.c.p. maps from $\mathcal{A}_1$ into $\mathcal{B}(\mathcal{H}), S \circ \alpha \prec T \circ \alpha$ and

$$\delta(S \circ \alpha : T \circ \alpha) = \tilde{U}(T, \alpha)^* \delta(S : T) \tilde{U}(T, \alpha) \qquad (3.1)$$

where $\tilde{U}(T, \alpha) : \mathcal{K}_{T \circ \alpha} \to \mathcal{K}_T$ is the unique unitary operator satisfying

$$\tilde{U}(T, \alpha) \left[\sum_i X_i \otimes u_i \right]_{T \circ \alpha} = \left[\sum_i \alpha(X_i) \otimes u_i \right]_T \qquad (3.2)$$

for all $X_i \in \mathcal{A}_1$, $u_i \in \mathcal{H}, i$ varying over any finite set.

Proof : Let $S \prec T$. Clearly $S \circ \alpha \prec T \circ \alpha$. From property (iii) of Proposition 3.1 we have

$$S \circ \alpha(X) = V_T^* \delta(S : T) \pi_T(\alpha(X)) V_T, \quad X \in \mathcal{A}_1.$$

Since $\tilde{U}(T, \alpha)$ defined by (3.2) as a map from $\mathcal{D}_{T \circ \alpha}$ onto $\mathcal{D}_T$ is scalar product preserving and $\mathcal{D}_{T \circ \alpha}$ and $\mathcal{D}_T$ are dense in $\mathcal{K}_{T \circ \alpha}$ and $\mathcal{K}_T$ respectively it follows that $\tilde{U}(T, \alpha)$ extends uniquely as a unitary operator from $\mathcal{K}_{T \circ \alpha}$ onto $\mathcal{K}_T$. By definition

$$\tilde{U}(T, \alpha) V_{T \circ \alpha} = V_T, \quad \tilde{U}(T, \alpha) \pi_{T \circ \alpha}(X) = \pi_T(\alpha(X)) \tilde{U}(T, \alpha)$$

for all $X \in \mathcal{A}_1$. Combining this with (3.2) we get

$$S \circ \alpha(X) = V_{T \circ \alpha}^* \tilde{U}(T, \alpha)^* \delta(S : T) \tilde{U}(T, \alpha) \pi_{T \circ \alpha}(X) V_{T \circ \alpha}.$$

40

Since $\pi_{T \circ \alpha}(X) = \tilde{U}(T, \alpha)^* \pi_T(\alpha(X)) \tilde{U}(T, \alpha)$ commutes with $\tilde{U}(T, \alpha)^* \delta(S : T) \tilde{U}(T, \alpha)$ we obtain (3.1) from property (iv) of Proposition 3.1. ∎

Remark : It is interesting to note that in Proposition 3.2 one has the relation

$$J(S, T)\tilde{U}(T, \alpha) = \tilde{U}(S, \alpha)J(S \circ \alpha, T \circ \alpha)$$

which implies (3.1).

Proposition 3.3 : Let $\mathcal{A}_0 \subset \mathcal{A}$ be a unital C^* subalgebra of $\mathcal{A}$ and let $S, T : \mathcal{A} \to \mathcal{B}(\mathcal{H})$ be unital c.p. maps such that $S \prec T$. Denote by S_0, T_0 respectively the restrictions of S, T to $\mathcal{A}_0$. Then $S_0 \prec T_0$. If $\mathcal{K}_0$ is the closed subspace of $\mathcal{K}_T$ spanned by $\{\pi_T(X)V_T u, \quad X \in \mathcal{A}_0, \quad u \in \mathcal{H}\}$, $V_0 = V_T$ considered as an isometry from $\mathcal{H}$ into $\mathcal{K}_0$, $\pi_0(X)$ is the restriction of $\pi_T(X)$ to $\mathcal{K}_0$ for $X \in \mathcal{A}_0$ and P_0 is the projection on $\mathcal{K}_0$ then $(\mathcal{K}_0, \pi_0, V_0)$ is a Stinespring dilation of T_0 and

$$\delta(S_0 : T_0)P_0 = P_0 \delta(S : T)P_0.$$

Proof : Immediate from definitions. ∎

Remark : If $\mathcal{A}_n$ is an increasing sequence of unital C^* subalgebras of $\mathcal{A}$ such that $\bigcup_n \mathcal{A}_n$ is dense in $\mathcal{A}$, S and T are unital c.p. maps from $\mathcal{A}$ into $\mathcal{B}(\mathcal{H})$ such that $S \prec T$, and S_n and T_n are the restrictions of S and T respectively to $\mathcal{A}_n$, then it follows from Proposition 3.3 that

$$s. \lim_{n \to \infty} \delta(S_n : T_n)P_n = s. \lim_{n \to \infty} P_n \delta(S : T)P_n = \delta(S : T)$$

where P_n is the projection on the subspace $\mathcal{K}_n$ generated by $\{\pi_T(X)V_T u, X \in \mathcal{A}_n, u \in \mathcal{H}\}$ in $\mathcal{K}_T$ and s.lim denotes strong limit. Here $\delta(S_n : T_n)$ is the Radon-Nikodym derivative in the Stinespring dilation $(\mathcal{K}_n, \pi_n, V_n)$, π_n being the restriction of π_T to $\mathcal{A}_n$ and V_n the isometry $V_T : \mathcal{H} \to \mathcal{K}_n$. This may be looked upon as a quantum probabilistic analogue of the classical theorem that

41

Radon-Nikodym derivatives of one probability measure Q with respect to a dominating probability measure P in an increasing filtration of sub σ-algebras constitute a convergent martingale with respect to P.

We conclude this section with a discussion on Radon-Nikodym derivatives concerning compositions of unital c.p. maps when the uniform domination property is fulfilled at every stage for a sequence of pairs of u.c.p. maps.

To begin with, consider a sequence $\mathcal{A}_1, \mathcal{A}_2, ..., \mathcal{A}_n$ of unital C^* algebras of operators in Hilbert spaces $\mathcal{H}_1, \mathcal{H}_2, ..., \mathcal{H}_n$ respectively. Let $\mathcal{A}_0 = \mathcal{B}(\mathcal{H})$ and $T_i : \mathcal{A}_i \to \mathcal{A}_{i-1}$ be a unital c.p. map, $1 \leq i \leq n$. Following Section 2 (and also the approach outlined in [BP]) consider the algebraic tensor product vector space $\mathcal{A}_n \otimes \mathcal{A}_{n-1} \otimes \cdots \otimes \mathcal{A}_1 \otimes \mathcal{H}$ with the nonnegative bilinear form $B_\mathbf{T}$ defined by

$$B_\mathbf{T} \left(\sum_i X_{ni} \otimes X_{n-1\,i} \otimes \cdots \otimes X_{1i} \otimes u_i, \ \sum_j Y_{nj} \otimes Y_{n-1\,j} \otimes \cdots \otimes Y_{1j} \otimes v_j \right)$$

$$= \sum_{i,j} \langle u_i, T_1(X_{1i}^* \cdots T_{n-1}(X_{n-1\,i}^* T_n(X_{ni}^* Y_{nj}) Y_{n-1\,j}) \ldots Y_{1j}) v_j \rangle. \tag{3.3}$$

Note that this is the analogue of (2.2). Exactly as in Section 2 this leads to the GNS Hilbert space $\mathcal{K}_\mathbf{T}^{(n)}$ by completing the prehilbert space of equivalence classes $[\xi]_\mathbf{T}$, $\xi = \sum_i X_{ni} \otimes X_{n-1\,i} \otimes \cdots \otimes X_{1i} \otimes u_i$. Define the isometry $V_\mathbf{T} : \mathcal{H} \to \mathcal{K}_\mathbf{T}$ by $V_\mathbf{T} u = [1 \otimes \cdots \otimes 1 \otimes u]_\mathbf{T}$ and the representation $\pi_\mathbf{T}^{(n)}$ of $\mathcal{A}_n$ by

$$\pi_\mathbf{T}^{(n)}(X) \left[\sum_i X_{ni} \otimes \cdots \otimes X_{1i} \otimes u_i \right]_\mathbf{T} = \left[\sum_i X X_{ni} \otimes X_{n-1\,i} \otimes \cdots \otimes X_{1i} \otimes u_i \right]_\mathbf{T}.$$

Consider the increasing sequence of subspaces $\mathcal{K}_\mathbf{T}^{(m)}$, $m = 0, 1, 2, .., n$ where $\mathcal{K}_\mathbf{T}^{(0)} = \mathcal{H}$ and $\mathcal{K}_\mathbf{T}^{(m)}$ is the span of

$$\{[1 \otimes \cdots \otimes 1 \otimes X_m \otimes X_{m-1} \otimes \cdots \otimes X_1 \otimes u]_\mathbf{T}, \ X_j \in \mathcal{A}_j, 1 \leq j \leq m, \ u \in \mathcal{H}\}.$$

Denote by $V_\mathbf{T}^{(m)}$ the inclusion isometry from $\mathcal{K}_\mathbf{T}^{(m)}$ into $\mathcal{K}_\mathbf{T}^{(m+1)}$. One has the representation $\pi_\mathbf{T}^{(m)}$ of $\mathcal{A}_m$ in $\mathcal{K}_\mathbf{T}^{(m)}$ defined by

$$\pi_{\mathbf{T}}^{(m)}(X)\left[\sum_i 1 \otimes \cdots \otimes 1 \otimes X_{mi} \otimes \cdots \otimes X_{1i} \otimes u\right]$$

$$= \left[\sum_i 1 \otimes \cdots \otimes 1 \otimes XX_{mi} \otimes X_{m-1i} \otimes \cdots \otimes X_{1i} \otimes u_i\right]_{\mathbf{T}}, \ X \in \mathcal{A}_m.$$

$\mathcal{A}$ simple computation leads to the relation

$$V_{\mathbf{T}}^{(m)*} \pi_{\mathbf{T}}^{(m+1)}(X) V_{\mathbf{T}}^{(m)} = \pi_{\mathbf{T}}^{(m)} \circ T_{m+1}(X), \ X \in \mathcal{A}_{m+1}.$$

It is easily seen that $(\mathcal{K}_{\mathbf{T}}^{(m+1)}, \pi_{\mathbf{T}}^{(m+1)}, V_{\mathbf{T}}^{(m)})$ is a Stinespring dilation of the unital c.p. map $\pi_{\mathbf{T}}^{(m)} \circ T_{m+1} : \mathcal{A}_{m+1} \to \mathcal{B}(\mathcal{K}_{\mathbf{T}}^{(m)})$, $m = 0, 1, 2, ..., n-1$.

Now suppose $S_i : \mathcal{A}_i \to \mathcal{A}_{i-1}$, $i = 1, 2, ..., n$ is another sequence of u.c.p maps such that $S_i \prec T_i$ for each i. Then $S_1 \circ S_2 \circ \cdots \circ S_n \prec T_1 \circ T_2 \circ \cdots \circ T_n$. As in Proposition 3.1 one can construct a bounded operator $J(\mathbf{S}, \mathbf{T}) : \mathcal{K}_{\mathbf{T}} \to \mathcal{K}_{\mathbf{S}}$ satisfying

$$J(\mathbf{S}, \mathbf{T})[\xi]_{\mathbf{T}} = [\xi]_{\mathbf{S}}, \ \xi \in \mathcal{A}_n \otimes \cdots \otimes \mathcal{A}_1 \otimes \mathcal{H}.$$

Note that $J(\mathbf{S}, \mathbf{T})$ maps $\mathcal{K}_{\mathbf{T}}^{(m)}$ into $\mathcal{K}_{\mathbf{S}}^{(m)}$. Denote this operator by $J_m(\mathbf{S}, \mathbf{T})$ and write $\delta_m(\mathbf{S} : \mathbf{T}) = J_m^*(\mathbf{S}, \mathbf{T}) J_m(\mathbf{S}, \mathbf{T})$, $m = 1, 2, ..., n$. Then
$$V_{\mathbf{T}}^{(m)*} \delta_{m+1}(\mathbf{S} : \mathbf{T}) \pi_{\mathbf{T}}^{(m+1)}(X) V_{\mathbf{T}}^{(m)}$$
$$= J_m(\mathbf{S}, \mathbf{T})^* \pi_{\mathbf{S}}^{(m)} \circ S_{m+1}(X) J_m(\mathbf{S}, \mathbf{T}), \ X \in \mathcal{A}_{m+1}.$$
Furthermore $\delta_{m+1}(\mathbf{S} : \mathbf{T})$ commutes with $\pi_{\mathbf{T}}^{(m+1)}(X)$. In other words $\delta_{m+1}(\mathbf{S} : \mathbf{T})$ is the Radon-Nikodym density of the (not necessarily unital) c.p. map $J_m(\mathbf{S}, \mathbf{T})^* \pi_{\mathbf{S}}^{(m)} \circ S_{m+1}(\cdot) J_m(\mathbf{S}, \mathbf{T})$ from $\mathcal{A}_{m+1}$ into $\mathcal{B}(\mathcal{K}_{\mathbf{T}}^{(m)})$ with respect to the unital c.p. map $\pi_{\mathbf{T}}^{(m)} \circ S_m$ for each $m = 1, 2, ..., n-1$. It may also be noted that $V_{\mathbf{T}} = V_{\mathbf{T}}^{(n-1)} V_{\mathbf{T}}^{(n-2)} \cdots V_{\mathbf{T}}^{(0)}$ and

$$V_{\mathbf{T}}^* \delta_n(\mathbf{S} : \mathbf{T}) \pi_{\mathbf{T}}^{(n)}(X) V_{\mathbf{T}} = S_1 \circ S_2 \circ \cdots \circ S_n(X), \ X \in \mathcal{A}_n.$$

It may be interesting to examine the continuous time analogue of $\delta_m(\mathbf{S} : \mathbf{T})$ and describe it through stochastic differential equations at least in some examples.

43

4 Lebesgue decomposition of one c.p. map with respect to another

Let S_1, S_2 be two unital c.p. maps from a unital C^* algebra $\mathcal{A}$ into $\mathcal{B}(\mathcal{H})$. Then S_1 and S_2 are uniformly dominated by the unital c.p. map $T = \frac{1}{2}(S_1 + S_2)$. Let

$$\delta_i = \delta(S_i : T), \quad i = 1, 2 \tag{4.1}$$

be the Radon-Nikodym density of S_i with respect to T, defined as in Proposition 3.1. Then

$$S_i(X) = V_T^* \, \delta_i \pi_T(X) V_T, \quad X \in \mathcal{A}, \quad i = 1, 2.$$

Taking the arithmetic mean on both the sides we have

$$T(X) = V_T^* \, \frac{1}{2}(\delta_1 + \delta_2) \pi_T(X) V_T, \quad X \in \mathcal{A}.$$

On the other hand from the definition of Stinespring dilation we have

$$T(X) = V_T^* \pi_T(X) V_T, \quad X \in \mathcal{A}.$$

Hence by property (iv) in Proposition 3.1 we get

$$\frac{1}{2}(\delta_1 + \delta_2) = 1,$$

1 being the identity operator in $\mathcal{K}_T$. In particular, δ_1 and δ_2 are positive commuting selfadjoint operators which commute with $\pi_T(X)$ for every X in $\mathcal{A}$. Denote by P_i the projection on the null space of δ_i and put $P_0 = 1 - P_1 - P_2$. Then P_1 and P_2 commute with each other and also commute with $\pi_T(X)$. Since $\delta_2 P_2 = 0$, $\delta_2 P_1 = 2 P_1$ it follows that $P_1 P_2 = 0$. Thus P_0, P_1, P_2 are projections satisfying $P_0 + P_1 + P_2 = 1$ and the representation π_T decomposes into a direct sum of three representations which are the restrictions of π_T to the ranges of

P_0, P_1, P_2. Denote the range of P_i by $\mathcal{K}_i$ and the restriction of π_T to $\mathcal{K}_i$ by π_i, $i = 0, 1, 2$. Then

$$\mathcal{K}_T = \mathcal{K}_0 \oplus \mathcal{K}_1 \oplus \mathcal{K}_2, \quad \pi_T = \pi_0 \oplus \pi_1 \oplus \pi_2. \tag{4.2}$$

Write

$$
\begin{aligned}
S_1^{ac}(X) &= V_T^* \delta_1 P_0 \pi_T(X) V_T = V_T^* \delta_1 (1 - P_2) \pi_T(X) V_T, & (4.3) \\
S_1^s(X) &= V_T^* \delta_1 P_2 \pi_T(X) V_T = 2 V_T^* P_2 \pi_T(X) V_T. & (4.4)
\end{aligned}
$$

Then S_1^{ac} and S_1^s are c.p. maps from $\mathcal{A}$ into $\mathcal{B}(\mathcal{H})$ and $S_1 = S_1^{ac} + S_1^s$. Following the analogy with classical integration theory we call S_1^{ac} and S_1^s the *absolutely continuous* and *singular part* of S_1 respectively with respect to S_2.

Theorem 4.1 : Let S_1, S_2 be unital c.p. maps from $\mathcal{A}$ into $\mathcal{B}(\mathcal{H})$. Define

$$
\begin{aligned}
\tilde{\mathcal{K}}_1 &= \mathcal{K}_0 \oplus \mathcal{K}_2, \quad \tilde{\mathcal{K}}_2 = \mathcal{K}_0 \oplus \mathcal{K}_1 \\
\tilde{\pi}_1(X) &= \pi_0(X) \oplus \pi_2(X), \quad \tilde{\pi}_2(X) = \pi_0(X) \oplus \pi_1(X), \ X \in \mathcal{A}
\end{aligned}
$$

where $\mathcal{K}_i$, π_i are as in (4.2). Let

$$\tilde{V}_i = \sqrt{\delta_i} \, V_T : \mathcal{H} \to \tilde{\mathcal{K}}_i, \quad i = 1, 2 \tag{4.5}$$

where δ_i is given by (4.1). Then there exist unitary operators $\Gamma_i : \tilde{\mathcal{K}}_i \to \mathcal{K}_{S_i}$, $i = 1, 2$ satisfying

$$
\begin{aligned}
\Gamma_i \sqrt{\delta_i} V_T &= V_{S_i}, \\
\Gamma_i \tilde{\pi}_i(X) &= \pi_{S_i}(X) \Gamma_i, \ X \in \mathcal{A}, \ i = 1, 2.
\end{aligned}
$$

Proof : Since δ_i is a positive selfadjoint operator the null spaces of δ_i and $\sqrt{\delta_i}$ are same. Thus $\tilde{V}_i$ maps $\mathcal{H}$ into $\tilde{\mathcal{K}}_i$. By definition the range of δ_i is contained in $\tilde{\mathcal{K}}_i$ as a dense linear manifold. Since $\{\pi_T(X) V_T u, \ X \in \mathcal{A}, \ u \in \mathcal{H}\}$ is total in $\mathcal{K}$

45

it follows that $\{\tilde{\pi}_i(X)V_T u, \; X \in \mathcal{A}, \; u \in \mathcal{H}\}$ is total in $\tilde{\mathcal{K}}_i$. Since the restriction of δ_i to $\tilde{\mathcal{K}}_i$ is an invertible selfadjoint operator (with probably an unbounded inverse) and $\sqrt{\delta_i}$ commutes with $\pi_T(X)$ it follows that $\{\tilde{\pi}_i(X)\tilde{V}_i u, \; X \in \mathcal{A}, \; u \in \mathcal{H}\}$ is total in $\tilde{\mathcal{K}}_i$. Furthermore

$$
\begin{aligned}
\langle \tilde{\pi}_i(X)\tilde{V}_i u, \; & \tilde{\pi}_i(Y)\tilde{V}_i v \rangle \\
&= \langle V_T u, \; \sqrt{\delta_i}\, \tilde{\pi}_i(X^*Y) \sqrt{\delta_i} V_T v \rangle \\
&= \langle V_T u, \; \sqrt{\delta_i}\, \pi_i(X^*Y) \sqrt{\delta_i} V_T v \rangle \\
&= \langle u, \; V_T^* \delta_i \pi_T(X^*Y) V_T v \rangle \\
&= \langle u, \; S_i(X^*Y)v \rangle \\
&= \langle [X \otimes u]_{S_i}, \; [Y \otimes v]_{S_i} \rangle.
\end{aligned}
$$

This shows that there exists a unique unitary operator $\Gamma_i : \tilde{\mathcal{K}}_i \to \mathcal{K}_{S_i}$ satisfying

$$
\Gamma_i \tilde{\pi}_i(X)\tilde{V}_i u = \pi_{S_i}(X)V_{S_i} u, \quad u \in \mathcal{H}, \; X \in \mathcal{A}.
$$

Putting $X = 1$ we get $\Gamma_i \tilde{V}_i = V_{S_i}$. Furthermore

$$
\begin{aligned}
\Gamma_i \tilde{\pi}_i(X) \sum_k & \tilde{\pi}_i(X_k)\tilde{V}_i u_k \\
&= \Gamma_i \sum_k \tilde{\pi}_i(XX_k)\tilde{V}_i u_k \\
&= \sum_k \pi_{S_i}(XX_k)V_{S_i} u_k \\
&= \pi_{S_i}(X)\Gamma_i \sum_k \tilde{\pi}_i(X_k)\tilde{V}_i u_k, \quad i = 1, 2.
\end{aligned}
$$

The arbitrariness of X_k and u_k implies that $\Gamma_i \tilde{\pi}_i(X) = \pi_{S_i}(X)\Gamma_i$. ∎

Corollary 4.2 : Let S_1, S_2 be unital c.p. maps from $\mathcal{A}$ into $\mathcal{B}(\mathcal{H})$. Then there exists a unique positive selfadjoint operator δ in $\mathcal{K}_{S_2}$ satisfying the following:

(i) δ commutes with $\pi_{S_2}(X)$ for every $X \in \mathcal{A}$;

(ii) $\mathcal{D}_{S_2} \subset \mathcal{D}(\sqrt{\delta})$;

(iii) $\sqrt{\delta}\, V_{S_2}$ is a bounded operator and

$$S_1^{ac}(X) = (\sqrt{\delta}V_{S_2})^* \, \pi_{S_2}(X)\sqrt{\delta}V_{S_2}, \; X \in \mathcal{A}.$$

Proof: Consider the operators δ_1, δ_2 and Γ_2 of Theorem 4.1. δ_1 and $\delta_2 = 2 - \delta_1$ are positive selfadjoint opeators in $\mathcal{K}_T$ with $T = \frac{1}{2}(S_1 + S_2)$, which leave the subspace $\tilde{\mathcal{K}}_2$ invariant and δ_2 is invertible in $\tilde{\mathcal{K}}_2$. Put $\delta = \Gamma_2 \, \delta_1 \delta_2^{-1} \Gamma_2^*$. Then δ is a positive selfadjoint operator on $\mathcal{K}_{S_2}$. From (4.3) we have for any $u, v \in \mathcal{H}$ and $X \in \mathcal{A}$,

$$
\begin{aligned}
\langle u, S_1^{ac}(X)v \rangle &= \langle u, V_T^* \delta_1 (1 - P_2)\pi_T(X)V_T v \rangle \\
&= \langle V_T u, \; \delta_1 \tilde{\pi}_2(X)V_T v \rangle \\
&= \langle \delta_1^{\frac{1}{2}} \delta_2^{-\frac{1}{2}} \delta_2^{\frac{1}{2}} V_T u, \tilde{\pi}_2(X)\delta_1^{\frac{1}{2}} \delta_2^{-\frac{1}{2}} \delta_2^{\frac{1}{2}} V_T v \rangle \\
&= \langle \sqrt{\delta}\, V_{S_2} u, \; \pi_{S_2}(X)\sqrt{\delta}V_{S_2} v \rangle. \qquad (4.6)
\end{aligned}
$$

Putting $X = 1$ we have

$$S_1^{ac}(1) = (\sqrt{\delta}V_{S_2})^* (\sqrt{\delta}V_{S_2}),$$

which shows that $\sqrt{\delta}V_{S_2}$ is a bounded operator. By definition δ commutes with $\pi_{S_2}(X)$ and hence $\pi_{S_2}(X)V_{S_2}u$ is in the domain of $\sqrt{\delta}$. Thus (i) and (ii) hold good. The second part of property (iii) follows from (4.6).

Now suppose that δ' is another positive selfadjoint operator satisfying (i) - (iii). Then we have from (4.6)

$$\langle \sqrt{\delta}\pi_{S_2}(X)V_{S_2}u, \; \sqrt{\delta}\pi_{S_2}(X)V_{S_2}v \rangle = \langle \sqrt{\delta'}\pi_{S_2}(X)V_{S_2}u, \; \sqrt{\delta'}\pi_{S_2}(X)V_{S_2}v \rangle$$

for all $X, Y \in \mathcal{A}, u, v \in \mathcal{H}$. This implies the existence of an isometry W from the range of $\sqrt{\delta}$ to the range of $\sqrt{\delta'}$ such that $W\sqrt{\delta} = \sqrt{\delta'}$. By the uniqueness of polar decomposition it follows that $\delta = \delta'$. ∎

Remark : According to the Lebesgue decomposition theorem in classical measure theory, given two σ-finite measures μ_1, μ_2 on a Borel space Ω there exists a measurable partition of Ω into three disjoint sets $\Omega_0, \Omega_1, \Omega_2$ satisfying the following:

(i) $\mu_i(\Omega_i) = 0, \quad i = 1, 2$;

(ii) The restrictions of μ_1 and μ_2 to Ω_0 are equivalent. The absolutely continuous and singular parts of μ_1 with respect to μ_2, denoted respectively by μ_1^{ac} and μ_1^s are given by $\mu_1^{ac}(E) = \mu_1(E \cap \Omega_0) = \mu_1(E \cap (\Omega \backslash \Omega_2))$ and $\mu_1^s(E) = \mu_1(E \cap \Omega_2)$. μ_1^{ac} admits a Radon-Nikodym density with respect to μ_2.

The situation above may be compared with the existence of the three projections P_0, P_1, P_2 such that $P_0 + P_1 + P_2 = 1$ satisfying (4.3), (4.4) and the properties mentioned in Theorem 4.1 and Corollary 4.2. The unique positive selfadjoint (but probably unbounded) operator δ can legitimately be called the Radon-Nikodym derivative of S_1^{ac} with respect to S_2 in the Stinespring dilation space of S_2. We say that S_1 is *absolutely continuous with respect to S_2* and $S_1 \ll S_2$ in symbols if $S_1^s = 0$. This is equivalent to $P_2 = 0$ in view of (4.4), i.e., the null space of $\delta(S_2 : \frac{S_1+S_2}{2})$ being trivial. Since by the definition of δ in Proposition 3.1,

$$\delta(S_2 : \frac{S_1 + S_2}{2}) = J(S_2, \frac{S_1 + S_2}{2})^* J(S_2, \frac{S_1 + S_2}{2})$$

it follows that $S_1 \ll S_2$ if and only if the null space of $J(S_2, \frac{1}{2}(S_1 + S_2))$ is trivial.

Proposition 4.3 : Let S_1, S_2 be unital c.p. maps from $\mathcal{A}$ into $\mathcal{B}(\mathcal{H})$. Then $S_1 \ll S_2$ if and only if the linear map $J_0(S_1, S_2) : \mathcal{D}_{S_2} \to \mathcal{D}_{S_2}$ defined by

$$J_0(S_1, S_2)[\xi]_{S_2} = [\xi]_{S_1}, \quad \xi \in \mathcal{A} \otimes \mathcal{H}$$

is a closable operator from $\mathcal{K}_{S_2}$ into $\mathcal{K}_{S_1}$.

48

Proof : Let $T = \frac{1}{2}(S_1 + S_2)$. For any $\xi = \sum_i X_i \otimes u_i$ in $\mathcal{A} \otimes \mathcal{H}$ we have

$$\|[\xi]_T\|^2 = \frac{1}{2}(\|[\xi]_{S_1}\|^2 + \|[\xi]_{S_2}\|^2). \tag{4.7}$$

Suppose $S_1 \ll S_2$. From the discussion preceding the statement of Proposition 4.3 it follows that the null space of $J(S_2, T)$ is trivial. Now consider a sequence $\{\xi_n\}$ in $\mathcal{A} \otimes \mathcal{H}$ for which $\lim_{n\to\infty} [\xi_n]_{S_2} = 0$, $\lim_{n\to\infty} [\xi_n]_{S_1} = \psi_1$. Then (4.7) implies that $\{[\xi_n]_T\}$ is a Cauchy sequence in $\mathcal{K}_T$ and hence there exists $\psi \in \mathcal{K}_T$ such that $\lim_{n\to\infty} [\xi_n]_T = \psi$. Again (4.7) implies that $\|\psi\|^2 = \frac{1}{2}\|\psi_1\|^2$. If $\psi_1 \neq 0$ then $\psi \neq 0$ and $J(S_2, T)\psi = \lim_{n\to\infty} [\xi_n]_{S_2} = 0$, a contradiction. In other words $J_0(S_1, S_2)$ with domain $\mathcal{D}_{S_2}$ is closable.

Conversely assume that $J_0(S_1, S_2)$ with domain D_{S_2} is closable. Suppose that $J(S_2, T)\psi = 0, \psi \in \mathcal{K}_T$. Since $J(S_2, T)$ is a bounded operator there exists a sequence $\{\xi_n\}$ in $\mathcal{A} \otimes \mathcal{H}$ such that $\lim_{n\to\infty} [\xi_n]_T = \psi$, $\lim_{n\to\infty} [\xi_n]_{S_2} = 0$. By (4.7) $\{[\xi_n]_{S_1}\}$ is a Cauchy sequence and hence $\psi_1 = \lim_{n\to\infty} [\xi_n]_{S_1}$ exists and $\|\psi\|^2 = \frac{1}{2}\|\psi_1\|^2$. Closability of $J_0(S_1, S_2)$ implies $\psi_1 = 0$ and hence $\psi = 0$. In other words the null space of $J(S_2, T)$ is trivial and $S_1 \ll S_2$. ∎

Remark If $S_1 \ll S_2$ and $S_2 \ll S_3$ then $J_0(S_1, S_2)J_0(S_2, S_3) = J_0(S_1, S_3)$. The closability of $J_0(S_1, S_2)$ and $J_0(S_2, S_3)$ do not immediately imply the closability of $J_0(S_1, S_3)$. However, if $S_1 \ll S_2$ and $S_2 \prec S_3$ then the closure of $J_0(S_2, S_3)$, which is $J(S_2, S_3)$, is bounded and hence $J_0(S_1, S_3)$ is closable and $S_1 \ll S_3$. Thus the problem of transitivity of absolute continuity calls for a deeper investigation.

5 Examples

Example 5.1 : Let S_1, S_2 be two representations of a unital C^* algebra $\mathcal{A}$ in $\mathcal{B}(\mathcal{H})$, $T = \frac{1}{2}(S_1 + S_2)$ and

$$\mathcal{H}_0 = \{u | (S_1(X) - S_2(X))u = 0 \text{ for all } X \in \mathcal{A}\}.$$

Then $\mathcal{H}_0$ is invariant under $S_1(X)$ and $S_2(X)$ for every X. Denote by P the projection on $\mathcal{H}_0$. S_1 and S_2 are unital c.p. maps and we claim that the absolutely continuous and singular parts of S_1 with respect to S_2 are given by

$$S_1^{ac}(X) = PS_1(X), \quad S_1^s(X) = (1 - P)S_1(X).$$

In particular, $S_1 \ll S_2$ if and only if $S_1 = S_2$.

Indeed, note that

$$T(X) = PS_1(X) + (1 - P)\frac{1}{2}(S_1 + S_2)(X)$$

where P commutes with $S_1(X)$ and $S_2(X)$ for every X. Write $\mathcal{H}_1 = \mathcal{H}_0^\perp$. Then $\frac{1}{2}(S_1 + S_2)(X)|_{\mathcal{H}_1}$ yields a c.p. map from $\mathcal{A}$ into $\mathcal{B}(\mathcal{H}_1)$. Consider the isometry $V : \mathcal{H}_1 \to \mathcal{H}_1 \oplus \mathcal{H}_1$ defined by $Vu = 2^{-\frac{1}{2}}(u \oplus u)$ and the representation $\pi(X) = S_1(X) \oplus S_2(X)$ restricted to $\mathcal{H}_1 \oplus \mathcal{H}_1$. We claim that $(\mathcal{H}_1 \oplus \mathcal{H}_1, \pi, V)$ is a Stinespring dilation of the c.p. map $X \to \frac{1}{2}(S_1 + S_2)(X)|_{\mathcal{H}_1}$. It is easy to check that $\frac{1}{2}(S_1(X)u + S_2(X)u) = V^*\pi(X)Vu$ for $u \in \mathcal{H}_1$. Suppose there exist $v_1, v_2 \in \mathcal{H}_1$ such that $v_1 \oplus v_2$ is orthogonal to the set

$$\{S_1(X)u \oplus S_2(X)u, \quad u \in \mathcal{H}_1, \quad X \in \mathcal{A}\}.$$

Then

$$\langle v_1, S_1(X)u \rangle + \langle v_2, S_2(X)u \rangle = 0 \text{ for all } u \in \mathcal{H}_1, X \in \mathcal{A}.$$

This implies $S_1(X)v_1 + S_2(X)v_2 = 0$ for every X. Putting $X = 1$ we get $v_1 + v_2 = 0$ and hence $(S_1(X) - S_2(X))v_1 = 0$. Thus $v_1 \in \mathcal{H}_0 \cap \mathcal{H}_1$. Hence $v_1 = v_2 = 0$, proving the claim.

Now let

$$\mathcal{K}' = \mathcal{H}_0 \oplus \mathcal{H}_1 \oplus \mathcal{H}_1$$

$$V'u = Pu \oplus 2^{-\frac{1}{2}}(1-P)u \oplus 2^{-\frac{1}{2}}(1-P)u, \ u \in \mathcal{H},$$
$$\pi'(X) = S_1(X)|_{\mathcal{H}_0} \oplus S_1(X)|_{\mathcal{H}_1} \oplus S_2(X)|_{\mathcal{H}_1}.$$

Then $(\mathcal{K}', V', \pi')$ is a Stinespring dilation of T. If $\mathcal{K}_0 = \mathcal{H}_0, \mathcal{K}_1$ and $\mathcal{K}_2$ are respectively the first and second copies of $\mathcal{H}_1$ in $\mathcal{K}'$ then the projections P_i on $\mathcal{K}_i$ satisfy the requirements of Theorem 4.1 from which our initial claim follows. In the decomposition $\mathcal{H} = \mathcal{H}_0 \oplus \mathcal{H}_1$ the projection P on $\mathcal{H}_0$ is the Radon-Nikodym density of S_1^{ac} with respect to S_2.

If S_1 and S_2 are finite or continuous convex combinations of representations of $\mathcal{A}$ in a fixed $\mathcal{B}(\mathcal{H})$ we are unable to get an expression for S_1^{ac}. Our next example is a mild attempt in this direction.

Example 5.2 : Consider a Borel space $(\Omega, \mathcal{F})$ and a strongly measurable map $x \to U_x$ from Ω into the unitary group of a separable Hilbert space $\mathcal{H}$. Let $\mathcal{A} = \mathcal{B}(\mathcal{H})$. For any probability measure λ on $(\Omega, \mathcal{F})$ consider the unital c.p. map S_λ from $\mathcal{B}(\mathcal{H})$ into itself defined by

$$S_\lambda(X) = \int U_x^* X U_x d\lambda(x), \quad X \in \mathcal{B}(\mathcal{H})$$

where the integral is in the strong Bochner sense. Given two probability measures λ and μ we do not know how to obtain the Lebesgue decomposition of S_λ with respect to S_μ. The difficulty seems to lie in achieving a convenient form of the Stinespring dilation of S_λ.

Suppose the map $x \to U_x$ satisfies the 'linear independence condition' :

$$\int \varphi(x) U_x d\lambda(x) = 0 \text{ if and only if } \varphi(x) = 0 \text{ a.e. } x(\lambda) \qquad (5.1)$$

for any $\varphi \in L^1(\lambda)$. Put $\mathcal{K} = L^2(\lambda, \mathcal{H})$, the Hilbert space of $\mathcal{H}$-valued λ-square integrable maps, $(Vu)(x) = u$, the constant vector u for each $u \in \mathcal{H}$ and

$$(\pi(X)v)(x) = U_x^* X U_x v(x), \quad v \in \mathcal{K}.$$

We claim that $(\mathcal{K}, \pi, V)$ is a Stinespring dilation of the unital c.p. map S_λ. Clearly, V is an isometry from $\mathcal{H}$ into $\mathcal{K}$ and $V^*\pi(X)V = S_\lambda(X)$. Suppose $v \in \mathcal{K}$ is orthogonal to every vector of the form $\pi(X)Vu$, $X \in \mathcal{B}(\mathcal{H})$, $u \in \mathcal{H}$. Then

$$\int \langle v(x),\ U_x^* X U_x u \rangle d\lambda(x) = 0$$

which implies

$$\int U_x^* X U_x v(x) d\lambda(x) = 0, \quad X \in \mathcal{B}(\mathcal{H}).$$

Putting $X = |u_1\rangle\langle u_2|$, $u_1, u_2 \in \mathcal{H}$ we get

$$\int \langle u_2,\ U_x v(x) \rangle U_x^* u_1 d\lambda(x) = 0.$$

By condition (5.1) we conclude that

$$\langle U_x v(x),\ u_2 \rangle = 0 \text{ a.e. } x(\lambda)$$

for each $u_2 \in \mathcal{H}$. Since $\mathcal{H}$ is separable this can hold only if $U_x v(x) = 0$ a.e. $x(\lambda)$, i.e., $v(x) = 0$ a.e. $x(\lambda)$, proving the claim.

Now suppose that μ, ν are two probability measures on $(\Omega, \mathcal{F})$, $\lambda = \frac{1}{2}(\mu + \nu)$ and condition (5.1) holds. Then $S_\lambda = \frac{1}{2}(S_\mu + S_\nu)$. In the Stinespring dilation $(\mathcal{K}, \pi, V)$ described in the preceding paragraph the Radon-Nikodym density $\delta(S_\mu : S_\lambda)$ turns out to be just multiplication by the function $\frac{d\mu}{d\lambda}$. This shows that the absolutely continuous part S_μ^{ac} of S_μ with respect to S_ν is given by

$$S_\mu^{ac}(X) = \int_{\{x:\frac{d\nu}{d\lambda}(x)\neq 0\}} U_x^* X U_x\, d\mu(x), \quad X \in \mathcal{B}(\mathcal{H}).$$

In particular, $S_\mu \ll S_\nu$ if and only if $\mu \ll \nu$.

Example 5.3 : Let $(\Omega, \mathcal{F}, \mu)$ be a σ-finite measure space, $A : x \to A_x$ a strongly measurable map from Ω into the space of bounded positive operators on the Hilbert space $\mathcal{H}$ such that $\int A_x d\mu(x) = 1$ and for any $\varphi \in L^\infty(\mu)$ define

$$S_A(\varphi) = \int \varphi(x) A_x d\mu(x).$$

Then S_A is a unital c.p. map from the abelian C^* algebra $L^\infty(\mu)$ into the nonabelian $\mathcal{B}(\mathcal{H})$.

Consider the Hilbert space $L^2(\mu, \mathcal{H})$ and the subspace $\mathcal{K}_A$ defined by

$$\mathcal{K}_A = \{u \in L^2(\mu, \mathcal{H}), \; u(x) \in N(A_x)^\perp \text{ a.e. } x(\mu)\}$$

where $N(A_x)$ denotes the null space of A_x. Define the isometry $V_A : \mathcal{H} \to \mathcal{K}_A$ by $(V_A u)(x) = A_x^{1/2} u$. That V_A is an isometry follows from

$$
\begin{aligned}
\langle V_A u, \; V_A v \rangle &= \int \langle A_x^{1/2} u, \; A_x^{1/2} v \rangle d\mu(x) \\
&= \langle u, \int A_x d\mu(x) v \rangle \\
&= \langle u, v \rangle.
\end{aligned}
$$

Since $N(A_x) = N(A_x^{1/2})$ it is clear that $V_A u \in \mathcal{K}_A$. Define the representation $\varphi \to \pi_A(\varphi)$ in $\mathcal{K}_A$ by

$$(\pi_A(\varphi)v)(x) = \varphi(x)v(x), \quad \varphi \in L^\infty(\mu).$$

It is easy to check that $V_A^* \pi_A(\varphi) V_A = S_A(\varphi)$. If $v \in \mathcal{K}_A$ is orthogonal to every element of the form $\pi_A(\varphi)V_A u$, $u \in \mathcal{H}$, $\varphi \in L^\infty(\mu)$ then

$$\int \langle v(x), \; \varphi(x) A_x^{1/2} u \rangle d\mu(x) = 0$$

implies $A_x^{1/2} v(x) = 0$ a.e. $x(\mu)$. By the definition of $\mathcal{K}_A, v = 0$. In other words $(\mathcal{K}_A, \pi_A, V_A)$ is a Stinespring dilation of S_A.

If S_B is another such unital c.p. map from $L^\infty(\mu)$ into $\mathcal{B}(\mathcal{H})$ then $S_A \ll S_B$ if and only if the following holds: for every sequence $v_n \in L^2(\mu, \mathcal{H})$ satisfying

(i) $\int \langle v_n(x), \; B_x v_n(x) \rangle d\mu(x) \to 0$ as $n \to \infty$;

(ii) $\int \langle v_n(x) - v_m(x), \; A_x(v_n(x) - v_m(x)) \rangle d\mu(x) \to 0$ as m and $n \to \infty$,

$$\lim_{n \to \infty} \int \langle v_n(x), \; A_x v_n(x) \rangle d\mu(x) = 0.$$

If each B_x is invertible with a bounded inverse and $\sup_x \|B_x^{-\frac{1}{2}} A_x B_x^{-\frac{1}{2}}\| < \infty$ then $S_A \prec S_B$ and

$$(\delta(S_A : S_B)v)(x) = B_x^{-\frac{1}{2}} A_x B_x^{-\frac{1}{2}} v(x),$$

$v \in \mathcal{K}_B$.

References

[BP] B.V.R. Bhat and K.R. Parthasarathy : Kolomogorov's existence theorem for Markov processes in C^* algebras, Proc. Ind. Acad. Sci. (Math. Sci.), 104 (1994), 253-262.

[BS] V.P. Belavkin and P. Staszewski : A Radon-Nikodym theorem for completely positive maps, Rep. Math. Phys. 24(1986) 49-55.

[H] P.R. Halmos : Measure Theory, van Nostrand, Princeton (1950).

[M] P.A. Meyer : Quantum Probability for Probabilists, Lecture Notes in Mathematics 1538, Springer-Verlag (1993).

[P1] K.R. Parthasarathy : Introduction to Probability and Measure, Macmillan India, Delhi (1977).

[P2] K.R. Parthasarathy : An Introduction to Quantum Stochastic Calculus, Birkhauser Verlag, Basel (1992).

ABEL EXPANSIONS AND GENERALIZED ABEL POLYNOMIALS IN STOCHASTIC MODELS

PHILIPPE PICARD[1]
CLAUDE LEFEVRE[2]

ABSTRACT To build expansions, the family of the Abel polynomials $\left\{ (x-a)(x-a-bn)^{n-1}/n! \; ; \; n \in \mathbf{N} \right\}$ can be used as a basis in place of the classical family of monomials $\left\{ x^n/n! \; ; \; n \in \mathbf{N} \right\}$. In that case we get Abel's expansions that generalize Taylor's ones. The purpose of the present paper is to show that these polynomials and expansions are present implicitly in several probability models, and that making explicit their hidden algebraic structure is very useful. More complex stochastic models can then also be considered, after extending the Abelian structure to more general polynomials.

AMS 1991   Subject classification Primary 60 G 17
                        Secondary 60 J 75, 60 G 40

Keywords   Appell polynomials, Busy period, M/D/1 queue, one sided test.

1   Introduction

1.1   *The Abel series.* In the fifth problem of his posthumous work "Sur les fonctions génératrices et leurs déterminantes" [ABE81], N. ABEL expands $\varphi(x + \alpha)$ in terms of the $\varphi^{(n)}(x + n\beta)$'s getting the (formal) expansion

$$\varphi(x + \alpha) = \sum_{n=0}^{\infty} \varphi^{(n)}(x + n\beta) \frac{\alpha(\alpha - n\beta)^{n-1}}{n!} . \qquad (1.1)$$

Except in the interpolation theory context, where (1.1) is frequently called the Abel *interpolation* series - a rather unfortunate qualification for such an original generalization of Taylor's series -, this expansion is rarely referred to in the literature. Actually only the case $\varphi(x) = e^x$ in which (1.1), after cancellation by

---

[1] Université de Lyon 1
[2] Université Libre de Bruxelles

e^x, is reduced to

$$e^\alpha = \sum_{n=0}^{\infty} e^{n\beta} \frac{\alpha(\alpha - n\beta)^{n-1}}{n!} \qquad (1.2)$$

seems to be popular. Notice that, when $|\beta| < 1$, (1.2) is an exact formula. Unfortunately, after [JEN02], (1.2) has usually been proved in using systematically the Lagrange expansion

$$f(t) = f(0) + \sum_{n=1}^{\infty} \frac{u^n}{n!} \left\{ f'(t)\varphi(t)^n \right\}_{t=0}^{(n-1)} ,$$

where $t = u\varphi(t)$, $f(t) = e^{\alpha t}$, $\varphi(t) = e^{-\beta t}$, the connection between (1.1) and (1.2) being therefore lost.

The originality of (1.1) appears more deeply if one notices that, in addition to giving an expansion of $\varphi(x + \alpha)$ in terms of the $\varphi^{(n)}(x + n\beta)$, it also gives its expansion in terms of the polynomials $\alpha(\alpha - n\beta)^{n-1}/n!$, α being now taken as the main variable. Since we shall be concerned mainly with this second interpretation, we change the notation writing b for β, a for x, $x - a$ for α, so that (1.1) will be presented in the equivalent form

$$\varphi(x) = \sum_{n=0}^{\infty} \varphi^{(n)}(u_n) \frac{(x - u_0)(x - u_n)^{n-1}}{n!} , \qquad (1.3)$$

where $u_n = a + bn$, $n \geq 0$. We call (1.3) the *Abel expansion* of $\varphi(x)$ with respect to the *Abel polynomials*

$$B_n(x) = (x - u_0)(x - u_n)^{n-1}/n! , \quad n \geq 0 . \qquad (1.4)$$

These polynomials being a generalization of the monomials $x^n/n!$, it is not surprising to meet them in several fields such as analysis, statistics or probability. Unfortunately in most fields, except again in interpolation theory, they are usually not identified, and the possibility of using their algebraic properties is ignored. The main purpose of this paper is to show on some stochastic models how an Abelian structure can be recognized and exploited.

1.2   *Abel versus Appell polynomials*. When in (1.4) we perform the exchange of u_0 and u_n, we get the polynomials

$$A_n(x) = (x - u_n)(x - u_0)^{n-1} / n! \ , \ n \geq 0 \ , \tag{1.5}$$

or, when $n > 0$,

$$A_n(x) = \frac{(x - u_0)^n}{n!} - b\frac{(x - u_0)^{n-1}}{(n-1)!} \ . \tag{1.6}$$

Let us compare the families $\{ B_n \ ; \ n \in \mathbf{N} \}$ and $\{ A_n \ ; \ n \in \mathbf{N} \}$. Clearly :

$$A_0(x) = B_0(x) = 1 \ . \tag{1.7}$$

$$B_n(u_0) = 0 \ \text{for} \ any \ n > 0 \ , \tag{1.8}$$

while

$$A_n(u_0) = 0 \ only \ \text{for} \ n > 1 \ . \tag{1.9}$$

$$A_n' = A_{n-1} \ \text{or} \ DA_n = A_{n-1} \ , \ n > 0 \ , \tag{1.10}$$

D being the differentiation operator. This last result means that $\{ A_n \ ; \ n \in \mathbf{N} \}$ is an *Appell family*. Similarly, when $n > 0$, one may write

$$B_n(x) = \frac{(x - u_n)^n}{n!} + b\frac{(x - u_n)^{n-1}}{(n-1)!}$$

the derivative of which is *not* $B_{n-1}(x)$ but $B_{n-1}(x - b)$ since $u_n = b + u_{n-1}$. Apparently this is less simple than (1.10) but, introducing the operator

$$\tilde{\Delta} = S^b D \ , \tag{1.11}$$

where S^b is the translation operator (i.e. $S^b f(x) = f(x + b)$), we can write

$$\tilde{\Delta}B_n(x) = S^b DB_n(x) = S^b B_{n-1}(x - b) = B_{n-1}(x) \ . \tag{1.12}$$

Consequently $\{ B_n \ ; \ n \in \mathbf{N} \}$ can be interpretated as a *generalized Appell family*, the operator $\tilde{\Delta}$ being used in place of D .

The results (1.10) and (1.12) are interesting but, with respect to expansions, the result (1.8) is more essential as one will see below.

1.3   Abel expansions and generalized Taylor expansions. Let R be a polynomial of degree n. Using convenient coefficients β_i we may write it

$$R(x) = \sum_{i=0}^{n} \beta_i B_i(x) . \qquad (1.13)$$

For $0 \le r \le n$, we have

$$\tilde{\Delta}^r R(a) = \sum_{i=r}^{n} \beta_i \tilde{\Delta}^r B_i(a) = \sum_{i=r}^{n} \beta_i B_{i-r}\left(u_0\right) = \beta_r$$

due to (1.7) and (1.8) . Therefore, working with $\tilde{\Delta}$ instead of Δ, we get the generalized Taylor formula

$$R(x) = \sum_{i=0}^{n} \tilde{\Delta}^i R(a) B_i(x) \qquad (1.14)$$

which is really the Abel expansion

$$R(x) = \sum_{i=0}^{n} R^{(i)}\left(u_i\right) B_i(x) . \qquad (1.15)$$

Since (1.9) is less powerful than (1.8), such beautiful simple results cannot be expected when the A_i's are used in place of the B_i's.

1.4   From Abel to Abel-Gontcharov polynomials. It is clear that in the preceding sub-section (1.15) could be proved from (1.13) by direct calculation of $D^r R(u_r)$, and without introducing the operator $\tilde{\Delta}$. This suggests that the concept of Abel expansion could be valuable even when the function $i \rightarrow u_i$ is not affine (in that case $\tilde{\Delta}$ would not exist and Abel expansions would not be reducible to generalized Taylor expansions).

To see that, we introduce the new notation

$$G_n\left(x|U\right) = B_n(x) = \left(x - u_0\right)\left(x - u_n\right)^{n-1} \Big/ n! \qquad (1.16)$$

where $U = \left\{ u_i ; i \in \mathbb{N} \right\}$. Instead of (1.7) and (1.8) we have

$$G_n\left(u_0|U\right) = \delta_{n0} \qquad (1.17)$$

and from (1.12)

$$G_n'(x|U) = \frac{(x - b - u_0)(x - b - u_{n-1})^{n-2}}{(n-1)!} = G_{n-1}(x|EU)$$

where

$$EU = \{ u_{1+i} \; ; \; i \in N \} \; .$$

More generally, for $r = 0, 1, ..., n$,

$$G_n^{(r)}(x|U) = G_{n-r}(x|E^r U) \tag{1.18}$$

where $E^r U = \{ u_{r+i} \; ; \; i \in N \}$ and consequently

$$G_n^{(r)}(u_0|U) = \delta_{nr} \; . \tag{1.19}$$

Now for a polynomial R of degree n we can write

$$R(x) = \sum_{i=0}^{n} \beta_i G_i(x|U)$$

with convenient β_i's. From (1.19) we get $R^{(r)}(u_r) = \beta_r$ so that (1.15) is proved again. In this proof only (1.19) is used and *no specification of the u_i's is required*. This means that Abel expansions can be defined not only with respect to Abel polynomials, but also with respect to polynomials defined by (1.18) and associated with any family $U = \{ u_i \; ; \; i \in N \}$ (affine or not). These polynomials are recognized to be Abel-Gontcharoff polynomials. For a brief presentation, the reader is referred to the appendix of [LEF96].

2   Simple applications of Abel polynomials

2.1   *The busy period of a M/D/1 queue.* Consider a queueing system for which any service time is of length 1, and new customers arrive according to a Poisson process with parameter λ and starting at time $-x$, $0 \le -x \le r$, r being the number of customers present in the queue initially. We denote by L the straight line defined by $t \to t - r$, by U the family $\{ u_i \; ; \; i \in N \}$ where $u_i = -r - i$ for any i, by $M = r + N$ the number of customers served during the first busy period and by K the number of new customers arriving during the time interval $(-x, -u_0)$. Clearly K has a Poisson distribution with parameter $\lambda(x - u_0)$ (see Figure 1) and N, the number of new customers served during the first busy period, can be identified with the level of the first crossing of the Poisson trajectory with the straight line L.

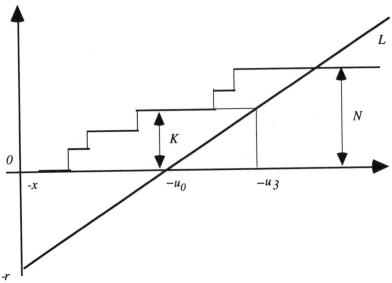

Fig. 1

To build $P(N=n)$ it will be convenient to stress its dependence on x and U, so we denote it by $P(N = n|x, U)$. First

$$P(N = 0|x, U) = P(K = 0) = e^{-\lambda\left(x - u_0\right)} . \tag{2.1}$$

When $n>0$,

$$P(N = n|u_0, U) = 0 \tag{2.2}$$

and, from a renewal argument,

$$P(N = n|x, U) = E\left(P\left(N = n - K|u_0, E^K U\right)\right), \tag{2.3}$$

where of course $E^K U = \{ u_{i+K} ; i \in \mathbf{N} \}$, i.e. explicitly,

$$P(N=n|x, U) = \sum_{k=0}^{n} e^{-\lambda\left(x-u_0\right)} \frac{\left(\lambda\left(x-u_0\right)\right)^k}{k!} P\left(N=n-k|u_0, E^k U\right). \tag{2.4}$$

Putting

$$P(N=n|x, U) = e^{-\lambda\left(x-u_n\right)}\lambda^n H_n(x|U) \tag{2.5}$$

we transform (2.1) and (2.2) respectively into

$$H_0(x|U) = 1, \tag{2.6}$$

$$H_n(u_0|U) = 0 , \quad n > 0 . \tag{2.7}$$

Since

$$P\left(N=n-k|u_0, E^kU\right) = e^{-\lambda\left(u_0-u_{k+n-k}\right)}\lambda^{n-k}H_{n-k}\left(u_0|E^kU\right),\qquad (2.8)$$

(2.4) is changed into

$$H_n\left(x|U\right) = \sum_{k=0}^{n}\frac{(x-u_0)^k}{k!}H_{n-k}\left(u_0|E^kU\right),\ n>0 .\qquad (2.9)$$

These formulae show that $H_n\left(x|U\right)$ is a polynomial of degree n with respect to x. Clearly $G_n\left(x|U\right)$ satisfies (2.6) and (2.7) and also (2.9) since, from Taylor's formula,

$$G_n\left(x|U\right) = \sum_{k=0}^{n}G_n^{(k)}\left(u_0|U\right)\frac{(x-u_0)^k}{k!} = \sum_{k=0}^{n}\frac{(x-u_0)^k}{k!}G_{n-k}\left(u_0|E^kU\right).$$

We conclude that $H_n\left(x|U\right)$ can be identified with $G_n\left(x|U\right), n>0$. This implies that

$$P\left(N=n|x, U\right) = e^{-\lambda\left(x-u_n\right)}\lambda^n\left(x-u_0\right)\left(x-u_n\right)^{n-1}\Big/n! ,\ n \geq 0 .\qquad (2.10)$$

Note that putting $x=0$, we get for $M=r+N$ the Borel-Tanner distribution (see e.g. [PRA65]). When $\lambda \leq 1$, $N < \infty$ a.s. so that we have also proved

$$\sum_{n=0}^{\infty}e^{-\lambda\left(x-u_n\right)}\lambda^nG_n\left(x|U\right) = 1$$

or equivalently

$$e^{\lambda x} = \sum_{n=0}^{\infty}\left(e^{\lambda x}\right)^{(n)}_{x=u_n}G_n\left(x|U\right) ,$$

that is, $e^{\lambda x}$ is then the sum of its Abel expansion.

2.2   *A non parametric one-sided test.* Let $X_1, X_2,..., X_n$ be a sample of n i.i.d. random variables uniformly distributed on $(0,1)$, let $X^{(1)} \leq X^{(2)} \leq...\leq X^{(n)}$ be their order statistics, and let

$$\mathbb{F}_n(x) = \frac{1}{n}\sum_{j=1}^{n}\mathbb{1}_{\left\{X^{(j)}\leq x\right\}} ,$$

the empirical distribution function.

An important problem is to compare $\mathbb{F}_n(x)$ with the exact distribution function. Following, for example, [BIR51], let us consider the probability

$$\Delta_n(\varepsilon) = P\left(\mathbb{F}_n(x) > x - \varepsilon, \; 0 \le x \le 1\right) , \tag{2.11}$$

where $0 < \varepsilon < 1$. Putting

$$v_i = Min\left(\varepsilon + \frac{i}{n}, 1\right) \quad \text{and} \quad V = \left\{v_i \; ; \; i = 0, 1, ..., n-1\right\},$$

we observe that

$$\Delta_n(\varepsilon) = P\left(X^{(i+1)} < v_i \; , \; i = 0, 1, ..., n-1\right)$$

$$= n! \int_0^{v_0} d\xi_0 \int_{\xi_0}^{v_1} d\xi_1 ... \int_{\xi_{n-2}}^{v_{n-1}} d\xi_{n-1} = n!(-1)^n G_n(0|V). \tag{2.12}$$

Thus it remains to evaluate $G_n(0|V)$. Applying Lemma 2.1 below when $w_i = 1$, $u_i = \varepsilon + \dfrac{i}{n}$ for any i, and $k = Max\left\{i \; ; \; \varepsilon + \dfrac{i}{n} < 1\right\}$, we get

$$G_{n-i}\left(u_i\middle|E^iW\right) = \left(u_i - 1\right)^{n-i} \middle/ (n-i)!,$$

$$G_i\left(0|U\right) = (-\varepsilon)(-\varepsilon - i/n)^{i-1} \middle/ i!$$

yielding

$$\Delta_n(\varepsilon) = \sum_{i=k+1}^{n} \binom{n}{i}\left(1 - u_i\right)^{n-i}\varepsilon\left(\varepsilon + i/n\right)^{i-1} , \; n \ge 1. \tag{2.13}$$

Lemma 3.1. Having $U = \left\{u_i \; ; \; i = 0, 1, ..., n-1\right\}$, $W = \left\{w_i \; ; \; i = 0, 1, ..., n-1\right\}$ and k natural, $0 \le k < n$, define $V = \left\{v_i \; ; \; i = 0, 1, ..., n-1\right\}$ where for any i,

$$v_i = u_i \mathbb{1}_{\{i \le k\}} + w_i \mathbb{1}_{\{i > k\}}.$$

Then

$$G_n\left(x|V\right) = \sum_{i=k+1}^{n} G_{n-i}\left(u_i\middle|E^iW\right)G_i\left(x|U\right) . \tag{2.14}$$

Proof: The Abel expansion of $G_n\left(x|V\right)$ with respect to the $G_i\left(x|U\right)$'s is

$$G_n(x|V) = \sum_{i=0}^{n} G_n^{(i)}(u_i|V) G_i(x|U) . \qquad (2.15)$$

But, when $i \leq k$, $v_i = u_i$ and $G_n^{(i)}(u_i|V) = G_{n-i}\left(v_i|E^iV\right) = \delta_{ni} = 0$, and

when $i > k$, $v_i = w_i$ and $E^iV = E^iW$. Therefore (2.15) reduces to (2.14).☐

2.3   The level of first meeting of two independent trajectories.

Consider a markovian death process with infinitesimal generator $\alpha_i = \mu i/(Ci + D)$, $C \geq 0$, $D > 0$, $\mu > 0$ and a Poisson process with parameter $\lambda > 0$. These processes are supposed to be independent. Initially they start at levels m and r respectively, m and r being natural numbers, $m \geq r$. We note by N the level of first meeting of the trajectories. N is a random variable with range $\{r, r+1, ..., m-1, m\}$ and we are going to build its generating function,

$$E\left(x^N\right) = \psi(x|m, r) .$$

Clearly

$$\psi(x|m, m) = x^m \qquad (2.16)$$

and, when $r < m$, a renewal argument leads to

$$\psi(x|m, r) = \frac{\alpha_m}{\lambda + \alpha_m} \psi(x|m-1, r) + \frac{\lambda}{\lambda + \alpha_m} \psi(x|m, r+1). \qquad (2.17)$$

(2.16) and (2.17) together determine $\psi(x|m, r)$ recursively for $r = m-1, m-2, ...$. Notice that (2.17) admits solutions of the special form $a_m(k)b_r(k)$, k being any natural number, where

$$b_r(k) = \left(1 + \alpha_k/\lambda\right)^{r-k} , \qquad (2.18)$$

$$a_m(k) = m_{[k]}\left(1 + Ck/D\right)^{m-k}, \qquad (2.19)$$

and

$$m_{[k]} = m(m-1)(m-2)...(m-k+1) .$$

Hence we may look for $\psi(x|m, r)$ of the following form

$$\psi(x|m, r) = \sum_{k=0}^{m} a_m(k)b_r(k)H_k(x) , \qquad (2.20)$$

$H_k(x)$ being, for any k, a polynomial of degree k with respect to x. Indeed (2.20) implies (2.17), and to satisfy (2.16) one must choose the H_k's such that

$$x^m = \sum_{k=0}^{m} m_{[k]} \left(\left(1 + Ck/D\right)\left(1 + \alpha_k/\lambda\right)\right)^{m-k} H_k(x) , \ m \geq 0 ,$$

or equivalently

$$x^m = \sum_{k=0}^{m} \left(x^m\right)^{(k)}_{x = u_k} H_k(x) , \ m \geq 0 \tag{2.21}$$

where

$$u_k = \left(1 + Ck/D\right)\left(1 + \alpha_k/\lambda\right) . \tag{2.22}$$

Clearly (2.21) determines the H_k's recursively and for any k, $H_k(x)$ can be identified with $G_k(x|U)$ where $U = \{ u_i \ ; \ i \in \mathbf{N} \}$ is defined by (2.22), since then (2.21) is true as being an Abel expansion.

Finally it has been proved that

$$E\left(x^N\right) = \sum_{k=0}^{m} a_m(k) b_r(k) \left(x - u_0\right)\left(x - u_k\right)^{k-1} / k!$$

where $a_m(k)$, $b_r(k)$, u_k are given respectively by (2.19), (2.18), (2.22) and $u_0 = 1.\Box$

3   Generalized Appell and Abel polynomials

Definition 3.1. From the real formal power series $g(s) = \sum_{i=1}^{\infty} \alpha_i s^i$, *we define the family of real polynomials* $\{ e_n \ ; \ n \in \mathbf{N} \}$ *by its generating function*

$$\sum_{n=0}^{\infty} e_n(x) s^n = exp\{xg(s)\} \tag{3.1}$$

and then the linear operator Δ *by*

$$\begin{cases} \Delta e_{n+1} = e_n , \ n \geq 0 , \\ \Delta e_0 = 0 , \end{cases} \tag{3.2}$$

Δ^r *being defined recursively by* $\Delta^r = \Delta\left(\Delta^{r-1}\right), \ r > 1.$ $\tag{3.3}$

Remarks : $e_0(x) = 1$; the degree of e_n is n ; $e_n(0) = 0$ when $n>0$. When $g(s)=s$, $e_n(x) = x^n/n!$, $\Delta = D$ the differentiation operator. Formulae (3.2), (3.3) mean that $\{ e_n ; n \in N \}$ is a family of *generalized Appell polynomials*, Δ having been substituted for D.

Proposition 3.2. Δ *is shift invariant (i.e.* $\forall c \in IR$, $\Delta S^c = S^c \Delta$).

Proof : $\Delta S^c \left(\sum\limits_{n=0}^{\infty} e_n(x)s^n \right) = \Delta S^c e^{xg(s)} = \Delta e^{(c+x)g(s)} = e^{cg(s)} \sum\limits_{n=1}^{\infty} e_{n-1}(x)s^n$

$= e^{cg(s)} s e^{xg(s)} = S^c \left(s e^{xg(s)} \right) = S^c \sum\limits_{n=0}^{\infty} e_{n-1}(x)s^n = S^c \Delta \left(\sum\limits_{n=0}^{\infty} e_n(x)s^n \right)$

which shows that $\Delta S^c = S^c \Delta .\square$

Proposition 3.3. Any polynomial R can be expanded with respect to Δ and the e_n's, according to a generalized Taylor expansion

$$R(x) = \sum_r \Delta^r R(c) e_r(x-c) , \ c \in IR , \ x \in IR . \tag{3.4}$$

Proof : We write

$$R(x) = \sum_{i=0}^{k} \gamma_i e_i(x-c)$$

with convenient γ_i's. Due to proposition 3.2. and the definition of the e_i's, for $r=0,1,...,k$ we have $\Delta^r e_i(c-c) = S^{-c} e_{i-r}(c) = e_{i-r}(0) = \delta_{ir}$ so that $\Delta^r R(c) = \gamma_r$ and the proposition is proved.$\square$

Proposition 3.4. $\quad \Delta\big(n e_n(x)/x\big) = (n-1)e_{n-1}(x)/x$, $n > 0$.

Proof : For $n=1$ the formula is obvious. For $n>1$, we start with

$$s \sum_{n=0}^{\infty} e_n(x) n s^{n-1} = s \frac{d}{ds} e^{xg(s)} = sxg'(s)e^{xg(s)}$$

and, identifying the terms in s^n, we get

$$n e_n(x) = x \sum_{j=1}^{n} j \alpha_j e_{n-j}(x) . \tag{3.5}$$

Hence, using (3.5) and this formula again after substitution of $n-1$ for n ,

$$\Delta\big(ne_n(x)/x\big) = \sum_{j=1}^{n-1} j\alpha_j e_{n-1-j}(x) = (n-1)e_{n-1}(x)/x. \;\Box$$

Definition 3.5. To $U=\{ u_i \; ; \; i \in \mathbf{N} \}$ where $u_i = a + bi$ for any i, we associate the generalized Abel polynomials

$$\overline{G}_n\big(x|U\big) = \frac{x - u_0}{x - u_n} e_n\big(x - u_n\big), \; n \geq 0 .$$

Remark : $\overline{G}_n\big(x|U\big)$ is clearly a polynomial of degree n with respect to x . To justify the designation of $\overline{G}_n$ as generalized Abel polynomial we have the following proposition.

Proposition 3.6. $\qquad \overline{G}_n\big(u_0|U\big) = \delta_{n0}$ and, for $0 \leq r \leq n$,

$$\Delta^r \overline{G}_n\big(x|U\big) = \overline{G}_{n-r}\big(x|E^r U\big) .$$

Proof : The first result is obvious. For the second, since

$$\Delta \overline{G}_n\big(x|U\big) = \Delta S^{-u_n} \left(\frac{x + u_n - u_0}{x} e_n(x) \right) = S^{-u_n} \Delta \left(e_n(x) + bn \frac{e_n(x)}{x} \right)$$

$$= S^{-u_n} \left(e_{n-1}(x) + b(n-1)e_{n-1}(x)/x \right) = S^{-u_n} \frac{x + u_n - u_1}{x} e_{n-1}(x)$$

$$= \frac{x - u_1}{x - u_{1+n-1}} e_{n-1}\big(x - u_{1+n-1}\big) = \overline{G}_{n-1}\big(x|EU\big) ,$$

the result follows for $r=1$ and then for any r . $\Box$

Proposition 3.7. Any polynomial R admits, with respect to the operator Δ and the polynomials $\overline{G}_i$'s , the generalized Abel expansion

$$R(x) = \sum_i \Delta^i R\big(u_i\big) \overline{G}_i\big(x|U\big) .$$

Proof : Writing $R(x) = \sum_i \gamma_i \overline{G}_i\big(x|U\big)$, we easily check that $\Delta^i R\big(u_i\big) = \gamma_i. \Box$

4   Application to the busy period of some queueing systems

The results in the following examples are most probably known but we think that the methology is new.

4.1. A M/D/1 queue with batch arrivals.

Coming back to sub-section 2.1. we introduce $\{ q_j ; j \in N^* \}$ as the distribution of the number of people arriving in the same batch. During the time interval $(-x, -u_0)$ the generating function of the number K of arrivals is

$$E\left(s^K\right) = exp\left\{-\lambda\left(x - u_0\right)\left(1 - \sum_{j=1}^{\infty} q_j s^j\right)\right\} = e^{-\lambda(x-u_0)}e^{(x-u_0)g(s)}$$

where $g(s) = \sum_{j=1}^{\infty} \lambda q_j s^j$. Using the notations of Section 3 we have

$$P(K = k) = e^{-\lambda\left(x-u_0\right)}e_k\left(x - u_0\right),\qquad(4.1)$$

hence as before

$$P\left(N = 0 | x, U\right) = e^{-\lambda\left(x-u_0\right)}\qquad(4.2)$$

and, when $n>0$,

$$P\left(N = n | u_0, U\right) = 0,\qquad(4.3)$$

$$P\left(N = n | x, U\right) = \sum_{k=0}^{n} e^{-\lambda\left(x-u_0\right)}e_k\left(x - u_0\right)P\left(N = n - k | u_0, E^k U\right).\qquad(4.4)$$

Putting

$$P\left(N = n | x, U\right) = e^{-\lambda\left(x-u_n\right)}K_n(x | U),\qquad(4.5)$$

we arrive at

$$K_0(x | U) = 1,\qquad(4.6)$$

$$K_n\left(u_0 | U\right) = 0 \quad\text{when } n>0,\qquad(4.7)$$

$$K_n\left(x | U\right) = \sum_{k=0}^{n} e_k\left(x - u_0\right)K_{n-k}\left(u_0 | E^k U\right), n>0,\qquad(4.8)$$

$K_n\left(x | U\right)$ being therefore a polynomial of degree n with respect to x that can be identified with $\overline{G}_n\left(x | U\right)$. (Notice that, for $\overline{G}_n\left(x | U\right)$, (4.8) is equivalent to

$$\overline{G}_n\left(x | U\right) = \sum_r \Delta^r \overline{G}_n\left(u_0 | U\right)e_r\left(x - u_0\right)$$

which is true due to (3.4)). Finally

$$P(N = n|x, U) = e^{-\lambda(x-u_n)} \frac{x - u_0}{x - u_n} e_n(x - u_n) . \tag{4.9}$$

Example : When $\forall i > 0$, $q_i = \rho \alpha^i / i$ where $\rho = -1/\ell n(1 - \alpha), 0 < \alpha < 1,$

$$P(N = n|0, U) = e^{\lambda u_n} \frac{u_0}{u_n} \binom{\lambda \rho u_n}{n} (-\alpha)^n .$$

4.2. A similar queue with a Bernoulli arrival process.

Here we suppose that $-x$ and $-u_0$ are integers, $-x \le -u_0$, a new customer being able to join the queue only at times $-x+1, -x+2, \ldots$. For any i , at time $-x+i$, there is an arrival with probability q , $0 < q < 1$, and no arrival with probability $1-q$. All arrivals are independent. The formulae (2.1) and (2.4) in sub-section 2.1 are turned into

$$P(N = 0|x, U) = (1 - q)^{x - u_0} , \tag{4.10}$$

$$P(N=n|x, U) = \sum_{k=0}^{n} \binom{x-u_0}{k} q^k (1-q)^{x-u_0-k} P\left(N=n-k|u_0, E^k U\right) . \tag{4.11}$$

Instead of (2.5) we put

$$P(N = n|x, U) = (1 - q)^{x-u_n} \left(\frac{q}{1-q}\right)^n Q_n(x|U) , \tag{4.12}$$

$Q_n(x|U)$ satisfying

$$Q_0(x|U) = 1 , \tag{4.13}$$

$$Q_n(u_0|U) = 0 , n > 0 , \tag{4.14}$$

$$Q_n(x|U) = \sum_{k=0}^{n} \binom{x - u_0}{k} Q_{n-k}(u_0|U) , n > 0 . \tag{4.15}$$

These last three conditions show that $Q_n(x|U)$ is a polynomial of degree n with respect to x , that can be identified with $\overline{G}_n(x|U)$ when the family $\{ e_k ; k \in \mathbf{N}\}$ is such that

$$e_k(x) = \binom{x}{k} \tag{4.16}$$

(the identification is performed as in sub-section 4.1). Notice that (4.16) means that here Δ is the usual forward difference operator. The conclusion is that

$$Q_n\left(x|U\right) = \frac{x - u_0}{x - u_n}\binom{x - u_n}{n}$$

and therefore

$$P\left(N = n|x, U\right) = (1 - q)^{x - u_n}\left(\frac{q}{1 - q}\right)^n \frac{x - u_0}{x - u_n}\binom{x - u_n}{n}. \tag{4.17}$$

The case of batch arrivals could be treated similarly.

5   References

[ABE81] N. ABEL. Sur les fonctions génératrices et leurs déterminantes. *Oeuvres complètes,* Sylow and Lie, Eds. Christiana, 2, 67-81, 1881.

[APP80] P.E. APPELL. Sur une classe de polynômes. *Ann. Scient. Ecole Normale Sup.* (2) 9, 119-144, 1880.

[BIR51] Z.W. BIRNBAUM and F.H. TINGEY. One-sided contours for probability distribution functions. *Ann. Math. Stat.,* 22, 592-596, 1951.

[JEN02] J.L. JENSEN. Sur une identité d'Abel et sur d'autres formules analogues. *Acta. Mathematica* 26, 307-318, 1902.

[LEF96] C. LEFEVRE and Ph. PICARD. On the first crossing of a Poisson process in a lower boundary, this book, 1996.

[PRA65] N.U. PRABHU. *Queues and inventories.* John Wiley and Sons Inc., New York, 1965.

Positive Dependence Orders: A Survey

Marco Scarsini
Moshe Shaked

ABSTRACT Notions of positive dependence of two random variables X and Y have been introduced in the literature in an effort to mathematically describe the property that "large (respectively, small) values of X go together with large (respectively, small) values of Y." Some of these notions are based on some comparison of the joint distribution of X and Y with their distribution under the theoretical assumption that X and Y are independent. Often such a comparison can be extended to general pairs of bivariate distributions with given marginals. This fact led researchers to introduce various notions of positive dependence orders. These orders are designed to compare the strength of positive dependence of the two underlying bivariate distributions. In this survey we describe some such notions.

1 Introduction

Notions of positive dependence of two random variables X and Y have been introduced in the literature in an effort to mathematically describe the property that "large (respectively, small) values of X go together with large (respectively, small) values of Y." The idea goes back to Gini (1915–16) who used it as the heuristics for measures of concordance. Many of the modern notions of positive dependence are defined by means of some comparison of the joint distribution of X and Y with their distribution under the theoretical assumption that X and Y are independent. Often such a comparison can be extended to general pairs of bivariate distributions with given marginals. This fact led researchers to introduce various notions of positive dependence orders. These orders are designed to compare the strength of positive dependence of the two underlying bivariate distributions. In this survey we describe some such notions.

In an effort to formalize the fundamental ideas that are associated with notions of positive dependence orders, Kimeldorf and Sampson (1987) have presented ten basic desirable properties (axioms) that such notions should satisfy. Some of the orders that are described in this survey do not satisfy all of Kimeldorf and Sampson's axioms. We will not list the axioms here, but in sequel, when we encounter an order that violates some of the axioms then we will point it out, and we will then describe the underlying axioms.

In this paper "increasing" and "decreasing" mean "non-decreasing" and "non-increasing," respectively. Also, whenever an expectation is written, it is implicitly assumed that it exists.

Most of the orders that we describe in this paper are defined on on the Fréchet class $\mathcal{M}(H_1, H_2)$ of bivariate distributions with fixed marginals H_1 and H_2. The upper bound of this class is the distribution defined by $\min\{H_1(x), H_2(y)\}$ (whose probability mass is concentrated on the set $\{(x, y) : H_1(x) = H_2(y)\}$). The lower bound of this class is the distribution defined by $\max\{H_1(x) + H_2(y) - 1, 0\}$ (whose probability mass is concentrated on the set $\{(x, y) : H_1(x) + H_2(y) = 1\}$).

2   The PQD Order

Let the random variables X and Y have the joint distribution G. Let G_X and G_Y denote, respectively, the marginal distributions of X and Y. Lehmann (1966) defines G (or X and Y) to be positive quadrant dependent (PQD) if

$$G(x, y) \geq G_X(x)G_Y(y), \qquad \text{for all } x \text{ and } y. \tag{2.1}$$

This notion led Yanagimoto and Okamoto (1969) and Tchen (1980) to introduce the order that is given in definition 2.1 below. Recall that the survival function $\overline{G}$, that corresponds to G above, is defined by $\overline{G}(x, y) \equiv 1 - G_X(x) - G_Y(y) + G(x, y)$ for all x and y.

Definition 2.1 *Let F and G be two bivariate distributions with the same marginals. Then F is said to be smaller than G in the* PQD *sense (denoted by $F \leq_{\text{PQD}} G$) if*

$$F(x, y) \leq G(x, y), \qquad \text{for all } x \text{ and } y. \tag{2.2}$$

Equivalently, $F \leq_{\text{PQD}} G$ if, and only if,

$$\overline{F}(x, y) \leq \overline{G}(x, y), \qquad \text{for all } x \text{ and } y. \tag{2.3}$$

If (U, V) has the distribution F and (X, Y) has the distribution G, and $F \leq_{\text{PQD}} G$, then we will sometimes write $(U, V) \leq_{\text{PQD}} (X, Y)$. If P_F and P_G are the measures associated with F and G, respectively (namely, $P_F\{(-\infty, x]\} = F(x)$ and $P_G\{(-\infty, y]\} = G(y)$), then we sometimes write $P_F \leq_{\text{PQD}} P_G$.

It is easily seen that G is PQD if, and only if, $G^I \leq_{\text{PQD}} G$, where $G^I(x, y) \equiv G_X(x)G_Y(y)$ for all x and y.

By Hoeffding's Lemma (see Lehmann, 1966, page 1139) we see that if (U, V) and (X, Y) are as in Definition 2.1 then

$$\text{Cov}(U, V) = \int_{-\infty}^{\infty} \int_{-\infty}^{\infty} [F(x, y) - F_U(x)F_V(y)]dxdy$$

and

$$\text{Cov}(X,Y) = \int_{-\infty}^{\infty} \int_{-\infty}^{\infty} [G(x,y) - G_X(x)G_Y(y)]dxdy,$$

provided the covariances are well-defined (here the notation F_U and F_V is obvious). It thus follows from (2.2) that if $(U,V) \leq_{\text{PQD}} (X,Y)$ then (recall that then $U =_{\text{st}} X$ and $V =_{\text{st}} Y$, where '$=_{\text{st}}$' denotes equality in law)

$$\text{Cov}(U,V) \leq \text{Cov}(X,Y), \tag{2.4}$$

and therefore

$$\rho_{U,V} \leq \rho_{X,Y},$$

where $\rho_{U,V}$ and $\rho_{X,Y}$ denote the correlation coefficients associated with (U,V) and (X,Y), respectively, provided the underlying variances are well-defined. Yanagimoto and Okamoto (1969) have shown that some other correlation measures, such as Kendall's τ, Spearman's ρ and Blomquist's q, are preserved under the PQD order. The inequality (2.4), and the monotonicity of other correlation measures under the PQD order, can also be obtained as corollaries from Theorem 2.5 below.

The PQD order enjoys some desirable preservation properties. Dabrowska (1985) and Kimeldorf and Sampson (1987) have shown the following result.

Theorem 2.2 *Suppose that the four random vectors* (U_1, V_1), (X_1, Y_1), (U_2, V_2) *and* (X_2, Y_2) *satisfy*

$$(U_1, V_1) \leq_{\text{PQD}} (X_1, Y_1) \quad and \quad (U_2, V_2) \leq_{\text{PQD}} (X_2, Y_2), \tag{2.5}$$

and assume that (U_1, V_1) *and* (U_2, V_2) *are independent and also that* (X_1, Y_1) *and* (X_2, Y_2) *are independent. Then*

$$(\phi(U_1, U_2), \psi(V_1, V_2)) \leq_{\text{PQD}} (\phi(X_1, X_2), \psi(Y_1, Y_2)),$$

$$for\ all\ increasing\ functions\ \phi\ and\ \psi.$$

In particular, if (2.5) holds then

$$(U_1 + U_2, V_1 + V_2) \leq_{\text{PQD}} (X_1 + X_2, Y_1 + Y_2),$$

that is, PQD order is closed under convolutions. From Theorem 2.2 it also follows that

$$(U,V) \leq_{\text{PQD}} (X,Y) \Longrightarrow (\phi(U), \psi(V)) \leq_{\text{PQD}} (\phi(X), \psi(Y)),$$

for all increasing functions ϕ and ψ. Combining this with (2.4) it is seen that

$$(U,V) \leq_{\text{PQD}} (X,Y) \Longrightarrow \text{Cov}(\phi(U), \psi(V)) \leq \text{Cov}(\phi(X), \psi(Y)),$$

for all increasing functions ϕ and ψ.

The following closure under mixtures property of the PQD order is easy to prove.

Theorem 2.3 *Consider the two random vectors* (U, V, Θ) *and* (X, Y, Θ). *Suppose that for every value* θ *in the support of* Θ *one has*

$$[(U, V)|\Theta = \theta] \leq_{\mathrm{PQD}} [(X, Y)|\Theta = \theta],$$

where $[(U, V)|\Theta = \theta]$ *and* $[(X, Y)|\Theta = \theta]$ *denote random variables whose distribution functions are the conditional distributions of* (U, V) *and* (X, Y), *respectively, given* $\Theta = \theta$. *Then*

$$(U, V) \leq_{\mathrm{PQD}} (X, Y).$$

The following closure property of the PQD order (which corresponds to one of the axioms of Kimeldorf and Sampson, 1987) is also easy to prove.

Theorem 2.4 *Let* $\{F_n\}$ *and* $\{G_n\}$ *be two sequences of bivariate distributions. If* $F_n \leq_{\mathrm{PQD}} G_n$, $n = 1, 2, \ldots$, *and if* $F_n \to F$ *and* $G_n \to G$ *(where* $\to$ *denotes convergence in law), then* $F \leq_{\mathrm{PQD}} G$.

A bivariate function ϕ is said to be supermodular if $\phi(x, y) + \phi(x', y') \geq \phi(x', y) + \phi(x, y')$ whenever $x \leq x'$ and $y \leq y'$. For example, every bivariate distribution function is a bounded supermodular function. Cambanis, Simons, and Stout (1976), Tchen (1980), Rüschendorf (1980), Cambanis and Simons (1982), and Mosler (1984) have shown the next result.

Theorem 2.5 *Let* (U, V) *and* (X, Y) *be two random vectors. If* (U, V) $\leq_{\mathrm{PQD}} (X, Y)$ *then*

$$E[\phi(U, V)] \leq E[\phi(X, Y)], \quad \text{for every bounded supermodular function } \phi.$$

Some extensions of this result can be found in the above mentioned references.

If F and G of (2.2) are the distribution functions of bivariate vectors with integer-valued components, then the comparison $F \leq_{\mathrm{PQD}} G$ is the same as a comparison of the partial sums of two matrices with non-negative entries (which sum up to 1). Nguyen and Sampson (1985a) studied the geometry of such matrices.

The PQD comparison can be used also to compare contingency tables that have the same row and column sums. Nguyen and Sampson (1985b) obtained some results regarding the number of such contingency tables that are more PQD than a given contingency table.

Block, Chhetry, Fang and Sampson (1990) found necessary and sufficient conditions (by means of orders of permutations) for two bivariate empirical distributions to be ordered according to the PQD order. Further results in this vein are given in Metry and Sampson (1993). Examples of pairs of bivariate distributions that are PQD ordered can be found in de la Horra and Ruiz-Rivas (1988) and in Joe (1993). Bassan and Scarsini (1994) characterized the PQD order by means of the usual stochastic ordering of some related stopping times. Some applications of the PQD order in reliability and queueing theory are described in Section 3.6 of Kalashnikov and Rachev (1990). Some related measures of association can be found in Kruskal (1958) which basically discovered the works of the Italian school that C. Gini founded.

Ebrahimi (1982) discussed negatively dependent distributions that are ordered according to the PQD order.

Fang, Hu and Joe (1994) apply the idea of positive dependence to stationary Markov chains and show that, if the process is stochastically increasing, then dependence (in the sense of $\leq_{\text{PQD}}$) is decreasing with the lag, namely, if $\{X_1, X_2, \ldots\}$ is a Markov chain and X_i is distributed according to F, (X_1, X_n) is distributed according to F_{1n}, then

$$F_{12} \geq_{\text{PQD}} F_{13} \geq_{\text{PQD}} \cdots \geq_{\text{PQD}} F_{1n} \geq_{\text{PQD}} \cdots \geq_{\text{PQD}} F^{(2)},$$

where $F^{(2)}(x, y) = F(x)F(y)$.

Joe (1990) extended the PQD order to the multivariate case (from the bivariate case given in Definition 2.1). He did it by letting F and G in Definition 2.1 each be a joint distribution of n random variables, where $n > 2$, and then required *both* (2.2) and (2.3) to hold. Explicitly, let $U = (U_1, U_2, \ldots, U_n)$ and $X = (X_1, X_2, \ldots, X_n)$ be two random vectors with distribution functions F and G, respectively. Denote $\overline{F}(u) = P\{U_1 > u_1, U_2 > u_2, \ldots, U_n > u_n\}$ and $\overline{G}(x) = P\{X_1 > x_1, X_2 > x_2, \ldots, X_n > x_n\}$. Then Joe (1990) defined F to be smaller than G in the concordant ordering if

$$F(x) \leq G(x) \quad \text{and} \quad \overline{F}(x) \leq \overline{G}(x), \quad \text{for all } x. \tag{2.6}$$

From (2.6) it is seen that F and G must have the same marginals. It is also seen that if $(U_1, U_2, \ldots, U_n)$ is smaller than $(X_1, X_2, \ldots, X_n)$ in the concordant ordering then also $(U_{i_1}, U_{i_2}, \ldots, U_{i_\ell})$ is smaller than $(X_{i_1}, X_{i_2}, \ldots, X_{i_\ell})$ in the concordant ordering for every $1 \leq i_1 < i_2 < \cdots < i_\ell \leq n$. See Joe (1990) for some further properties and applications of this order. Brown and Rinott (1988) showed that some pairs of multivariate infinitely divisible distributions satisfy (2.6).

If one requires (2.2) or (2.3) to hold, without assuming that F and G have the same marginals, then one obtains orderings which compare (U, V) and (X, Y) by size (in addition to the comparison by their strength of positive dependence). Such orders are described in Subsection 4.G.1 of Shaked and Shanthikumar (1994). These orders can be easily extended for the comparison of vectors of dimension $n \geq 2$; see, for example, Section 4.6 in Samorodnitsky and Taqqu (1994).

3 The PRD Order

Let the random variables X and Y have the joint distribution G. Let $G_{Y|X}(\cdot|x)$ denote the conditional distribution of Y given $X = x$, and let

$G_{Y|X}^{-1}(\cdot|x)$ denote its right-continuous inverse. Lehmann (1966) defines G (or X and Y) to be positive regression dependent (PRD) if Y is stochastically increasing in X, that is, if

$$G_{Y|X}(y|x) \geq G_{Y|X}(y|x'), \qquad \text{for all } x \leq x' \text{ and } y,$$

or, equivalently, if

$$G_{Y|X}^{-1}(u|x) \leq G_{Y|X}^{-1}(u|x'), \qquad \text{for all } x \leq x' \text{ and } u \in [0,1].$$
(3.1)

Note that when $G_{Y|X}(u|x)$ is continuous in u for all x then (3.1) can be equivalently written as

$$G_{Y|X}\left[G_{Y|X}^{-1}(u|x)\Big|x'\right] \leq u, \qquad \text{for all } x \leq x' \text{ and } u \in [0,1].$$
(3.2)

This notion led Yanagimoto and Okamoto (1969) to introduce the order that is given in the following definition.

Definition 3.1 *Let F and G be two bivariate distributions with the same marginals, and let (U,V) and (X,Y) be two random vectors having the distributions F and G, respectively. Then F is said to be smaller than G in the* PRD *sense (denoted by $F \leq_{\text{PRD}} G$ or $(U,V) \leq_{\text{PRD}} (X,Y)$ or $P_F \leq_{\text{PRD}} P_G$) if, for any $x \leq x'$,*

$$F_{V|U}^{-1}(u|x) \leq F_{V|U}^{-1}(v|x') \Longrightarrow G_{Y|X}^{-1}(u|x) \leq G_{Y|X}^{-1}(v|x'),$$

$$\textit{for any } u,v \in [0,1], \quad (3.3)$$

where the notations $F_{V|U}$ and $F_{V|U}^{-1}$ are obvious.

Note that (3.3) can be equivalently written as

$$G_{Y|X}\left[G_{Y|X}^{-1}(u|x)\Big|x'\right] \leq F_{V|U}\left[F_{V|U}^{-1}(u|x)\Big|x'\right],$$

$$\text{for all } x \leq x' \text{ and } u \in [0,1]. \quad (3.4)$$

Yanagimoto and Okamoto (1969) have shown that $F \leq_{\text{PRD}} G$ if, and only if, for any $x \leq x'$,

$$F_{V|U}(y|x) \geq G_{Y|X}(y'|x) \Longrightarrow F_{V|U}(y|x') \geq G_{Y|X}(y'|x'),$$

$$\text{for any } y \text{ and } y'. \quad (3.5)$$

Note that (3.5) can be equivalently written as

$$G_{Y|X}^{-1}\left[F_{V|U}(y|x)\Big|x\right] \leq G_{Y|X}^{-1}\left[F_{V|U}(y|x')\Big|x'\right],$$

$$\text{for all } x \leq x' \text{ and } y, \quad (3.6)$$

that is, $G_{Y|X}^{-1}\left[F_{V|U}(y|x)\big|x\right]$ is increasing in x for all y (Fang and Joe, 1992). Another characterization of the PRD order can be found in Rüschendorf (1986).

Yanagimoto and Okamoto (1969) have also shown that G is PRD if, and only if, $G^I \leq_{\text{PRD}} G$, where G^I is defined in Section 2. In the continuous case, this is immediate from (3.2) and (3.4). Yanagimoto and Okamoto (1969) also showed that

$$(U,V) \leq_{\text{PRD}} (X,Y) \Longrightarrow (U,V) \leq_{\text{PQD}} (X,Y).$$

The PRD order is not symmetric in the sense that $(U,V) \leq_{\text{PRD}} (X,Y)$ does not necessarily imply that $(V,U) \leq_{\text{PRD}} (Y,X)$. Thus, this order violates one of the axioms of Kimeldorf and Sampson (1987). However, it satisfies the following closure under monotone transformations property; this property is equivalent to another axiom of Kimeldorf and Sampson (1987).

Proposition 3.2 *Let* (U,V) *and* (X,Y) *be two random vectors. If* (U,V) $\leq_{\text{PRD}} (X,Y)$ *then* $(\phi(U),\psi(V)) \leq_{\text{PRD}} (\phi(X),\psi(Y))$ *for all increasing functions* ϕ *and* ψ.

Example 3.3 Let U and V be any independent random variables, each having a continuous distribution. Define

$$X = U, \quad Y_\rho = \rho U + (1-\rho^2)^{1/2}V, \qquad \text{for } -1 \leq \rho \leq 1.$$

Then $(X,Y_{\rho_1}) \leq_{\text{PRD}} (X,Y_{\rho_2})$ whenever $\rho_1 \leq \rho_2$. A bivariate normal distribution is a particular case of this example when U and V are normally distributed.

Example 3.4 Let U and V be any independent random variables, each having a continuous distribution. Define

$$X = U, \quad Y_\alpha = \alpha U + V, \qquad \text{for } -\infty \leq \alpha \leq \infty.$$

Then $(X,Y_{\alpha_1}) \leq_{\text{PRD}} (X,Y_{\alpha_2})$ whenever $\alpha_1 \leq \alpha_2$.

Example 3.5 Let U and V be any independent random variables, each having a continuous distribution, such that U is distributed on $(0,1)$, while V is non-negative. Define

$$X = U, \quad Y_\alpha = (1 + \alpha U)V, \qquad \text{for } \alpha \geq -1.$$

Then $(X,Y_{\alpha_1}) \leq_{\text{PRD}} (X,Y_{\alpha_2})$ whenever $\alpha_1 \leq \alpha_2$.

Many other examples of pairs of random vectors that are PRD ordered can be found in Fang and Joe (1992).

Block, Chhetry, Fang and Sampson (1990) found necessary and sufficient conditions (by means of orders of permutations) for two bivariate empirical distributions to be ordered according to the PRD order. Some variations of the PRD order are discussed in Fang and Joe (1992).

We close this section by mentioning some related orders. Hollander, Proschan and Sconing (1990) briefly discussed the order according to which (U, V) is smaller than (X, Y) if

$$G_{Y|X}(y|x) - F_{V|U}(y|x) \quad \text{is increasing in } x \text{ for all } y,$$

where the notation here is the same as in Definition 3.1. They also briefly considered orders that correspond to the right tail increasing (RTI) and left tail decreasing (LTD) notions of positive dependence.

4   The PLRD Order

Let the random variables X and Y have the joint distribution G. For any two intervals I_1 and I_2 of the real line, let us denote $I_1 \leq I_2$ if $x_1 \in I_1$ and $x_2 \in I_2$ imply that $x_1 \leq x_2$. For any two intervals I and J of the real line let $G(I, J)$ represent the probability assigned by G to the rectangle $I \times J$. Block, Savits and Shaked (1982) essentially define G (or X and Y) to be positive likelihood ratio dependent if

$$G(I_1, J_1)G(I_2, J_2) \geq G(I_1, J_2)G(I_2, J_1),$$
$$\text{whenever } I_1 \leq I_2 \text{ and } J_1 \leq J_2. \quad (4.1)$$

In fact, Block, Savits and Shaked (1982) call G totally positive of order 2 (TP$_2$) if (4.1) holds. When G has a (continuous or discrete) density g, then (4.1) is equivalent to the condition that g is TP$_2$, that is,

$$g(x_1, y_1)g(x_2, y_2) \geq g(x_1, y_2)g(x_2, y_1), \quad \text{whenever } x_1 \leq x_2 \text{ and } y_1 \leq y_2.$$

Then (4.1) is the same as the condition for the positive dependence notion that Lehmann (1966) called positive likelihood ratio dependence (PLRD). This notion led Kimeldorf and Sampson (1987) to introduce the order that is described in the following definition.

Definition 4.1 *Let F and G be two bivariate distributions with the same marginals, and let (U, V) and (X, Y) be two random vectors having the distributions F and G, respectively. Then F is said to be smaller than G in the PLRD sense (denoted by $F \leq_{\mathrm{PLRD}} G$ or $(U, V) \leq_{\mathrm{PLRD}} (X, Y)$ or $P_F \leq_{\mathrm{PLRD}} P_G$) if*

$$F(I_1, J_1)F(I_2, J_2)G(I_1, J_2)G(I_2, J_1)$$
$$\leq F(I_1, J_2)F(I_2, J_1)G(I_1, J_1)G(I_2, J_2),$$
$$\textit{whenever } I_1 \leq I_2 \textit{ and } J_1 \leq J_2. \quad (4.2)$$

where the generic notation $F(I, J)$ is obvious. When F and G have (continuous or discrete) densities f and g, then (4.2) is equivalent to

$$f(x_1, y_1)f(x_2, y_2)g(x_1, y_2)g(x_2, y_1)$$
$$\leq f(x_1, y_2)f(x_2, y_1)g(x_1, y_1)g(x_2, y_2),$$
$$\text{whenever } x_1 \leq x_2 \text{ and } y_1 \leq y_2.$$

It is easy to see (see, for instance, Kimeldorf and Sampson, 1987) that G is PLRD if, and only if, $G^I \leq_{\text{PLRD}} G$, where G^I is defined in Section 2. Kimeldorf and Sampson (1987) also showed that

$$(U, V) \leq_{\text{PLRD}} (X, Y) \Longrightarrow (U, V) \leq_{\text{PQD}} (X, Y).$$

We do not know whether $(U, V) \leq_{\text{PLRD}} (X, Y) \Longrightarrow (U, V) \leq_{\text{PRD}} (X, Y)$.

The PLRD order satisfies all the axioms of Kimeldorf and Sampson (1987). In particular, we have the following closure property.

Proposition 4.2 *Let (U, V) and (X, Y) be two random vectors. If (U, V) $\leq_{\text{PLRD}} (X, Y)$ then $(\phi(U), \psi(V)) \leq_{\text{PLRD}} (\phi(X), \psi(Y))$ for all increasing functions ϕ and ψ.*

Example 4.3 Let H and K be two continuous univariate distribution functions. For $-1 \leq \alpha \leq 1$, define the following Farlie-Gumbel-Morgenstern distribution function

$$F_\alpha(x, y) = H(x)K(y)\{1 + \alpha[1 - H(x)][1 - K(y)]\}, \qquad \text{for all } x \text{ and } y.$$

Then $F_{\alpha_1} \leq_{\text{PLRD}} F_{\alpha_2}$ whenever $\alpha_1 \leq \alpha_2$.

Yanagimoto (1990) introduced a collection of 16 orders based on the idea of (4.2). He did it by requiring (4.2) to hold for special choices of intervals I_1, I_2, J_1 and J_2. The PQD order is one of the 16 orders in the collection of Yanagimoto. Metry and Sampson (1991) extended Yanagimoto's idea and presented a more general approach for generating positive dependence orderings. That approach makes it fairly easy to study the properties of the resulting orders and the interrelationships among them. Yanagimoto (1990) also introduced an order that is similar to the PLRD order, and which applies to random vectors of dimension $n \geq 2$.

We close this section by mentioning a related order. Kemperman (1977) and Karlin and Rinott (1980) suggested an ordering according to which the bivariate distribution F (with density f) is smaller than the bivariate distribution G (with density g) if

$$f(x_1, y_1)g(x_2, y_2) \geq f(x_1, y_2)g(x_2, y_1), \quad \text{whenever } x_1 \leq x_2 \text{ and } y_1 \leq y_2.$$

This ordering has not been studied in the literature as a positive dependence order. In fact, Kimeldorf and Sampson have noticed that it does not satisfy some of the basic axioms that they introduced.

5   Association Orders

The random variables X and Y are said to be associated if

$$\mathrm{Cov}(\phi(X,Y),\psi(X,Y)) \geq 0$$

for all increasing functions ϕ and ψ for which the covariance is well-defined; see Esary, Proschan and Walkup (1967). This notion led Schriever (1987) to introduce the order that is described in the following definition.

Definition 5.1 *Let F and G be two bivariate distributions, and let (U,V) and (X,Y) be two random vectors having the distributions F and G, respectively. Then F is said to be smaller than G in the association sense (denoted by $F \leq_{\mathrm{ASSOC}} G$ or $(U,V) \leq_{\mathrm{ASSOC}} (X,Y)$ or $P_F \leq_{\mathrm{ASSOC}} P_G$) if*

$$(X,Y) =_{\mathrm{st}} (K_1(U,V), K_2(U,V)),$$

for some increasing functions K_1 and K_2 which satisfy

$$K_1(x_1,y_1) < K_1(x_2,y_2),\ K_2(x_1,y_1) > K_2(x_2,y_2)$$
$$\Longrightarrow x_1 < x_2,\ y_1 > y_2. \quad (5.1)$$

The restriction (5.1) on the functions K_1 and K_2 is for the purpose of making the ASSOC order applicable in situations which are not symmetric in the X and Y variables. [In case (5.1) is dropped, $(X,Y) \geq_{\mathrm{ASSOC}} (Y,X) \geq_{\mathrm{ASSOC}} (X,Y)$.] Fang and Joe (1992) showed that if K_1 and K_2 are partially differentiable increasing functions then (5.1) is equivalent to

$$\frac{\partial}{\partial x}K_1(x,y) \cdot \frac{\partial}{\partial y}K_2(x,y) \geq \frac{\partial}{\partial y}K_1(x,y) \cdot \frac{\partial}{\partial x}K_2(x,y), \quad \text{for all } x \text{ and } y.$$

It is easy to see that if $G^I \leq_{\mathrm{ASSOC}} G$ then G is the distribution function of associated random variables, where G^I is defined in Section 2.

The following closure property is easy to prove (see Schriever, 1987).

Proposition 5.2 *Let (U,V) and (X,Y) be two random vectors. If $(U,V) \leq_{\mathrm{ASSOC}} (X,Y)$ then $(\phi(U),\psi(V)) \leq_{\mathrm{ASSOC}} (\phi(X),\psi(Y))$ for all strictly increasing functions ϕ and ψ.*

Note that in Definition 5.1 it is not assumed that F and G have the same marginals. But see Remark 5.6 below.

If $(U,V) \leq_{\mathrm{PRD}} (X,Y)$) then (5.1) holds with $K_1(x,y) = x$. Thus, it follows that

$$(U,V) \leq_{\mathrm{PRD}} (X,Y) \Longrightarrow (U,V) \leq_{\mathrm{ASSOC}} (X,Y).$$

Example 5.3 Let U and V be any independent random variables. Define

$$X_\alpha = (1-\alpha)U + \alpha V, \quad Y = U, \quad \text{for } \alpha \in [0,1].$$

Then $(X_{\alpha_1},Y) \leq_{\mathrm{ASSOC}} (X_{\alpha_2},Y)$ whenever $\alpha_1 \leq \alpha_2$.

Example 5.4 Let U and V be any independent random variables. Define

$$X_\alpha = (1 - \alpha)U + \alpha V, \quad Y = \alpha U + (1 - \alpha)V, \quad \text{for } \alpha \in [0, \tfrac{1}{2}].$$

Then $(X_{\alpha_1}, Y) \leq_{\text{ASSOC}} (X_{\alpha_2}, Y)$ whenever $\alpha_1 \leq \alpha_2$.

Example 5.5 Let (U, V) and (X, Y) have bivariate normal distributions with correlation coefficients ρ_1 and ρ_2, respectively. Then $(U, V) \leq_{\text{ASSOC}} (X, Y)$ if, and only if, $-1 \leq \rho_1 \leq \rho_2 \leq 1$.

Other examples of pairs of random vectors that are ASSOC ordered can be found in Fang and Joe (1992).

Block, Chhetry, Fang and Sampson (1990) found necessary and sufficient conditions (by means of orders of permutations) for two bivariate empirical distributions to be ordered according to the ASSOC order. Some variations of the ASSOC order are discussed in Fang and Joe (1992).

We close this section by mentioning a related order. Kimeldorf and Sampson (1989) and Hollander, Proschan and Sconing (1990) discuss briefly an order according to which (U, V) is smaller than (X, Y) if

$$\text{Cov}(\phi(U, V), \psi(U, V)) \leq \text{Cov}(\phi(X, Y), \psi(X, Y)), \tag{5.2}$$

for all increasing functions ϕ and ψ for which the covariance is well-defined. Kimeldorf and Sampson (1989) showed that this order does not satisfy one of their axioms. This order can clearly be extended to the case in which the dimension of the random vectors described in (5.2) is $n \geq 2$.

Remark 5.6 Most of the dependence orders considered in this paper are invariant with respect to strictly increasing monotone transformations of the random variables, namely if $\leq_*$ is a dependence order and

$$(U, V) \leq_* (X, Y)$$

then for strictly increasing functions ϕ and ψ,

$$(\phi(U), \psi(V)) \leq_* (\phi(X), \psi(Y)).$$

This fact provides a way of extending the dependence orders from the Fréchet class of distributions with fixed marginals to the whole class of bivariate distributions, by assuming that $(X, Y) \sim_* (\phi(X), \psi(Y))$ whenever ϕ and ψ are strictly increasing. Another way to achieve the same result is by imposing that the orders be defined in terms of the copulas and not in terms of the distribution functions (given a distribution function F with marginals H_1 and H_2, there always exists a function C_F, called copula, such that $F(x, y) = C_F(H_1(x), H_2(y))$). The equivalence of the two approaches is due to the fact that the copula is invariant with respect to strictly monotone transformation of the random variables (see, for example, Scarsini, 1984).

Thus it is seen, for example, that by not assuming in Proposition 5.2 that F and G have the same marginals, we do not really increase the generality of that proposition.

6   The PDD Order

Let the random variables X and Y have the symmetric (or exchangeable, or interchangeable) joint distribution G. Shaked (1979) defines G (or X and Y) to be positive definite dependent (PDD) if G is a positive definite kernel on $\mathbb{S} \times \mathbb{S}$, where $\mathbb{S}$ is the support of X (and therefore, by symmetry, $\mathbb{S}$ is also the support of Y). Shaked (1979) has shown that X and Y are PDD if, and only if,

$$\operatorname{Cov}(h(X), h(Y)) \geq 0, \qquad \text{for every real function } h. \qquad (6.1)$$

Based on this notion, Rinott and Pollak (1980) introduced the order that is given in the next definition.

Definition 6.1 *Let F and G be two symmetric bivariate distributions with the same common marginals, and let (U, V) and (X, Y) be two random vectors having the distributions F and G, respectively. Then F is said to be smaller than G in the* PDD *sense (denoted by $F \leq_{\mathrm{PDD}} G$ or $(U, V) \leq_{\mathrm{PDD}} (X, Y)$ or $P_F \leq_{\mathrm{PDD}} P_G$) if*

$$\operatorname{Cov}(h(U), h(V)) \leq \operatorname{Cov}(h(X), h(Y)), \quad \text{for every real function } h.$$
$$(6.2)$$

It is easily seen that G is PDD if, and only if, $G^I \leq_{\mathrm{PDD}} G$, where G^I is defined in Section 2. Rinott and Pollak (1980) have noted the following characterization of the PDD order.

Proposition 6.2 *Let F and G be two symmetric bivariate distributions with the same marginals. Then $F \leq_{\mathrm{PDD}} G$ if, and only if, $G(x, y) - F(x, y)$ is a positive definite kernel.*

The PDD order enjoys some desirable preservation properties. Using the method of proof of Theorem 3.1 in Shaked (1979) one can obtain the following result.

Theorem 6.3 *Suppose that the four random vectors (U_1, V_1), (X_1, Y_1), (U_2, V_2) and (X_2, Y_2) satisfy*

$$(U_1, V_1) \leq_{\mathrm{PDD}} (X_1, Y_1) \quad and \quad (U_2, V_2) \leq_{\mathrm{PDD}} (X_2, Y_2), \qquad (6.3)$$

and assume that (U_1, V_1) and (U_2, V_2) are independent and also that (X_1, Y_1) and (X_2, Y_2) are independent. Then

$$(\phi(U_1, U_2), \phi(V_1, V_2)) \leq_{\mathrm{PDD}} (\phi(X_1, X_2), \phi(Y_1, Y_2)),$$
$$\text{for all increasing functions } \phi.$$

In particular, if (6.3) holds then the PDD order is closed under convolutions, that is,

$$(U_1 + U_2, V_1 + V_2) \leq_{\mathrm{PDD}} (X_1 + X_2, Y_1 + Y_2).$$

The PDD order is also preserved under mixtures and under limits in distributions. These are stated in the next two theorems.

Theorem 6.4 *Consider the two random vectors (U, V, Θ) and (X, Y, Θ). Suppose that for every value θ in the support of Θ one has*

$$[(U, V)|\Theta = \theta] \leq_{\text{PDD}} [(X, Y)|\Theta = \theta] ,$$

where $[(U, V)|\Theta = \theta]$ and $[(X, Y)|\Theta = \theta]$ are described in Theorem 2.3. Then

$$(U, V) \leq_{\text{PDD}} (X, Y).$$

Theorem 6.5 *Let $\{F_n\}$ and $\{G_n\}$ be two sequences of bivariate distributions. If $F_n \leq_{\text{PDD}} G_n$, $n = 1, 2, \ldots$, and if $F_n \to F$ and $G_n \to G$ (where $\to$ denotes convergence in law), then $F \leq_{\text{PDD}} G$.*

Example 6.6 Let (U, V) and (X, Y) have bivariate normal distributions with common marginals and correlation coefficients ρ_1 and ρ_2, respectively. If $0 \leq \rho_1 \leq \rho_2 \leq 1$ then $(U, V) \leq_{\text{PDD}} (X, Y)$.

Rinott and Pollak (1980) have essentially shown that if $(U, V) \leq_{\text{PDD}} (X, Y)$ then some of the first-passage times of related Gaussian processes are ordered in the usual stochastic order.

If F and G of Definition 6.1 are not symmetric, but still have the same common marginals (that is, U, V, X and Y of Definition 6.1 are all identically distributed) then the PDD order can still be defined on the symmetrizations $\tilde{F}(x, y) = \frac{1}{2}[F(x, y) + F(y, x)]$ and $\tilde{G}(x, y) = \frac{1}{2}[G(x, y) + G(y, x)]$ of F and G; this is the approach taken by Rinott and Pollak (1980).

The defining inequality (6.2) can be written as

$$Eh(U)h(V) \leq Eh(X)h(Y), \quad \text{for every real function } h, \qquad (6.4)$$

because (U, V) and (X, Y) have the same common marginals. Tong (1989) has extended (6.4) for the case when $n \geq 2$. Explicitly, let $\boldsymbol{U} = (U_1, U_2, \ldots, U_n)$ and $\boldsymbol{X} = (X_1, X_2, \ldots, X_n)$ have distribution functions with common marginals (but we do not necessarily assume that they are exchangeable; see Section 7). Then, according to Tong (1989), $\boldsymbol{U}$ is less positively dependent than $\boldsymbol{X}$ if

$$E \prod_{i=1}^{n} h(U_i) \leq E \prod_{i=1}^{n} h(X_i), \quad \text{for every real function } h. \qquad (6.5)$$

Tong (1989) lists some examples of vectors $\boldsymbol{U}$ and $\boldsymbol{X}$ that satisfy (6.5), and shows some applications of this order.

7 Ordering Exchangeable Distributions

Let $U = (U_1, U_2, \ldots, U_n)$ and $X = (X_1, X_2, \ldots, X_n)$ be two exchangeable random vectors (that is, the joint distributions F and G are permutation symmetric). Let $U_{(1)} \leq U_{(2)} \leq \cdots \leq U_{(n)}$ and $X_{(1)} \leq X_{(2)} \leq \cdots \leq X_{(n)}$ be the corresponding order statistics. Intuitively, if X is "more positively dependent" than U (or, alternatively, X is "less dispersed" than U) then we can expect the X_i's to "hang together" more than the U_i's. For example, we can expect quantities such as $U_{(n)} - U_{(1)}$ or $U_{(n)} + U_{(n-1)} - U_{(2)} - U_{(1)}$ to be stochastically larger than $X_{(n)} - X_{(1)}$ or $X_{(n)} + X_{(n-1)} - X_{(2)} - X_{(1)}$, respectively (a random variable Z is said to be stochastically larger than a random variable W, denoted by $W \leq_{\mathrm{st}} Z$, if $P\{W \leq x\} \geq P\{Z \leq x\}$ for all x). This observation led Shaked and Tong (1985) to introduce the first two orders given in this section.

Definition 7.1 *Let U and X be two n-dimensional exchangeable random vectors with the same common marginals. We write $U \leq_{\mathrm{pd\text{-}1}} X$ if*

$$\left| \sum_{i=1}^{n} c_i U_{(i)} \right| \geq_{\mathrm{st}} \left| \sum_{i=1}^{n} c_i X_{(i)} \right|, \qquad whenever \sum_{i=1}^{n} c_i = 0. \qquad (7.1)$$

When the interest is in the unordered components of the random vectors then the following definition is useful.

Definition 7.2 *Let U and X be two n-dimensional exchangeable random vectors with the same common marginals. We write $U \leq_{\mathrm{pd\text{-}2}} X$ if*

$$\left| \sum_{i=1}^{n} c_i U_i \right| \geq_{\mathrm{st}} \left| \sum_{i=1}^{n} c_i X_i \right|, \qquad whenever \sum_{i=1}^{n} c_i = 0. \qquad (7.2)$$

Recall from Marshall and Olkin (1979) that a vector $a = (a_1, a_2, \ldots, a_n)$ is said to be smaller in the majorization order than the vector $b = (b_1, b_2, \ldots, b_n)$ (denoted $a \prec b$) if $\sum_{i=1}^{n} a_i = \sum_{i=1}^{n} b_i$ and if $\sum_{i=1}^{j} a_{[i]} \leq \sum_{i=1}^{j} b_{[i]}$ for $j = 1, 2, \ldots, n-1$, where $a_{[i]}$ $[b_{[i]}]$ is the ith largest element of a $[b]$, $i = 1, 2, \ldots, n$. For any random variable W, let F_W denote the distribution function of W.

Definition 7.3 *Let U and X be two n-dimensional exchangeable random vectors with the same common marginals. We write $U \leq_{\mathrm{pd\text{-}3}} X$ if*

$$(F_{U_{(1)}}(x), F_{U_{(2)}}(x), \ldots, F_{U_{(n)}}(x)) \succ (F_{X_{(1)}}(x), F_{X_{(2)}}(x), \ldots, F_{X_{(n)}}(x)),$$
$$for\ all\ x. \qquad (7.3)$$

Marshall and Olkin (1979, page 350) showed that (7.3) is equivalent to

$$(E\phi(U_{(1)}), E\phi(U_{(2)}), \ldots, E\phi(U_{(n)})) \succ (E\phi(X_{(1)}), E\phi(X_{(2)}), \ldots, E\phi(X_{(n)}))$$

for all monotone functions ϕ for which the expectations exist. A farther insight into the meaning of (7.3) can be obtained by rewriting it as the set of inequalities

$$E\left[\sum_{i=1}^{j} I_{(-\infty,x]}(U_{(i)})\right] \geq E\left[\sum_{i=1}^{j} I_{(-\infty,x]}(X_{(i)})\right],$$

for $j = 1, 2, \ldots, n$, and all x,   (7.4)

with equality holding for $j = n$. That is, for each j, the expected value of the number of order statistics which are less than or equal to x among the first k ordered U_i's is at least as large as the corresponding expected value based on the ordered X_i's.

When one is concerned only with the expectations of the order statistics then the following definition is useful.

Definition 7.4 *Let U and X be two n-dimensional exchangeable random vectors with the same common marginals. We write $U \leq_{\text{pd-4}} X$ if*

$$(EU_{(1)}, EU_{(2)}, \ldots, EU_{(n)}) \succ (EX_{(1)}, EX_{(2)}, \ldots, EX_{(n)}).$$
(7.5)

Shaked and Tong (1985) have shown the following interrelationships among the orders $\leq_{\text{pd-}k}$, $k = 1, 2, 3, 4$.

$$\begin{array}{ccc} \leq_{\text{pd-1}} & \Rightarrow & \leq_{\text{pd-2}} \\ & \Downarrow & \\ \leq_{\text{pd-3}} & \Rightarrow & \leq_{\text{pd-4}} \end{array}$$

They also proved the closure properties that are given in the next theorem.

Theorem 7.5 (i) *For each m, let $U^{(m)}$ and $X^{(m)}$ be two n-dimensional exchangeable random vectors with the same common marginals such that $U^{(m)} \leq_{\text{pd-}k} X^{(m)}$. If $U^{(m)} \to U$ and $X^{(m)} \to X$ (where $\to$ denotes convergence in law), then $U \leq_{\text{pd-}k} X$, $k = 1, 2, 3$.*
(ii) *Let $(U_1, U_2, \ldots, U_n)$ and $(X_1, X_2, \ldots, X_n)$ be two n-dimensional exchangeable random vectors with the same common marginals. If $(U_1, U_2, \ldots, U_n) \leq_{\text{pd-}k} (X_1, X_2, \ldots, X_n)$ then $(U_{i_1}, U_{i_2}, \ldots, U_{i_\ell}) \leq_{\text{pd-}k} (X_{i_1}, X_{i_2}, \ldots, X_{i_\ell})$ for every $1 \leq i_1 < i_2 < \cdots < i_\ell \leq n$, $k = 1, 2, 3, 4$.*
(iii) *Let $(U_1, U_2, \ldots, U_n)$ and $(X_1, X_2, \ldots, X_n)$ be as in part (ii). If $(U_1, U_2, \ldots, U_n) \leq_{\text{pd-}k} (X_1, X_2, \ldots, X_n)$ then $(aU_1 + b, aU_2 + b, \ldots, aU_n + b) \leq_{\text{pd-}k} (aX_1 + b, aX_2 + b, \ldots, aX_n + b)$ for any constants a and b, $k = 1, 2, 3, 4$.*
(iv) *Let $(U_1, U_2, \ldots, U_n)$ and $(X_1, X_2, \ldots, X_n)$ be as in part (ii), let Θ be a random variable independent of U and X, and let ϕ be any function. If $(\phi(U_1, \theta), \phi(U_2, \theta), \ldots, \phi(U_n, \theta)) \leq_{\text{pd-}k} (\phi(X_1, \theta), \phi(X_2, \theta), \ldots, \phi(X_n, \theta))$ for every θ in the support of Θ, then $(\phi(U_1, \Theta), \phi(U_2, \Theta), \ldots, \phi(U_n, \Theta)) \leq_{\text{pd-}k} (\phi(X_1, \Theta), \phi(X_2, \Theta), \ldots, \phi(X_n, \Theta))$, $k = 1, 2, 3, 4$.*

It follows from parts (iii) and (iv) of Theorem 7.5 that if $(U_1, U_2, \ldots, U_n)$ $\leq_{\text{pd-}k} (X_1, X_2, \ldots, X_n)$ and Θ is a random variable independent of U and X, then

$$(U_1 + \Theta, U_2 + \Theta, \ldots, U_n + \Theta) \leq_{\text{pd-}k} (X_1 + \Theta, X_2 + \Theta, \ldots, X_n + \Theta)$$

and

$$(\Theta U_1, \Theta U_2, \ldots, \Theta U_n) \leq_{\text{pd-}k} (\Theta X_1, \Theta X_2, \ldots, \Theta X_n),$$

for $k = 1, 2, 3, 4$.

For the bivariate exchangeable case Shaked and Tong have shown that

$$(U, V) \leq_{\text{PDD}} (X, Y) \Longrightarrow (U, V) \leq_{\text{pd-3}} (X, Y).$$

Many examples of pairs of exchangeable vectors that satisfy the orders $\leq_{\text{pd-}k}$, $k = 1, 2, 3, 4$, are listed in Shaked and Tong (1985). Gupta and Richards (1992) give examples of pairs of multivariate Liouville distributions that are ordered according to $\leq_{\text{pd-1}}$ and therefore also according to $\leq_{\text{pd-2}}$ and $\leq_{\text{pd-4}}$.

Shaked and Tong (1985) have noted that, intuitively, exchangeable random vectors are "more positively dependent" if, and only if, they are "less dispersed." Thus they suggested to define orderings according to which $(U_1, U_2, \ldots, U_n)$ is smaller than $(X_1, X_2, \ldots, X_n)$ if

$$E\phi(U_1, U_2, \ldots, U_n) \geq E\phi(X_1, X_2, \ldots, X_n), \tag{7.6}$$

for every ϕ which belongs to some properly chosen class of permutation symmetric functions. In addition to the classes defined in (7.1), (7.2) and (7.4) [there exists also a class under which (7.6) gives (7.5)] a natural choice of such a class is the class of all Schur-convex functions. Chang (1992) considered some orderings that are defined by (7.6) for several classes of permutation symmetric functions. His paper contains a rich bibliography regarding several stochastic majorization orders.

8 Nonmonotone Dependence Orders

A suitable modification of some monotone dependence orders leads to the definition of nonmonotone dependence orders (with respect to some pre-specified function ϕ). Scarsini and Venetoulias (1993) developed this idea with the aim of defining parametric families of bivariate distributions with nonmonotone dependence structure. Given the considerations of Remark 5.6, we can restrict our attention to the class $\mathcal{M}_2$ of probability measures on the Borel sets of $[0, 1]^2$ with uniform marginals.

Consider a measure preserving function $\phi : [0, 1] \to [0, 1]$ (ϕ is measure preserving if $\text{Leb}\phi^{-1}(A) = \phi(A)$ for all Borel sets A, where Leb is the Lebesgue measure). The function ϕ will be the theoretical benchmark, namely, under maximal dependence all the probability measure will be concentrated on the graph of ϕ. For every $x \in [0,1]$ define $A_x^\phi = \phi^{-1}([0,x])$. For $(x,y) \in [0,1]^2$ and $P \in \mathcal{M}_2$, let $C_P^\phi(x,y) = P(A_x^\phi, [0,y])$. Given $P, Q \in \mathcal{M}_2$, we define

$$P \leq_{\phi\text{-PQD}} Q \quad \text{if} \quad C_P^\phi(x,y) \leq C_Q^\phi(x,y) \quad \text{for all} \quad (x,y) \in [0,1]^2.$$

Therefore the order $\leq_{\phi\text{-PQD}}$ is just the PQD order applied to the functions C_P^ϕ and C_Q^ϕ rather than to the distribution functions.

By using results of Scarsini (1989) it is possible to prove that for all $P \in \mathcal{M}_2$,

$$P^{1-\phi} \leq_{\phi\text{-PQD}} P \leq_{\phi\text{-PQD}} P^\phi,$$

where P^ϕ ($P^{1-\phi}$) is the unique probability measure in $\mathcal{M}_2$ whose support coincides with the graph of ϕ ($1 - \phi$). The measure P^ϕ ($P^{1-\phi}$) plays the role of the upper (lower) Fréchet bound in the ϕ-order of dependence.

It is not difficult to see that, if U, V, X and Y are random variables with uniform marginal distributions on $[0,1]$, then

$$(U,V) \leq_{\phi\text{-PQD}} (X,Y) \iff (\phi(U), V) \leq_{\text{PQD}} (\phi(X), Y).$$

Given this consideration, ϕ-versions of all the previously examined dependence orders for bivariate distributions can easily be constructed. An extension to nonfunctional dependence allows to generalize also the multivariate orderings. Let ϕ_1 and ϕ_2 be measure preserving functions on $[0,1]$. Define

$$C_P^{\phi_1,\phi_2}(x,y) = P(A_x^{\phi_1}, A_y^{\phi_2}), \qquad \text{for all } (x,y) \in [0,1]^2.$$

For $P, Q \in \mathcal{M}_2$ define $P \leq_{\phi_1,\phi_2\text{-PQD}} Q$ if

$$C_P^{\phi_1,\phi_2}(x,y) \leq C_Q^{\phi_1,\phi_2}(x,y), \qquad \text{for all } (x,y) \in [0,1]^2.$$

The following equivalence is easy to prove:

$$(U,V) \leq_{\phi_1,\phi_2\text{-PQD}} (X,Y) \iff (\phi_1(U), \phi_2(V)) \leq_{\text{PQD}} (\phi_1(X), \phi_2(Y)).$$

Furthermore, for every $P \in \mathcal{M}_2$,

$$P^{\phi_1,1-\phi_2} \leq_{\phi_1,\phi_2\text{-PQD}} P \leq_{\phi_1,\phi_2\text{-PQD}} P^{\phi_1,\phi_2},$$

where P^{ϕ_1,ϕ_2} ($P^{\phi_1,1-\phi_2}$) is a measure whose support is contained in the set $\{(x,y) : \phi_1(x) = \phi_2(y)\}$ ($\{(x,y) : \phi_1(x) = 1 - \phi_2(y)\}$). In general the measures P^{ϕ_1,ϕ_2} and $P^{\phi_1,1-\phi_2}$ are not unique.

Given n measure preserving functions $\phi_1, \phi_2, \ldots, \phi_n$ on $[0,1]$, it is possible to define an order $\leq_{\phi_1,\phi_2,\ldots,\phi_n\text{-*}}$ on the class $\mathcal{M}_n$ of probability measures on $[0,1]^n$ with uniform marginals as follows

$$(U_1, U_2, \ldots, U_n) \leq_{\phi_1,\phi_2,\ldots,\phi_n\text{-*}} (X_1, X_2, \ldots, X_n)$$

if

$$(\phi_1(U_1), \phi_2(U_2), \ldots, \phi_n(U_n)) \leq_* (\phi_1(X_1), \phi_2(X_2), \ldots, \phi_n(X_n)),$$

where $\leq_*$ is any multivariate dependence order.

A completely different approach to nonmonotone dependence orders is the one taken by Silvey (1964), Ali and Silvey (1965a, 1965b), Joe (1985, 1987) and Scarsini (1990). Given a measure $P \in \mathcal{M}_2$, the density f of P (with respect to the Lebesgue measure) is such that the closer f is to the product measure, the more concentrated f will be around 1. The further P is from the product measure, the more dispersed f will be. Therefore a dispersion order for f can be seen as a dependence order for the measure P. In particular, for P and $Q \in \mathcal{M}_2$, with respective densities f and g, we can define

$$P \leq_{\mathrm{DEP}} Q$$

if

$$\int_{[0,1]^2} \phi(f(x,y)) \, dx dy \leq \int_{[0,1]^2} \phi(g(x,y)) \, dx dy,$$

for every convex function ϕ.

The order $\leq_{\mathrm{DEP}}$ satisfies the following property. For all $P \in \mathcal{M}$,

$$\mathrm{Leb} \leq_{\mathrm{DEP}} P \leq_{\mathrm{DEP}} \mathrm{Leb}^{\perp},$$

where $\mathrm{Leb}^{\perp}$ is any measure orthogonal to the Lebesgue measure.

The definition of the order $\leq_{\mathrm{DEP}}$ for measures on $\mathbb{R}^2$ with fixed arbitrary marginals is easily obtained by choosing f as the density of P with respect to the product measure of its marginals.

Acknowledgments: The research of Marco Scarsini was partially supported by MURST. The research of Moshe Shaked was supported by NSF Grant DMS 9303891.

9   References

[1] Ali, S. M. and Silvey, S. D. (1965a), Association between random variables and the dispersion of a Radon-Nikodym derivative, *Journal of the Royal Statistical Society, Series B* **27**, 100–107.

[2] Ali, S. M. and Silvey, S. D. (1965b), A further result about the relevance of the dispersion of a Radon-Nikodym derivative to the problem of measuring association, *Journal of the Royal Statistical Society, Series B* **27**, 108–110.

[3] Bassan, B. and Scarsini, M. (1994), Positive dependence orderings and stopping times, *Annals of the Institute of Statistical Mathematics* **46**, 333–342.

[4] Block, H. W., Chhetry, D., Fang, Z. and Sampson, A. R. (1990), partial orders on permutations and dependence orderings on bivariate empirical distributions, *Annals of Statistics* **18**, 1840–1850.

[5] Block, H. W., Savits, T. H. and Shaked, M. (1982), Some concepts of negative dependence, *Annals of Probability* **10**, 765–772.

[6] Brown, L. D. and Rinott, Y. (1988), Inequalities for multivariate infinitely divisible processes, *Annals of Probability* **16**, 642–657.

[7] Cambanis, S. and Simons, G. (1982), Probability and expectation inequalities, *Zeitschrift für Wahrscheinlichkeitstheorie und verwandte Gebiete* **59**, 1–25.

[8] Cambanis, S., Simons, G. and Stout, W. (1976), Inequalities for $Ek(X,Y)$ when the marginals are fixed, *Zeitschrift für Wahrscheinlichkeitstheorie und verwandte Gebiete* **36**, 285–294.

[9] Chang, C.-S. (1992), A new ordering for stochastic majorization: theory and applications, *Advances in Applies Probability* **24**, 604–634.

[10] Dabrowska, D. (1985), Descriptive parameters of location, dispersion and stochastic dependence, *Statistics* **16**, 63–88.

[11] Ebrahimi, N. (1982), The ordering of negative quadrant dependence, *Communications in Statistics—Theory and Methods* **11**, 2389–2399.

[12] Esary, J. D., Proschan, F. and Walkup, D. W. (1967), Association of random variables with applications, *Annals of Mathematical Statistics* **38**, 1466–1474.

[13] Fang, Z., Hu, T. and Joe, H. (1994), On the decrease in dependence with lag for stationary Markov chains, *Probability in the Engineering and Informational Sciences* **8**, 385–401.

[14] Fang, Z. and Joe, H. (1992), Further developments on some dependence orderings for continuous bivariate distributions, *Annals of the Institute of Statistical Mathematics* **44**, 501–517.

[15] Gini, C. (1915–16), Sul criterio di concordanza tra due caratteri, *Atti del Reale Istituto Veneto di Scienze, Lettere ed Arti* **75**, 309–331.

[16] Gupta, R. D. and Richards, D. St. P. (1992), Multivariate Liouville Distributions, III, *Journal of Multivariate Analysis* **43**, 29–57.

[17] Hollander, M., Proschan, F. and Sconing, J. (1990), Information, censoring, and dependence, *Topics in Statistical Dependence*, (edited by Block, H. W., Sampson, A. R. and Savits, T. H.), IMS Lecture Notes—Monograph Series **16**, Hayward, CA, 257–268.

[18] de la Horra, J. and Ruiz-Rivas, C. (1988), A Bayesian method for inferring the degree of dependence for a positively dependent distributions, *Communications in Statistics—Theory and Methods* **17**, 4357–4370.

[19] Joe, H. (1985), An ordering of dependence for contingency tables, *Linear Algebra and its Applications* **70**, 89–103.

[20] Joe, H. (1987), Majorization, randomness and dependence for multivariate distributions, *Annals of Probability* **15**, 1217–1225.

[21] Joe, H. (1990), Multivariate concordance, *Journal of Multivariate Analysis* **35**, 12–30.

[22] Joe, H. (1993), Parametric families of multivariate distributions with given marginals, *Journal of Multivariate Analysis* **46**, 262–282.

[23] Kalashnikov, V. V. and Rachev, S. T. (1990), *Mathematical Methods for Construction of Queueing Models*, Wadsworth and Brooks, Pacific Grove, CA.

[24] Karlin, S. and Rinott, Y. (1980), Classes of orderings of measures and related correlation inequalities. I. Multivariate totally positive distributions, *Journal of Multivariate Analysis* **10**, 467–498.

[25] Kemperman, J. H. B. (1977), On the FKG inequality for measures on a partially ordered space, *Indagationes Mathematicae* **39**, 313–331.

[26] Kimeldorf, G. and Sampson, A. R. (1987), Positive dependence orderings, *Annals of the Institute of Statistical Mathematics* **39**, 113–128.

[27] Kimeldorf, G. and Sampson, A. R. (1989), A framework for positive dependence, *Annals of the Institute of Statistical Mathematics* **41**, 31–45.

[28] Kruskal, W. H. (1958), Ordinal measures of association, *Journal of the American Statistical Association* **53**, 814–861.

[29] Lehmann, E. (1966), Some concepts of dependence, *Annals of Mathematical Statistics* **37**, 1137–1153.

[30] Marshall, A. W. and Olkin, I. (1979), *Inequalities: Theory of Majorization and Its Applications*, Academic Press, New York, NY.

[31] Metry, M. H. and Sampson, A. R. (1991), A family of partial orderings for positive dependence among fixed marginal bivariate distributions, *Advances in Probability Distributions with Given Marginals*, (edited by Dall'Aglio, G., Kotz, S. and Salinetti, G.), Kluwer Academic Publishers, Boston, MA, 129–138.

[32] Metry, M. H. and Sampson, A. R. (1993), Orderings for positive dependence on multivariate empirical distributions, *Annals of Applied Probability* **3**, 1241–1251.

[33] Mosler, K. C. (1984), Stochastic dominance decision rules when the attributes are utility independent, *Management Science* **30**, 1311–1322.

[34] Nguyen, T. T., and Sampson, A. R. (1985a), The geometry of a certain fixed marginal probability distributions, *Linear Algebra and Its Applications* **70**, 73–87.

[35] Nguyen, T. T., and Sampson, A. R. (1985b), Counting the number of $p \times q$ integer matrices more concordant than a given matrix, *Discrete Applied Mathematics* **11**, 187–205.

[36] Rinott, Y. and Pollak, M. (1980), A stochastic ordering induced by a concept of positive dependence and monotonicity of asymptotic test sizes, *Annals of Statistics* **8**, 190–198.

[37] Rüschendorf, L. (1980), Inequalities for the expectation of Δ-monotone functions, *Zeitschrift für Wahrscheinlichkeitstheorie und verwandte Gebiete* **54**, 341–349.

[38] Rüschendorf, L. (1986), Monotonicity and unbiasedness of tests via a. s. construction, *Statistics* **17**, 221–230.

[39] Samorodnitsky, G. and Taqqu, M. S. (1994), *Stable Non-Guassian Random Processes*, Chapman and Hall, New York, NY.

[40] Scarsini, M. (1984), On measures of concordance, *Stochastica* **8**, 201–218.

[41] Scarsini, M. (1989), Copulae of probability measures on product spaces, *Journal of Multivariate Analysis*, **31**, 201–219.

[42] Scarsini, M. (1990), An ordering of dependence. *Topics in Statistical Dependence*, (edited by Block, H. W., Sampson, A. R. and Savits, T. H.), IMS Lecture Notes—Monograph Series **16**, Hayward, CA, 403–414.

[43] Scarsini, M. and Venetoulias, A. (1993), Bivariate distributions with nonmonotone dependence structure, *Journal of the American Statistical Association* **88**, 338–344.

[44] Schriever, B. F. (1987), An ordering for positive dependence, *Annals of Statistics* **15**, 1208–1214.

[45] Shaked, M. (1979), Some concepts of positive dependence for bivariate interchangeable distributions, *Annals of the Institute of Statistical Mathematics* **31**, 67–84.

[46] Shaked, M. and Tong, Y. L. (1985), Some partial orderings of exchangeable random variables by positive dependence, *Journal of Multivariate Analysis* **17**, 333–349.

[47] Shaked, M. and Shanthikumar, J. G. (1994), *Stochastic Orders and Their Applications*, Academic Press, New York, NY.

[48] Silvey, S. D. (1964), On a measure of association, *Annals of Mathematical Statistics* **35**, 1157–1166.

[49] Tchen, A. H. (1980), Inequalities for distributions with given marginals, *Annals of Probability* **8**, 814–827.

[50] Tong, Y. L. (1989), Inequalities for a class of positively dependent random variables with a common marginal, *Annals of Statistics* **17**, 429–435.

[51] Yanagimoto, Y. (1990), Dependence ordering in statistical models and other notions, *Topics in Statistical Dependence*, (edited by Block, H. W., Sampson, A. R. and Savits, T. H.), IMS Lecture Notes—Monograph Series **16**, Hayward, CA, 489–496.

[52] Yanagimoto, Y. and Okamoto, M. (1969), Partial orderings of permutations and monotonicity of a rank correlation statistic, *Annals of the Institute of Statistical Mathematics* **21**, 489–506.

A POISSON LIMIT THEOREM ON THE NUMBER OF APPEARANCES OF A PATTERN IN A MARKOV CHAIN

Ourania Chryssaphinou, University of Athens
Stavros Papastavridis, University of Athens

Abstract

A sequence of Markov dependent trials is performed, each one of them producing a letter from a given finite alphabet. Under quite general conditions we prove that the number of non–overlapping occurrences of long patterns approximates a Poisson distribution.

Key words and phrases: Markov chains, sequences of patterns, Poisson approximation, DNA sequences.

AMS 1980 subject classification. 60C05

1 Introduction

Let us consider a sequence of Markov dependent trials taking values in a given finite alphabet $\Omega = \{\omega_1, \ldots, \omega_q\}$, $q \geq 2$. A pattern is a specific finite sequence of letters from the alphabet Ω. Recently there is a great interest about problems concerning patterns derived under the above scheme. Many of these problems are higly motivated by analysis of DNA sequences (see Bishop, D. T, Williamson, J. A. and Skolnick, M. H. [3], Fousler and Karlin [8]). In the case of independent trials a lot of work has been done. Consider Guibas and Odlyzko [11], [12], Gerber and Li [9], Breen, Waterman and Zhang [4] and the references contained therein. Some work has been also done for the case of Markov dependent trials. For background material see Biggins [1], Biggins and Cannings [1], Godbole [10], Pittinger [13], Rajarshi [14], Schwager [15].

In this paper we consider a sequence of patterns $\{B_n\}$. For any pattern B, let $|B|$ denote its length, or the number of elements. We denote by $P(B_n)$ the probability that B_n occurrs at the $|B_n|$–th trial and by μ_n the mean time for the first occurrence of B_n. We are interested in the number N_n of non–overlapping occurrences of B_n among the n first trials.

In Theorem 2.1 we prove that if $n/\mu_n \to \lambda$ as $n \to \infty$ $(0 \leq \lambda < \infty)$ then the r.v. N_n converges in distribution to a Poisson distributed r.v. with parameter λ.

The proof is based on a result in Chryssaphinon and Papastavridis [7] concerning the probability generating function (p.g.f.) of the waiting time for the first occurrence of B_n say W_{n1}.

Using this result will prove in Propositions 2.1 and 2.2 that the r.v.'s W_{n1} and W_{ni}, $i = 2, 3, \ldots$ are approximately exponential. Then the proof of Theorem 2.1 comes out easily.

The corresponding results for the case of independent trials as well as the case of the total number of (overlapping) appearances of the pattern B_n have been studied by the Chryssaphinon and Papastavridis in [5] and [6] respectively.

2 The main result

Let us consider a discrete Markov chain $\{X_n, n \geq 1\}$ with state space $\Omega = \{\omega_1, \ldots, \omega_q\}$, $2 \leq q < \infty$, stationary transition probabilities p_{ij} and initial probabilities p_k. We assume that $0 < p_{ij} < 1$, $i, j = 1, \ldots, q$. Let $\{B_n\}$ be a sequence of patterns of lengths $\{|B_n|\}$ where $|B_n| \to \infty$ as $n \to \infty$ and $P(B_n)$ be the probability that B_n occurs at the $|Bn|$–th trial.

Let us assume $B_n = (\beta_{n,1}, \ldots, \beta_{n,|B_n|})$. We define $B_{n,i} = (\beta_{n,1}, \ldots, \beta_{n,i})$ and $B'_{n,i} = (\beta_{n,|B_n|-i+1}, \ldots, \beta_{n,|B_n|})$, $0 \leq i \leq |B_n|$ with $B_{n,0} = B'_{n,0} = \varnothing$

We put

$$e_{ni} = \begin{cases} 1 & \text{if } B_{n,i} = B'_{n,i} \\ 0 & \text{otherwise} \end{cases} , e_{n_0} = 1$$

and we define the conditional probability of a pattern say $Y = (y_1, \ldots, y_{|Y|})$ given the pattern $H = (h_1, \ldots, h_{|H|})$ as

$$P(Y|H) = P((X_{n+1}, \ldots, X_{n+|Y|}) = Y | X_n = \ell(H))$$

where $\ell(H)$ is the last element of the pattern H, if $H \neq \varnothing$.

Clearly $P(Y|H)$ is independent of n. Furthermore, set

$$P(Y|H) = P(X_1 = y_1, \ldots, X_{|Y|} = y_{|Y|})$$

if $H = \varnothing$. Finally, we introduce the polynomial

$$(B_n * B_n)(z) = \sum_{i=1}^{|B_n|} e_{ni} P(B'_{n,|B_n|-i} B_{n,i}) z^{|B_n|-i}$$

Following the same notation as in Chryssaphinou and Papastavridis [7] we consider the probabilities

$$a_n(m) = P(B_n \text{ is a subpattern of } Y_m \text{ but not of } Y_{m-1}), m \geq 1$$

where $Y_n = (X_1, \ldots, X_n)$. Denote by W_{n1} the waiting time for the first occurrence of B_n. Clearly it is $P(W_{n1} = m) = a_n(m)$ and the probability generating function (p.g.f.)

$$A_n(z) = E(z^{W_{n1}}) = \sum_{m=1}^{\infty} a_n(m) z^m$$

93

is fully determined by the result of Theorem 2 p. 171 of Chryssaphinou and Papastavridis [7]. The p.g.f. $A_n(z)$ has been derived under the assumption that the last element of B_n is ω_q (i.e. $\ell(B_n) = \beta_{n,|B_n|} = \omega_q$). Solving the system of Theorem 2 mentioned above we get the following expression for the p.g.f.

$$A_n(z) = \frac{z^{|B_n|} f_n(z)}{(1 - z)\phi_n(z) + z^{|B_n|} g_n(z)} \tag{1}$$

where

$$
\begin{aligned}
f_n(z) &= P(\widetilde{B}_n | \beta_{n,1}) F(z) \\
\phi_n(z) &= (B_n * B_n)(z) \Phi(z) \\
g_n(z) &= P(\widetilde{B}_n | \beta_{n,1}) G(z)
\end{aligned} \tag{2}
$$

$\widetilde{B}_n = (\beta_{n,2}, \ldots, \beta_{n,|B_n|})$ and

$$
F(z) = \begin{vmatrix}
P(\beta_{n,1}) & -P(\beta_{n,1}|\omega_1) & -P(\beta_{n,1}|\omega_2) & \cdots & -P(\beta_{n,1}|\omega_q) \\
z & (1-z) & (1-z) & \cdots & (1-z) \\
zP_1 & 1 - zP_{11} & -zP_{21} & \cdots & -zP_{q1} \\
zP_2 & -zP_{12} & 1 - zP_{22} & \cdots & -zP_{q2} \\
\cdot & \cdot & \cdot & \cdots & \cdot \\
\cdot & \cdot & \cdot & \cdots & \cdot \\
\cdot & \cdot & \cdot & \cdots & \cdot \\
zP_{q-1} & -zP_{1,q-1} & -zP_{2,q-1} & \cdots & -zP_{q,q-1}
\end{vmatrix}
$$

$$
\Phi(z) = \begin{vmatrix}
1 & 1 & 1 & \cdots & 1 \\
1 - zP_{11} & -zP_{21} & -zP_{31} & \cdots & -zP_{q1} \\
-zP_{12} & 1 - zP_{22} & -zP_{32} & \cdots & -zP_{q2} \\
\cdot & \cdot & \cdot & \cdots & \cdot \\
\cdot & \cdot & \cdot & \cdots & \cdot \\
\cdot & \cdot & \cdot & \cdots & \cdot \\
-zP_{1,q-1} & -zP_{2,q-1} & -zP_{3,q-1} & \cdots & -zP_{q,q-1}
\end{vmatrix}
$$

$$
G(z) = \begin{vmatrix}
P(\beta_{n,1}|\omega_1) & P(\beta_{n,1}|\omega_2) & P(\beta_{n,1}|\omega_3) & \cdots & P(\beta_{n,1}|\omega_q) \\
1 - zP_{11} & -zP_{21} & -zP_{31} & \cdots & -zP_{q1} \\
-zP_{12} & 1 - zP_{22} & -zP_{32} & \cdots & -zP_{q2} \\
\cdot & \cdot & \cdot & \cdots & \cdot \\
\cdot & \cdot & \cdot & \cdots & \cdot \\
\cdot & \cdot & \cdot & \cdots & \cdot \\
-zP_{1,q-1} & -zP_{2,q-1} & -zP_{3,q-1} & \cdots & -zP_{q,q-1}
\end{vmatrix}
$$

Since

$$f_n(1) = g_n(1) \tag{3}$$

we have that the mean waiting time of the first occurrence of B_n is given by

$$A'_n(z)|_{z=1} = \mu_n = \frac{f'(1) + \phi_n(1) - g'_n(1)}{g_n(1)} \qquad (4)$$

(the symbol $'$ denotes the derivative with respect to z). Obviously μ_n is the mean occurence time under initial distribution and it is not the mean of the other inter–event times.

The following lemma will be useful in the sequel.

Lemma 2.1 *It is*

$$\Phi(1) \neq 0 \text{ and } F(1) = G(1) \neq 0 \qquad (5)$$

Proof. $F(z)$ and $G(z)$ do not depend on n, and only depend on the first element of B_n. Relation (5) follows from determinant theory arguments.

Our main result is the following

Theorem 2.1 *If $|B_n| \to \infty$ and $\frac{n}{\mu_n} \to \lambda$ as $n \to \infty$ $(0 \leq \lambda < \infty)$, then*

$$\lim_{n \to \infty} P(N_n = k) = e^{-\lambda} \frac{\lambda^k}{k}, \quad k = 0, 1, \dots$$

The proof of this theorem is based on the following lemma and propositions

Lemma 2.2 *If $|B_n| \to \infty$ as $n \to \infty$, then $\frac{|B_n|}{\mu_n} \to 0$ as $n \to \infty$.*

Proof. We distinguish the following two cases:

(i) The elemetns $\ell(B_n)$ and $\beta_{n,1}$ are the same for all n. Then using (2) and (4) we have

$$\frac{|B_n|}{\mu_n} = \frac{|B_n| P(\tilde{B}_n|\beta_{n,1}) G(1)}{P(\tilde{B}_n|\beta_{n,1}) F'(1) - P(\tilde{B}_n|\beta_{n,1}) G'(1) + (B_n * B_n)(1)\Phi(1)}.$$

We observe that $B_n * B_n(1) \geq 1$ and $P(\tilde{B}_n|\beta_{n,1}) \leq p^{|B_n|-1}$ where $p = \max\{p_{ij} : i, j = 1, \dots, q\}$. By our assumption $0 < p < 1$, we get that $|B_n| P(\tilde{B}_n|\beta_{n,1}) \to 0$ as $|B_n| \to \infty$ which together with (5) gives the result.

(ii) The general case follows easily since there are only a finite number of choices, i.e. q^2, for $\ell(B_n)$ and $\beta_{n,1}$ so the sequence under consideration splits into a finite number of subsequences each one of those having $\ell(B_n)$ and $\beta_{n,1}$ constant, so case (i) is applicable.

Proposition 2.1 *If $|B_n| \to \infty$ as $n \to \infty$, then*

$$\lim_{n \to \infty} P(W_{n,1}/\mu_n > x) = e^{-x}, \quad x > 0$$

95

Proof. We will assume that $\ell(B_n)$, $\beta_{n,1}$ are the same for all n. The general case follows easily as in the previous lemma. We consider the Laplace transform for the r.v. W_{n1}/μ_n which according to relation (1) is

$$E(e^{-sW_{n1}/\mu_n}) = A_n(e^{-s/\mu_n}) \tag{6}$$

$$= \frac{e^{-s|B_n|/\mu_n} f_n(e^{-s/\mu_n})}{(1 - e^{-s/\mu_n})\phi_n(e^{-s/\mu_n}) + e^{-s|B_n|/\mu_n} g_n(e^{-s/\mu_n})}$$

$$= \left\{ \frac{1 - e^{-s/\mu_n}}{s/\mu_n} s e^{s|B_n|/\mu_n} \frac{\phi_n(e^{-s/\mu_n})}{\mu_n f_n(e^{-s/\mu_n})} + \frac{g_n(e^{-s/\mu_n})}{f_n(e^{-s/\mu_n})} \right\}^{-1}$$

Lemma 2.2 guarantees that

$$\lim_{n\to\infty} \frac{1 - e^{-s/\mu_n}}{s/\mu_n} = 1 \text{ and } \lim_{n\to\infty} e^{s|B_n|/\mu_n} = 1 \tag{7}$$

Using Lemma 2.1,

$$\lim_{n\to\infty} \frac{g_n(e^{-s/\mu_n})}{f_n(e^{-s/\mu_n})} = \lim_{n\to\infty} \frac{G(e^{-s/\mu_n})}{F(e^{-s/\mu_n})} = \frac{G_n(1)}{F_n(1)} = 1 \tag{8}$$

Finally, using (4)

$$\lim_{n\to\infty} \frac{\phi_n(e^{-s/\mu_n})}{\mu_n f_n(e^{-s/\mu_n})} = \lim_{n\to\infty} \frac{g_n(1)}{f_n(e^{-s/\mu_n})} \left\{ \frac{f_n'(1) - g_n'(1)}{\phi_n(e^{-s/\mu_n})} + \frac{\phi_n(1)}{\phi_n(e^{-s/\mu_n})} \right\}^{-1} = 1 \tag{9}$$

since

$$\lim_{n\to\infty} \frac{f_n'(1) - g_n'(1)}{\phi_n(e^{-s/\mu_n})} = \lim_{n\to\infty} \frac{P(\tilde{B}_n|\beta_{n,1})[F'(1) - G'(1)]}{\phi_n(e^{-s/\mu_n})} = 0$$

(see proof of Lemma 2.2, case (i)).
Relations (6), (7), (8) and (9) give us that

$$\lim_{n\to\infty} E(e^{-sW_{n1}/\mu_n}) = \frac{1}{1 + s}$$

as required.

Defining W_{ni} being the waiting time from the $(i - 1)$th occurrence of B_n till the ith one, $i = 2, 3, \ldots$ the following result is valid.

Proposition 2.2 If $|B_n| \to \infty$ as $n \to \infty$, then

$$\lim_{n\to\infty} P(W_{ni}/\mu_n > x) = e^{-x}, \quad x > 0$$

Proof. The p.g.f. of W_{ni}, $i = 1, 2, 3, \ldots$ is slightly different than the p.g.f of W_{n1}. Namely, it is

$$E(z^{W_{ni}}) = \frac{z^{|B_n|} f_n^*(z)}{(1-z)\phi_n(z) + z^{|B_n|} g_n(z)}$$

where

$$f_n^*(z) = \begin{vmatrix} P(B_n|\omega_q) & -P(B_n|\omega_1) & \cdots & -P(B_n|\omega_q) \\ z & 1-z & \cdots & 1-z \\ zP_{q1} & 1-zP_{11} & \cdots & -zP_{q1} \\ \cdot & \cdot & \cdot & \cdot \\ \cdot & \cdot & \cdot & \cdot \\ \cdot & \cdot & \cdot & \cdot \\ zP_{q,q-1} & -zP_{1,q-1} & \cdots & -zP_{q,q-1} \end{vmatrix}$$

The same procedure as in Proposition 2.1 does the trick.

Proof of Theorem 2.1. Using Propositions 2.1 and 2.2 we derive the result of Theorem 2.1 immediately, since we have a delayed renewal process with asymtotically exponential waiting times. So the number of renewals is asymptotically Poisson.

References

[1] Biggins, J.D. (1987), *A note on repeated sequences in Markov chains.* Adv. Appl. Prob. 19, 740–743.

[2] Biggins, J. D. and Cannings C. (1987), *Markov renewal processes counters and repeated sequences in Markov chains.* Adv. Appl. Prob. 19, 521–545.

[3] Bishop, D.T., Williamson, J.A. and Skolnick, M.H. (1983), *A model for restgriction fragment length distributions.* Amer. J. Hum. Genet. 35, 795–815.

[4] Breen, S., Waterman, M.S. and Zhang, N. (1985), *Renewal theory for several patterns.* J. Appl. Prob. 22 , 228–234.

[5] Chryssaphinou, O. and Papastavridis S. (1988a), *A limit theorem for the number of non-overlapping occurrences of a pattern in a sequence of independent trials.* J. Appl. Prob. 25, 428–431.

[6] Chryssaphinou, O. and Papastavridis, S. (1988b), *A limit theorem for the number of over-lapping appearances of a pattern in a sequence of independent trials.* Prob. Th. Rel. Fields 79, 129–143.

[7] Chryssaphinou, O. and Papastavridis, S.(1990), *The occurrence of sequence patterns in repeated dependent experiments Theory.* Prob. Appl. vol. 35, 167–173.

[8] Fousler, O.E. and Karlin, S. (1987), *Maximal success duration for a semi–Markov process.* Stoch. Proc. Appl. 24, 203–224.

[9] Gerber, H. U. and Li, S.–Y. R. (1981), *The occurrence of sequence patterns in repeated experiments and hitting times in a Markov chain.* Stoch Proc. Appl. 11, 101–108.

[10] Godbole, A.P. *Poisson approximations for runs and patterns of rare events.* Preprint.

[11] Guibas, L.J. and Odlyzko, A.M. (1980), *Long repetitive pattern in random sequences.* Z. Wahrsch. 53, 241–262.

[12] Guibas, L.J. and Odlyzko, A.M. (1981), *String overlaps, patterns matching and nontransitive games.* J. Comb. Th. (A) 30, 183–208.

[13] Pittenger, A.O. (1987), *Hitting time of sequences.* Stoch. Proc. Appl. 24, 225–240.

[14] Rajarshi M.B. (1974), *Success runs in a two-state Markov chain.* J. Appl. Prob. 11, 190–192.

[15] Schwager, S.J. (1983), *Run probabilities in sequences of Markov–dependent trials.* J. Amer. Stat. Assoc. 78, 168–175.

[16] Solov'ev, A.O. (1966), *A combinatorial identity and its application to the problem concerninig the first occurrence of a rare event.* Theory Prob. Appl. 11 276–282.

Palindromes in Random Letter Generation: Poisson Approximations, Rates of Growth, and Erdős-Rényi Laws

Debashis Ghosh
Anant P. Godbole

ABSTRACT Consider a sequence $\{X_j\}_{j=1}^n$ of i.i.d. uniform $\{0, 1, \ldots, d - 1\}$-valued random variables, and let $M_{n,k}$ be the number of *palindromes* of length k counted in an overlapping fashion; a palindrome is any word that is symmetric about its center. We prove that the distribution of $M_{n,k}$ can be well-approximated by that of a Poisson random variable. Similar approximations are obtained for various other random quantities of interest. We also obtain maximal and minimal rates of growth for the length L_n of the longest palindrome; an Erdős-Rényi law is derived as a corollary: the length of the longest palindrome is, almost surely, of order $\log_a n$, where a is the *square root* of the alphabet size d. Analogous results on *partial palindromes*, i.e., words in which a certain (non-zero) number of "mismatches" prevent symmetry about the center, have been presented by Revelle [R] in a sequel to this paper.

1   Introduction and Motivation

Palindromic segments are those that read the same way from either direction. They have had a long history in linguistics and recreational mathematics (see, e.g., [B]), and in molecular genetics, where palindromes often signal the existence, in a nucleotide sequence, of a binding site for a regulatory protein (see, e.g., Waterman [W]). Examples of palindromes in linguistics include "Was it a rat I saw?" (with an obvious bending of the rules) and "saippuakivikauppias"; the latter is listed in the *Guinness Book of World Records* as being the longest palindromic *word* known. It means "a dealer in lye", in the Finnish language. Richard Arratia (personal communication) has informed the authors of the longest palindromic segment

> T. Eliot, top bard, notes putrid tang emanating. Is sad. I'd
> assign it a name: "gnat dirt upset on drab pot toilet"

known in the English language. We note that, in molecular biology, DNA segments such as ACTTGATCAAGT are considered to be palindromic, since the bases A and T are complementary, as are C and G; in this paper, however, we use the regular convention for defining palindromic segments.

Throughout this paper, we consider a sequence $\{X_j\}_{j=1}^n$ of i.i.d. letters uniformly chosen from a finite alphabet of size d; there are obvious possible extensions of our results to the case where the letters are generated in an independent but non-i.i.d. fashion, and to various kinds of Markov-dependent situations. We focus our attention on results involving perfect palindromes; in a sequel to this paper, Revelle [R] has considered the case where word segments are allowed to have a fraction α of mismatches that prevent symmetry about the center (partial palindromes), using somewhat different methods to obtain results that are more complicated, but of essentially the same form. Revelle's paper bears the same relation to this one as does the work of Arratia et al. [AGW] to the classical theorems of Erdős and Rényi [ER70] and Erdős and Révész [ER75]. Our primary goal in this paper is thus to obtain results that mirror those of Erdős, Rényi and Révész on the length L_n of the longest head run in n independent coin tosses. We remind the reader of the Erdős-Rényi laws: Given a sequence $\{X_j\}_{j=1}^n$ of i.i.d. (p) coin-tosses, we have

$$\lim_{n\to\infty} \frac{L_n}{\log_{1/p} n} = 1 \quad \text{a.s.,}$$

and, more significantly, the hierarchy of results

$$\limsup_{n\to\infty} \frac{(L_n - \log_{1/p} n)}{\log_{1/p} \log_{1/p} n} = 1 \quad \text{a.s.;}$$

$$\limsup_{n\to\infty} \frac{(L_n - \log_{1/p} n - \log_{1/p} \log_{1/p} n)}{\log_{1/p} \log_{1/p} \log_{1/p} n} = 1 \quad \text{a.s.}$$

etc. holds, whereas we have the rather sharp minimal growth rate criterion

$$\alpha \le \liminf_{n\to\infty} (L_n - \log_{1/p} n + \log_{1/p} \log_{1/p} \log_{1/p} n) \le \beta$$

for two constants α and β. In Section 3, we derive almost sure results that exhibit the parallel between longest pure head runs, on the one hand, and longest palindromic segments, on the other. In particular, it is shown that the longest palindrome in n i.i.d. letters is approximately *twice as long* as the longest pure run of a fixed letter, being (asymptotically) of order $\log_a n$ almost surely, where $a = \sqrt{d}$. This result is probably not too surprising from an intuitive point of view (since one would expect a word segment with half the "degrees of freedom" to be twice as long) but several interesting ideas go into its proof, which uses, as a vehicle, the Stein-Chen method of Poisson approximation - in a fashion reminiscent of the development in

Arratia et al. [AGW]. Our results on Poisson approximation are presented in Section 2, and are of interest in their own right; of special significance in their proofs are the following:

(i) The introduction of the critically important variable $M'_k = M'_{n,k}$, defined as the number of *maximal* palindromes of length k or more (the precise definition follows); and

(ii) The (unexpected?) pairwise independence of several systems of indicator variables.

There are obvious similarities between our distributional and almost sure results, on the one hand, and those obtained by Móri in a series of papers written during the last decade, on the other (see, e.g., Móri [M] and the references therein). In particular, the study of the waiting time W until one of several word patterns appears, when specialized to the case where the patterns consist of the set of maximal palindromes of length k or more, yields estimates for the point probabilities $\mathbf{P}(M'_{n,k} = 0)$. Our development is quite different from Móri's however, since the pairwise independence alluded to earlier allows one to effectively employ the Stein-Chen method to obtain an entire distributional approximation for the law of M'_k. This is a technique that fails in the absence of any significant information about the nature of the overlaps between the various words in question.

2   Poisson Approximations

We begin with some definitions. Let $M = M_{n,k}$ and $N = N_{n,k}$ denote, respectively, the number of overlapping and non-overlapping occurrences of palindromes of length k in the sequence $\{X_j\}_{j=1}^n$. Non-overlapping occurrences are counted, as is customary, in the recurrent fashion advocated by Feller [F]. Also, $L = L_n$ will denote the length of the longest palindrome, and the variable $M^*_\alpha = M^*_{n,\alpha}$ is defined as the number of *maximal* palindromes of length α, i.e., segments such that $\{X_j, \ldots, X_{j+\alpha-1}\}$ *is* palindromic, but $\{X_{j-1}, \ldots, X_{j+\alpha}\}$ *isn't*. Note that occurrences of palindromes that involve "edge effects" can be incorporated into this definition quite easily; for example, $\{X_1, X_2, X_3\}$ can be defined to be a maximal palindrome of length three iff it is a palindrome. Finally, let $M'_k = M'_{n,k} = \sum_{\alpha=k}^n M^*_\alpha$; we occasionally refer to M'_k, somewhat loosely, as the "number of maximal palindromes of length k or more". We give a couple of examples to fix these ideas: Given the sequence ATATAGCTCG, the values of $M_{10,3}$ and $N_{10,3}$ are 4 and 2, respectively. Next consider the sequence GATTAATTAC: the longest palindrome is of length 8, whereas the values of $M_{10,4}$ and $N_{10,4}$ are 3 and 2 respectively. Finally, $M'_{10,4} = 3$.

We shall deal with each of the variables introduced above, but the almost sure results of Section 3 will depend most critically on the Poisson approximation that we establish for $\mathcal{L}(M'_{n,k})$, where we will use the notation $\mathcal{L}(Z)$ for the probability distribution of the r.v. Z. We begin by studying the approximate distribution of the number $M = M_{n,k}$ of overlapping occurrences of palindromes of length k, and first establish a somewhat surprising result:

Lemma 2.1 *Fix an integer k. Define the indicators I_j by letting $I_j = 1$ if $X_j, \ldots, X_{j+k-1}$ constitutes a palindrome. Then $\{I_j\}_{j=1}^{n-k+1}$ consists of a sequence of pairwise independent random variables.*

Proof. We employ elementary combinatorial methods. If $\{i, i+1, \ldots, i+k-1\} \cap \{j, j+1, \ldots, j+k-1\} = \emptyset$, then it is clear that $\mathbf{P}(I_i I_j = 1) = \mathbf{P}(I_i = 1)\mathbf{P}(I_j = 1)$. Assume, then, that $|\{i, i+1, \ldots, i+k-1\} \cap \{j, j+1, \ldots, j+k-1\}| = r;$ $(1 \leq r \leq k-1)$, and let k be even. The first case we consider is when $r = k/2$. In this case, one can freely choose the characters in the overlap, but once the overlap is determined, there only exists one arrangement such that both words are palindromes. It follows that

$$\mathbf{P}(I_i I_j = 1) = \frac{d^r}{d^{2k-r}} = \frac{1}{d^k} = \mathbf{P}(I_i = 1)\mathbf{P}(I_j = 1).$$

Assume next that $r < k/2$. After having chosen freely in the overlap, one still has to choose $(k/2) - r$ characters for each word until only one arrangement of the letters exist such that both words are palindromes. Thus,

$$\mathbf{P}(I_i I_j = 1) = \frac{d^r d^{2(\frac{k}{2}-r)}}{d^{2k-r}} = \frac{1}{d^k} = \mathbf{P}(I_i = 1)\mathbf{P}(I_j = 1).$$

Finally, we let r be larger than $k/2$. One chooses a certain number of characters inside the overlap before a unique arrangement of letters exists such that both words are palindromes; the number of characters that can be freely chosen in this fashion may be verified to be $|i - j| = k - r$. It follows that

$$\mathbf{P}(I_i I_j = 1) = \frac{d^{k-r}}{d^{2k-r}} = \frac{1}{d^k} = \mathbf{P}(I_i = 1)\mathbf{P}(I_j = 1).$$

If k is odd, on the other hand, we first let $r = (k+1)/2$ and note that one can freely choose r characters before only one arrangement of the remaining letters exists such that both words are palindromes, so that

$$\mathbf{P}(I_i I_j = 1) = \frac{d^r}{d^{2k-r}} = \frac{d^{k+1}}{d^{2k}} = \frac{1}{d^{k-1}} = \mathbf{P}(I_i = 1)\mathbf{P}(I_j = 1).$$

If $r < (k+1)/2$, after having chosen freely in the overlap, one still has to choose $\{(k+1)/2\} - r$ characters for each word until a unique arrangement of the remaining letters exist such that both words are palindromes. Thus,

$$\mathbf{P}(I_i I_j = 1) = \frac{d^r d^{2 \frac{(k+1)}{2} - r)}}{d^{2k-r}} = \frac{1}{d^{k-1}} = \mathbf{P}(I_i = 1)\mathbf{P}(I_j = 1).$$

Finally, if $r > (k+1)/2$, one must choose a certain number of characters in the overlap before a unique arrangement of the remaining letters exist such that both words are palindromes. The number of characters that can be arbitrarily picked may be checked as before to be $|i - j| + 1 = k - r + 1$. It follows that

$$\mathbf{P}(I_i I_j = 1) = \frac{d^{k-r+1}}{d^{2k-r}} = \frac{1}{d^{k-1}} = \mathbf{P}(I_i = 1)\mathbf{P}(I_j = 1);$$

this concludes the proof of Lemma 2.1; in conjunction with the Stein-Chen method of Poisson approximation (Barbour et al. [BHJ]), the lemma yields the following

Proposition 2.2 *Let $M = M_{n,k}$ denote the number of overlapping occurrences of k-palindromes in the sequence $\{X_j\}_{j=1}^n$, and let $\mathrm{Po}(\lambda)$ denote the Poisson r.v. with mean $\lambda = \mathbf{E}(M)$, where*

$$\mathbf{E}(M) = \begin{cases} \frac{(n-k+1)}{d^{k/2}} & \text{if } k \text{ is even,} \\ \frac{(n-k+1)}{d^{(k-1)/2}} & \text{if } k \text{ is odd.} \end{cases}$$

Then, with $\mathcal{L}(Z)$ representing the probability distribution of the r.v. Z, and d_{TV} the usual variation distance, we have

$$d_{\mathrm{TV}}(\mathcal{L}(M), \mathrm{Po}(\lambda)) \le 4\sqrt{d} \frac{k}{d^{k/2}}.$$

Proof. Note that $M = \sum_{j=1}^{n-k+1} I_j$, where $I_j = 1$ iff $X_j, \ldots, X_{j+k-1}$ is a palindrome, and $I_j = 0$ otherwise. Since the I_j's are dissociated, it follows by Corollary 2.C.5 in Barbour et al. [BHJ] that

$$d_{\mathrm{TV}}(\mathcal{L}(M), \mathrm{Po}(\lambda)) \le \frac{(1 - e^{-\lambda})}{\lambda} \bullet$$

$$\sum_{j=1}^{n-k+1} \left\{ \mathbf{P}^2(I_j = 1) + \sum_{r=1}^{k-1} \sum_i [\mathbf{P}^2(I_i = 1) + \mathbf{P}(I_i I_j = 1)] \right\}, (1.1)$$

where the third sum above is over the two indicators I_i whose "neighbourhoods of dependence" have an overlap of magnitude r with that of the indicator I_j. Thus by Lemma 2.1, (1.1), and the fact that $\mathbf{P}(I_j = 1) \le 1/d^{(k-1)/2}$, we have

$$d_{\mathrm{TV}}(\mathcal{L}(M), \mathrm{Po}(\lambda)) \le \frac{(1 - e^{-\lambda})}{\lambda}(n + k - 1)(4k - 3)\mathbf{P}^2(I_1 = 1)$$

$$\le (4k - 3)\mathbf{P}(I_1 = 1) \le 4\sqrt{d} \frac{k}{d^{k/2}}, \qquad (1.2)$$

as asserted.   □

Remark. Note that the duality relation $P(M_{n,k} = 0) = P(L_n \leq k-1)$ *is not true*, as evidenced by setting $k = 3$ in the simple example "ATTA". It is this fact that leads us, later in this section, to consider Poisson approximations for the random variable $\mathcal{L}(M'_{n,k})$. Revelle [R] has used different methods to deal with this problem.

An elementary coupling technique, together with Proposition 2.2, leads to the following result on the Poisson approximation of N, the number of non-overlapping palindromes of length k.

Proposition 2.3 *Let $N = N_{n,k}$ denote the number of non-overlapping occurrences of k-palindromes in the alphabet-valued sequence $\{X_j\}_{j=1}^n$. Then,*

$$d_{\mathrm{TV}}(\mathcal{L}(N), \mathrm{Po}(\lambda)) \leq \frac{2nkd}{d^k} + 4\sqrt{d}\frac{k}{d^{k/2}}$$

where $\lambda = E(M_{n,k})$; note that a comparison is made between $\mathcal{L}(N)$ and a Poisson r.v. with a different mean.

Proof. Since

$$d_{\mathrm{TV}}(\mathcal{L}(N), \mathrm{Po}(\lambda)) \leq d_{\mathrm{TV}}(\mathcal{L}(N), \mathcal{L}(M)) + d_{\mathrm{TV}}(\mathcal{L}(M), \mathrm{Po}(\lambda)),$$

and since, by (1.2),

$$d_{\mathrm{TV}}(\mathcal{L}(M), \mathrm{Po}(\lambda)) \leq 4\sqrt{d}\frac{k}{d^{k/2}},$$

it remains to estimate $d_{\mathrm{TV}}(\mathcal{L}(N), \mathcal{L}(M))$. We have

$$
\begin{aligned}
d_{\mathrm{TV}}(\mathcal{L}(N), \mathcal{L}(M)) &\leq P(M \neq N) \\
&\leq P\left(\bigcup_{j=1}^{n-k+1} \bigcup_{i=j-k+1}^{j+k-1} \{I_i I_j = 1\} \right) \\
&\leq \frac{2nkd}{d^k},
\end{aligned}
$$

by Lemma 2.1. This proves the proposition.   □

We next turn to the critical quantity in this paper, *viz.* L_n, the length of the longest palindromic segment in n random letters. A fundamental auxiliary variable that facilitates the understanding of the growth rate of the longest palindromic sequence is $M'_k = M'_{n,k} = \sum_{\alpha=k}^n M^*_{n,\alpha}$, where we recall that $M^*_{n,\alpha}$ is the number of *maximal* palindromes of length α [i.e., occurrences of situations where $X_j, \ldots, X_{j+\alpha-1}$ constitutes a palindrome, but $X_{j-1}, \ldots, X_{j+\alpha}$ doesn't]. We shall use the basic duality relation $L_n \leq k-1$ iff $M'_{n,k} = 0$ (whose proof is straightforward) together with a Poisson approximation for $\mathcal{L}(M'_k)$, to estimate the probability that the longest palindrome is of length at most $k-1$; this in turn will enable us to prove results on the almost sure behaviour of L_n.

The following lemma, which states that occurrences of palindromes of lengths α and β are pairwise independent, is crucial:

Lemma 2.4 *Fix two integers α, β. Define the indicators $\{I_{j,r} : 1 \leq j \leq n-k+1; 1 \leq r \leq n-j+1\}$ by letting $I_{j,r} = 1$ if $X_j, \ldots, X_{j+r-1}$ constitutes a palindrome. Then for $2j + \alpha \neq 2i + \beta$,*

$$\mathbf{P}(I_{j,\alpha}I_{i,\beta} = 1) = \mathbf{P}(I_{j,\alpha} = 1)\mathbf{P}(I_{i,\beta} = 1);$$

in other words, occurrences of palindromes of (possibly) different lengths are pairwise independent, provided that one of the corresponding "windows" is not contained within the other in a perfectly symmetric fashion.

Proof. We offer here simply a partial proof of the lemma. If $\alpha = \beta$, then we get back Lemma 2.1. Assume that $\alpha > \beta$, that α and β are both even, and that $i > j$. Let r denote the overlap between the two words.

CASE 1: $r = \frac{\beta}{2}, i > j + \frac{\alpha}{2} - 1$. After picking any characters in the overlap, only one arrangement of the remaining letters exists such that the second word is a palindrome. For the first word, we can still pick $i - (j + \frac{\alpha}{2})$ characters before a unique arrangement of the remaining letters exists such that the first word is a palindrome. Thus,

$$\mathbf{P}(I_{j,\alpha}I_{i,\beta} = 1) = \frac{1}{d^{\frac{\beta}{2}}} \frac{d^{i-(j+\frac{\alpha}{2})}}{d^{i-j}} = \frac{1}{d^{\frac{\alpha+\beta}{2}}} = \mathbf{P}(I_{j,\alpha} = 1)\mathbf{P}(I_{i,\beta} = 1),$$

establishing pairwise independence in this case.

CASE 2: $r > \frac{\beta}{2}, i < j + \frac{\alpha}{2} - 1$. One can choose $i + \frac{\beta}{2} - (j + \frac{\alpha}{2})$ characters in the overlap freely before a unique arrangement of the remaining letters exists such that both words are palindromes. It follows that

$$\mathbf{P}(I_{j,\alpha}I_{i,\beta} = 1) = \frac{d^{i-j+\frac{\beta}{2}-\frac{\alpha}{2}}}{d^{i+\beta-j}} = \frac{1}{d^{\frac{\alpha+\beta}{2}}} = \mathbf{P}(I_{j,\alpha} = 1)\mathbf{P}(I_{i,\beta} = 1),$$

as required. We consider just one more case

CASE 3: $r < \frac{\beta}{2}, i > j + \frac{\alpha}{2} - 1$. In this case, one can freely choose the letters for positions $j + \frac{\alpha}{2}$ through $i + \frac{\beta}{2} - 1$ until a unique arrangement of the remaining letters exists such that both words are palindromes. Therefore,

$$\mathbf{P}(I_{j,\alpha}I_{i,\beta} = 1) = \frac{d^{i+\frac{\beta}{2}-j-\frac{\alpha}{2}}}{d^{i+\beta-j}} = \frac{1}{d^{\frac{\alpha+\beta}{2}}} = \mathbf{P}(I_{j,\alpha} = 1)\mathbf{P}(I_{i,\beta} = 1).$$

The reader may verify that pairwise independence holds for all the remaining cases; the method of proof is exactly the same as before. The only situation where pairwise independence fails, therefore, is when $2j + \alpha = 2i + \beta$; in this case one clearly has $\mathbf{P}(I_{j,\alpha}I_{i,\beta} = 1) = \mathbf{P}(I_{j,\alpha} = 1)$ [or $\mathbf{P}(I_{j,\alpha}I_{i,\beta} = 1) = \mathbf{P}(I_{i,\beta} = 1)$.] □

Next consider the variable $M'_{n,k} = \sum_{\alpha=k}^{n} M^*_{n,\alpha} = \sum_{\alpha=k}^{n} \sum_{j=1}^{n-\alpha+1} I^*_{j,\alpha} = \sum_{j=1}^{n-k+1} \sum_{\alpha=k}^{n-j+1} I^*_{j,\alpha}$, where $I^*_{j,\alpha} = 1$ if the letters $\{X_j, \ldots, X_{j+\alpha-1}\}$ constitute a maximal palindrome. It is clear that $\mathbf{P}(I^*_{j,\alpha}I^*_{i,\beta} = 1) = 0$ if $2j + \alpha = 2i + \beta$, and thus by Lemma 2.4 that

$$\mathbf{P}(I^*_{j,\alpha}I^*_{i,\beta} = 1) \leq \mathbf{P}(I_{j,\alpha} = 1)\mathbf{P}(I_{i,\beta} = 1)$$

for all choices of j, i, α, and β. We use this observation as a crucial component in establishing a Poisson approximation for $\mathcal{L}(M'_{n,k})$; the failure of pairwise independence for the indicators $\{I_{j,\alpha} : 1 \leq j \leq n - k + 1; k \leq \alpha \leq n - j + 1\}$ leads to a rather poor Poisson approximation for the variable $\sum_{\alpha=k}^{n} M_{n,\alpha}$. We next compute $\lambda' = \mathbf{E}(M'_{n,k})$:

Lemma 2.5

$$\lambda' = \mathbf{E}(M'_{n,k}) = \frac{(n - k + 1)}{d^{\lfloor k/2 \rfloor}} + \frac{(n - k)}{d^{\lceil k/2 \rceil}}.$$

Proof. The above discussion implies that

$$\lambda' = \mathbf{E}(M'_{n,k}) = \sum_{\alpha=k}^{n} \sum_{j=1}^{n-\alpha+1} \mathbf{P}(I^*_{j,\alpha} = 1).$$

Assume first that k is even and that n is odd. Notice that palindromes of length $\alpha (\geq k)$ that start at the first position or end at the last position *cannot* be maximal in the regular sense and are defined to be maximally palindromic, we recall, simply if they are palindromic. There are two such word segments of length α for each $\alpha = k, \ldots, n - 1$, but only one possible "non-regular" maximal palindrome of length n. Since the probability of a palindrome in a segment of length α is $1/d^{\lfloor \alpha/2 \rfloor}$, we see that the contribution, to λ', of the indicators corresponding to the non-regular occurrences of maximal palindromes is given by

$$2\left(\frac{2}{d^{k/2}} + \frac{2}{d^{(k+2)/2}} + \cdots + \frac{2}{d^{(n-3)/2}}\right) + \frac{3}{d^{(n-1)/2}}. \tag{1.3}$$

Next consider "regular" maximal palindromes, i.e. ones for which there is at least one letter on each side of the palindromic occurrence. It is easy to verify that there are $(n - k - 1)$ possible regular maximal palindromes of length k, $(n - k - 2)$ of length $k + 1, \ldots$, two of length $(n - 3)$, and one of length $(n - 2)$. Moreover, it is evident that $\mathbf{E}(I^*_{j,\alpha}) = 1/d^{\lfloor \alpha/2 \rfloor} -$

$1/d^{\lfloor(\alpha+2)/2\rfloor}$ for such indicators, so that the contribution, to λ', of the indicators corresponding to the regular occurrences of maximal palindromes is given by

$$(n-k-1)(\frac{1}{d^{k/2}} - \frac{1}{d^{(k+2)/2}}) + (n-k-2)(\frac{1}{d^{k/2}} - \frac{1}{d^{(k+2)/2}}) +$$

$$(n-k-3)(\frac{1}{d^{(k+2)/2}} - -\frac{1}{d^{(k+4)/2}}) + (n-k-4)(\frac{1}{d^{(k+2)/2}} - -\frac{1}{d^{(k+4)/2}})$$

$$+\ldots+$$

$$2(\frac{1}{d^{(n-3)/2}} - -\frac{1}{d^{(n-1)/2}}) + (\frac{1}{d^{(n-3)/2}} - -\frac{1}{d^{(n-1)/2}}),$$

which may be checked to simplify to

$$\frac{(2n-2k-3)}{d^{k/2}} - \frac{4}{d^{(k+2)/2}} - \frac{4}{d^{(k+4)/2}} \cdots - \frac{4}{d^{(n-3)/2}} - \frac{3}{d^{(n-1)/2}}. \quad (1.4)$$

(1.3) and (1.4) may now be combined to give the required result, i.e., $\lambda' = (2n-2k+1)/d^{k/2}$. If k is odd or n is even, a similar argument may be employed to prove the lemma. $\qquad\square$

Theorem 2.6 Let $M'_k = M'_{n,k} = M^*_{n,k} + \ldots M^*_{n,n}$, where $M^*_{n,\alpha}$ denotes the number of maximal palindromes of length α in the sequence $\{X_j\}^n_{j=1}$. Then

$$d_{\mathrm{TV}}(\mathcal{L}(M'_k), \mathrm{Po}(\lambda')) \leq D\frac{k}{d^{k/2}}$$

where $D = D_d$ is an explicit constant, and $\lambda' = \mathbf{E}(M'_k)$ is given by Lemma 2.5.

Proof. By Theorem 1 in Arratia et al. [AGG],

$$d_{\mathrm{TV}}(\mathcal{L}(M'_k), \mathrm{Po}(\lambda')) \leq \left[(b_1+b_2)\frac{(1-e^{-\lambda'})}{\lambda'} + b_3\right] \quad (1.5)$$

where, given the "neighbourhoods of dependence" $\{\mathcal{B}_{j,\alpha} : 1 \leq j \leq n-k+1; k \leq \alpha \leq n-j+1\}$,

$$b_1 = \sum_{j=1}^{n-k+1}\sum_{\alpha=k}^{n-j+1}\sum_{(i,\beta)\in\mathcal{B}_{j,\alpha}} P(I^*_{j,\alpha} = 1)P(I^*_{i,\beta} = 1), \quad (1.6)$$

$$b_2 = \sum_{j=1}^{n-k+1}\sum_{\alpha=k}^{n-j+1}\sum_{(j,\alpha)\neq(i,\beta)\in\mathcal{B}_{j,\alpha}} P(I^*_{j,\alpha}I^*_{i,\beta} = 1), \quad (1.7)$$

and

$$b_3 = \sum_{j=1}^{n-k+1}\sum_{\alpha=k}^{n-j+1} s_{j,\alpha} \quad (1.8)$$

where

$$s_{j,\alpha} = \mathbf{E}\big|\mathbf{E}\{I_{j,\alpha} - \mathbf{P}(I_{j,\alpha} = 1)|\sigma(I_{i,\beta} : (i,\beta) \notin \mathcal{B}_{j,\alpha})\}\big|. \qquad (1.9)$$

We use the non-rectangular neighbourhoods of dependence given by

$$\begin{aligned}
\mathcal{B}_{j,\alpha} = &\ \{(i,\beta) : j - k - 1 \le i \le j + \alpha + 1, k \le \beta \le n - i + 1\} \\
&\cup \{(i,\beta) : i \le j - k - 2, (j - i - 1) \le \beta \le n - i + 1\};
\end{aligned}$$

due to this choice, it follows that $b_3 = 0$. On the other hand, b_1, as defined by (1.6), is given by

$$\begin{aligned}
b_1 =&\ \sum_{j=1}^{n-k+1} \sum_{\alpha=k}^{n-j+1} \mathbf{P}(I_{j,\alpha}^* = 1) \bullet \\
&\left(\sum_{i=j-k-1}^{j+\alpha+1} \sum_{\beta=k}^{n-i+1} \mathbf{P}(I_{i,\beta}^* = 1) + \sum_{i=1}^{j-k-2} \sum_{\beta=j-i-1}^{n-i+1} \mathbf{P}(I_{i,\beta}^* = 1) \right) \\
\le&\ \sum_{j=1}^{n-k+1} \sum_{\alpha=k}^{n-j+1} \mathbf{P}(I_{j,\alpha}^* = 1) \bullet \\
&\left((\alpha + k + 3) \sum_{\beta=k}^{n} \frac{\sqrt{d}}{d^{\beta/2}} + \sum_{i=1}^{j-k+2} \frac{d^{3/2}}{d^{(j-i)/2}(\sqrt{d}-1)} \right) \\
\le&\ \sum_{j=1}^{n-k+1} \sum_{\alpha=k}^{n-j+1} \frac{\sqrt{d}}{d^{\alpha/2}} \left(\frac{(\alpha+k+3)d}{d^{k/2}(\sqrt{d}-1)} + \frac{d^3}{d^{k/2}(\sqrt{d}-1)^2} \right), \\
=&\ \sum_{j=1}^{n-k+1} \sum_{\alpha=k}^{n-j+1} \frac{A\alpha + B}{d^{\alpha/2}}, \qquad\qquad (1.10)
\end{aligned}$$

where

$$A = A_{k,d} = \frac{d^{3/2}}{d^{k/2}(\sqrt{d}-1)} \qquad (1.11)$$

and

$$B = B_{k,d} = \frac{(k+3)d^{3/2}}{d^{k/2}(\sqrt{d}-1)} + \frac{d^{7/2}}{d^{k/2}(\sqrt{d}-1)^2}. \qquad (1.12)$$

Now it is easy to verify that

$$\sum_{\alpha=k}^{\infty} \frac{\alpha}{d^{\alpha/2}} \le 3\frac{(k+3)}{d^{k/2}},$$

so that (1.10) reduces to

$$b_1 \le \frac{n-k+1}{d^{k/2}} \left\{ 3A(k+3) + \frac{B\sqrt{d}}{\sqrt{d}-1} \right\}$$

$$\leq \ \lambda' \left\{ 3A(k+3) + \frac{B\sqrt{d}}{\sqrt{d}-1} \right\}. \tag{1.13}$$

Since $\mathbf{P}(I_{j,\alpha}^* I_{i,\beta}^* = 1) \leq \mathbf{P}(I_{j,\alpha} = 1)\mathbf{P}(I_{i,\beta} = 1)$, the above calculation shows that b_2, too, may be bounded above by the right hand side of (1.13), so that by (1.13) and (1.5),

$$d_{\mathrm{TV}}\big(\mathcal{L}(M_k'), \mathrm{Po}(\lambda')\big) \ \leq \ 2 \left\{ 3A(k+3) + \frac{B\sqrt{d}}{\sqrt{d}-1} \right\}$$

$$\leq \ D_d \frac{k}{d^{k/2}},$$

as claimed. This proves Theorem 2.6.   □

Corollary 2.7 *(An extreme value limit.)* *Let $a = \sqrt{d}$, and set $\nu = d^{1/2} + d^{-1/2}$. Assume that $\gamma_n \to \infty$ is arbitrary. Then*

$$\lim_{n\to\infty} \sup_{x\in\Gamma_n} \left| \mathbf{P}(L_n - \log_a 2n < x) - \exp\{-\frac{1}{d^{x/2}}\} \right| = 0,$$

where $\Gamma_n = \{x \in \mathbf{R} : x \geq -\log_a 2n + \log_a \log_a 2n + \gamma_n,\ x + \log_a 2n = 0, 2, 4, \ldots\}$. Also,

$$\lim_{n\to\infty} \sup_{x\in\Delta_n} \left| \mathbf{P}(L_n - \log_a \nu n < x) - \exp\{-\frac{1}{d^{x/2}}\} \right| = 0,$$

where $\Delta_n = \{x \in \mathbf{R} : x \geq -\log_a \nu n + \log_a \log_a \nu n + \gamma_n,\ x + \log_a \nu n = 1, 3, 5, \ldots\}$

Proof. By Theorem 2.6,

$$\left| \mathbf{P}(L_n \leq k-1) - e^{-\lambda'} \right| \leq D_d \frac{k}{d^{k/2}} \tag{1.14}$$

where $\lambda' = (n-k+1)/d^{\lfloor k/2 \rfloor} + (n-k)/d^{\lceil k/2 \rceil}$ satisfies

$$\frac{2(n-k)}{d^{k/2}} \leq \lambda' \leq \frac{n(\sqrt{d} + (1/\sqrt{d}))}{d^{k/2}} = \frac{n\nu}{d^{k/2}}$$

and thus stays away from zero and infinity iff $k - \log_a n$ stays bounded, where $a = \sqrt{d}$. Let $k - \log_a n = x; x \in \mathbf{R}$. Then the error bound $D_d k/d^{k/2}$ given by (1.14) is of magnitude

$$\frac{D_d(x + \log_a n)}{d^{(x+\log_a n)/2}} = \frac{D_d x}{d^{x/2}n} + \frac{D_d \log_a n}{d^{x/2}n} = \zeta_1 + \zeta_2, \text{ say,}$$

where ζ_1 and ζ_2 may both be checked to tend to zero as $n \to \infty$ provided that $x \geq -\log_a n + \log_a \log_a n + \gamma_n$, where $\gamma_n \to \infty$ is arbitrary.

Assume next that $x + \log_a 2n$ is even, $x \geq -\log_a 2n + \log_a \log_a 2n + \gamma_n$. Then

$$\left| \mathbf{P}(L_n - \log_a 2n < x) - \exp\{-\frac{1}{d^{x/2}}\} \right|$$

$$\leq \left| \mathbf{P}(L_n - \log_a 2n < x) - \exp\{-\lambda'\} \right| + \left| \exp\{-\frac{1}{d^{x/2}}\} - -\exp\{-\lambda'\} \right|$$

$$= \varepsilon_1 + \varepsilon_2, \quad \text{say.}$$

Now $\varepsilon_1 \to 0$ $(n \to \infty)$ uniformly in $x \geq -\log_a 2n + \log_a \log_a 2n + \gamma_n$ by the above discussion. Consider ε_2 next. We have

$$\varepsilon_2 = \left| \exp\{\frac{-2(n - x - \log_a 2n) - 1}{2nd^{x/2}}\} - \exp\{-\frac{1}{d^{x/2}}\} \right|$$

$$= \exp\{-\frac{1}{d^{x/2}}\} \left| \exp\left\{\frac{x + \log_a 2n - 1/2}{nd^{x/2}}\right\} - 1 \right|$$

$$\leq \left| \exp\left\{\frac{x + \log_a 2n - 1/2}{nd^{x/2}}\right\} - 1 \right|,$$

which again tends to zero uniformly in $x \geq -\log_a 2n + \log_a \log_a 2n + \gamma_n$. This proves the first part of the corollary. Now if $x + \log_a \nu n$ is odd, $x \geq -\log_a \nu n + \log_a \log_a \nu n + \gamma_n$, then

$$\mathbf{P}(L_n - \log_a \nu n < x) \approx \exp\left\{ -\frac{n\sqrt{d}}{d^{(x+\log_a \nu n)/2}} - \frac{n}{\sqrt{dd^{(x+\log_a \nu n)/2}}} \right\}$$

$$= \exp\left\{ -\frac{1}{d^{x/2}} \right\},$$

where the error once again goes to zero. This establishes the corollary. $\square$

3   Rates of Growth of the Longest Palindromic Segment

Our goal in this section is to prove almost sure results on the rate of growth of the longest palindromic sequence. We start by establishing an Erdős-Rényi law for L_n, showing that the longest palindromic segment is asymptotically *twice as long* as the longest pure run of any single letter, being of order $\log_a n$, where $a = \sqrt{d}$ as before. This result will be refined in Theorems 3.2 and 3.3, which address the almost sure growth rate of the sequence $\{L_n - \log_a n\}$.

Proposition 3.1 *Let L_n be the longest palindromic segment generated when n letters are independently and uniformly selected from a d-letter* *alphabet. Set $a = \sqrt{d}$. Then*

$$\lim_{n \to \infty} \frac{L_n}{\log_a n} = 1 \quad a.s.$$

Proof. The proof will be a straightforward application of the Borel-Cantelli lemma. Note first that $L_n/\log_a n \to 1$ a.s. iff $L_{a^n}/n \to 1$ a.s.; to see this, we sandwich any integer r between successive powers of a, and thus obtain

$$\frac{L_{a^m}}{m}\frac{m}{m+1} \le \frac{L_r}{\log_a r} \le \frac{L_{a^{m+1}}}{m+1}\frac{m+1}{m},$$

which establishes the claim. Now, given any $\varepsilon > 0$, define the events A_n and B_n by

$$A_n = \{L_{a^n} > (1+\varepsilon)n\}$$

and

$$B_n = \{L_{a^n} < (1-\varepsilon)n\}$$

respectively. We shall show that $\mathbf{P}(A_n \text{ i.o.}) = \mathbf{P}(B_n \text{ i.o.}) = 0$; this will establish the Erdős-Rényi law. We have, for some constants K_1 and K_2,

$$
\begin{aligned}
\sum_{n=1}^{\infty} \mathbf{P}(B_n) &= \sum_{n=1}^{\infty} \mathbf{P}\left(L_{a^n} < \lceil (1-\varepsilon)n \rceil\right) \\
&\le \sum_{n=1}^{\infty} \exp\left\{-K_1 \frac{a^n}{a^{(1-\varepsilon)n}}\right\} + K_2 \cdot D_d \sum_{n=1}^{\infty} \frac{(1-\varepsilon)n}{a^{(1-\varepsilon)n}} \\
&< \infty,
\end{aligned}
$$

which proves, via Borel-Cantelli, that $\mathbf{P}(B_n \text{ i.o.}) = 0$. Next, note that the inequality $1 - \lambda' \le \exp\{-\lambda'\}$ implies, that for some constants K_3 and K_4,

$$
\begin{aligned}
\sum_{n=1}^{\infty} \mathbf{P}(A_n) &= \sum_{n=1}^{\infty} \left(1 - \mathbf{P}(L_{a^n} \le \lfloor (1+\varepsilon)n \rfloor)\right) \\
&\le K_3 \sum_{n=1}^{\infty} \frac{a^n}{a^{(1+\varepsilon)n}} + D_d K_4 \sum_{n=1}^{\infty} \frac{(1+\varepsilon)n}{a^{(1+\varepsilon)n}} \\
&< \infty,
\end{aligned}
$$

proving that $\mathbf{P}(A_n \text{ i.o.}) = 0$, and completing the proof. $\square$

Theorem 3.2 *Let L_n denote the length of the longest palindrome in n i.i.d. letters, drawn uniformly from the d-letter alphabet $\{0,1,\ldots,d-1\}$. Then*
(a)

$$\limsup_{n \to \infty} \frac{L_n - \log_a n}{\log_a \log_a n} = 1 \quad almost\ surely,$$

which is a result that can be refined to

(b)

$$\limsup_{n\to\infty} \frac{L_n - \log_a n - \log_a \log_a n}{\log_a \log_a \log_a n} = 1 \quad almost\ surely,$$

and further generalized to

(c) Given a monotone increasing function f, with $f(n) \sim \log_a \log_a n$, we have $\mathbf{P}(L_n \geq \log_a n + f(n)$ i.o.) = 0 (or 1) according as the series $\sum_{n=1}^{\infty} (1/a)^{f(a^n)}$ converges (or diverges).

Proof. Throughout this proof (and the rest of the paper), K, L, etc. will denote generic constants, possibly depending on d, whose values might change from line to line. We first observe that

$$\limsup \frac{(L_n - \log_a n)}{\log_a \log_a n} = 1 \text{ a.s.} \iff \limsup \frac{(L_{a^n} - n)}{\log_a n} = 1 \text{ a.s.};$$

this is proved, as before, by sandwiching an integer r between two successive powers of a. For a fixed $\varepsilon > 0$, set

$$E_n = \{L_{a^n} > n + (1 + \varepsilon)\log_a n\}$$

and

$$G_n = \{L_{a^n} > n + (1 - \varepsilon)\log_a n\}.$$

We have

$$\sum_{n=1}^{\infty} \mathbf{P}(E_n) = \sum_{n=1}^{\infty} \left(1 - \mathbf{P}(L_{a^n} \leq \lfloor n + (1 + \varepsilon)\log_a n \rfloor)\right)$$

$$\leq K \sum_{n=1}^{\infty} \frac{a^n}{a^{n+(1+\varepsilon)\log_a n}} + L \sum_{n=1}^{\infty} \frac{n + (1 + \varepsilon)\log_a n}{a^{n+(1+\varepsilon)\log_a n}}$$

$$< \infty,$$

so that $\mathbf{P}(E_n$ i.o.) = 0. We need next to show that $\mathbf{P}(G_n$ i.o.) = 1, and must address the fact that the events $\{G_n\}_{n=1}^{\infty}$ are not independent. Towards this end, we define $L'_{a,n}$ to be the length of the longest palindrome in the sequence of letters $\{a^{n-1} + 1, \ldots, a^n\}$, and set $G'_n = \{L'_{a,n} > n + (1 - \varepsilon)\log_a n\}$. Note that $G'_n \subseteq G_n$ and that $\{G'_n\}_{n=1}^{\infty}$ is an ensemble of independent events. Finally, observe that there are constants A and B such that $A \leq \mathbf{P}(G'_n)/\mathbf{P}(G_n) \leq B$ ($n \to \infty$), since $\mathbf{P}(G'_n) \sim \mathbf{P}(L_{a^{n-1}(a-1)} > n + (1-\varepsilon)\log_a n) \sim 1 - \exp\{-K(a-1)/an^{1-\varepsilon}\} \sim K(a-1)/an^{1-\varepsilon}$, whereas $\mathbf{P}(L_{a^n} > n + (1 - \varepsilon)\log_a n) \sim L/n^{1-\varepsilon}$ by the same argument. Now

$$\sum_{n=1}^{\infty} \mathbf{P}(G'_n) \geq K \sum_{n=1}^{\infty} \mathbf{P}(L_{a^{n-1}(a-1)} \geq n + (1 - \varepsilon)\log_a n)$$

$$\geq K \sum_{n=1}^{\infty} \frac{(a - 1)}{an^{1-\varepsilon}} - -L \sum_{n=1}^{\infty} \frac{n + (1 - \varepsilon)\log_a n}{a^{n+(1-\varepsilon)\log_a n}}$$

$$= \infty,$$

so that $\mathbf{P}(G_n'$ i.o.$) = 1$, proving that $\mathbf{P}(G_n$ i.o.$) = 1$. This establishes part (a) of the theorem; parts (b) and (c) can be proved in an entirely analogous fashion.   □

Our final result shows, that in analogy with the pure head run situation, the longest palindromic segment exhibits a very sharp minimal rate of growth:

Theorem 3.3 *Almost surely,*

$$(\log_a \log_a e^2) - 2 \leq \liminf\{L_n - \log_a n + \log_a \log_a \log_a n\} \leq (\log_a \log_a e^\nu) + 1,$$

where $\nu = \sqrt{d} + (1/\sqrt{d})$.

Proof. For a fixed $\delta > 0$, set $M = (\log_a \log_a e^{2-\delta}) - 1$. We shall, as before, argue along the exponential subsequence $\{a^n\}$, but will be more careful than before with the analysis. Given $\varepsilon > 0$, let

$$H_n = \{L_{a^n} < M - \varepsilon - \log_a \log_a n + n\}.$$

We have, since $2(n-k)/d^{k/2} \leq \lambda' \leq n\nu/d^{k/2}$,

$$
\begin{aligned}
\sum_{n=1}^{\infty} \mathbf{P}(H_n) &= \sum_{n=1}^{\infty} \mathbf{P}\big(L_{a^n} < \lceil M - \varepsilon - \log_a \log_a n + n\rceil\big) \\
&\leq \sum_{n=1}^{\infty} \exp\left\{--2\frac{a^n - \lceil M - \varepsilon - \log_a \log_a n + n\rceil}{a^{\lceil M-\varepsilon-\log_a \log_a n+n\rceil}}\right\} \\
&\quad + \sum_{n=1}^{\infty} D_d \frac{\lceil M - \varepsilon - \log_a \log_a n + n\rceil}{a^{\lceil M-\varepsilon-\log_a \log_a n+n\rceil}} \\
&\leq \sum_{n=1}^{\infty} \exp\left\{--2\frac{a^n - M + \varepsilon + \log_a \log_a n - n - 1}{a^{M-\varepsilon-\log_a \log_a n+n+1}}\right\} \\
&\quad + \sum_{n=1}^{\infty} D_d \frac{\lceil M - \varepsilon - \log_a \log_a n + n\rceil}{a^{\lceil M-\varepsilon-\log_a \log_a n+n\rceil}} \\
&\leq \sum_{n=1}^{\infty} \exp\left\{-(2-\delta)\frac{a^n}{a^{M-\varepsilon-\log_a \log_a n+n+1}}\right\} \\
&\quad + \sum_{n=1}^{\infty} D_d \frac{\lceil M - \varepsilon - \log_a \log_a n + n\rceil}{a^{\lceil M-\varepsilon-\log_a \log_a n+n\rceil}} \\
&= \sum_{n=1}^{\infty} \frac{1}{n^{a^\varepsilon}} + \sum_{n=1}^{\infty} D_d \frac{\lceil M - \varepsilon - \log_a \log_a n + n\rceil}{a^{\lceil M-\varepsilon-\log_a \log_a n+n\rceil}} \\
&< \infty,
\end{aligned}
$$

so that $(\log_a \log_a e^2) - 1 \leq \liminf\{L_{a^n} - n + \log_a \log_a n\}$, since $\delta > 0$ is arbitrary. It follows that $\liminf\{L_n - \log_a n + \log_a \log_a \log_a n\} \geq (\log_a \log_a e^2) -$

2 as claimed. We need next to show that

$$\liminf\{L_n - \log_a n + \log_a \log_a \log_a n\} \leq (\log_a \log_a e^\nu) + 1 := R.$$

Towards this end, let, for a given $\varepsilon > 0$,

$$J_n = \{L_n < R + \varepsilon - \log_a \log_a \log_a n + \log_a n\};$$

we wish to show that $\mathbf{P}(J_n \text{ i.o.}) = 1$, and will argue along the superexponential subsequence $\{a^{n^a}\}$ this time, proving that

$$\mathbf{P}(J_n' \text{ i.o.}) := \mathbf{P}\left(L_{a^{n^a}} < n^a - \log_a \log_a n - 1 + R + \varepsilon \text{ i.o.}\right) = 1.$$

Let $L_{a,n}''$ denote the length of the longest palindrome in the sequence of letters $\{a^{(n-1)^a} + 1, \ldots, a^{n^a}\}$ and set

$$J_n'' = \left\{L_{a,n}'' < n^a - \log_a \log_a n - 1 + R + \varepsilon\right\},$$

where $\{J_n''\}_{n=1}^\infty$ is an ensemble of independent events. We have

$$\sum_{n=1}^\infty \mathbf{P}(J_n'') = \sum_{n=1}^\infty \mathbf{P}\left(L_{a^{n^a}-a^{(n-1)^a}} < n^a - \log_a \log_a n + R - 1 + \varepsilon\right)$$

$$= \sum_{n=1}^\infty \mathbf{P}\left(L_{a^{n^a}-a^{(n-1)^a}} < \lceil n^a - \log_a \log_a n + R - 1 + \varepsilon\rceil\right)$$

$$\geq \sum_{n=1}^\infty \exp\left\{-\nu\frac{(a^{n^a} - a^{(n-1)^a})}{a^{\lceil n^a - \log_a \log_a n + R - 1 + \varepsilon\rceil}}\right\}$$

$$- \sum_{n=1}^\infty D_d \frac{\lceil n^a - \log_a \log_a n + R - 1 + \varepsilon\rceil}{a^{\lceil n^a - \log_a \log_a n + R - 1 + \varepsilon\rceil}}$$

$$\geq \sum_{n=1}^\infty \exp\left\{-\nu\frac{a^{n^a}}{a^{n^a - \log_a \log_a n + R - 1 + \varepsilon}}\right\}$$

$$- \sum_{n=1}^\infty D_d \frac{\lceil n^a - \log_a \log_a n + R - 1 + \varepsilon\rceil}{a^{\lceil n^a - \log_a \log_a n + R - 1 + \varepsilon\rceil}}$$

$$= \sum_{n=1}^\infty \frac{1}{n^{a-\varepsilon}} - \sum_{n=1}^\infty D_d \frac{\lceil n^a - \log_a \log_a n + R - 1 + \varepsilon\rceil}{a^{\lceil n^a - \log_a \log_a n + R - 1 + \varepsilon\rceil}}$$

$$= \infty,$$

so that $\mathbf{P}(J_n'' \text{ i.o.}) = 1$. Note next that

$$\mathbf{P}\left(L_{a^{n^a}} \neq L_{a,n}''\right)$$

$$\leq \mathbf{P}\left(\bigcup_{j=1}^{a^{(n-1)^a}} \{\text{the longest palindrome starts at the } j\text{th position}\}\right)$$

$$\leq K \frac{a^{(n-1)^a}}{a^{n^a}}$$

$$\leq \frac{K}{a^{Ln^{a-1}}},$$

so that $\mathbf{P}\left(L_{a^{n^a}} \neq L''_{a,n} \text{ i.o.}\right) = 0$. It follows that $\mathbf{P}(J'_n \text{ i.o.}) = 1$, proving that $\liminf\{L_{a^{n^a}} - n^a + 1 + \log_a \log_a n\} \leq R$, and thus that $\liminf\{L_n - \log_a n + \log_a \log_a \log_a n\} \leq R$. This establishes Theorem 3.3. $\qquad \square$

Acknowledgments: The research of both authors was partially supported by National Science Foundation Grant DMS-9322460, and has benefited from useful discussions with David Revelle.

4 References

[AGG] Arratia, R., Goldstein, L., and Gordon, L. (1989) Two moments suffice for Poisson approximations: The Chen-Stein method. *Ann. Probab.* **17** 9–25.

[AGW] Arratia, R., Gordon, L., and Waterman, M. (1990) The Erdős-Rényi law in distribution, for coin tossing and sequence matching. *Ann. Statist.* **18** 539–570.

[B] Bergerson, H. (1973) *Palindromes and Anagrams.* Dover Publications, Inc., New York.

[BHJ] Barbour, A. D., Holst, L., and Janson, S. (1992) *Poisson Approximation.* Oxford University Press.

[ER70] Erdős, P. and Rényi, A. (1970) On a new law of large numbers. *J. Anal. Math.* **22** 103–111.

[ER75] Erdős, P. and Révész, P. (1975) On the length of the longest head run. *Colloq. Math. Soc. János Bolyai* **16** 219–228.

[F] Feller, W. (1971) *An Introduction to Probability Theory and its Applications, Vol. 1, 3rd Edition.* John Wiley and Sons, Inc., New York.

[M] Móri, T. (1992) Maximum waiting times are asymptotically independent. *Comb. Probab. Computing* **1**, 251–264.

[R] Revelle, D. (1995) The length of the longest partial palindrome. Preprint.

[W] Waterman, M. (1989) *Mathematical Methods for DNA Sequences.* CRC Press, Inc., Boca Raton.

Direct analytical methods for determining quasistationary distributions for continuous-time Markov chains

A.G. Hart
P.K. Pollett

ABSTRACT We shall be concerned with the problem of determining the quasistationary distributions of an absorbing continuous-time Markov chain directly from the transition-rate matrix Q. We shall present conditions which ensure that any finite μ-invariant probability measure for Q is a quasistationary distribution. Our results will be illustrated with reference to birth and death processes.

1 Introduction

The most useful conditions to date, which guarantee that a μ-invariant probability distribution m be a quasistationary distribution, stipulate that μ should be equal to the probability flux into the absorbing state; see for example [2], [7], [12], [13] and [15]. However, although these conditions have proved useful in practice (see for example [10], [11] and [14]), they are deficient in so far as μ and m are interrelated; indeed, there is usually a one-parameter family of quasistationary distributions indexed by μ. Here, we address this problem by presenting conditions, solely in terms of the transition rates, which guarantee that *any* finite μ-invariant measure (or, more generally, any which is finite with respect to the absorption probabilities) can be normalized to produce a quasistationary distribution.

We begin by reviewing existing work on the relationship between μ-invariant measures and quasistationary distributions.

2 Quasistationary distributions

Let $S = \{0, 1, \ldots\}$ and let $Q = (q_{ij}, \ i, j \in S)$ be a stable, conservative and regular q-matrix of transition rates over S, let $(X(t), \ t \geq 0)$ be the unique Markov chain associated with Q and denote its transition function

116

by $P(\cdot) = (p_{ij}(\cdot),\ i,j \in S)$.

Let C be a subset of S and μ some fixed non-negative real number. Then, the measure $m = (m_j,\ j \in C)$ is called a μ-*invariant measure for P* if

$$\sum_{i \in C} m_i p_{ij}(t) = e^{-\mu t} m_j, \qquad j \in C,\ t \geq 0. \tag{1.1}$$

In contrast, m is called a μ-*invariant measure for Q* if

$$\sum_{i \in C} m_i q_{ij} = -\mu m_j, \qquad j \in C. \tag{1.2}$$

We shall take $C = \{1, 2, \ldots\}$ and for simplicity we shall suppose that C is irreducible; this guarantees that all non-trivial μ-invariant measures m satisfy $m_j > 0$ *for all $j \in C$*. We shall also assume that 0 is an absorbing state, that is $q_{00} = 0$, and, that $q_{i0} > 0$ for at least one $i \in C$, a condition which guarantees a positive probability of absorption starting in i.

We shall use van Doorn's [18] definition of a quasistationary distribution.

Definition 1. Let $m = (m_j,\ j \in C)$ be a probability distribution over C and define $h(\cdot) = (h_j(\cdot),\ j \in S)$ by

$$h_j(t) = \sum_{i \in C} m_i p_{ij}(t), \qquad j \in S,\ t \geq 0. \tag{1.3}$$

Then, m is a *quasistationary distribution* if, for all $t > 0$ and $j \in C$,

$$\frac{h_j(t)}{\sum_{i \in C} h_i(t)} = m_j.$$

That is, if the chain has m as its initial distribution, then m is a quasistationary distribution if the state probabilities at time t, conditional on the chain being in C at t, are the same for all t.

The relationship between quasistationary distributions and the transition probabilities of the chain is made more precise in the following proposition [7]:

Proposition 1. Let $m = (m_j,\ j \in C)$ be a probability measure over C. Then, m is a quasistationary distribution if and only if, for some $\mu > 0$, m is a μ-invariant measure for P.

Thus, in a way which mirrors the theory of *stationary distributions*, one can interpret quasistationary distributions as eigenvectors of the transition function. However, the transition function is available explicitly in only a few simple cases, and so one requires a means of determining quasistationary distributions directly from transition rates of the chain. Since q_{ij} is the right-hand derivative of $p_{ij}(\cdot)$ near 0, an obvious first step is to rewrite (1.1) as

$$\sum_{i \in C: i \neq j} m_i p_{ij}(t) = \left((1 - p_{jj}(t)) - (1 - e^{-\mu t})\right) m_j, \qquad j \in C,\ t \geq 0.$$

Then, proceeding formally, dividing this expression by t and letting $t \downarrow 0$, we arrive at (1.2). This argument can be justified rigorously (see Proposition 2 of [17]), and so, in view of Proposition 1, we have the following simple result:

Proposition 2. If m is a quasistationary distribution then, for some $\mu > 0$, m is a μ-invariant measure for Q.

The more interesting question of when a positive solution m to (1.2) is also a solution to (1.1) was answered in [8, 9]:

Proposition 3. A μ-invariant measure m for Q is μ-invariant for P if and only if the equations

$$\sum_{i \in C} y_i q_{ij} = \nu y_j, \qquad 0 \le y_j \le m_j, \ j \in C, \tag{1.4}$$

have no non-trivial solution for some (and then for all) $\nu > -\mu$.

If we seek a quasistationary distribution then the μ-invariant measure m for Q can be taken to be finite, in which case simpler conditions obtain. The following result can be deduced from Theorems 3.2, 3.4 and 4.1 of [7]:

Proposition 4. Let m be a probability measure over C and suppose that m is μ-invariant for Q. Then, m is a quasistationary distribution if and only if $\mu = \sum_{i \in C} m_i q_{i0}$.

3 The Reuter FE conditions

Our main result gives conditions on Q which guarantee that *any* probability distribution over C which is μ-invariant for Q is a quasistationary distribution:

Theorem 1. If the equations

$$\sum_{i \in C} y_i q_{ij} = \nu y_j, \quad j \in C, \tag{1.5}$$

have no non-trivial, non-negative solution such that $\sum_{i \in C} y_i < \infty$, for some (and then all) $\nu > 0$, then *all* μ-invariant probability measures for Q are μ-invariant for P.

Remarks. (1) Many of the assumptions we have made can be relaxed. First, Q need not be regular or even conservative; the conclusions of the theorem are valid taking P to be the minimal transition function. Next, C need not be irreducible. For example, we can take C to be the whole of the state space S, which itself need not be of any particular form. More generally, C can be the union of irreducible classes (that is, irreducible with respect to the minimal chain), provided we impose some extra conditions, as follows. If C_1 and C_2 are two such classes with $C_1 \prec C_2$, that is, C_2 is accessible from C_1 (again, for the minimal chain), then we require that there be no class C' of states *outside* C with $C_1 \prec C' \prec C_2$: all paths leading from $i \in C_1$ to $j \in C_2$ must be wholly contained in C. (See Theorem 2 of [8] and the remarks at the end of Section 5 of that paper.)

(2) We call our invariance conditions the *Reuter FE conditions*, because they are G.E.H. Reuter's familiar necessary and sufficient conditions for the minimal transition function to be the unique solution to the forward equations when Q is not regular (see Section 6 of [16]); under our assumption that Q be regular, the Reuter FE conditions have no bearing on the forward equations.

Proof. Let m be a μ-invariant probability measure for Q. If the Reuter FE conditions hold, then any non-trivial, non-negative solution y to (1.4), for say $\nu = 1$, must satisfy $\sum_{i \in C} y_i = \infty$. However, since $\sum_{i \in C} m_i < \infty$, such a solution cannot satisfy $y_i \leq m_i$ for all i. Thus, by Proposition 3, m is μ-invariant for P.

4   When absorption is not certain

When the absorption probabilities are less than 1, we cannot use Theorem 1, because, under the conditions we have imposed (specifically, the regularity of Q), the μ-invariant measure cannot be finite. To see this, first observe that if m is a finite μ-invariant measure for P, with $\mu > 0$, then, for all $i \in C$,

$$\lim_{t \to \infty} \sum_{j \in C} p_{ij}(t) = 0,$$

since from (1.1) we have that

$$m_i p_{ij}(t) \leq e^{-\mu t} m_j, \qquad j \in S.$$

But P is honest, and so the probability of absorption starting in i, given by

$$a_i = \lim_{t \to \infty} p_{i0}(t),$$

is equal to 1 for all $i \in C$.

When the absorption probabilities *are* less than 1, the natural premise is that the μ-invariant measure m be finite *with respect to* $a = (a_i, \ i \in S)$, that is, $\sum_{i \in C} m_i a_i < \infty$, and, as we shall see, the conditions of Theorem 1 can be relaxed accordingly. (Since m is the subject of our attention, we prefer this to the measure-theoretic statement that a is an m-measurable function.) The premise arises in connection with the more general definition of a quasistationary distribution:

Definition 2. Let $m = (m_j, \ j \in C)$ be a measure over C such that $\sum_{j \in C} m_j a_j < \infty$ and define $h(\cdot) = (h_j(\cdot), \ j \in S)$ by (1.3) and $p = (p_j, \ j \in C)$ by

$$p_j = \frac{m_j a_j}{\sum_{i \in C} m_i a_i}, \qquad j \in C.$$

Then, p is a *quasistationary distribution* if, for all $t > 0$ and $j \in C$,

$$\frac{h_j(t)a_j}{\sum_{i \in C} h_i(t)a_i} = p_j.$$

Our next result, which is a generalization of Theorem 1, provides a means of determining quasistationary distributions directly from the q-matrix in cases where absorption occurs with probability less than 1.

Theorem 2. If the equations

$$\sum_{i \in C} y_i q_{ij} = \nu y_j, \quad j \in C, \tag{1.6}$$

have no non-trivial, non-negative solution such that $\sum_{i \in C} y_i a_i < \infty$, for some (and then all) $\nu > 0$, then all μ-invariant measures for Q which are finite with respect to a are also μ-invariant for P.

Proof. Suppose that the Reuter FE conditions (1.6) hold. Then, any non-trivial, non-negative solution to (1.4) satisfies $\sum_{i \in C} y_i a_i = \infty$. But, since $\sum_{i \in C} m_i a_i < \infty$, we cannot have $y_i \leq m_i$ for every i and so, again by Proposition 3, m must be μ-invariant for P.

An alternative proof, which also provides a useful way of interpreting Theorem 2, is based on the dual of Q, namely the q-matrix $\bar{Q} = (\bar{q}_{ij}, i, j \in S)$ given by

$$\bar{q}_{ij} = q_{ij}a_j/a_i, \quad i, j \in S.$$

This dual q-matrix is *conservative* because $a = (a_j, j \in S)$ is an invariant vector for Q, and, by Lemma 3.3 (ii) of [9], the corresponding minimal transition function $\bar{P}$ bears the same duality relationship with P:

$$\bar{P}_{ij}(t) = P_{ij}(t)a_j/a_i, \quad i, j \in S, t \geq 0.$$

Since a is an invariant vector for P, $\bar{P}$ is honest ($\bar{Q}$ regular) and is hence the unique $\bar{Q}$-transition function. Now, on setting $\bar{m}_j = m_j a_j, j \in C$, we see that m is μ-invariant for Q if and only if $\bar{m}$ is μ-invariant for $\bar{Q}$, and, that m is μ-invariant for P if and only if $\bar{m}$ is μ-invariant for $\bar{P}$. It is then a simple matter to check that Theorem 2 follows by applying Theorem 1 to $\bar{Q}$.

In applications involving chains for which absorption occurs with probability less than 1, it frequently easier to construct the dual transition rates from the absorption probabilities and then apply Theorem 1. We shall use this approach in Section 5.

5 Birth and death processes

We shall illustrate the results of the previous section with reference to absorbing birth and death processes. Some further applications are described in [4].

Van Doorn [18] has given a complete treatment of questions concerning the existence of quasistationary distributions for absorbing birth and death processes in cases where the probability of absorption is 1. We shall explain how his conditions for the existence of quasistationary distributions arise in the context of Theorem 1 and then extended these results to cases where absorption occurs with probability less than 1.

An absorbing birth and death process on $S = \{0, 1, \ldots\}$ has transition rates given by

$$
q_{ij} = \begin{cases} \lambda_i, & \text{if } j = i + 1, \\ -(\lambda_i + \mu_i), & \text{if } j = i, \\ \mu_i, & \text{if } j = i - 1, \\ 0, & \text{otherwise,} \end{cases}
$$

where the birth rates (λ_i, $i \geq 0$) and the death rates (μ_i, $i \geq 0$) satisfy λ_i, $\mu_i > 0$, for $i \geq 1$, and $\lambda_0 = \mu_0 = 0$. Thus, 0 is an absorbing state and $C = \{1, 2, \ldots\}$ is an irreducible class. We shall assume that

$$
\sum_{i=1}^{\infty} \frac{1}{\lambda_i \pi_i} \sum_{j=1}^{i} \pi_j = \infty, \tag{1.7}
$$

where $\pi_1 = 1$ and

$$
\pi_i = \prod_{j=2}^{i} \frac{\lambda_{j-1}}{\mu_j}, \qquad i \geq 2,
$$

a condition which is necessary and sufficient for Q to be regular (see Section 3.2 of [1]).

The classical Karlin and McGregor theory of birth and death processes involves the recursive construction of a sequence of orthogonal polynomials using the equations for an x-invariant vector (see [18]): define ($\phi_i(\cdot)$, $i \in C$), where $\phi_i : \mathrm{R} \to \mathrm{R}$, by $\phi_1(x) = 1$,

$$
\lambda_1 \phi_2(x) = \lambda_1 + \mu_1 - x,
$$

$$
\lambda_i \phi_{i+1}(x) - (\lambda_i + \mu_i)\phi_i(x) + \mu_i \phi_{i-1}(x) = -x\phi_i(x), \qquad i \geq 2,
$$

and let

$$
m_i = \pi_i \phi_i(x), \qquad i \in C, \ x \in \mathrm{R}. \tag{1.8}
$$

It can be shown [18] that $\phi_i(x) > 0$ for x in the range $0 \leq x \leq \lambda$, where λ (≥ 0) is the decay parameter of C (see [5]). Since π is a subinvariant measure for Q, that is $\sum_{i \in S} \pi_i q_{ij} \leq 0$, it follows, from Theorem 4.1 b(ii) of [9], that, for each fixed x in the above range, $m = (m_i, \ i \in C)$ is an x-invariant measure for Q; specifically, m satisfies (1.2) with $\mu = x$. Moreover, m is uniquely determined up to constant multiples. Something which is not at all obvious, is that when the probability of absorption is 1, that is

$$
A := \sum_{i=1}^{\infty} \frac{1}{\lambda_i \pi_i} = \infty, \tag{1.9}
$$

as well as $\lambda > 0$, we have that $\sum_{i=1}^{\infty} \pi_i \phi_i(x) < \infty$ for all x in the range $0 < x \le \lambda$ (see [3] and the proof of Theorem 3.2 of [18]). Thus, when $A = \infty$, each x-invariant measure for x in this range is finite, and so in order to apply Theorem 1 it remains only to check the Reuter FE conditions. (Note that $A = \infty$ implies (1.7) and hence the regularity of Q). It is well known (see Section 3.2 of [1]) that the Reuter FE conditions hold whenever

$$D := \sum_{i=1}^{\infty} \frac{1}{\lambda_i \pi_i} \sum_{j=i+1}^{\infty} \pi_j = \infty. \qquad (1.10)$$

So, in the case where absorption occurs with probability 1, $D = \infty$ is sufficient to ensure that for every x in the range $0 < x \le \lambda$, $m = (m_i, i \in C)$, given by $m_i = \pi_i \phi_i(x)$, is a (finite) x-invariant measure for P and, hence, can be normalized to produce a quasistationary distribution. Thus, we have proved the following result, which is encapsulated by Theorem 3.2 (i) of [18] (see also Theorem 3.5 (i) of [6]):

Theorem 3. Consider a birth and death process with $A = \infty$ (which of necessity is regular). Then, for every μ in the range $0 < \mu \le \lambda$, the essentially unique μ-invariant measure m for Q is finite. Furthermore, if $D = \infty$, then m is always μ-invariant for P and $p = (p_j, j \in C)$, given by

$$p_j = \frac{\pi_j \phi_j(\mu)}{\sum_{i \in C} \pi_i \phi_i(\mu)}, \qquad j \in C,$$

is a quasistationary distribution.

Let us now deal with the case where absorption occurs with probability less than 1, that is, $A < \infty$. The absorption probabilities are given by $a_0 = 1$ and

$$a_j = \frac{\mu_1}{1 + \mu_1 A} \sum_{i=j}^{\infty} \frac{1}{\lambda_i \pi_i}, \qquad j \in C. \qquad (1.11)$$

As suggested at the end of Section 4 we shall construct $\bar{Q}$ the dual of Q and apply Theorem 1 to $\bar{Q}$. This is clearly the q-matrix of an absorbing birth and death process on S whose birth rates and death rates are given (in an obvious notation) by

$$\bar{\lambda}_j = \lambda_j a_{j+1}/a_j \quad \text{and} \quad \bar{\mu}_j = \mu_j a_{j-1}/a_j, \qquad j \in C,$$

with $\bar{\lambda}_0 = \bar{\mu}_0 = 0$. The corresponding potential coefficients are given by $\bar{\pi}_j = \pi_j a_j^2/a_1^2$, $j \in C$, and the polynomials by $\bar{\phi}_j = \phi_j a_1/a_j$, $j \in C$. It follows that $\bar{m} = (\bar{m}_j, j \in C)$, the essentially unique μ-invariant measure for $\bar{Q}$, is given by $\bar{m}_j = m_j a_j/a_1$, $j \in C$. It is easily shown that the counterparts, $\bar{A}$ and $\bar{D}$, of the series in (1.9) and (1.10) are both divergent. Hence $\bar{m}$ is finite and, by Theorem 1, $\bar{m}$ is μ-invariant for $\bar{P}$. We have therefore proved the following result, which can be compared with Theorem 3.5 (ii) of [6]:

Theorem 4. Consider a regular birth and death process with $A < \infty$. Then, for every μ in the range $0 < \mu \leq \lambda$, the essentially unique μ-invariant measure m for Q is finite with respect to a. Furthermore, m is always μ-invariant for P and $p = (p_j, \ j \in C)$, given by

$$p_j = \frac{\pi_j \phi_j(\mu) a_j}{\sum_{i \in C} \pi_i \phi_i(\mu) a_i}, \qquad j \in C,$$

is a quasistationary distribution.

Remark. What has happened to the invariance condition $D = \infty$? This condition is not needed when $A < \infty$, for is the regularity of Q which is making the result work. Indeed, in a more general setting, where Q is not regular and our attention is focused on the minimal transition function, we can show that if $A < \infty$, then $\bar{D} = \infty$ if and only if the series in (1.7) diverges. It is also worth remarking that, in this more general context, Theorem 3 remains valid with P being interpreted as the minimal process.

6   Some concluding remarks

For birth and death processes, the transition rates are *reversible* with respect to the subinvariant measure π, that is,

$$\pi_i q_{ij} = \pi_j q_{ji}, \qquad i, j \in C, \tag{1.12}$$

and, since the Reuter FE conditions are expressed simply and explicitly in terms of the divergence of certain series, one might expect a simplification of the Reuter FE conditions in the general case of reversible Markov chains. We shall content ourselves with the following simple result:

Theorem 5. Suppose that Q is reversible with respect a subinvariant measure $\pi = (\pi_i, \ i \in S)$, that is, (1.12) holds. Then, every μ-invariant measure $m = (m_i, \ i \in C)$ for Q satisfying $\sup_{i \in C}\{m_i/\pi_i\} < \infty$ is μ-invariant for P.

Remarks. (1) Neither the subinvariant measure π nor the μ-invariant measure m need be finite. We require only that m be bounded above by π.

(2) Our assumption that Q be regular cannot be relaxed.

Proof. Let m be a μ-invariant measure which is bounded above by π and suppose that m *is not* μ-invariant for P. Then, by Proposition 3, the equations (1.4) have a non-trivial solution y, certainly for $\nu > 0$. On substituting (1.12) into (1.4) we find that $z = (z_j, \ j \in C)$, given by $z_j = y_j/\pi_j$, satisfies

$$\sum_{j \in C} q_{ij} z_j = \nu z_j, \tag{1.13}$$

with $0 < z_j \leq m_j/\pi_j, \ j \in C$. But, $\sup_{i \in C}\{m_i/\pi_i\} < \infty$, and so we have found a bounded, non-trivial, non-negative solution to (1.13), which, by Theorem 2.2.7 of [1], contradicts our assumption that Q is regular.

We shall conclude with a tantalizing conjecture. For birth and death process with absorption probability 1, Theorem 3 identifies a one-parameter family of quasistationary distributions under the condition that $D = \infty$. As mention earlier, this result is contained in the first part of Theorem 3.2 of [18]. The second part of van Doorn's result states that if $D < \infty$, then there is *only one* quasistationary distribution: in the notation of Theorem 3, if $D = \infty$, then m is μ-invariant for P only when $\mu = \lambda$, and $p = (p_j, \ j \in C)$, given by

$$p_j = \frac{\pi_j \phi_j(\lambda)}{\sum_{i \in C} \pi_i \phi_i(\lambda)}, \qquad j \in C,$$

is the only quasistationary distribution.

Since, for birth and death processes, the Reuter FE conditions hold whenever the D series (1.10) diverges, we arrive at the following conjecture, which would extend Theorems 1 and 2:

Conjecture. If the Reuter FE conditions fail, that is, the equations

$$\sum_{i \in C} y_i q_{ij} = \nu y_j, \quad j \in C, \tag{1.14}$$

have a non-trivial, non-negative solution satisfying $\sum_{i \in C} y_i a_i < \infty$, for some (and then all) $\nu > 0$, then μ-invariant probability measures for Q which are finite with respect to a are μ-invariant for P *only* when μ is the decay parameter of C.

7   Acknowledgments

The authors are grateful to Laird Breyer and David Walker for valuable conversations on this work. The work of the first author was carried out during the period he spent as a Ph.D. student in the Department of Mathematics of the University of Queensland.

8   References

[1] W.J. Anderson. *Continuous-time Markov chains: an applications oriented approach.* Springer-Verlag, New York, 1991.

[2] S. Elmes, P.K. Pollett, and D. Walker. Further results on the relationship between μ-invariant measures and quasistationary distributions for continuous-time Markov chains. Submitted for publication, 1995.

[3] P. Good. The limiting behaviour of transient birth and death processes conditioned on survival. *J. Austral. Math. Soc.*, 8:716–722, 1968.

[4] A.G. Hart. *Quasistationary distributions for continuous-time Markov chains.* Ph.D. Thesis, The University of Queensland, 1996.

[5] J.F.C. Kingman. The exponential decay of Markov transition proba-
     bilities. *Proc. London Math. Soc.*, 13:337–358, 1963.

[6] M. Kijima, M.G. Nair, P.K. Pollett, and E. van Doorn. Limiting con-
     ditional distributions for birth-death processes. Submitted for publi-
     cation, 1995.

[7] M.G. Nair and P.K. Pollett. On the relationship between μ-
     invariant measures and quasistationary distributions for continuous-
     time Markov chains. *Adv. Appl. Probab.*, 25:82–102, 1993.

[8] P.K. Pollett. On the equivalence of μ-invariant measures for the min-
     imal process and its q-matrix. *Stochastic Process. Appl.*, 22:203–221,
     1986.

[9] P.K. Pollett. Reversibility, invariance and μ-invariance. *Adv. Appl.
     Probab.*, 20:600–621, 1988.

[10] P.K. Pollett. Analytical and computational methods for modelling the
     long-term behaviour of evanescent random processes. In D.J. Sutton,
     C.E.M. Pearce, and E.A. Cousins, editors, *Decision Sciences: Tools for
     Today, Proceedings of the 12th National Conference of the Australian
     Society for Operations Research*, pages 514–535, Adelaide, 1993. Aus-
     tralian Society for Operations Research.

[11] P.K. Pollett. Modelling the long-term behaviour of evanescent eco-
     logical systems. In M. McAleer, editor, *Proceedings of the Interna-
     tional Congress on Modelling and Simulation*, volume 1, pages 157–
     162, Perth, 1993. Modelling and Simulation Society of Australia.

[12] P.K. Pollett. Recent advances in the theory and application of qua-
     sistationary distributions. In S. Osaki and D.N.P. Murthy, editors,
     *Proceedings of the Australia-Japan Workshop on Stochastic Models in
     Engineering, Technology and Management*, pages 477–486, Singapore,
     1993. World Scientific.

[13] P.K. Pollett. The determination of quasistationary distributions di-
     rectly from the transition rates of an absorbing Markov chain. *Math.
     Computer Modelling*, 1995. To appear.

[14] P.K. Pollett. Modelling the long-term behaviour of evanescent ecolog-
     ical systems. *Ecological Modelling*, 75, 1995. To appear.

[15] P.K. Pollett and D. Vere-Jones. A note on evanescent processes. *Aus-
     tral. J. Statist.*, 34:531–536, 1992.

[16] G.E.H. Reuter. Denumerable Markov processes and the associated
     contraction semigroups on l. *Acta Math.*, 97:1–46, 1957.

[17] R.L. Tweedie. Some ergodic properties of the feller minimal process. *Quart. J. Math. Oxford*, 25:485–495, 1974.

[18] E.A. van Doorn. Quasi-stationary distributions and convergence to quasi-stationarity of birth-death processes. *Adv. Appl. Probab.*, 23:683–700, 1991.

Explosions in Markov Processes
and Submartingale Convergence.

Kersting G. and Klebaner F. C.*
University of Frankfurt and University of Melbourne.

December 21, 1995

Abstract

Conditions for nonexplosions and explosions in Markov pure jump processes are given in terms of the rate of change in the process. We show how these conditions follow from a submartingale convergence theorem. As a corollary, new conditions for nonexplosions in Birth-Death processes in terms of the survival rate are obtained.

1 Introduction

Let $Z = \{Z_t\}$ be a time homogeneous Markov Jump process in continuous time with the state-space $S \subset R$. The process can be described as follows. If the process is in state z then it stays there for an exponential length of time with mean $\lambda^{-1}(z)$ after which it jumps from z. For a general description and construction of such processes see e.g. Breiman 1968, pp. 328-338, Chung 1967, Ch.2. Let T_n be the time of the n-th jump of the process, $T_n = \inf\{t > T_{n-1} : Z_t \neq Z_{t-}\}$, $T_0 = 0$. If $\lambda(z) \to \infty$ in some domain of the state-space S the expected duration of stay in state z, $\lambda^{-1}(z) \to 0$, and time spent in z becomes shorter and shorter. It can happen that the process jumps infinitely many times in a finite time interval, then $\lim_n T_n < \infty$, this phenomenon is called explosion. This terminology comes from the case when S represents integers, and $\lambda(z)$ can go to ∞ only when $z \to \infty$. In this case we can have infinitely many jumps only if the process can reach ∞ in finite time.

When the process does not explode it is called *regular*. The question of explosion is an important one, since when there are no explosions the forwards and backwards equations are satisfied and the solutions are unique.

Martingale convergence and related theorems are basic results in probability. We show how results on explosions and nonexplosions follow as an application of these basic results.

*Research supported by the Australian Research Council

127

To shorten notations denote by Z_n the embedded jump chain, $Z_n = Z(T_n)$. The well-known necessary and sufficient condition for nonexplosion is

$$\sum_{n=1}^{\infty} \frac{1}{\lambda(Z_n)} = \infty \ a.s. \tag{1}$$

see e.g. Chung 1967, p.259-260, Breiman 1968, p. 337. This condition follows from the fact that for a sequence of independent random variables τ_n, which are exponentially distributed with parameter λ_n, $\sum_{n=0}^{\infty} \tau_n < \infty$ a.s. if and only if $\sum_{n=0}^{\infty} 1/\lambda_n < \infty$. Condition (1), however, is hard to check in general, as it involves the embedded chain.

Sufficient conditions for nonexplosion and explosion given below are formulated in terms of the parameters of the process: $\lambda(z)$ and moments of jumps from z. Let $X(z)$ denote the r.v. with the distribution of the jump from z, i.e. the conditional distribution of $Z_1 - z$ given that $Z_0 = z$. Let $m(z) = EX(z)$. The rate of change in the process when the process is in state z is given by $\lambda(z)m(z)$. It is desirable to have conditions in terms of this rate. In Birth-Death processes, for example, $\lambda(z)m(z) = b(z) - d(z)$, the birth rate minus the death rate, which is the survival rate. For a large class of processes, essentially divergence of $\int_0^{\infty} \frac{dz}{\lambda(z)m(z)}$ is necessary and sufficient for nonexplosions.

2   Results

In this section we give results together with a discussion of motivation and conditions.

The state-space S is taken to be a subset of nonnegative reals, and we assume throughout that $\lambda(z)$, $z \geq 0$ is bounded on bounded intervals. By $\mathcal{F}_t$ we denote the natural filtration of the process Z_t and $\mathcal{F}_n = \mathcal{F}_{T_n}$. In what follows C_1, C_2 etc. stands for positive constants, and C is a non specified positive constant.

Theorem 1 *Suppose there exists a positive, increasing function $f(z)$ such that for all $z \geq 0$, $m(z)\lambda(z) \leq f(z)$ and*

$$\int_0^{\infty} \frac{dz}{f(z)} = \infty. \tag{2}$$

Then the process Z does not explode.

Thus according to Theorem 1 condition (2) is sufficient for nonexplosions. The question now is whether it is also necessary. Or in other words, does the convergence of the integral in (2) imply explosions? The following example shows that for some processes condition (2) is necessary and sufficient for nonexplosions.

EXAMPLE 1. Let $\lambda(z)$ and $m(z)$, $z \geq 0$, be positive. Define $0 = s_0 < s_1 < s_2 < \cdots$ by $s_{n+1} = s_n + m(s_n)$. Let Z_t be the jump process which jumps from s_n to s_{n+1} after exponential waiting time with parameter $\lambda(s_n)$, $n = 0, 1, 2 \ldots$.
Then

$$\sum_{n=0}^{\infty} \frac{1}{\lambda(Z_n)} = \sum_{n=0}^{\infty} \frac{1}{\lambda(s_n)} = \sum_{n=0}^{\infty} \frac{s_{n+1} - s_n}{\lambda(s_n)m(s_n)}$$

128

If $m(z)$ is continuous then $s_n \to \infty$. If moreover it satisfies the growth condition $s_{n+1} - s_n < C(s_n - s_{n-1})$ for some constant $C > 0$ (which holds for e.g. $m(z) = z^\alpha$, $\alpha < 1$) then divergence of the above series is equivalent to $\int_0^\infty \frac{dz}{m(z)\lambda(z)} = \infty$, e.g. Knopp 1956.

It is, however, naive to expect that the convergence of the integral $\int_0^\infty \frac{dz}{m(z)\lambda(z)} < \infty$ is sufficient for explosions in general. If $\lambda(z)$ are bounded, then there is no explosion, no matter what growth condition is placed on $m(z)$. Turning to unbounded $\lambda(z)$, the following example shows that we can have $\lambda(z)$ increasing to infinity as fast as we like, but there is no explosion. This is mainly due to the fact that explosions can occur only on the set of divergence of the process to infinity, and the rates alone can not guarantee that this set has a positive probability.

EXAMPLE 2. Let the process be on integers $1, 2, \ldots$ and jump from z to 1 or $4z - 1$ with probability $1/2$. Then $m(z) = z$ and we can take $\lambda(z)$ increasing to infinity as fast as we like, nevertheless $P(Z_n = 1 \text{ i.o.}) = 1$ and (1) holds a.s., implying that there is no explosion.

This example also demonstrates that if the process is allowed to jump all the way down, explosion will not occur. Hence in the following we shall introduce conditions to control the jumps downwards.

We also know that if $\lambda(z)$ are bounded from above, then $1/\lambda(Z_n)$ are bounded away from zero, making the series in (1) diverge, and explosion cannot occur. A similar effect holds if $\lambda(z)$ are bounded from above on a subsequence, and the process Z_n takes values along this subsequence. Thus we are led to the condition $\lambda(z) \to \infty$. However, instead of $\lambda(z) \to \infty$ we assume a much weaker condition that $\lambda(z)$ are bounded away from zero for all z sufficiently large, namely: for $z \geq C_1$

$$(A) \qquad\qquad \inf_{z \geq C_1} \lambda(z) > 0.$$

The next result treats the case of bounded jumps.

Theorem 2 *Suppose (A) and also that for all z sufficiently large, $z \geq C$, $-\infty < -C_2 < X(z) < C_1 < \infty$, and $m(z)\lambda(z) \geq f(z)$, f is monotone nondecreasing, and*

$$\int_0^\infty \frac{dz}{f(z)} < \infty. \tag{3}$$

Let $m^+(z) = EX^+(z)$ and $m^-(z) = EX^-(z)$, and suppose that

$$(B) \qquad\qquad \liminf_{z \to \infty} \frac{1}{m(z)} \left(\frac{f(z)}{f(z+C_1)} m^+(z) - \frac{f(z)}{f(z-C_2)} m^-(z) \right) > 0.$$

Then the process Z explodes on the set $\{Z_n \to \infty\}$.

An important corollary is the case of Birth-Death Processes.

129

Corollary 1 *Let Z_t be a Birth-Death process with birth and death rates $b(z)$ and $d(z)$ respectively. If $b(z) - d(z)$ is non decreasing and*

$$\sum_{z=0}^{\infty} \frac{1}{b(z) - d(z)} = \infty,$$

then the process does not explode. If

$$\liminf_{z \to \infty} \left(\frac{b(z)}{b(z+1) - d(z+1)} - \frac{d(z)}{b(z-1) - d(z-1)} \right) > 0,$$

and

$$\sum_{z=0}^{\infty} \frac{1}{b(z) - d(z)} < \infty$$

then the process explodes on the set of divergence to infinity.

Note that conditions are given in terms of the survival rate $\lambda(z)m(z) = b(z) - d(z)$.

Proof of the Corollary. It is easy to see that $\lambda(z) = b(z)+d(z)$, $m^+(z) = b(z)/(b(z)+d(z))$, $m^-(z) = d(z)/(b(z) + d(z))$, $m(z) = (b(z) - d(z))/(b(z) + d(z))$, $\lambda(z)m(z) = b(z) - d(z)$. Nonexplosion follows from Theorem 1; explosion from Theorem 2. $\square$

In the following result the jumps are unbounded and such that the mean jump $m(z)$ increases to infinity, but the contribution of the negative part of the jump becomes negligible in comparison to the positive part.

Processes with jumps bounded from below occur frequently in applications as population models. In such models a parent particle is replaced by its offspring and the jumps are bounded by -1.

Theorem 3 *Suppose (A) and also that there exists a positive, increasing function $f(z)$ such that $m(z)\lambda(z) \geq f(z)$ for all z sufficiently large, $z \geq C$, and (3) holds. Suppose in addition assumptions (a) - (d):*

(a) $X(z) \geq -h(z)$, for all z, $h(z) \geq 0$ and satisfies $\lim_{z \to \infty} z - h(z) = +\infty$.

(b) There exist $0 < \delta < 1$ and a positive function $k(z)$ such that for all z large enough
$$E(X^+(z)I(X^+(z) < k(z))) > \delta m(z).$$

(c) $\liminf_{z \to \infty} \frac{f(z)}{f(z+k(z))} > 0$, $\limsup_{z \to \infty} \frac{f(z)}{f(z-h(z))} < \infty$,

(d) $\lim_{z \to \infty} \frac{EX^-(z)}{EX^+(z)} = 0$.

Then the process Z explodes on the set $\{Z_n \to \infty\}$.

If the jumps are bounded away from zero then the integral condition on growth of $\lambda(z)$ is sufficient for explosions.

130

Theorem 4 *Suppose (A) and also that for all z sufficiently large, $z \geq C$, $X(z) > C_3 > 0$, and $\lambda(z)$ is monotone nondecreasing, and*

$$\int_0^\infty \frac{dz}{\lambda(z)} < \infty. \qquad (4)$$

Then the process Z explodes on the set $\{Z_n \to \infty\}$.

The next two results are from Kersting and Klebaner (1995). We give them here for completeness. The proofs given here are particularly suited for these results. The conditions involve positive and negative moments of the jumps.

Denote for $\alpha > 0$,

$$n_\alpha(z) = E((\frac{Z_1}{z})^\alpha - 1 | Z_0 = z),$$

$$n_{-\alpha}(z) = E(1 - (\frac{z}{Z_1})^\alpha | Z_0 = z).$$

Theorem 5 *Suppose there exists a positive function $f(z)$ such that for some $\alpha > 0$ $f(z)z^{\alpha-1}$ is increasing and*

$$zn_\alpha(z)\lambda(z) \leq f(z), \quad \text{and} \quad \int_0^\infty \frac{dz}{f(z)} = \infty. \qquad (5)$$

Then the process Z does not explode.

Note that Theorem 1 is a special case of Theorem 5 with $\alpha = 1$.

Theorem 6 *Suppose (A) and also that there exists a positive function $f(z)$ such that for some $\alpha > 0$, $f(z)z^{-\alpha-1}$ is decreasing and for all z sufficiently large*

$$zn_{-\alpha}(z)\lambda(z) \geq f(z), \quad \text{and} \quad \int_0^\infty \frac{dz}{f(z)} < \infty.$$

Then the process Z explodes on the set $\{Z_n \to \infty\}$.

REMARK.
For processes with state-space in R conditions for nonexplosions can be obtained by considering the variation process $|Z|_t$ of Z_t, defined as

$$|Z|_t = |Z_0| + \sum_{s \leq t} |\Delta Z_s|.$$

Then $|Z_t| \leq |Z|_t$, and $|Z_t|$ and $|Z|_t$ jump at the same times. For example, $|m|(z) = E(|X(z)|)$ should replace $m(z)$ in the conditions of Theorems 1 and 2. Similar generalizations are available for the rest of theorems with obvious modifications in conditions by replacing z with $|z|$.

131

3   PROOFS.

We derive the results from the following version of the submartingale convergence theorem, given in a textbook on Probability by Shiryaev p.485, see also convergence of special semimartingales in Liptser and Shiryaev (1989), p 137. In this theorem the Doob's decomposition of a submartingale is used. If S is a submartingale then its Doob's decomposition is the representation $S_n = M_n + A_n$ where M_n is a martingale and A_n is an adapted, predictable, increasing process. Such decomposition is unique. Since A is increasing, $\lim_n A_n = A_\infty$ exists. In what follows by the event $\{S_n$ converges$\}$ we understand the event $\{-\infty < \lim_{n\to\infty} S_n < \infty\}$.

Theorem 7 (Convergence sets of a submartingale.) *Let S be a submartingale and $S_n = M_n + A_n$ be its Doob decomposition.*

(a) If $S \geq 0$, or $\sup_n ES_n^- < \infty$, then $\{A_\infty < \infty\} \subseteq \{S_n$ converges$\} \subseteq \{\sup S_n < \infty\}$.

(b) If $S \in C^+$, then $\{S_n$ converges$\} = \{\sup S_n < \infty\} \subseteq \{A_\infty < \infty\}$.

By definition, S belongs to class C^+ if $E(\Delta S_{\tau_a})I(\tau_a < \infty) < \infty$ for all $a > 0$. Here $\tau_a = \inf\{n : S_n > a\}$, is the first time S goes above level a, and $\Delta S_n = S_n - S_{n-1}$.

A sufficient condition to belong to C^+ is

$$S \in C^+ \text{ if } |\Delta S_n| \leq C < \infty.$$

We give two Lemmas first.

Lemma 1 *Let $U_n = F(Z_n)$, where F is a positive increasing function with $F(\infty) = \infty$. Suppose that for some $C > 0$*

$$E(U_{n+1}|\mathcal{F}_n) \leq U_n + C/\lambda(Z_n). \tag{7}$$

Then $\sum_{k=0}^\infty 1/\lambda(Z_n) = \infty$ a.s.

Proof of Lemma 1. Adding up and rearranging the terms in (7) we obtain

$$U_{n+1} \leq M_{n+1} + \sum_{i=1}^n C/\lambda(Z_i) = M_{n+1} + A_{n+1},$$

where $M_n = \sum_{i=1}^n U_i - E(U_i|\mathcal{F}_{i-1})$ is a martingale and $A_n = \sum_{i=1}^{n-1} C/\lambda(Z_i)$ is a predictable and increasing process. Consequently, $M_n + A_n$ is a submartingale, moreover it is positive since $U_n > 0$. Thus by the first part of Theorem 7 it converges on the set $\{A_\infty < \infty\}$. Since $\lambda(z)$ is bounded on finite intervals, $\{A_\infty < \infty\} \subseteq \{Z_n \to \infty\}$, because on the set $\{Z_n \not\to \infty\}$ there is a subsequence on which $1/\lambda(Z_i) > C > 0$ making $A_\infty = \infty$. By the assumption of the theorem $F(\infty) = \infty$, hence $U_n \to \infty$ on $\{Z_n \to \infty\}$. This implies that $P(\{A_\infty < \infty\}) = 0$, or $\sum_{i=1}^\infty 1/\lambda(Z_i) = \infty$ a.s. $\square$

132

Lemma 2 *Let $U_n = F(Z_n)$, where F is a positive increasing function with $F(\infty) < \infty$. Suppose that condition (A) holds, and for all sufficiently large Z_n, $Z_n > C_1$ the following bound holds: for some $C > 0$*

$$E(U_{n+1}|\mathcal{F}_n) \geq U_n + C/\lambda(Z_n). \tag{8}$$

Then $\sum_{k=0}^{\infty} 1/\lambda(Z_n)$ converges on the set $\{Z_n \to \infty\}$.

Proof of Lemma 2. It follows from (8) that

$$E(U_{n+1}|\mathcal{F}_n) \geq U_n + C/\lambda(Z_n)I(Z_n > C_1) + I(Z_n \leq C_1)(E(U_{n+1}|\mathcal{F}_n) - U_n). \tag{9}$$

Hence, it follows from (9) that W_n defined by (10) is a submartingale,

$$W_{n+1} = U_{n+1} - \sum_{k=0}^{n}[C/\lambda(Z_k)I(Z_k > C_1) + I(Z_k \leq C_1)(E(U_{k+1}|\mathcal{F}_k) - U_k)]. \tag{10}$$

Now, by using $U_n \leq F(\infty) < \infty$ and Assumption (A)

$$|\Delta W_n| \leq |\Delta U_n| + I(Z_{n-1} > C_1)C/\lambda(Z_{n-1}) + F(\infty) < 2F(\infty) + C/\inf_{z > C_1}\lambda(z) < \infty.$$

Hence W is in C^+. Thus by Theorem 7 (b),

$$\{W_n \text{ converges}\} = \{\sup W_n < \infty\}. \tag{11}$$

It is clear from (10) that

$$W_n \leq U_n + \sum_{k=0}^{n-1} I(Z_k \leq C_1)U_k, \tag{12}$$

and since $U_n \leq F(\infty) < \infty$, from (12) and (11) we have

$$\{Z_n \to \infty\} \subseteq \{\sup W_n < \infty\} = \{W_n \text{ converges}\}.$$

Hence from (10), $U_n - \sum_{k=0}^{n-1} C/\lambda(Z_k)$ converges on $\{Z_n \to \infty\}$. But $U_n \to F(\infty) < \infty$ on this set, hence $\sum_{k=0}^{\infty} 1/\lambda(Z_k)$ converges on this set. $\square$

In the proofs of the theorems we define

$$F(z) = \int_0^z \frac{dx}{f(x)},$$

where f is the function appearing in the conditions.

Proof of Theorem 5. It is enough to show how (7) follows from conditions of the theorem. The following inequality is obtained by monotonicity of $f(z)z^{\alpha-1}$. For $y > x$

$$\frac{1}{f(y)} \leq \frac{1}{f(x)}(\frac{y}{x})^{\alpha-1}.$$

133

Using this inequality for positive and negative x separately we obtain for *any* $z > 0$ and $x > -z$,

$$F(z + x) - F(z) = \int_z^{z+x} \frac{1}{f(y)} dy \le \frac{z}{\alpha f(z)} ((\frac{z+x}{z})^\alpha - 1).$$

Letting now $z = Z_n$ and $x = Z_{n+1} - Z_n$ in the above inequality and taking conditional expectations it is easily checked that (7) follows. (In Theorem 1 it is easier to see (7) from concavity of $F(z)$.) The proof is completed by applying Lemma 1. $\square$

To prove Theorems 2, 3 and 6 on explosions it is enough to show how (8) follows from the conditions of the theorems and then apply Lemma 2. The proof of Theorem 6 is the shortest, we give it first.

Proof of Theorem 6. Since $f(z)z^{-\alpha-1}$ is decreasing we have the inequality for $y > x$

$$\frac{1}{f(y)} \ge \frac{1}{f(x)} (\frac{y}{x})^{-1-\alpha},$$

from which we obtain for *any* $z > 0$ and $x > -z$,

$$F(z + x) - F(z) \ge \frac{1}{f(z)} z^{1+\alpha} \int_z^{z+x} y^{-1-\alpha} dy = \frac{z}{\alpha f(z)} (1 - (\frac{z}{z+x})^\alpha).$$

Letting $z = Z_n$ and $x = Z_{n+1} - Z_n$ in the above inequality and taking conditional expectations it is easily checked that (8) follows from the conditions of the theorem. $\square$

Proof of Theorem 2. The bound in (8) is obtained from the inequality: for any a, b

$$\int_a^{a+b} \frac{dx}{f(x)} > \frac{b}{f(a+b)}, \tag{13}$$

which holds since f is increasing. So that

$$E(\Delta U_{n+1} | Z_n = z) \ge E \frac{X(z)}{f(z + X(z))} = E \frac{X^+(z)}{f(z + X^+(z))} - E \frac{X^-(z)}{f(z - X^-(z))}. \tag{14}$$

Now, since $-C_2 < X(z) < C_1$

$$E \frac{X^+(z)}{f(z + X^+(z))} \ge \frac{m^+(z)}{f(z + C_1)},$$

and

$$E \frac{X^-(z)}{f(z - X^-(z))} < \frac{m^-(z)}{f(z - C_2)}.$$

Hence

$$E \frac{X(z)}{f(z + X(z))} \ge \frac{m^+(z)}{f(z + C_1)} - \frac{m^-(z)}{f(z - C_2)}. \tag{15}$$

134

It now follows from the Condition (B) of the theorem and (15), (14) that

$$E(\Delta U_{n+1}|Z_n = z) \geq \frac{C}{\lambda(z)},$$

provided z is large enough. (8) now follows. □

Proof of Theorem 3. We have the bounds (13) and (14). The positive term is bounded as follows:

$$E \frac{X^+(z)}{f(z + X^+(z))} \geq E\left(\frac{X^+(z)I(X^+(z) < k(z))}{f(z + X^+(z))}\right).$$

By monotonicity of f and by the Conditions (b) and (c) of the theorem we obtain that there is a positive constant C such that

$$E \frac{X^+(z)}{f(z + X^+(z))} \geq \frac{EX^+(z)I(X^+(z) < k(z))}{f(z + k(z))} \geq \frac{Cm(z)}{f(z)} \geq \frac{C}{\lambda(z)}, \tag{16}$$

if only z is large enough. The negative term is bounded as follows. By monotonicity of f and condition (a)

$$E \frac{X^-(z)}{f(z - X^-(z))} < \frac{EX^-(z)}{f(z - h(z))}.$$

By Condition (c) we obtain

$$\frac{EX^-(z)}{f(z - h(z))} < C \frac{EX^-(z)}{f(z)} < \frac{C}{\lambda(z)} \frac{EX^-(z)}{m(z)},$$

where the last inequality holds for all z large enough. It follows from Condition (d) $\frac{EX^-(z)}{m(z)} = o(1)$, $z \to \infty$, hence

$$\frac{EX^-(z)}{f(z - h(z))} = o\left(\frac{1}{\lambda(z)}\right), \; z \to \infty. \tag{17}$$

From (16) and (17) it follows that for some C

$$E(\Delta U_{n+1}|Z_n = z) \geq \frac{C}{\lambda(z)},$$

for all z large enough, which is the desired bound in (8). □

Proof of Theorem 4. From (13) we obtain, with f replaced by λ,

$$E(\Delta U_{n+1}|\mathcal{F}_n) \geq E\left(\frac{X(Z_n)}{\lambda(Z_{n+1})}|\mathcal{F}_n\right) \geq C_3 E\left(\frac{1}{\lambda(Z_{n+1})}|\mathcal{F}_n\right), \tag{18}$$

for all Z_n sufficiently large. Using Lemma 2 with $E\left(\frac{1}{\lambda(Z_{n+1})}|\mathcal{F}_n\right)$ instead of $\frac{1}{\lambda(Z_{n+1})}$, we obtain that

$$\sum_{n=0}^{\infty} E\left(\frac{1}{\lambda(Z_{n+1})}|\mathcal{F}_n\right) < \infty$$

135

on the set $\{Z_n \to \infty\}$. Now

$$\sum_{k=0}^{n} \frac{1}{\lambda(Z_{k+1})} = \sum_{k=0}^{n} \frac{1}{\lambda(Z_{k+1})} - E(\frac{1}{\lambda(Z_{k+1})}|\mathcal{F}_k) + \sum_{k=0}^{n} E(\frac{1}{\lambda(Z_{k+1})}|\mathcal{F}_k)$$

is a positive submartingale in C^+, hence by Theorem 7 the following sets are equal

$$\{\sum_{k=0}^{n} \frac{1}{\lambda(Z_{k+1})} < \infty\} = \{\sum_{k=0}^{n} E(\frac{1}{\lambda(Z_{k+1})}|\mathcal{F}_k) < \infty\},$$

and the proof is complete. $\square$

References

[1] Breiman L.(1968). *Probability*. Addison-Wesley, Reading, Ma.

[2] Chung, K.L. (1967). *Markov Chains with Stationary Transition Probabilities*. Springer, New York.

[3] Kersting, G. and Klebaner, F.C. (1995) Sharp conditions for nonexplosions and explosions in Markov jump processes. *Ann. Prob.* 23, 268-272.

[4] Knopp K. (1956). *Infinite sequences and series*. Dover, New York.

[5] Liptser, R. and Shiryaev, A.N. (1989). *Theory of Martingales*. Kluwer, Dordrecht.

[6] Shiryaev, A.N. (1984). *Probability*. Springer, New York.

Probability Bounds for Product Poisson Processes

Joong Sung Kwon
Ronald Pyke

ABSTRACT Consider processes formed as products of independent Poisson and symmetrized Poisson processes. This paper provides exponential bounds for the tail probabilities of statistics representable as integrals of bounded functions with respect to such product processes. In the derivations, tail probability bounds are also obtained for product empirical measures. Such processes arise in tests of independence. A generalization of the Hanson-Wright inequality for quadratic forms is used in the symmetric case. The paper also provides some reasonably tractable approximations to the more general bounds that are derived first.

Prologue

The second author met Joe and Ruth Gani for the first time when they arrived at a New York airport with their son Jonathan to spend the calendar year 1959 at Columbia University. Joe was a tremendously stimulating and energetic addition to the group of active young people that were there at the time. We shared an office during that visit, and two joint papers grew out of these early discussions. We have kept close contacts over the years during which times Joe has been a constant friend and stimulating and supportive colleague. We have served together in particular on many committees, often involving (not surprisingly) publications. To me, Joe is unique because of his "can do" attitude. He has always been willing to bring his full energies and insights to bear upon important problems, societal and mathematical, to work toward solutions and to see that good ideas, very often his own, are in fact carried through for the benefit of the discipline and its scientific community.

At the Athens conference, my oral contribution was entitled, "Brownian Motion: A plied and applied process". Its purpose was to overview early developments in the study of Brownian motion in order to emphasize the close ties between applications and theory that it shows. Although the modelling by Einstein(1905) of the movement of particles required in the molecular theory of heat is best known, my emphasis was upon the first use of Brownian motion in the modelling by Bachelier(1900) of the movement of bond prices in the French Bourse. From these independent and very different beginnings, Brownian motion has become one of the most significant components of probability, both pure and applied. Certainly the study of the fine and global structures of Brownian and related processes is a significant activity plied by many probabilists today, at the same time when both the scope and the complexity of their applications continue their dramatic increase.

This emphasis upon the inseparability of theory and application has always been central to Joe's outlook on probability. For example, I recalled a pair of talks given by Joe and I some 20 years earlier at a Conference on Future Directions for Mathematical Statistics organized by Sudhish Ghurye and held in 1974 in Edmonton, Canada. Joe spoke on "Theory and practice in applied probability", (cf. Gani(1975)) while I spoke on "Applied probability; an editor's dilemma",(cf. Pyke(1975)). Both talks focused on concerns about preserving the inherent unity of theory and applications in the presence of the growing tendency to separate both probability and probabilists into a pure and applied dichotomy.

The dichotomy in and of itself need not be viewed as bad, for it often aids communication about emphases and coverage (as in the names of societies, journals and departments.) The unhealthy aspect of the dichotomy arises when the descriptive adjectives "pure" and "applied" become subjective labels of quality and importance, and when the separating walls become increasingly inhibiting to the essential intellectual interchanges by which both camps can so greatly benefit.

I view the 1900 paper by Bachelier as one of the finest examples of a major contribution blending both theory and application. Not only does the paper appear very modern in its approach and flavor (one may see in it the Chapman-Kolmogorov equations, random walks, martingales, stopping times, the heat equation and first crossing probabilities among other topics) it is exemplary in the attention (over 40% of this long paper) given by the author to comparisons between his proposed Brownian model and actual bond prices during the five years, 1894–98. More about this and the other parts of the review of Brownian motion will be included in Pyke (1996).

In the following written portion of our contribution to Joe, we pick up from the product Brownian processes mentioned briefly at the end of the Athen's talk, and look at the products of their Poisson and empirical precursors.

1   Introduction

The purpose of this paper is to derive exponential bounds for the tail probabilities of product processes that have Poisson and related processes for factors. We define constructively the factor processes as follows. Let

$(S, \mathcal{S}, Q)$ be a finite measure space. Let $\mathcal{F}$ be a set of S - measurable functions on S. A process $Y = \{Y(f) : f \in \mathcal{F}\}$ is said to be a *Poisson process with parameter Q* if it has independent increments in the sense that $Y(A_1), Y(A_2), \ldots, Y(A_k)$ are independent whenever $A_1, A_2, \ldots, A_k$ are disjoint subsets (equivalently, indicator functions) of S, and the marginal distributions are Poisson with parameters $Q(A_i)$. Write $\lambda = Q(S)$. We can represent a version of such a Poisson process as follows : Write $F(A) = Q(A)/Q(S)$. Let $\{U_i\}$ be a sequence of independent identically F-distributed S-valued random variables (the locations of points) and $N = Y(S)$ (the number of points) be an independent Poisson r.v. with parameter $\lambda = Q(S)$. Then, for a function f on S set

$$Y(f) = \sum_{i=1}^{N} f(U_i), \qquad (1.1)$$

interpreting the empty sum as zero. When we replace (1.1) by

$$\overline{Y}(f) = \sum_{i=1}^{N} \varepsilon_i f(U_i) \qquad (1.2)$$

where $\{\varepsilon_i\}$ is a sequence of independent symmetric Bernoulli (or Rademacher) random variables $(P(\varepsilon_i = 1) = 1/2 = P(\varepsilon_i = -1))$ that is independent of $\{U_i\}$ and N, we refer to the $\overline{Y}$ - process as a *symmetric Poisson process with parameter Q*. Similarly, if we replace $\{\varepsilon_i\}$ by $\{X_i\}$, a general sequence of independent identically distributed random variables, independent of all other r.v.'s, we obtain a *compound Poisson process* with marginal distributions being those of

$$\hat{Y}(f) = \sum_{i=1}^{N} X_i f(U_i). \qquad (1.3)$$

We begin by stating exponential bounds for the one-dimensional case. Let $\mathcal{P}(\lambda)$ denote the ordinary Poisson distribution with parameter λ. If N is $\mathcal{P}(\lambda)$ we refer to the law of $\overline{N} = \sum_{i=1}^{N} \varepsilon_i \equiv \overline{Y}(1)$ as *symmetrized $\mathcal{P}(\lambda)$*. Alternatively, $\overline{N}$ may be represented by $\overline{N} = N_1 - N_2$ with N_1 and N_2 being independent $\mathcal{P}(\lambda/2)$ r.v.'s.

Theorem 1.1 (C.f. Pyke(1983) and Bass and Pyke(1984)) *Let N be $\mathcal{P}(\lambda)$. Then*

$$
\begin{aligned}
P(N > \eta) \quad &< \quad \exp(-\lambda - \eta[\ln(\eta/\lambda) - 1]) \quad && \textit{if} \quad \eta \geq \lambda \\
&< \quad \exp(-\lambda - \eta) \quad && \textit{if} \quad \eta \geq e^2 \lambda. \qquad (1.4)
\end{aligned}
$$

The form of the second bound is chosen for convenience in application. Since $g(u) := 1 + u(lnu - 1)$ is (convex) increasing on $(1, \infty)$, it follows that $g(u) > 1 + \varepsilon u$ whenever $u > e^{1+\varepsilon}$ for any $\varepsilon > 0$. Thus one also has

$$P(N > \eta) \leq \exp(-\lambda - \varepsilon\eta) \quad \text{if} \quad \eta \geq e^{1+\varepsilon}\lambda. \tag{1.5}$$

Corollary 1.1 *If N is $\mathcal{P}(\lambda)$ and $v > 0$, then*

$$P\left(\frac{N - \lambda}{\sqrt{\lambda}} > v\right) \leq \exp\left(-\frac{v^2}{2}\left(1 - \frac{v}{\sqrt{\lambda}}\right)\right). \tag{1.6}$$

Proof. Use the fact that $ln(1 + x) \geq x - x^2/2$ if $x > 0$, and substitute $\eta = \lambda + v\sqrt{\lambda}$ in Theorem 1.1. $\square$

Notice that the bound in (1.6) is useful only when $v < \sqrt{\lambda}$; i.e. $\eta = \lambda + v\sqrt{\lambda} < 2\lambda$. By using higher degree polynomials as lower bounds for $ln(1 + x)$, one can expand the range of v, but the improvement is small. For example, the fourth degree bound, $ln(1+x) \geq x - x^2/2 + x^3/3 - x^4/4$, allows one to improve the exponent in (1.6) to

$$-\frac{v^2}{2}\left(1 - \frac{1}{3}\left(\frac{v}{\sqrt{\lambda}}\right) + \frac{1}{6}\left(\frac{v}{\sqrt{\lambda}}\right)^2 - \frac{1}{2}\left(\frac{v}{\sqrt{\lambda}}\right)^3\right)$$

which remains negative at least for $v \leq \frac{7}{6}\sqrt{\lambda}$, a modest improvement.

Though the emphasis of this paper is upon upper tails, bounds for lower tails may also be obtained. For example, one may show analogously to Theorem 1.1 that when N is $\mathcal{P}(\lambda)$,

$$P(N \leq \eta) \leq \exp(-\lambda - \eta(\ln(\eta/\lambda) - 1) \quad \text{if} \quad 0 < \eta < \lambda.$$

It should be remarked that some lower tail bounds for the distributions of $Y(f)$ and for the product $Z(f)$ defined in (1.9) below are derived in Kallenberg and Szulga(1989).

Theorem 1.2 (Bass and Pyke(1987)) *If $\overline{N}$ is symmetrized $\mathcal{P}(\lambda)$, then, for $\eta > 0$,*

$$P(\overline{N} > \eta) \leq \exp\left(-\frac{\eta^2}{2\lambda + 2\eta/3}\right). \tag{1.7}$$

Furthermore, if in the representation (1.9), $X_i f(U_i)$ has mean 0 and is bounded by b in absolute value, then

$$P\left(\widehat{Y}(f) > \eta\right) \leq \exp\left(-\frac{\eta^2}{2\lambda\sigma^2 + 2\eta b/3}\right) \tag{1.8}$$

where $\sigma^2 = E[X_i f(U_i)]^2$.

Proof. The second bound is an immediate consequence of Lemma 2.3 of Bass and Pyke (1984). The first bound is then a special case with $\sigma^2 = 1 = b$. The first bound can also be derived as a corollary of Proposition 3.3 and Lemma 3.1 of Bass and Pyke (1987) because $\overline{N}$ is subpoissonian with parameters $(\lambda/2, 1)$. □

When $\mathcal{F}$ is restricted to the indicator functions, $f = 1_B$ for $B \in \mathcal{S}$, (1.1), (1.2) and (1.3) define set functions, Y, $\overline{Y}$ and $\hat{Y}$, that are σ-finite measures with a finite number of atoms almost surely. Moreover, these well defined atomic random signed measures are independently scattered, which is to say, that they are countably additive set functions which assign independent values to disjoint sets. However, the product random measures studied in this paper, which have Y, $\overline{Y}$ and $\hat{Y}$ as factors, will *not* be independently scattered.

In general, suppose that for each $i = 1, 2$, $(\mathbf{S}_i, \mathcal{S}_i, Q_i)$ is a finite measure space, and Z_i is a random σ-finite signed measure on $(\mathbf{S}_i, \mathcal{S}_i)$. Then the product process, $Z := Z_1 \times Z_2$, is well defined on the product space $(\mathbf{S}_1 \times \mathbf{S}_2, \sigma(\mathcal{S}_1 \times \mathcal{S}_2))$ by Fubini's theorem. Our interest here is to obtain probability bounds for such product processes when the factors are Poisson. In this case the atomic structure of Poisson processes enables one to write, for example,

$$Z(f) := (Y_1 \times Y_2)(f) = \sum_{i=1}^{N_1} \sum_{j=1}^{N_2} f(U_i, V_j), \qquad (1.\,9)$$

where Y_1 and Y_2 are Poisson processes with respective parameters Q_1 and Q_2, f on $\mathbf{S}_1 \times \mathbf{S}_2$ is $\sigma(\mathcal{S}_1 \times \mathcal{S}_2)$-measurable and $N_i = Y_i(\mathbf{S}_i)$. Products of symmetrized and compound Poisson processes are defined similarly. (The constructive definitions used here are expeditious for this paper since we focus entirely on probability bounds. More extensive studies into the definitions of product processes can be found in Perez-Abreu(1985, Section 3.2) for the L_2 - approach, and in Kallenberg and Szulga(1989) for the direct sample path approach.) These product processes arise naturally in the context of testing for independence between Poisson processes. They are also useful as building blocks in the study of products of set-indexed infinitely divisible processes just as (1.1), (1.2) and (1.3) were in the construction of the processes themselves; see Adler and Feigin(1984) and Bass and Pyke(1984).

2   Bounds for products of independent Poisson processes

Note first that for any f with finite mean $\delta_f := Ef(U, V)$, we may write $Z(f) = Z(f - \delta_f) + N_1 N_2 \delta_f$. Thus, for any $\eta, x > 0$,

$$
\begin{aligned}
P(Z(f) > \eta) &= P(Z(f - \delta_f) > \eta - N_1 N_2 \delta_f, N_1 N_2 > x) \\
&\quad + P(Z(f - \delta_f) > \eta - N_1 N_2 \delta_f, N_1 N_2 \le x) \\
&\le P(N_1 N_2 > x) + P\left(\frac{Z(f - \delta_f)}{N_1 N_2} > \frac{\eta}{x} - \delta_f \right) \quad (2.\,1)
\end{aligned}
$$

where we interpret $Z/N_1 N_2$ to be zero when $N_1 N_2 = 0$. Therefore, tail bounds for general f's will follow once bounds are obtained for the two special cases implicit in the right hand side of (2.1), namely $f \equiv 1$ and f's with $\delta_f = 0$. We begin with the first case by deriving the following bound for the simple product of Poisson r.v.'s.

Proposition 2.1 *If N_1 and N_2 are independent $\mathcal{P}(\lambda)$ and $\mathcal{P}(\mu)$ r.v.'s, respectively, then for all $\eta > 0$,*

$$
\begin{aligned}
P(N_1 N_2 > \eta) &< \exp(-\lambda - \mu - 2\sqrt{\eta}(\log(2\sqrt{\eta}/(\lambda + \mu)) - 1)) \\
&\qquad\qquad \textit{if }\ \eta > (\lambda + \mu)^2/4,
\end{aligned}
$$

$$
< \exp(-\lambda - \mu - 2\sqrt{\eta}) \qquad \textit{if }\ \eta > e^4(\lambda + \mu)^2/4. \quad (2.\,2)
$$

Proof. Note that
$$
N_1 N_2 \le (N_1 + N_2)^2/4. \quad (2.\,3)
$$

Hence,
$$
P(N_1 N_2 > \eta) \le P(N_1 + N_2 > 2\sqrt{\eta})
$$
and the result follows from Theorem 1.1 since $N_1 + N_2$ is $\mathcal{P}(\lambda + \mu)$. □

Another way to obtain a bound for a product is to observe simply that for any $a > 0$

$$
P(N_1 N_2 > \eta) \le P(N_1 > \sqrt{\eta}/a) + P(N_2 > a\sqrt{\eta}).
$$

An application of Theorem 1.1 to each of these terms leads to the bound

$$
\begin{aligned}
P(N_1 N_2 > \eta) &< \exp\left(-\lambda - \sqrt{\eta}/a\right) + \exp(-\mu - a\sqrt{\eta}) \\
&\qquad \textit{if }\ \eta > \max\{a^2 e^4 \lambda^2, e^4 \mu^2/a^2\}.
\end{aligned}
$$

By choosing for a the value that equates the two exponents on the right, namely $a = \{(\lambda - \mu)/2 + \sqrt{((\lambda - \mu)/2)^2 + \eta}\}/\sqrt{\eta}$, one obtains

$$P(N_1 N_2 > \eta) \quad < \quad 2\exp\left(-\frac{\lambda+\mu}{2} - \sqrt{(\lambda-\mu)^2/4 + \eta}\right)$$

$$\textit{if} \quad \eta > (e^4 + e^2)(\mu \vee \lambda)^2 - e^2\mu\lambda \ . \quad (2.\ 4)$$

A third approach is to observe that

$$P(N_1 N_2 > \eta) \le P(N_1 + N_2 > 2\sqrt{\eta}) \le P(N_1 > x) + P(N_2 > y)$$

where x and y are positive real numbers satisfying $x + y = 2\sqrt{\eta}$. In this case, Theorem 1.1 yields

$$
\begin{aligned}
P(N_1 N_2 > \eta) \quad &< \quad \exp(-\lambda - x) + \exp(-\mu - y) \\
&\qquad \textit{if} \quad x + y = 2\sqrt{\eta}, \quad x \ge e^2\lambda \quad \textit{and} \quad y \ge e^2\mu \\
&< \quad 2\exp\{-\frac{1}{2}(\lambda + \mu + 2\sqrt{\eta})\} \qquad\qquad (2.\ 5) \\
&\qquad \textit{if} \quad \eta > \max\{[(e^2 + \frac{1}{2})\lambda - \frac{\mu}{2}]^2, \ [(e^2 + \frac{1}{2})\mu - \frac{\lambda}{2}]^2\}
\end{aligned}
$$

where the second bound is again obtained by choosing x and y to equate the exponents.

Remark   The bound given in Proposition 2.1 is the strongest of the three discussed above. To see this, assume without loss of generality that $\mu \ge \lambda$. Then the negative exponents for the bounds of (2.1), (2.3) and (2.4) are, respectively, $\lambda + \mu + 2\sqrt{\eta}$, $\frac{\lambda+\mu}{2} + \sqrt{\left(\frac{\lambda-\mu}{2}\right)^2 + \eta} - ln2$ and $\frac{\lambda+\mu}{2} + \sqrt{\eta} - ln2$. It is easy to show that the first is always larger than the second, which is clearly larger than the third. Moreover, the restrictions placed upon η are, respectively, $\sqrt{\eta} > e^2(\lambda+\mu)/2$, $\eta > (e^4+e^2)\mu^2 - e^2\mu\lambda$ and $\sqrt{\eta} > e^2\mu + \frac{\mu-\lambda}{2}$. These are the same if $\mu = \lambda$, but when $\mu > \lambda$, the second is clearly less restrictive than the third, and straightforward algebra shows that the first is least restrictive of all. Thus the bound of Proposition 2.1, which is derived using (2.2), is the best of the three. However, it should be noted that only the first bound requires the assumption of independence. Thus the second bound would be the best for arbitrary Poisson products.

We now return to the original problem of obtaining bounds for general integrals with respect to a product Poisson process.

Theorem 2.1. *For $i = 1, 2$, let Y_i be independent Poisson processes with parameters Q_i on measurable spaces (S_i, S_i). Set $N_i = Y_i(S_i)$, $Z = Y_1 \times Y_2$,*

$Q = Q_1 \times Q_2$, $\lambda = Q_1(S)$, $\mu = Q_2(S_2)$ and $P = Q/\lambda\mu$. For any $\sigma(S_1 \times S_2)$-measurable function f with $|f| \le 1$, set

$$\delta_f = \int f dP \quad and \quad \sigma_f^2 = \int f^2 dP - \delta_f^2.$$

Then if $\delta_f > 0$,

$$P(Z(f) > \eta) \le P(N_1 N_2 > x) + 2\exp\{-\frac{1}{2}(\lambda \wedge \mu)C(\eta/x - \delta_f)\} \quad (2.6)$$

for any $\eta > 0$ and $x > 0$, where

$$C(u) = C(u; f) := \begin{cases} \frac{u^2}{2(\sigma_f^2 + u/3)} & for \quad 0 \le u < 1 - \delta_f, \\ +\infty & for \quad u \ge 1 - \delta_f. \end{cases} \quad (2.7)$$

Proof. In view of (2.1), we need only verify the bound for the normalized integral $Z(f)/N_1 N_2$ when f is bounded with $\delta_f = 0$. However, in order to keep the role of the bound for f separate from that of the centering, we will not set $\delta_f = 0$. Note first that when $mn > 0$,

$$P\left[\frac{Z(f - \delta_f)}{N_1 N_2} > u | N_1 = m, N_2 = n\right] = P[T_{mn}(f) > u + \delta_f] \quad (2.8)$$

where

$$T_{mn} = T_{mn}(f) = \frac{1}{mn}\sum_{i=1}^{m}\sum_{j=1}^{n} f(U_i, V_j) = P_m \times Q_n(f) \quad (2.9)$$

and P_m (Q_n) is the empirical measure based on $U_1, U_2, \cdots, U_m$ ($V_1, V_2, \cdots, V_n$) and $\mathcal{L}(U) = P := Q_1(\cdot)/Q_1(S_1)$, $\mathcal{L}(V) = Q := Q_2(\cdot)/Q_2(S_2)$. Thus, we need to obtain a bound for the tail probabilities of a product of independent empirical measures of possibly different sample sizes. Since this part of the proof is of independent interest we state it separately as:

Proposition 2.2 *In the context of Theorem 2.1 and (2.9), for $mn > 0$,*

$$P(T_{mn}(f) > \delta_f + \gamma) \le \exp\{-(m \wedge n)C(\gamma; f)\} \quad for \ 0 < \gamma < 1 - \delta_f \quad (2.10)$$

where $\delta_f = Ef(U, V)$ *and* $\sigma_f^2 = \mathrm{Var}(f(U, V))$ *are as before.*

Proof. The case of $\gamma > 1 - \delta_f$ follows directly from the boundedness of f. For the other, assume that $m \wedge n = m$ and following Hoeffding (1963) write $T_{mn} = n^{-1}\sum_{k=1}^{n} T_{mn}^{(k)}$, where

$$T_{mn}^{(k)} = m^{-1}\sum_{i=1}^{m} f(U_i, V_{i+k-1})$$

and $V_j = V_{j-n}$ when $j > n$. Thus, each $T_{mn}^{(k)}$ is a sum of m-independent identically distributed r. v.'s and each $T_{mn}^{(k)}$ has the same distribution. Hence,

$$E \exp(sT_{mn}) \le E\{n^{-1} \sum_{k=1}^{n} \exp(sT_{mn}^{(k)})\} \le E(\exp(sT_{mn}^{(1)})),$$

so that

$$E(\exp(sT_{mn} - s\gamma - s\delta_f)) \le E(\exp(sT_{mn}^{(1)} - s\gamma - s\delta_f)).$$

The result then follows by an application of the proof of Bernstein's inequality.   □

To complete the proof of Theorem 2.1, apply Proposition 2.2 to (2.8) to obtain

$$P[T_{mn} > u + \delta_f] \le \begin{cases} \exp(-(m \wedge n)C(u)) & \text{if } 0 < u < 1 - \delta_f \\ & \text{and} \quad m \wedge n > 0, \\ 0 & \text{if } u \ge 1 - \delta_f \text{ or } m \wedge n = 0. \end{cases}$$

Thus, the second term in (2.1), with $u = \eta/x - \delta > 0$, satisfies

$$P[\frac{Z(f - \delta_f)}{N_1 N_2} > u] \le E\{\exp(-(N_1 \wedge N_2)C(u)) : N_1 N_2 > 0\}$$

$$\le E\{\exp(-N_1 C(u)) + \exp(-N_2 C(u)) : N_1 N_2 > 0\} \quad (2.\,11)$$

Now, if N is a $\mathcal{P}(\lambda)$ r.v., one may easily show that for $s > 0$,

$$E(e^{-sN} : N > 0) = e^{-\lambda}\{\exp(\lambda e^{-s}) - 1\}. \quad (2.\,12)$$

Thus by setting $s = C(u)$, one may evaluate the right hand side of (2.11). To obtain the simpler bound stated in the theorem, observe that when $u < 1 - \delta_f$,

$$C(u) = \frac{3}{2} \frac{u^2}{3\sigma_f^2 + u^2} \le 3/2 \,.$$

Since over the interval $(0, 3/2)$, the curve e^{-s} lies below the line $1 - \frac{2}{3}(1 - e^{-3/2})s$ that passes through $(0,1)$ and $(3/2, e^{-3/2})$, it is easy to see that the right hand side of (2.12) is bounded above by $\exp(-\frac{2}{3}(1 - e^{-3/2})\lambda s)$. Since $\frac{2}{3}(1 - e^{-3/2}) > 1/2$, this is in turn bounded by $\exp -\lambda s/2$ for $0 \le s \le 3/2$. The proof is completed by setting $s = C(u)$.   □

To obtain the optimal bound for the tails of $Z(f)$, one would choose a value of x that minimizes the right hand side of (2.5). Though manageable numerically for given η, λ and μ, it does not seem possible to find useful closed forms for these optimal value in general. Observe that the first term in (2.5) is a nonincreasing function of x, as is the bound given for it in (2.1). On the other hand, since $C(u)$ is a nondecreasing function in u, it follows that the second term in (2.5) is nondecreasing in x, as is the term it bounds, namely $E\{\exp(-(N_1 \wedge N_2)C(\eta/x - \delta)) : N_1 N_2 > 0\}$. By checking their ranges, one sees that the functions always cross, and an optimal value of x exists. It follows, moreover, that only values $x \geq \eta$ need be considered. However, though manageable numerically for given η, λ and μ, it does not seem possible to find useful general expressions for these optimal values.

The reader should note the relationship between the statistics T_{mn} and coup led versions of U-statistics; a bivariate U-statistic is the integral of f over the complement of the diagonal with respect to $P_n \times P_n$, whereas a decoupled U-statistic is the same integral with respect to $P_n \times Q_n$, in which Q_n is an independent copy of P_n. In recent years, there has been considerable activity in the study of tail inequalities for U-statistics; see e.g. Ossiander (1992), Arcones and Giné (1994) and references therein. This tie-in, however, does not result in improved inequalities for T_{mn}, even when $m = n$.

3   Bounds for products of symmetric Poisson processes

Before deriving bounds for the product processes themselves, we first apply Theorem 1.2 to obtain a probability bound for the product of individual symmetric Poisson r. v.'s.

Proposition 3.1   *Let $\overline{N}_1$ and $\overline{N}_2$ be independent symmetric Poisson r. v.'s with parameters λ and μ, respectively. Then for all $\eta > 0$,*

$$P(\overline{N_1}\,\overline{N_2} > \eta) < 2\exp\left(-\frac{(\sqrt{\eta}/a)^2}{2\lambda + 2\sqrt{\eta}/3a}\right) = 2\exp\left(-\frac{\eta a^2}{2\mu + 2\sqrt{\eta}a/3}\right)$$

where a is the unique positive solution of $x^4\lambda + x^3\sqrt{\eta}/3 - x\sqrt{\eta}/3 - \mu = 0$.
Proof. Since $\overline{N}_1$ and $\overline{N}_2$ are symmetric random variables,

$$P(\overline{N}_1\,\overline{N}_2 > \eta) \ \le\ 2P(\overline{N}_1\overline{N}_2 > \eta,\ \overline{N}_1 > 0 \text{ and } \overline{N}_2 > 0)$$

$$\le\ \inf_{x>0} 2\{P(\overline{N}_1 > \sqrt{\eta}/x,\ \overline{N}_1 > 0 \text{ and } \overline{N}_2 > 0)$$

$$+\ P(\overline{N}_2 > x\sqrt{\eta},\ \overline{N}_1 > 0 \text{ and } \overline{N}_2 > 0)\}.$$

By independence,

$$P(\overline{N}_1 > \sqrt{\eta}/x, \overline{N}_1 > 0 \text{ and } \overline{N}_2 > 0) \ =\ P(\overline{N}_1 > \sqrt{\eta}/x)P(\overline{N}_2 > 0)$$
$$\le\ P(\overline{N}_1 > \sqrt{\eta}/x)/2.$$

and

$$P(\overline{N}_2 > x\sqrt{\eta} : \overline{N}_1 > 0 \text{ and } \overline{N}_2 > 0) \ =\ P(\overline{N}_2 > x\sqrt{\eta})/2.$$

Thus, by Theorem 1.2,

$$P(\overline{N}_1\overline{N}_2 > \eta) \le \inf_{x>0}\{P(\overline{N}_1 > \sqrt{\eta}/x) + P(\overline{N}_2 > x\sqrt{\eta})\}$$

$$\le \inf_{x>0}\left\{ \exp\left(-\frac{\eta/x^2}{2\lambda + 2\sqrt{\eta}/3x}\right) + \exp\left(-\frac{\eta x^2}{2\mu + 2x\sqrt{\eta}/3}\right)\right\}. \qquad (3.\,1)$$

To equate the two exponents of equation (3.1) one would have to solve

$$x^4\lambda + x^3\sqrt{\eta}/3 - x\sqrt{\eta}/3 - \mu = 0$$

which, by elementary calculus, can be shown to have a unique positive solution which we have called a. The proof is complete.   □

When $\lambda = \mu$, the solution in the above is $a = 1$, yielding the bound $2\exp(-\sqrt{\eta}/(2\lambda/\sqrt{\eta}+2/3))$. This may be compared with what one gets by writing $\overline{N}_1\overline{N}_2 = (N_{11}-N_{12})(N_{21}-N_{22})$, in which the N_{ij} are independent $\mathcal{P}(\lambda/2)$ r.v.'s, observing

$$P(\overline{N}_1\overline{N}_2 > \eta) \le 2P(N_{11}N_{21} > \eta/2)$$

and applying (2.1). The bounds are of similar orders, with the advantage to Proposition 3.1 coming from the fact that it applies for all $\eta > 0$.

A key inequality for the study of product processes is the one given by Hanson and Wright (1971) for the tail probabilities of quadratic forms. Moreover, it is the first inequality to show the way in which this tail behavior depends upon *two* norms. In this paper, we derive a variant of this inequality that covers the case in which the vectors in the quadratic form are independent and of different dimensions. Let $X = (X_1, \cdots, X_n)$ be a (row) vector of independent symmetric (about zero) r.v.'s that are *uniformly sub-Gaussian* in that there exists a constant $\beta > 0$ such that for all reals,

$$\max_{1 \leq i \leq n} E(e^{sX_i}) \leq e^{\beta s^2}. \tag{3.2}$$

A consequence of this (cf. Lemma II.5.2 of Jain and Marcus (1978)) is that these r.v.'s satisfy the assumption of Hanson and Wright (1971), namely

$$\max_{1 \leq i \leq n} P(|X_i| > x) \leq MP(|\xi| > cx), \tag{3.3}$$

for some constants M and c, where ξ is a standard $N(0,1)$ r.v. We first state the Hanson-Wright inequality: If $A = (a_{ij})$ is an $n \times n$ matrix and $Q = XAX'$, then for any $\eta > 0$,

$$P(Q \geq \eta) \leq \exp\left(-\min\left\{C_1 \frac{\eta}{\|A\|}, \ C_2 \frac{\eta^2}{\Lambda_A^2}\right\}\right) \tag{3.4}$$

where C_1 and C_2 are constants depending only on the constants M and c, and where $\Lambda_A = (\Sigma_{i,j} a_{ij}^2)^{1/2}$ is the usual ℓ_2-norm on $\mathbb{R}^{n^2}$ and $\|A\| = \sup\{|xA| : x \in \mathbb{R}^n, |x| \leq 1\}$ is the norm of A considered as an operator on $\mathbb{R}^n$. (Note that the definitions of the norms Λ_A and $\|A\|$ do not require A to be a square matrix, and so in what follows the same notation will be used for general dimensions.)

To set up the variation needed in this paper, let $W = (W_1, W_2, \cdots, W_m)$ and $Z = (Z_1, \cdots, Z_n)$ be independent vectors of independent symmetric sub-Gaussian r.v.'s satisfying (3.2) for the same (without loss of generality) constants M and c. Consider the quadratic form

$$Q_{mn} := WBZ' \text{ with } B = (b_{ij}; 1 \leq i \leq m, 1 \leq j \leq n).$$

This differs in structure from Q in that the factors W and Z are now independent and of different dimensions. However, it is easy to put it into the Hanson-Wright framework to establish

Theorem 3.1. *Under the above assumptions, for all $\eta > 0$*

$$P(Q_{mn} \geq \eta) \leq \exp\left(-\min\left\{K_1 \frac{\eta}{\|BB'\|^{1/2}}, \ K_2 \frac{\eta^2}{\Lambda_B^2}\right\}\right) \tag{3.5}$$

where $K_1 = 2C_1$ and $K_2 = 2C_2$ are constants depending only on the constants M and c.

Proof. Let O_k and I_k denote $k \times k$ zero and identity matrices, respectively. Define

$$A = \frac{1}{2}\begin{pmatrix} O_m & B \\ B' & O_n \end{pmatrix}, \ X = (W, Z)$$

so that

$$Q = XAX' = (W, Z)A(W, Z)'$$

$$= \tfrac{1}{2}(ZB', WB)(W, Z)' = \tfrac{1}{2}ZB'W' + \tfrac{1}{2}WBZ'$$

$$= WBZ' = Q_{m,n} .$$

Thus the Hanson-Wright inequality may be applied directly to bound $P(Q_{mn} \leq \eta)$. It remains only to express Λ_A and $\| A \|$ in terms of B. For this, recall (see e.g. Rao (1973), p. 68) that

$$|I_{m+n} - \zeta A| = \begin{vmatrix} I_m & -\tfrac{\zeta}{2}B \\ -\tfrac{\zeta}{2}B' & I_n \end{vmatrix} = |I_m| \, |I_n - (-\tfrac{\zeta}{2}B')(-\tfrac{\zeta}{2}B)|$$

$$= |I_n - \tfrac{\zeta^2}{4}B'B|,$$

or similarly, $|I_{m+n} - \zeta A| = |I_m - \tfrac{\zeta^2}{4}BB'|$. This shows the relationship between the eigenvalues of A and those of BB' or $B'B$, implying that

$$\| A \| = \frac{1}{2} \| BB' \|^{1/2} = \frac{1}{2} \| B'B \|^{1/2} .$$

On the other hand, $\Lambda_A^2 = \tfrac{1}{2}\Lambda_B^2$ which completes the proof. $\square$

One may improve upon the non-linear term in the above inequality by using a different consequence of the sub-Gaussian property together with a preliminary result in the proof of the Hanson-Wright inequality, rather than the inequality itself. Though uses of these inequalities, being primarily for large η, place most importance upon the linear term, the proof of the following improvement is of interest.

Theorem 3.2. If $W = (W_1, \ldots, W_m)$ and $Z = (Z_1, \ldots, Z_n)$ are independent vectors of independent symmetric uniformly sub-Gaussian r.v.'s, there exist constants K_1 and K_2 such that for all $\eta > 0$

$$P(Q_{mn} \geq \eta) \leq \exp\left(-\min\left\{K_1\frac{\eta}{\|BB'\|^{1/2}}, K_2\frac{\eta^{4/3}}{\Lambda_{BB'}^{2/3}}\right\}\right) \qquad (3.6)$$

where $Q_{mn} = WBZ'$ for an $m \times n$ real matrix B.

Proof. By conditioning on the Z_j's, we use independence and the sub-Gaussian property (3.2) to obtain

$$E(e^{sQ_{mn}}|Z) = E\left\{\exp\left(\sum_{i=1}^{m} W_i\left(s\sum_{j=1}^{n} b_{ij}Z_j\right)\right)|Z\right\}$$

$$\leq \exp\left\{\beta s^2 \sum_{i=1}^{m}\left(\sum_{j=1}^{n} b_{ij}Z_j\right)^2\right\} \qquad (3.7)$$

for some constant $\beta > 0$. Now write

$$\sum_{i=1}^{m} \left(\sum_{j=1}^{n} b_{ij} Z_j \right)^2 = \sum_{j=1}^{n} \sum_{k=1}^{n} Z_j Z_k a_{jk} := Q$$

where $a_{jk} = \sum_{i=1}^{m} b_{ij} b_{ik}$. Thus $A = (a_{jk}) = B'B$ in this proof. From equation (13) of Hanson-Wright (1971) it follows then that there exist positive constants c_1 and τ, depending only upon the constant β in the sub-Gaussian bound for the Z_j's, such that

$$E(e^{\theta Q}) \le \exp(c_1 \theta^2 \Lambda_A^2) \tag{3.8}$$

for all $\theta \in [0, \tau/\|A\|]$. Upon applying this to (3.7) we obtain, for $\theta = \beta s^2$,

$$E(e^{s Q_{mn}}) \le E(e^{\beta s^2 Q}) \le \exp(c_1 \beta^2 s^4 \Lambda_A^2) \tag{3.9}$$

for $s^2 \le \tau/\beta\|A\|$. Thus by Chebyshev's inequality,

$$P(Q_{mn} > \eta) \le \exp(-s\eta + c_2 s^4 \Lambda_A^2) \tag{3.10}$$

for $0 \le s \le (\tau/\beta\|A\|)^{1/2}$ and $c_2 = c_1 \beta^2$. To minimize the right hand side of (3.10), observe that the minimum of $\phi(s) := -s\eta + c_2 s^4 \Lambda_A^2$, a concave function, occurs at $s_1 = (\eta/4c_2 \Lambda_A^2)^{1/3}$. Thus, upon setting $s_0 = \min\{s_1, (\tau/\beta\|A\|)^{1/2}\}$, we have

$$P(Q_{mn} > \eta) \le \exp\{-s_0(\eta - c_2 s_1^3 \Lambda_A^2)\}$$

$$= \exp(-\tfrac{3}{4} \eta s_0)$$

$$= \exp\left(-\min\left\{ C_1 \frac{\eta}{\|A\|^{1/2}}, C_2 \frac{\eta^{4/3}}{\Lambda_A^{2/3}} \right\}\right)$$

where $C_1 = \tfrac{3}{4}(\tau/\beta)^{1/2}$ and $C_2 = \tfrac{3}{4}(4c_2)^{-1/3}$. Since $A = B'B$, use $K_1 = C_1$ and $K_2 = 2^{1/3} C_2$ to complete the proof, noting as well that $\|BB'\| = \|B'B\|$ and $\Lambda_{BB'} = \Lambda_{B'B}$. □

Though the bounds in (3.4) and (3.5) are, importantly, functions of both norms, it is possible to obtain from them bounds that involve only the more tractable norm, Λ_B: Recall (Rao (1973), p. 34) that for a square matrix A, $\| A \| \le \{\text{trace } (AA')\}^{1/2} = \Lambda_A$ so that

$$\| BB' \| \le \sqrt{tr(BB'BB')} = \Lambda_{BB'} .$$

Also, $\Lambda_{BB'} \le \Lambda_B^2$, by Cauchy-Schwarz. It is then a consequence of Theorem 3.2 that for all $\eta > 0$

$$P(Q_{mn} \ge \eta) \le \exp\left(-K \min\left\{ \frac{\eta}{\sqrt{\Lambda_{BB'}}}, \left(\frac{\eta}{\sqrt{\Lambda_{BB'}}}\right)^{4/3} \right\}\right) \tag{3.11}$$

$$\le \exp\left(-K \min\left\{ \frac{\eta}{\Lambda_B}, \left(\frac{\eta}{\Lambda_B}\right)^{4/3} \right\}\right) \tag{3.12}$$

where $K = K_1 \wedge K_2$. Notice that the variance of Q_{mn} is $\Sigma_{i,j} \mathrm{var}(W_i)\mathrm{var}(Z_j)b_{ij}^2$, which by (3.2) is bounded by a constant times Λ_B^2. Thus, it is appropriate to record that (3.12) implies

$$P(Q_{mn} \geq x\Lambda_B) \leq \exp(-Kx), \quad \text{for } x > 1 . \tag{3.13}$$

The reader may check that the bound of Theorem 3.2 is uniformly better than that of Theorem 3.1, except possibly for the constants.

Consider now the product process $\overline{Z} = \overline{Y}_1 \times \overline{Y}_2$ in which the factors $\overline{Y}_i$ ($i = 1, 2$) are independent symmetric Poisson processes with parameters Q_i defined on $(\mathbf{S}_i, \mathcal{S}_i)$. We will use the representation (1.2) for each of them to give us

$$\overline{Z}(f) = (\overline{Y}_1 \times \overline{Y}_2)(f) = \sum_{i=1}^{N_1}\sum_{j=1}^{N_2} \epsilon_i \epsilon_j' f(U_i, V_j) \tag{3.14}$$

for any $\sigma(\mathcal{S}_1 \times \mathcal{S}_2)$-measurable f, where N_1, ϵ and U are associated with $\overline{Y}_1$ and N_2, ϵ', and V are with $\overline{Y}_2$. As in Section 2, set $\lambda = Q_1(\mathbf{S}_1)$, $\mu = Q_2(\mathbf{S}_2)$, $P_1 = \lambda^{-1}Q_1$ and $P_2 = \mu^{-1}Q_2$. Assume $\lambda \vee \mu \geq 1$. Let $\mathcal{F} = \sigma(U, V)$ be the σ-field generated by the "locations" $\{U_i\}$ and $\{V_j\}$. For integers $m, n \geq 0$, write

$$S_{mn} = \sum_{i=1}^{m}\sum_{j=1}^{n} \epsilon_i \epsilon_j' f(U_i, V_j) .$$

Since symmetric Bernoulli r.v.'s, being bounded, are sub-Gaussian, (cf. Lemma 5.3 of Jain and Marcus (1978)), we may apply (3.11) to S_{mn} conditionally given $\mathcal{F}$, with $b_{ij} = f(U_i, V_j)$, to obtain

$$P(\overline{Z}(f) > \eta | \mathcal{F}, N_1 = m, N_2 = n) = P(S_{mn} > \eta | \mathcal{F})$$

$$\leq \exp\left\{-K\left\{\frac{\eta}{\|BB'\|^{1/2}} \wedge \frac{\eta^{4/3}}{\Lambda_{BB'}^{2/3}}\right\}\right\} \tag{3.15}$$

$$\leq \exp\left\{-K\left\{\frac{\eta}{\sqrt{\Lambda_{mn}}} \wedge \frac{\eta^{4/3}}{\Lambda_{mn}^{2/3}}\right\}\right\} \tag{3.16}$$

$$\leq \exp\left\{-K\left\{\frac{\eta}{\sqrt{L_{mn}}} \wedge \frac{\eta^{4/3}}{L_{mn}^{2/3}}\right\}\right\} \tag{3.17}$$

since $0 \leq \Lambda_{mn} \leq L_{mn}$, where we write

$$\Lambda_{mn}^2 = \sum_{j=1}^{n}\sum_{k=1}^{n}\left(\sum_{i=1}^{m} f(U_i, V_j)f(U_i, V_k)\right)^2 \quad and \tag{3.18}$$

$$L_{mn} = \sum_{i=1}^{m}\sum_{j=1}^{n} f^2(U_i, V_j) \tag{3.19}$$

for $\Lambda_{BB'}^2$ and Λ_B^2, respectively, to emphasize the dependence upon m and n. We will also write Λ and L for $\Lambda_{N_1N_2}$ and $L_{N_1N_2}$, respectively. The following theorem is thus an immediate consequence of (3.16) and (3.17).

Theorem 3.3 *Under the above assumptions, there exists a constant* $K > 0$ *such that for all* $\eta > 0$

$$P(\overline{Z}(f) > \eta) \;\leq\; E\exp\left\{-K\left\{\frac{\eta}{\Lambda^{1/2}} \wedge \frac{\eta^{4/3}}{\Lambda^{2/3}}\right\}\right\} \qquad (3.20)$$

$$\leq\; E\exp\left\{-K\left\{\frac{\eta}{L^{1/2}} \wedge \frac{\eta^{4/3}}{L^{2/3}}\right\}\right\}. \qquad (3.21)$$

Since $0 \leq L_{mn} \leq mn$ whenever $|f| \leq 1$, (3.21) has the following immediate consequence.

Corollary 3.1 *If* $|f| \leq 1$, *then under the above assumptions, there exists a constant* $K > 0$ *such that for all* $\eta > 0$ *and* $z > 0$,

$$P(\overline{Z}(f) > \eta) \leq E\exp\left\{-K\left\{\frac{\eta}{\sqrt{N_1N_2}} \wedge \frac{\eta^{4/3}}{(N_1N_2)^{2/3}}\right\}\right\} \qquad (3.22)$$

and

$$P\left\{\frac{\overline{Z}(f)}{\sqrt{N_1N_2}} > z\right\} \leq e^{-K(z \wedge z^{4/3})}. \qquad (3.23)$$

Moreover, if $2K\eta > e^2(e^2+1)(\lambda+\mu)^2$ *and* $\eta > K/2$, *then*

$$P(\overline{Z}(f) > \eta) \leq 2e^{-\sqrt{2K\eta}}. \qquad (3.24)$$

Proof. Although (3.22) and (3.23) follow directly from (3.21), the proof of (3.24) requires further steps as follows. For any $\nu > 0$, write

$$P(\overline{Z}(f) > \eta) \;\leq P(\overline{Z}(f) > \eta, N_1N_2 \leq \nu^2) + P(N_1N_2 > \nu^2) \qquad (3.25)$$
$$\leq P(\overline{Z}(f) > (\eta/\nu)\sqrt{N_1N_2}) + P(N_1N_2 > \nu^2).$$

By (2.1) and (3.23) it then follows that

$$P(\overline{Z}(f) > \eta) \leq \exp\{-K(\tfrac{\eta}{\nu} \wedge (\tfrac{\eta}{\nu})^{4/3}\} + \exp(-\lambda - \mu - 2\nu) \qquad (3.26)$$

if $2\nu > e^2(\lambda + \mu)$. Now choose ν to equate the two terms. If $\eta \geq \nu$, this means solving $K\eta/\nu = \lambda + \mu + 2\nu$, for which the solution is

$$2\nu = \sqrt{2K\eta + (\frac{\lambda+\mu}{2})^2} - \frac{\lambda+\mu}{2}$$

and the common bound is

$$\exp(-\lambda - \mu - 2\nu) = \exp\left\{-\sqrt{2K\eta + (\frac{\lambda+\mu}{2})^2} - \frac{\lambda+\mu}{2}\right\}. \qquad (3.27)$$

Now check that the restriction on ν in (3.26) is equivalent to

$$2K\eta > (e^4 + e^2)(\lambda + \mu)^2, \qquad (3.28)$$

while the assumption $\eta/\nu \geq 1$ is satisfied if $2\eta \geq (K - \lambda - \mu)^+$. It remains to observe that the common exponent in (3.27) does not exceed $-\sqrt{2K\eta}$ as desired. In fact, under the restriction on η the approximation is very good since $\sqrt{2K\eta} \leq \sqrt{2K\eta + (\frac{\lambda+\mu}{2})^2} + \frac{\lambda+\mu}{2} \leq 1.066\sqrt{2K\eta}$. $\qquad\square$

4 Refinements

There is a big gap between the rather precise bounds of Theorem 3.3 and the more tractable crude ones of its corollary. Though the latter are useful, they are not sharp enough for all purposes since the only property of f that they utilize is its boundedness. For example, the bounds in Corollary 3.1 would be the same for all indicator functions without regard to the size or shape of their support. To make this more precise, the reader should consider the following prototypical examples involving the product of two independent standard symmetric Poisson random measures on the unit interval. That is, let $S_1 = S_2 = I := [0, 1]$ and take $Q_1 = \lambda P, Q_2 = \mu Q$ with $P = Q =$ Lebesgue measure. Consider $f = 1_A$ for $A \subset I^2$. Let $|A|$ denote the Lebesgue measure of A. Consider three types of sets A in the product space I^2, namely, a vertical strip, $A_V = J \times I$; a horizontal strip, $A_H = I \times J$ and a neighborhood of the diagonal $A_D = \{(x, y) : x \in I, |x - y| \leq \epsilon\}$. For given δ, one may choose J and ϵ so that all 3 sets have the same measure δ, and the same probability bounds. However, one expects rather distinct behavior in each of these cases when μ and λ are not equal. The reader will note the differences in Λ and L between these cases. The lack of dependence upon f that shows up in the bounds of Corollary 3.1 is due to the very crude substitution of 1 for f.

In the sharper bounds of Theorem 3.3, the inequalities $\|BB'\| \leq \Lambda_{BB'} \leq \Lambda_B^2$ were used for fixed m and n to obtain the conditional bounds (3.6), (3.11) and (3.12). The latter, using Λ_B^2, gives the weakest bound, but it is the most tractable one for study. $\Lambda_{BB'}$ and $\|BB'\|$ in turn give better bounds but their distributions are harder to analyse. Since the norm $\|BB'\|$ is the better discriminator between the functions f, it would be useful to have good bounds for its tail probabilities. In the absence of such, however, we now derive a tractable bound to (3.21), sharper than (3.22) and (3.24), that utilizes more information about f than just its boundedness. To begin, apply Proposition 2.2 to the function $h = f^2$ to get

$$P\left(L_{mn} > mn(\delta_h + \gamma)\right) \leq \exp\left\{-(m \wedge n)C(\gamma; h)\right\} \qquad (4.1)$$

for $\gamma > 0$ where $C(\gamma; h)$ is defined in (2.6). Combining (3.17) and (4.1) gives

$$P\left(\overline{Z}(f) > \eta | N_1 = m, N_2 = n\right) \leq E\{P(S_{mn} > \eta | \mathcal{F}) : L_{mn} \leq mn(\delta_h + \gamma)\}$$
$$+E\{P(S_{mn} > \eta | \mathcal{F}) : L_{mn} > mn(\delta_h + \gamma)\}$$

$$\leq \exp\left\{-K\left(\frac{\eta}{\sqrt{mn(\delta_h + \gamma)}} \wedge \frac{\eta^{4/3}}{(mn(\delta_h + \gamma))^{2/3}}\right)\right\}$$
$$+\exp\left\{-K\left(\frac{\eta}{\sqrt{mn}} \wedge \frac{\eta^{4/3}}{(mn)^{2/3}}\right) - \frac{(m \wedge n)\gamma^2}{2(\sigma_h^2 + \gamma/3)}\right\} \quad (4.2)$$

for all $0 < \gamma < 1 - \delta_h$, with the last term using $h \leq 1$. The object now is to choose that value of the free variable γ that minimizes, at least approximately, the sum of the two terms. For tractability in application, we choose to equate the exponents of the two terms rather than to minimize their sum; formally, of course, the latter is no more difficult to handle. To this end, set

$$g(\gamma) \equiv g(\gamma, v) = \exp\left\{-K\left(\frac{v}{\sqrt{\delta_h + \gamma}} \wedge \frac{v^{4/3}}{(\delta_h + \gamma)^{2/3}}\right)\right\}, \quad (4.3)$$
$$G(\gamma) \equiv G(\gamma, v, m) = \exp\left\{-K(v \wedge v^{4/3}) - mC(\gamma; h)\right\}.$$

Observe that over $[0, 1 - \delta_h]$, g is an increasing function of γ, whereas G is a decreasing function of γ. Since $g(1 - \delta_h) = G(0)$, with common value $\exp\{-K(v \wedge v^{4/3})\}$, there exists a unique value of γ, say $\gamma_m \equiv \gamma_m(v)$, at which the continuous functions are equal. This establishes

Theorem 4.1 *Under the assumptions of this section, with $|f| \leq 1$, $h = f^2$ and $M = N_1 \wedge N_2$,*

$$P(\overline{Z}(f) > v\sqrt{N_1 N_2}) \leq 2Eg(\Gamma, v) = 2EG(\Gamma, v, M) \quad (4.4)$$

for any $v > 0$ where $\Gamma := \gamma_M$ is the unique solution in $[0, 1 - \delta_h]$ of $g(\cdot, v) = G(\cdot, v, M)$. If $v > 1$, this is equivalent to

$$P(\overline{Z}(f) > v\sqrt{N_1 N_2}) \leq 2E\left(\exp\left\{-\frac{Kv}{\sqrt{\delta_h + \Gamma}}\right\}\right) \quad (4.5)$$

$$= 2e^{-Kv} E\left(\exp\left\{-\frac{1}{2}M\frac{\Gamma^2}{\sigma_h^2 + \Gamma/3}\right\}\right). \quad (4.6)$$

These bounds as desired depend upon more than just the boundedness of f and h, thereby improving on those of (3.23); notice that the bound of (3.23) is (twice) the common *maximum* values for g and G of $g(1 - \delta_h)$ and $G(0)$. Moreover, the new bounds decrease as δ_h (and hence σ_h^2) decreases.

Although (4.6) could be evaluated in any specific case, it is not suitable for general analysis. In the following, we modify the bound to give more tractability. Suppose $v \geq 1$, so that the simpler forms of the bound, (4.5) and (4.6), apply. Equivalently, for $v \geq 1 \geq \sqrt{\delta_h + \gamma}$, the sentence following (4.2) allows

$$P(\overline{Z}(f) > v\sqrt{N_1 N_2}) \leq E\left\{\min_{0 \leq \gamma < 1 - \delta_h}\left(\exp\left\{-\frac{Kv}{\sqrt{\delta_h + \gamma}}\right\}\right.\right.$$
$$\left.\left. + \exp\left\{-Kv - MC(\gamma; h)\right\}\right)\right\}. \tag{4.7}$$

Since the optimal value of γ conditional on $N_1 = m$ and $N_2 = n$ should be large relative to the standard deviation of L_{mn}/mn, whose mean is δ_h, note that for any function $h = f^2$ with the requisite integrability,

$$\text{var}(L_{mn}/mn) = \frac{1}{m}\left(1 - \frac{1}{n}\right)\text{cov}\left\{h(U,V), h(U,V')\right\}$$
$$+ \frac{1}{n}\left(1 - \frac{1}{m}\right)\text{cov}\left\{h(U,V), h(U',V)\right\}$$
$$+ \frac{1}{mn}\text{var}\left\{h(U,V)\right\}$$
$$\leq \frac{m+n}{mn}\sigma_h^2$$

where (U', V') is an independent copy of (U, V). Thus, since $\frac{1}{2}(m \wedge n) \leq mn/(m+n) \leq m \wedge n$, substitute $\gamma = x\sigma_h/\sqrt{m \wedge n}$, with the restriction $0 \leq \gamma \leq 1 - \delta_h$ translating into $0 \leq x \leq \sqrt{m}(1 - \delta_h)/\sigma_h$, when $m = m \wedge n$. Thus (4.7) becomes

$$P\left(\overline{Z}(f) > v\sqrt{N_1 N_2}\right) \leq E\left\{\min_{0 \leq x \leq \sqrt{M}(1-\delta_h)/\sigma_h}\left(\exp\left\{-\frac{Kv}{\sqrt{\delta_h + x\sigma_h M^{-1/2}}}\right\}\right.\right.$$
$$\left.\left. + \exp\left\{-Kv - \frac{x^2\sigma_h}{2(\sigma_h + x/3\sqrt{M})}\right\}\right)\right\} \tag{4.8}$$

Now denote the right hand side of (4.8) by $B(v)$. An application of (3.25) then yields

$$P(\overline{Z}(f) > \eta) \leq B(\eta/\nu) + P(N_1 N_2 > \nu^2)$$
$$\leq B(\eta/\nu) + \exp(-\lambda - \mu - 2\nu)$$

if $2\nu > e^2(\lambda + \mu)$ by (2.1). Thus

$$P(\overline{Z}(f) > \eta) \leq \min_{\nu > e^2(\lambda+\mu)/2}\{B(\eta/\nu) + \exp(-\lambda - \mu - 2\nu)\}. \tag{4.9}$$

Though the above procedure is clear, the expectation in (4.8) is difficult making the resultant bound tractable in specific cases only. However, an application of Fatou's lemma to (4.8) allows one to use an analysis similar to the above to obtain the following; the details are omitted.

Theorem 4.2. *In the context of this section,*

$$P(\overline{Z}(f) > \eta) \leq \min_{\nu > e^2(\lambda+\mu)/2} \{B_1(\eta/\nu) + \exp(-\lambda - \mu - 2\nu)\}$$

for all $\eta > 0$, where

$$B_1(v) = \min_{0 \leq \gamma < 1-\delta_h} \left\{ \exp\left(-\frac{Kv}{\sqrt{\delta_h + \gamma}}\right) + \exp\left(-Kv - \frac{\lambda}{2}\frac{\gamma^2}{\sigma_h^2 + \gamma/3}\right) \right\}.$$

One may further show that for each $\gamma \geq \sigma_h$,

$$B_1(v) \leq \exp\left(-\frac{Kv\delta_h^{-1/4}}{\sqrt{1 + \gamma/\sigma_h}}\right) + \exp\left(-Kv - \frac{\lambda}{4}\gamma\right); \qquad (4.\,10)$$

Since in most cases it is the second term in (4.19) that is the larger of the two, choose ν to equate the second term to the bound for $P(N_1 N_2 > \nu^2)$, namely $\exp(-\lambda - \mu - 2\nu)$. That is, solving $K\eta/\nu + \lambda\gamma/4 = \lambda + \mu + 2\nu$ leads to

Corollary 4.1. *If*

$$2\eta > \max\left\{(e^4 + e^2)(\lambda + \mu)^2/K,\ K(\delta_h + 8\sigma_h) - \sqrt{\delta_h + 8\sigma_h}\,(\lambda + \mu - 2\lambda\sigma_h)\right\},$$

then

$$\begin{aligned} P(\overline{Z}(f) > \eta) \ &\leq \exp\left\{-\frac{1}{3\delta_h^{1/4}}\left(\left\{2K\eta + \left(\tfrac{\lambda+\mu}{2} - \lambda\sigma_h\right)^2\right\}^{1/2} + \tfrac{\lambda+\mu}{2} - \lambda\sigma_h\right)\right\} \\ &+ 2\exp\left\{-\left(\left\{2K\eta + \left(\tfrac{\lambda+\mu}{2} - \lambda\sigma_h\right)^2\right\}^{1/2} + \tfrac{\lambda+\mu}{2} + \lambda\sigma_h\right)\right\}. \end{aligned}$$

$$(4.\,11)$$

The above proofs extend immediately to give bounds for the product, $\widehat{Z}$, of symmetric compound Poisson processes of (1.3) when they have uniformly bounded jumps $\{X_i, X_j'\}$. For example, when these jumps are symmetric and bounded by 1, the proof of Corollary 3.1 applies exactly to yield

$$P(\widehat{Z}(f) := \widehat{Y}_1 \times \widehat{Y}_2(f) > \eta) \leq 2\exp\left\{-\sqrt{2K\eta}\right\} \qquad (4.\,12)$$

if $2K\eta > (e^4 + e^2)(\lambda + \mu)^2$ and $2\eta > K$, where in this case, the constant K depends upon the distributions of the jumps.

Acknowledgements

We are grateful to Professor Mina Ossiander for suggesting the approach used in the proof of Proposition 2.2. We wish to thank her also for sending us early drafts of her papers cited herein and for her continued interest and support.

References

ALDER, R.J. and FEIGIN, P.D. (1984), On the cadlaguity of random measures. *Ann. Prob.* **12**, 615–630.

ARCONES, M.A. and GINÉ, E. (1994), Limit theorems for U-processes. *Ann. Prob.* **21**, 1494–1542.

BACHELIER, L., Théorie de la speculation, *Ann. Sci. École Norm. Sup.* **3**, 21–86 (English translation by A. J. Boness: in *The Random Character of Stock Market Prices*, 17–78, P. H. Cootner, Editor, M.I.T. Press, Cambridge, Mass. (1967).

BASS, R.F. and PYKE, R. (1984), The existence of set-indexed Lévy processes. *Z. Wahrsch. verw. Gebiete.* **66**, 157–172.

BASS, R.F. and PYKE, R. (1987), A central limit theorem for $\mathcal{D}(A)$-valued processes. *Stochastic Processes and their Applications.* **24**, 109–131.

BENNETT, G. (1962), Probability inequalities for the sums of bounded random variables. *J.A.S.A.* **57**, 33–45.

EINSTEIN, A. (1905). On the movement of small particles suspended in a stationary liquid demanded by the molecular-kinetic theory of heat. *Ann. Phys.* **17**, 549–560.

HANSON, D.L. and WRIGHT, F.T. (1971), A bound on tail probabilities for quadratic forms in independent random variables. *Ann. Math. Statist.* **42**, 1079-1083.

HOEFFDING, W. (1963), Probability inequalities for sums of bounded random variables. *J.A.S.A.* **57**, 33–45.

JAIN, N. and MARCUS, M. (1978), Continuity of subgaussian processes. In *Probability on Banach Spaces*, 81–196. Edited by J. Kuelbs. Marcel Dekker, NY.

OSSIANDER, M. (1991), Product measure and partial sums of Bernoulli random variables. In preparation.

OSSIANDER, M. (1992), Bernstein inequalities for U-statistics. To appear.

PYKE, R. (1983), A uniform central limit theorem for partial-sum processes indexed by sets. Probab. Statist. and Anal. (Edited by J. F. C. Kingman and G. E. H. Reuter) *London Math. Soc. Lect. Notes. Ser.* 79, 219–240.

PYKE, R. (1996). The early history of Brownian motion. In preparation.

RAO, C. R. (1973), *Linear Statistical Inference and its Applications.* (2nd Edition). J. Wiley and Sons, New York.

DEPARTMENT OF MATHEMATICS, SUN MOON UNIVERSITY, ASAN, CHUNG-NAM, REPUBLIC OF KOREA 337–840

DEPARTMENT OF MATHEMATICS, UNIVERSITY OF WASHINGTON, SEATTLE, WA 98195–4350

ON THE FIRST-CROSSING OF A POISSON PROCESS IN A LOWER BOUNDARY

CLAUDE LEFEVRE [1]
PHILIPPE PICARD [2]

ABSTRACT This paper is concerned with the first-crossing of a Poisson process in a general lower boundary. The statistic under investigation is the level of first-crossing N. It is shown that the distribution of N can be expressed in a simple and compact way in terms of a particular family of Abel-Gontcharoff polynomials. Attention is then paid to that specific law, called the Poisson-Gontcharoff distribution. It is pointed out that the law keeps some nice properties of the usual Poisson distribution. Finally, two related first-crossing problems that can be tackled using again Abel-Gontcharoff polynomials are examined.

Key words: First-crossing level; Poisson process; lower boundary; Abel-Gontacharoff polynomials; Poisson distribution; linear birth process.

AMS 1991 subject classifications: 60G17; 60G40; 60J75; 60E99.

1 Introduction

The problem of first-crossing of a Poisson process trajectory in a fixed boundary has a wide field of applications ranging from queueing, storage and risk theories to Kolmogorov-Smirnov statistics and sequential analysis. For such a counting process, it is crucial to distinguish between lower and upper boundaries. Indeed, in the lower case, the first-crossing occurs necessarily on a continuous part of the trajectory, while in the upper case, this can also happen at a jump time at the trajectory.

The literature on the subject goes back roughly to the fifties. Questions of direct relevance are the probability of absorption and the time and level of first-crossing. One of the earliest works is [PYK59] who considered the special case of linear boundaries, lower or upper. [WHI61] discussed several

---

[1] Université Libre de Bruxelles
[2] Université de Lyon 1

particular non-linear upper boundaries, and his method was followed by [GAL66] for general upper boundaries. The case of general lower or upper boundaries was examined by [DAN63] in a pioneering paper that gave rise to the research hereafter. [DUR71] dealt with the two-boundary situation. Recently, [ZAC91] reinvestigated the case of linear boundaries and [STA93] the case of general upper boundaries.

We mention that very few works have been devoted to the first-crossing of counting processes which are non-Poissonian. We refer to [GAL93], [PL94, PL96c] and [LP95] for the compound Poisson process with lower or upper boundaries, and to [PL96b] for various basic counting processes with lower boundaries.

In the present paper, we focus our attention mainly on the first-crossing of a Poisson process in a general lower boundary. The statistic under study is the first-crossing level N.

Our approach relies on the use of the remarkable family of the Abel-Gontcharoff (in short, A-G) polynomials. This family is relatively little known. It was introduced by [GON37] for the interpolation of entire functions, and it provides a natural extension of the classical family of Abel polynomials (see, e.g., [PL96a]). To our knowledge, the first references to these polynomials in probability and statistics are [DAN63] indicated earlier and [DAN67] who was concerned with the final size in the general epidemic model. In recent years, we made an extensive exploitation of the A-G polynomials within that framework of epidemic theory (see, e.g., [LP90], [PL90]. This time thus, we tackle the other application. A short presentation of these polynomials is given in the Appendix.

In Section 2, we show that the distribution of the level N can be expressed in a simple and compact way in terms of a particular family of A-G polynomials. When the boundary is linear, the law becomes the Lagrangian Poisson distibution (see, e.g., [CON89]); this reduces to the usual Poisson distribution when the boundary is a vertical line. We also mention briefly that with upper boundaries, the analysis is based on a family, quite distinct, of Appell polynomials.

Section 3 deals with that specific law obtained for N. For obvious reasons, it is called the Poisson-Gontcharoff distribution. This law keeps some nice properties related to the Poisson law. We first establish that it obeys a decomposition formula of the Raikov type, which generalizes a result of [GS74] for the Poisson law. Then, we construct a strongly connected law, named the binomial-Gontcharoff distribution, that generalizes the usual binomial law and the quasi-binomial law (see [CON89]).

In Section 4, we examine two other first-crossing problems for which the distribution of the level N can still be written in terms of A-G polynomials. The former variant substitutes a linear birth process with immigration for the original Poisson process. The latter considers the first-crossing of a Poisson process in an upper birth process with linearly decreasing rates.

2   Distribution of the first-crossing level

Consider a Poisson process with rate λ and initiated at time $-x, x$ being positive or negative. Later, we will put $x = 0$. The number of events generated in the interval $(-x, t), t > -x$, is a Poisson r.v. with parameter $\lambda(x + t)$. In other words, the level of first-crossing of the Poisson process in the vertical line at t has just that Poisson law. Now, instead of a vertical line, let us choose any continuous lower boundary $b(t), t \geq -x$. By lower, we mean, of course, that $b(-x) < 0$. We aim to determine the exact distribution of the first-crossing level N. Note that N is a positive integer valued r.v., with a possibly defective law.

Without loss of generality, we may suppose that the boundary $b(t)$ is increasing (some parts being vertical). Moreover, $b(t)$ can be reduced to the set $\mathcal{B}$ of points that are eligible as levels of first-crossing (i.e. with ordinates in $\mathbb{N}$). It is convenient to adopt the following unusual notation :

$$\mathcal{B} = \{(-u_i, i), i \geq 0\}, \tag{1.1}$$

where

$$0 < -u_0 \leq -u_1 \leq -u_2 \leq \ldots \text{ and } - x < -u_0. \tag{1.2}$$

This is illustrated in Figure 1. We observe that the time of first-crossing T is directly connected with N through the relation $T = -u_N$.

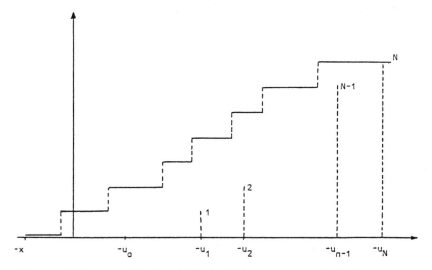

Figure 1. First-crossing level N of a Poisson process in the lower boundary $\mathcal{B} = \{-u_i, i), i \geq 0\}$.

Let U denote the family $\{u_i, i \geq 0\}$. For $r \geq 1$, $E^r U = \{u_{r+i}, i \geq 0\}$ will be the same family without its first r elements. To U we attach the family of A-G polynomials $G_n(x|U), n \geq 0$, of degree n in x. The distribution of N can then be expressed as follows (its dependence on λ, x, U being marked plainly).

Proposition 2.1.

$$P(N = n|\lambda, x, U) = \lambda^n e^{-\lambda(x-u_n)} G_n(x|U), n \geq 0. \tag{1.3}$$

Proof. Obviously,

$$P(N = 0|\lambda, x, U) = e^{-\lambda(x-u_0)}, \tag{1.4}$$

and for $n \geq 1$, we get

$$P(N = n|\lambda, x, U) = \int_0^{x-u_0} P(N = n - 1|\lambda, t - x, EU)\lambda e^{-\lambda t} \, dt. \tag{1.5}$$

Indeed, the realization of the event $[N = n|\lambda, x, U]$ demands that the first jump of the Poisson process occurs at time $T_0 = t < x - u_0$; T_0 is exponentially distributed with parameter λ. Furthermore, the process starting now at point $(t - x, 1)$ has to cross $\mathcal{B}$ for the first time at the level n, which is equivalent to requiring that a Poisson process starting at point $(t - x, 0)$ crosses the boundary $\{(-u_{1+i}, i), i \geq 0\}$ for the first-time at the level $n - 1$ (i.e., $[N = n - 1|\lambda, t - x, EU]$). Let us put

$$P(N = n|\lambda, x, U) = \lambda^n e^{-\lambda(x-u_n)} H_n(x|U), n \geq 0, \tag{1.6}$$

where $H_n(x|U)$ represents some appropriate function. From (1.4), (1.5) and (1.6), we obtain

$$H_0(x|U) = 1, \tag{1.7}$$

$$H_n(x|U) = \int_{u_0}^x H_{n-1}(\tau|EU) \, d\tau, n \geq 1. \tag{1.8}$$

Comparing (1.7), (1.8) with (1.48), (1.49), we see that for $n \geq 0$, $H_n(x|U)$ corresponds to the A-G polynomial $G_n(x|U)$. Thus, (1.6) yields (1.3). $\diamond$

We observe that as expected, the parameters in (1.3) are in fact λx and $\{\lambda u_i, i \geq 0\}$. From now on, we suppose that the Poisson process is of rate $\lambda = 1$ and starts at $x = 0$. These arguments are omitted in the notation.

Special case. Formula (1.3) becomes explicit when the boundary $\mathcal{B}$ is (increasing) linear. Le $U^l = \{u_i^l, i \geq 0\}$, where

$$u_i^l = -a - bi, \text{ with } a > 0 \text{ and } b \geq 0. \tag{1.9}$$

Since $G_n(x|U^l)$, $n \geq 0$, is given by the Abel formula (1.53), we get from (1.3) that

$$P(N = n|U^l) = e^{-(a+bn)} a(a + bn)^{n-1}/n!, n \geq 0. \tag{1.10}$$

The distribution (1.10) is recognized to be the Lagrangian or generalized Poisson law; see the book by Consul (1989). When $b = 0$, that is if the boundary is vertical, we have, of course, that N is a Poisson r.v. with parameter a.

Remark. In the general case, U may represent any sequence of real numbers satisfying (1.2), so that the law of N might depend on a large , even infinite, number of parameters. To evaluate (1.3) numerically, it suffices to apply the recurrence relations (1.55) defining te A-G polynomials. For satistical purposes, it is suitable to specify to the u_i's a simple parametric form. In a forthcoming work, [DEN95] shows that in automobile insurance, (1.3) with the parabolic sequence $u_i = -a - bi - ci^2$, $i \geq 0$, provides a good fit, better than those usually obtained, for the annual number of accidents incurred by a motorist.

An application to queueing theory. Consider a M/D/1 queueing system in which customers arrive according to a Poisson process with rate 1, service times are equal to the constant d and m customers are present intially in the queue. It is clear that during the first busy period of the server, the number N of new customers served is exactly given by the first-crossing level of the Poisson process in the linear boundary constructed from $u_i^l = -(m + i)d$, $i \geq 0$. Thus, the law of N is provided by (1.10) with $a = md$ and $b = d$.

We can now generalize this model by supposing that the service times are fixed but differ from a customer to another. Specifically, let $-u_0$ be the total service time of the m initial customers, and for $i \geq 1$, let $u_{i-1} - u_i$ denote the service time of the ith new customer. We directly see that the number N of new customer served during the first busy period is of law (1.3) where $U = \{u_i, i \geq 0\}$.

Property. *A sufficient condition guaranteeing $N < \infty$ a.s. is that the boundary B is above some straight line with slope at least 1, i.e. when u_i, $i \geq 0$, can be bounded as*

$$u_i \geq -a - bi, \text{ with } a > 0 \text{ and } 1 \geq b > 0. \tag{1.11}$$

Proof. We observe that under (1.11), N is smaller in distribution than the first-crossing level of the Poisson process in the linear boundary built from $u_i^l = -a - bi$, $i \geq 0$. The law of this level is given by (1.10) and it is known that (1.10) is a proper law if and only if $b \leq 1$, hence the result. $\Diamond$

First-crossing in an upper boundary. The problem with upper boundaries is quite different. Here, the first-crossing can occur at a jump time of the Poisson process. It does not correspond necessarily to a first-meeting and the time of first-crossing $\hat{T}$ has a continuous (instead of discrete) distribution.

As indicated in the Introduction, the law of $\hat{T}$ was investigated in previous works. A major point is that it can be expressed using this time a family of Appell polynomials. This family, much more standard, is sometimes confused with the family of A-G polynomials. We underline briefly in the Appendix that both families are locally connected but their underlying mathematical structures are really distinct.

We are content now with recalling the following basic result from [GAL66]. Let $g(t)$, $t \geq 0$, be an upper boundary, assumed to be strictly positive and increasing. Define

$$\hat{u}_i = g^{-1}(i), i \geq 0, \tag{1.12}$$

implying

$$0 = \hat{u}_0 \leq \hat{u}_1 \leq \hat{u}_2 \leq \ldots \tag{1.13}$$

Put $\hat{U} = \{\hat{u}_i, i \geq 0\}$ and attach to $\hat{U}$ the family of Appell polynomials $A_n(x|\hat{U}), n \geq 0$, of degree n in x. It can then be proved that for $x \in (\hat{u}_j, \hat{u}_{j+1}], j \geq 0$,

$$P(\tilde{T} > x|\hat{U}) = e^{-x} \sum_{n=0}^{j} A_n(x|\hat{U}). \tag{1.14}$$

3   A new generalization of the Poisson law

The distribution (1.3) constitutes a generalization of the Poisson and Lagrangian laws constructed from a family of A-G polynomials. For that reason, putting again $\lambda = 1$ and $x = 0$, we call it the Gontcharoff-Poisson distibution with parameters U (denoted by $\mathcal{G}\text{-}\mathcal{P}(U)$). As pointed out before, this law can be defective.

Definition 3.1. *The $\mathcal{G}\text{-}\mathcal{P}(U)$ distribution is the family*

$$\{e^{u_n} G_n(0|U), n \geq 0\}, \tag{1.15}$$

where $U = \{u_i, i \geq 0\}$ with $0 > u_0 \geq u_1 \geq u_2 \geq \ldots$

The $\mathcal{G}\text{-}\mathcal{P}(U)$ law keeps various nice properties of the Poisson distribution. Two of them are presented below.

3.1   A decomposition formula of the Raikov type

We remind the reader that [RAI37] showed that if X, Y are two independent positive non-degenerate r.v.'s whose sum $Z = X + Y$ follows a Poisson law, then both X and Y must themselves be Poisson r.v.'s.

A directly related problem of general interest is the possible decomposition of a Poisson r.v. into the sum of two non-independent r.v.'s. That question was studied by [GS74] who established the following result.

Gani-Shanbhag's theorem. *Let Z be a Poisson r.v. with parameter λ and $\lambda > \mu_0 \geq \mu_1 \geq \mu_2 \geq \ldots \geq 0$ any set of positive real numbers. It is always possible to decompose Z into the sum $X + Y$ of two r.v.'s X and Y such that the distribution of Y given $X = j$, $j \geq 0$, is Poisson with parameter μ_j, and X has a unique distribution $\{a_i, i \geq 0\}$ depending on the μ_j's. If the μ_j's are all equal to a single value μ, then X is Poisson with parameter $\lambda - \mu$, and X, Y are independent.*

Surprisingly enough, these authors obtained the theorem by exploiting a very particular property derived in the context of epidemic theory. Their proof is rather intricate and the result appears to be somewhat unexpected, if not mysterious.

Now, we are going to extend that decomposition formula to the case of $\mathcal{G}$-$\mathcal{P}(U)$ laws. Using functional properties of the A-G polynomials, the proof will become quite direct. Furthermore, we will give a simple interpretation of the formula within the previous framework of first-crossing problem.

Proposition 3.2. *Let $U = \{u_i, i \geq 0\}$ and $V = \{v_i, i \geq 0\}$ be two families satisfying*

$$0 > u_0 \geq u_1 \geq u_2 \geq \ldots, \quad 0 > v_0 \geq v_1 \geq v_2 \geq \ldots \quad \text{and } v_i > u_i, \, i \geq 0.$$
$$(1.16)$$

Any r.v. Z with $\mathcal{G}$-$\mathcal{P}(U)$ law can always be decomposed into the sum $X + Y$ of two r.v.'s X, Y such that X follows the $\mathcal{G}$-$\mathcal{P}(V)$ law and the law of Y given $X = j$, $j \geq 0$, is $\mathcal{G}$-$\mathcal{P}[U(j)]$, where $U(j) = \{u_i(j), i \geq 0\}$ with

$$u_i(j) = u_{j+i} - v_j. \tag{1.17}$$

Proof. Let $\{G_n(x|U), n \geq 0\}$ and $\{G_n(x|V), n \geq 0\}$ be the families of A-G polynomials attached to any families U and V. By the Abel expansion (1.54) of $G_n(0|U)$, $n \geq 0$, with respect to the family $\{G_n(0|V), n \geq 0\}$, we get

$$G_n(0|U) = \sum_{j=0}^{n} G_n^{(j)}(v_j|U)G_j(0|V),$$

which becomes by (1.50), (1.56) and (1.17)

$$G_n(0|U) = \sum_{j=0}^{n} G_{n-j}[0|U(j)]G_j(0|V), \, n \geq 0. \tag{1.18}$$

Multiplying (1.18) by $\exp(u_n)$, and noting that

$$u_n = (u_{j+n-j} - v_j) + v_j = u_{n-j}(j) + v_j,$$

we find

$$e^{u_n}G_n(0|U) = \sum_{j=0}^{n} e^{u_{n-j}(j)}G_{n-j}[0|U(j)]e^{v_j}G_j(0|V), \, n \geq 0. \tag{1.19}$$

Now, since the families U and V satisfy the conditions (1.16), we may define associated r.v.'s Z, X and Y that follow the $\mathcal{G}\text{-}\mathcal{P}$ laws indicated in the statement. Therefore, we see that (1.19) can then be rewritten as

$$P(Z = n) = \sum_{j=0}^{n} P(Y = n - j|X = j)P(X = j),\ n \geq 0,$$

hence the announced decomposition formula. $\Diamond$

Special case. To obtain Gani-Shanbhag's theorem, it suffices to take

$$u_i = -\lambda,\ i \geq 0,\ \text{and}\ v_i = \mu_i - \lambda, i \geq 0, \tag{1.20}$$

yielding for $j \geq 0$,

$$u_i(j) = -\mu_j,\ i \geq 0. \tag{1.21}$$

Indeed, the families U and $U(j)$, $j \geq 0$, being constant, the distributions $\mathcal{G}\text{-}\mathcal{P}(U)$ and $\mathcal{G}\text{-}\mathcal{P}[U(j)]$, $j \geq 0$, reduce to the Poisson laws with parameters λ and μ_j respectively. We add that the distribution $\{a_i, i \geq 0\}$ of X is recognized to be $\mathcal{G}\text{-}\mathcal{P}(V)$.

Interpretation. Returning to the first-crossing problem, we can easily visualize the decomposition formula. Consider the boundaries $\mathcal{B}_U = \{(-u_i, i),$ $i \geq 0\}$ and $\mathcal{B}_V = \{(-v_i, i), i \geq 0\}$ associated with U and V. By hypothesis, these boundaries are positive increasing, and $\mathcal{B}_V$ is above $\mathcal{B}_U$. Let Z and X be the first-crossing levels of the Poisson process in $\mathcal{B}_U$ and $\mathcal{B}_V$ respectively. By (1.3), the laws of Z and X are $\mathcal{G}\text{-}\mathcal{P}(U)$ and $\mathcal{G}\text{-}\mathcal{P}(V)$. Obviously, Z is larger that X. In fact, as illustrated in Figure 2, Z can be decomposed into $X + Y$ where the r.v. Y given $X = j$, $j \geq 0$, represents the first-crossing level of the Poisson process starting now at $(-v_j, j)$ in the boundary $\mathcal{B}_U$. The law of Y given $X = j$, $j \geq 0$, is the same as the distribution of the first-crossing level of the Poisson process starting at $(0, 0)$ in the boundary $\{[-(u_{j+i} - v_j), i], i \geq 0\}$, which is $\mathcal{G}\text{-}\mathcal{P}[U(j)]$ by (1.3).

We note that in the Poisson case, both boundaries are vertical, while in the Gani-Shanbhag case, only the farthest boundary $\mathcal{B}_U$ is vertical.

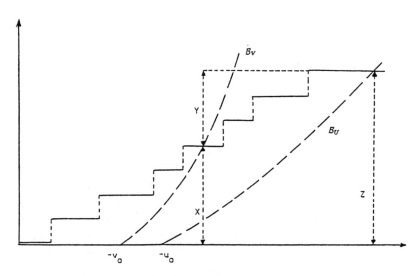

Figure 2. First-crossing levels Z and X of a Poisson process in the lower boundaries $\mathcal{B}_U$ and $\mathcal{B}_V$ respectively.

3.2   An associated distribution of binomial type

It is well-known that the Poisson law is connected with the binomial law through various properties. To show that in a simple way, let us observe again a Poisson process with rate 1 during the time intervals $(0, t)$ and $(0, t+s)$, $t, s > 0$. Given that the number of events generated during $(0, t+s)$ is equal to $n (\geq 0)$, we then obtain that the number of events during $(0, t)$ has a binomial law with exponent n and parameter $t/(t + s)$.

We first point out that similarly, it is possible to associate with Gontcharoff-Poisson distributions new distributions of binomial type. Consider two families U and V satisfying the conditions (1.16), let $\mathcal{B}_U$ and $\mathcal{B}_V$ be the corresponding boundaries and denote by Z and X the first-crossing levels of the Poisson process in $\mathcal{B}_U$ and $\mathcal{B}_V$ respectively. Then, the distribution of X given $Z = n$, $n \geq 0$, is as follows (in obvious notation).

Proposition 3.3.

$$P(X = j|n, U, V) = G_j(0|V)G_{n-j}[0|U(j)]/G_n(0|U), \quad 0 \leq j \leq n, \quad (1.22)$$

where $U(j) = \{u_i(j), i \geq 0\}$ is defined by (1.17).

Proof. We have

$$P(X = j|Z = n) = P(X = j)P(Z = n|X = j)/P(Z = n), \; 0 \leq j \leq n. \quad (1.23)$$

Arguing exactly as above in the interpretation of Proposition 3.2, we then deduce (1.22) from (1.23). ◊

Special case. Formula (1.22) becomes explicit when the boundaries $\mathcal{B}_U$ and $\mathcal{B}_V$ are (increasing) linear. Let $U^l = \{u_i^l, i \geq 0\}$ and $V^l = \{v_i^l, i \geq 0\}$, where

$$u_i^l = -a - bi \text{ and } v_i^l = -c - di, \text{ with } c > a > 0, \, b, d \geq 0, c + dn > a + bn. \tag{1.24}$$

From (1.53) we get

$$P(X = j | n, U^l, V^l) = \binom{n}{j} c(c + dj)^{j-1}(a - c + bj - dj) \tag{1.25}$$

$$(a - c + bn - dj)^{n-j-1}/a(a + bn)^{n-1}, 0 \leq j \leq n.$$

When $b = 0$, that is if $\mathcal{B}_U$ is vertical, (1.25) reduces to the quasi-binomial law of kind I; see [CON89]. When $b = d$, that is if $\mathcal{B}_U$ and $\mathcal{B}_V$ are parallel, (1.25) becomes the quasi-binomial law of kind II ([CON89]). When $b = d = 0$, that is if $\mathcal{B}_U$ and $\mathcal{B}_V$ are vertical, (1.25) reduces, of course, to the binomial law with exponent n and parameter c/a.

Definition 3.4. Given $n \geq 0$ and U, V satisfying (1.16), the family

$$\{G_j(0|V)G_{n-j}[0|U(j)]/G_n(0|U), 0 \leq j \leq n\} \tag{1.26}$$

is called the Gontcharoff-binomial distribution with exponent n and parameters U, V (denoted by $\mathcal{G}$-b(n, U, V)).

Now, let us come back to the decomposition problem of $\mathcal{G}$-$\mathcal{P}(U)$ laws discussed in Paragraph 3.1. That will allow us to reinforce Proposition 3.3 as follows.

Proposition 3.5. Let U and V be two families satisfying (1.16). Suppose that Z is a non-defective $\mathcal{G}$-$\mathcal{P}(U)$ r.v. that can be decomposed into the sum $X + Y$ of two r.v.'s X, Y. Then, X follows the $\mathcal{G}$-$\mathcal{P}(V)$ law if and only if the law of X given $Z = n$, $n \geq 0$, is $\mathcal{G}$-b(n, U, V).

Proof. We begin with the necessity part. By Proposition 3.2, the law of Y given $X = j$, $j \geq 0$, is $\mathcal{G}$-$\mathcal{P}[U(j)]$. Therefore, the result follows, as (1.22), directly from (1.23). For the sufficiency part, we obtain from (1.15) and (1.26) that

$$P(X = j) = \sum_{n=j}^{\infty} P(Z = n)P(X = j | Z = n)$$

$$= \sum_{n=j}^{\infty} e^{u_n} G_j(0|V)G_{n-j}[0|U(j)]$$

$$= e^{v_j} G_j(0|V) \sum_{n=0}^{\infty} e^{u_n(j)} G_n[0|U(j)], \, j \geq 0. \tag{1.27}$$

Since the $\mathcal{G}\text{-}\mathcal{P}[U(j)]$ laws, $j \geq 0$, are non-defective by hypothesis, we deduce from (1.27) that X is of $\mathcal{G}\text{-}\mathcal{P}(V)$ law as required. $\Diamond$

Several other properties can be established within that framework. In particular, we announce that the characterization of the Poisson distribution due to [RR64] can be generalized to $\mathcal{G}\text{-}\mathcal{P}(U)$ distributions under certain stringent conditions.

4 Two related first-crossing problems

We are going to present two variants of the original first-crossing problem that can be still analyzed using A-G polynomials. We indicate that by way of a generalization of these polynomials, it becomes possible to tackle a number of other situations [PL96b].

4.1 Extension to a linear birth proces with immigration

Instead of a Poisson process, let us consider a linear birth process with immigration, the rates λ_i, $i \geq 0$, being of the form

$$\lambda_i = \lambda + \beta i, \text{ with } \lambda > 0 \text{ and } \beta \geq 0; \tag{1.28}$$

let $\wedge = \{\lambda_i, i \geq 0\}$. We suppose that the process is initiated in state 0 at time $-x(\in \mathbb{R})$. As before, we denote by $\mathcal{B}$ the lower boundary (1.1) constructed from a family U satisfying (1.2). Let N be the first-crossing level of the process in $\mathcal{B}$.

Proposition 4.1.

$$P(N = n | \wedge, x, U) = [\lambda_0 \lambda_1 \ldots \lambda_{n-1}/(-\beta)^n] e^{-\lambda_0 x + \lambda_n u_n} G_n(e^{-\beta x} | \tilde{U}), \; n \geq 0, \tag{1.29}$$

where $\tilde{U}$ is the family $\{\tilde{u}_i = e^{-\beta u_i}, i \geq 0\}$.

Proof. The argument is quite similar to that followed for Proposition 2.1. By looking at the instant of the first jump of the process, we have

$$P(N = 0 | \wedge, x, U) = e^{-\lambda_0(x - u_0)}, \tag{1.30}$$

and for $n \geq 1$,

$$P(N = n | \wedge, x, U) = \int_0^{x - u_0} P(N = n - 1 | E \wedge, t - x, EU) \lambda_0 e^{-\lambda_0 t} \, dt, \tag{1.31}$$

with $E \wedge = \{\lambda_{1+i}, i \geq 0\}$. Let us put

$$P(N = n | \wedge, x, U) = [\lambda_0 \lambda_1 \ldots \lambda_{n-1}/(\lambda_0 - \lambda_1)^n] K_n[e^{(\lambda_0 - \lambda_1)x} | \wedge, \tilde{U}], \; n \geq 0. \tag{1.32}$$

where $\tilde{U}$ is defined as above and $K_n(.)$ represents some appropriate function. Inserting (1.32) in (1.30) and (1.31), and since $\lambda_i - \lambda_{i+1} = -\beta, i \geq 0$, we obtain that

$$K_0(e^{-\beta x}|\wedge, \tilde{U}) = 1, \tag{1.33}$$

$$K_n(e^{-\beta x}|\wedge, \tilde{U}) = \int_{u_0}^{x}(-\beta)K_{n-1}(e^{-\beta \tau}|E\wedge, E\tilde{U})e^{-\beta \tau}\, d\tau, \, n \geq 1. \tag{1.34}$$

Writing $y = e^{-\beta x}$ and $z = e^{-\beta \tau}$, (1.33) and (1.34) become

$$K_0(y|\wedge, \tilde{U}) = 1, \tag{1.35}$$

$$K_n(y|\wedge, \tilde{U}) = \int_{\tilde{u}_0}^{y} K_{n-1}(z|E\wedge, E\tilde{U})\, dz, \, n \geq 1. \tag{1.36}$$

From (1.35), (1.36), we observe that $K_n(y|\wedge, \tilde{U})$, $n \geq 0$, does not depend on $\wedge$ and is a polynomial of degree n in y. By (1.48),(1.49), we then see that this polynomial is just $G_n(y|\tilde{U})$, so that (1.32) yields (1.29). $\Diamond$

We note that the proposition remains valid when $\beta < 0$ provided that there exists some natural i_0 such that $\lambda_{i_0} = 0$. Then, the process stops at i_0 and the probabilities (1.29) are equal to 0 for $n \geq i_0 + 1$.

Special cases. It is easily checked that formula (1.29) generalizes (1.3). It also covers some other standard results.

Let $\{X_t, t \geq 0\}$ be a pure linear birth process with parameter λ and initiated by 1 individual at time 0. Then, the process $\{\hat{X}_t \equiv X_t - 1, t \geq 0\}$ falls within the preceding theory with $x = 0$ and $\beta = \lambda$ in (1.28). Therefore, the marginal law of X_t, for example, is given by

$$P(X_t = n) = P[N = n - 1|\{\lambda(1 + i), i \geq 0\}, 0, \{u_i = -t, i \geq 0\}], \, n \geq 1. \tag{1.37}$$

Using (1.29) and (1.53), we deduce from (1.37) that X_t does have a geometric law with parameter $e^{-\lambda t}$.

Let $\{Y_t, t \geq 0\}$ be a pure linear death process with parameter μ and initiated by i_0 individuals at time 0. Then, the process $\{\hat{Y}_t \equiv i_0 - Y_t, t \geq 0\}$ corresponds to the particular case where $x = 0$ and in (1.28), $\lambda = \mu i_0$ and $\beta = -\mu(< 0)$. Thus, we get, for example, that

$$P(Y_t = n) = P[N = i_0 - n|\{\mu(i_0 - i), i \geq 0\}, 0, \{u_i = -t, i \geq 0\}], 0 \leq n \leq i_0, \tag{1.38}$$

so that by (1.29) and (1.53), Y_t does have a binomial law with exponent n and parameter $e^{-\mu t}$.

4.2 First-crossing of a Poisson process in an upper birth process with linearly decreasing rates

We here examine the first-crossing between a Poisson process starting at time 0 in level $r(\geq 0)$ and with rate λ, and an upper birth process starting at time 0 in level $m(\geq r)$, stopping in level $k(\geq m)$ and with linearly decreasing rates λ_i, $i = m, m+1, \ldots, k$, of the form

$$\lambda_i = \alpha(k - i), \text{ with } \alpha > 0; \qquad (1.39)$$

put $l = k - m(\geq 0)$. We underline that this model is thus rather different from the original one. Now, let N be the first-crossing level between the two processes, and consider the r.v. $k - N$ which is valued in $\{0, 1, \ldots, l\}$. Its p.g.f. is given below (with $l_{[n]}$ denoting $\binom{l}{n} n!$, $n \geq 0$).

Proposition 4.2. *For $x \in \mathbb{R}$,*

$$E(x^{k-N}) = \sum_{n=0}^{l} l_{[n]} u_n^{k-r-n} G_n(x|U), \qquad (1.40)$$

where U is the family $\{u_i = \lambda/(\lambda + \alpha i), i \geq 0\}$.

Proof. Clearly, we may write, for $x \in \mathbb{R}$,

$$E(x^{k-N}) = \varphi(x|r, l), \qquad (1.41)$$

with $0 \leq r \leq m$ and $l \geq 0$. Let U be defined as above. If $r \leq m - 1$ and $l \geq 1$, the first jump that occurs after time 0 is concerned either with the Poisson process with probability u_n, or with the birth process with probability $1 - u_n$, which leads to

$$\varphi(x|r, l) = u_n \varphi(x|r + 1, l) + (1 - u_n)\varphi(x|r, l - 1). \qquad (1.42)$$

If $l = 0$, then $k - N = 0$, while if $r = m$, then $k - N = k - m = l$, yielding

$$\varphi(x|r, 0) = 1 \text{ and } \varphi(x|m, l) = x^l. \qquad (1.43)$$

It is easy to derive a family of particular solutions to (1.42), namely

$$l_{[n]} u_n^{k-r-n}, 0 \leq n \leq l. \qquad (1.44)$$

Since $\varphi(x|r, l)$ is a polynomial in x of degree l, we look for an expression of the form

$$\varphi(x|r, l) = \sum_{n=0}^{l} l_{[n]} u_n^{k-r-n} L_n(x), \qquad (1.45)$$

where $L_n(x)$ is some appropriate polynomial of degree n that does not depend on r and l. By construction and using (1.44), we see that (1.42) is automatically satisfied. Moreover, (1.43) will hold true provided that the $L_n(x)$'s satisfy the relations

$$L_0(x) = 1, \tag{1.46}$$

$$\sum_{n=0}^{l} l_{[n]} u_n^{l-n} L_n(x) = x^l, \, l \geq 0. \tag{1.47}$$

Comparing (1.46), (1.47) with (1.48), (1.55), we deduce that $L_n(x)$, $n \geq 0$, is just $G_n(x|U)$, and thus (1.45) becomes (1.40). ◊

Appendix

Let U be any given family of real numbers $\{u_i, i \geq 0\}$.

Abel-Gontcharoff polynomials

To U is attached a unique family of Abel-Gontcharoff polynomials $G_n(x|U)$ of degree n in x, $n \geq 0$, defined by

$$G_0(x|U) = 1, \tag{1.48}$$

$$G_n(x|U) = \int_{u_0}^{x} \int_{u_1}^{\xi_0} \int_{u_1}^{\xi_1} \cdots \int_{u_{n-1}}^{\xi_{n-2}} d\xi_1 \, d\xi_2 \, d\xi_3 \ldots d\xi_{n-1}, n \geq 1. \tag{1.49}$$

Therefore, putting $EU = \{u_{1+i}, i \geq 0\}$, we obtain by differentiation that they are characterized equivalently by

$$G_n^{(1)}(x|U) = G_{n-1}(x|EU), n \geq 1, \tag{1.50}$$

$$G_n(u_0|U) = \delta_{n,0}, n \geq 0. \tag{1.51}$$

The only particular family which can be written explicitely is the sequence of Abel polynomials. These correspond to the special case where $u_i, i \geq 0$, is linear in i, say

$$u_i = a + bi, \text{ with } a, b \in \mathbf{R}, \tag{1.52}$$

and they are given by

$$G_n(x|U) = (x - a)(x - a - bn)^{n-1}/n!, n \geq 0. \tag{1.53}$$

Of course, when $b = 0$, then $G_n(x|U) = (x - a)^n/n!$, $n \geq 0$.

A crucial advantage of the A-G polynomials is that they allow to generalize, under certain conditions, Taylor series expansions into Abel type expansions. In particular, any polynomial $R(x)$ of degree l in x, $l \geq 1$, can be expanded as

$$R(x) = \sum_{n=0}^{l} R^{(n)}(u_n) G_n(x|U). \tag{1.54}$$

As a special case, the A-G polynomials can be constructed recursively by

$$\frac{x^l}{l!} = \sum_{n=0}^{l} \frac{u_n^{l-n}}{(l-n)!} G_n(x|U), \ l \geq 1, \tag{1.55}$$

together with $G_0(x|U) = 1$.

A simple operational property is that

$$G_n(a + bx | a + bU) = b^n G_n(x|U), n \geq 0, \tag{1.56}$$

where $a, b \in \mathbb{R}$ and $a + bU = \{a + bu_i, i \geq 0\}$.

Appell polynomials

To U is attached a unique family of Appell polynomials $A_n(x|U)$ of degree n in x, $n \geq 0$, defined by

$$A_0(x|U) = 1, \tag{1.57}$$

$$A_n(x|U) = \int_{u_{n-1}}^{x} \int_{u_{n-2}}^{\xi_0} \int_{u_{n-3}}^{\xi_1} \cdots \int_{u_0}^{\xi_{n-2}} d\xi_0 \, d\xi_1 \, d\xi_2 \ldots d\xi_{n-1}. \tag{1.58}$$

Thus, they are characterized by

$$A_n^{(1)}(x|U) = A_{n-1}(x|0), n \geq 1, \tag{1.59}$$

$$A_n(u_{n-1}|U) = \delta_{n,0}, n \geq 0, \tag{1.60}$$

u_{-1} being arbitrary. Note that contrary to (1.50), differentiation in (1.59) does not interfere with U. That explains why most often the argument U is omitted in the notation of $A_n(x|U)$.

We underline the apparent similarity between the A-G and Appell polynomials. In fact, the two families are directly connected since

$$G_n(x | \{u_o, u_1, \ldots, u_{n-1}\}) = A_n(x | \{u_{n-1}, u_{n-2}, \ldots, u_o\}), n \geq 1. \tag{1.61}$$

This identification, however, does depend on the value of n. That has no real consequence on local properties of the polynomials, but the distinction becomes important for global properties involving the whole families.

Appell polynomials have received much more attention in the literature. They cover standard subfamilies such as Hermite, Laguerre, Bernoulli and Euler polynomials. For a brief presentation see, e.g., [KAZ88].

5 References

[CON89] P.C. CONSUL. *Generalized Poisson distributions. Properties and Applications.* Marcel Dekker, New York, 1989.

[DAN63] H.E. DANIELS. The Poisson process with a curved absorbing boundary. In *Bull. Intern. Statist. Inst. 40, 994-1008*, 1963.

[DAN67] H.E. DANIELS. The distribution of the total size of an epidemic. In *Proc. 5th Berkeley Symp. Math. Statist. Prob. 4, 281-293*, 1967.

[DEN95] M. DENUIT. A new distribution of Poisson type for the number of claims. Submitted, 1995.

[DUR71] J. DURBIN. Boundary-crossing probabilities for the Brownian motion and Poisson processes and techniques for computing the power of the Kolmogorov-Smirnov. In *J. Appl. Prob. 8, 431-453*, 1971.

[GAL66] S.F.L. GALLOT. Asymptotic absorption for a Poisson process. In *J. Appl. Prob. 3, 445-452*, 1966.

[GAL93] S.F.L. GALLOT. Absorption and first-passage times for a compound poisson process in a general upper boundary. In *J. Appl. Prob. 30, 835-850*, 1993.

[GON37] W. GONTCHAROFF. *Détermination des Fonctions Entières par Interpolation.* Hermann, Paris, 1937.

[GS74] J. GANI and D.N. SHANBHAG. An extension of Raikov's theorem derivable from a result in epidemic theory. In *Z. Wahrscheinlichkeitstheorie verw. Gebiete 29, 33-37*, 1974.

[KAZ88] Y.A. KAZ'MIN. Appell polynomials. In *Encyclopedia of Mathematics 1.* Reidel Kluwer, Dordrecht, 209-210, 1988.

[LP90] Cl. LEFEVRE and Ph. PICARD. A non standard family of polynomials and the final size distribution of Reed-Frost epidemic processes. In *Adv. Appl. Prob. 22, 25-48*, 1990.

[LP95] Cl. LEFEVRE and Ph. PICARD. First busy period for general D/M/1 and M/D/1 queues. Working paper, 1995.

[PL90] Ph. PICARD and Cl. LEFEVRE. A unified analysis of the final size and severity distribution in collective Reed-Frost epidemic processes. In *Adv. Appl. Prob. 22, 269-294*, 1990.

[PL94] Ph. PICARD and Cl. LEFEVRE. On the first crossing of the surplus process with a given upper barrier. In *Insurance : Math. Econ. 14, 163-179*, 1994.

[PL96a] Ph. PICARD and Cl. LEFEVRE. Abel expansions and generalized Abel polynomials in stochastic models. In *This volume*, 1996.

[PL96b] Ph. PICARD and Cl. LEFEVRE. First crossing of basic counting processes with lower non-linear boundaries : a unified approach with pseudopolynomials (I). In *Adv. Appl. Prob.*, to appear, 1996.

[PL96c] Ph. PICARD and Cl. LEFEVRE. The probability of ruin in finite time with discrete claim size distribution. In *Scand. Actuarial J.*, to appear, 1996.

[PYK59] R. PYKE. The supremum and infimum of the Poisson process. In *Ann. Math. Statist. 30, 568-576*, 1959.

[RAI37] D. RAIKOV. On the decomposition of Poisson laws. In *C.R. (Doklady). Acad. Sci. U.R.S.S. 14, 9-11*, 1937.

[RR64] C.R. RAO and H. RUBIN. On the characterization of the Poisson distribution. In *Sankhya A 26, 295-298*, 1964.

[STA93] W. STADJE. Distribution of first-exit times for empirical counting and Poisson processes with moving boundaries. In *Commun. Stat.-Stoch. Models 9, 91-103*, 1993.

[WHI61] P. WHITTLE. Some exact results for one-sided distribution tests of the Kolmogorov-Smirnov type. In *Ann. Math. Statist. 61, 499-505*, 1961.

[ZAC91] S. ZACKS. Distributions of stopping times for Poisson processes with linear boundaries. In *Commun. Stat.-Stoch. Models 7, 233-242*, 1991.

Explicit Rates of Convergence of Stochastically Ordered Markov Chains *

D.J. Scott and R. L. Tweedie[†]

ABSTRACT Let $\boldsymbol{\Phi} = \{\Phi_n\}$ be a Markov chain on a half-line $[0, \infty)$ that is stochastically ordered in its initial state. We find conditions under which there are explicit bounds on the rate of convergence of the chain to a stationary limit π: specifically, for suitable rate functions r which may be geometric or subgeometric and "moments" $f \geq 1$, we find conditions under which

$$r(n) \sup_{|g| \leq f} |\mathsf{E}_x[g(\Phi_n)] - \pi(g)| \leq M(x),$$

for all n and all x. We find bounds on $r(n)$ and $M(x)$ both in terms of geometric and subgeometric "drift functions", and in terms of behaviour of the hitting times on $\{0\}$ and on compact sets $[0, c]$ for $c > 0$. The results are illustrated for random walks and for a multiplicative time series model.

1   Introduction

It is with respect that we dedicate this paper to Joe Gani and to the memory of Ted Hannan. Both of us were taught as undergraduates by Ted Hannan; both of us were given early chances in academic and research environments by Joe Gani. We both owe much to them for our introduction to the world of applied probability and time series, and it is a pleasure to be able to repay these debts a little in this volume.

Markov chains provide the underlying structure for most of the models in epidemiology and other phenomena studied by Joe Gani over many years; and through the use of state space models it is becoming more and more clear that Markovian systems also provide an appropriate vehicle for studying broad aspects of the time series models in whose development

---

[*]Work supported in part by NSF Grants DMS-9205687 and DMS-9504561

[†]Postal Address: Department of Statistics, Colorado State University, Fort Collins CO 80523, USA

[0]Keywords: Irreducible Markov chains, invariant measures, geometric ergodicity, subgeometric ergodicity, random walks, state space models, multiplicative time series

AMS Subject classification: 60J10, 60K05

Ted Hannan stood so pre-eminent. The results here add, we hope, one more small building block to the structure of such models.

In this paper we will study some convergence properties of a discrete–time Markov chain $\mathbf{\Phi} = \{\Phi_n : n \in \mathbb{Z}_+\}$ evolving on a half-line $\mathsf{X} := [0, \infty)$. Our notation will follow that of Meyn and Tweedie [8], even though our space is not general as in [8]. We denote the transition probabilities of the chain by $P^n(x, A)$, $x \in \mathsf{X}$, $A \in \mathcal{B}(\mathsf{X})$ where $\mathcal{B}(\mathsf{X})$ is the usual Borel σ–field: that is, $P^n(x, A) = \mathsf{P}_x\{\Phi_n \in A\}$, $n \in \mathbb{Z}_+ = \{0, 1, \ldots\}$ where P_x and E_x denote respectively the probability law and expectation of the chain under the initial condition $\Phi_0 = x$. When the initial distribution is a general probability measure λ, the corresponding quantities are denoted by $\mathsf{P}_\lambda, \mathsf{E}_\lambda$. For any non-negative function f, we write Pf and $P^n f$ for the functions $\int P(x, dy) f(y)$ and $\int P^n(x, dy) f(y)$, and for any (signed) measure μ we write $\mu(f)$ for $\int \mu(dy) f(y)$. We write also $\tau_C = \min\{n \geq 1 : \Phi_n \in C\}$ for the hitting time of the chain on any set C, and for singleton sets we write τ_j for $\tau_{\{j\}}$.

Our central assumption is that the chain is *stochastically ordered in its initial state*. Here, we say that a random variable X is stochastically larger than a random variable Y if $\mathsf{P}(X \leq x) \leq \mathsf{P}(Y \leq x)$ for all $x \in \mathsf{X}$; and we require that for two copies of the chain Φ_n^1, Φ_n^2, whenever $\Phi_0^1 = x$ and $\Phi_0^2 = y$ and $x > y$, then Φ_n^1 is also stochastically larger than Φ_n^2. It then follows [14, 4] that this can be extended to *pathwise ordering*: we can construct Φ_n^1, Φ_n^2 so that in fact the sample paths themselves are ordered in the sense that $\Phi_n^1 \geq \Phi_n^2$ for all n when $\Phi_0^1 \geq \Phi_0^2$. Many chains satisfy these assumptions, especially in the queueing literature [14, 4].

We need an appropriate irreducibility and aperiodicity structure, and we assume throughout that for every x there is some n such that $P^n(x, 0) > 0$. If we put $\psi := \sum_n P^n(0, \cdot)$ then it is clear that the chain is ψ-irreducible and (because of the stochastic monotonicity) aperiodic (see [8, Chapters 4, 5]). We also assume that for every y we have $\psi(y, \infty) > 0$; if this does not hold then we can truncate the space to get essentially all of the results below, at least on the support of ψ.

Our goal is to find explicit bounds on rates of convergence of the transition probabilities P^n to π, which denotes a (necessarily unique) invariant or limiting probability measure for the chain. These extend (and render much sharper) the existence results developed in, for example, [8, Chapters 15, 16] and [15, 12, 10, 11, 7, 16] where one typically finds conditions under which there exists some "rate function" $r(n)$ such that

$$r(n)\|P^n(x, \cdot) - \pi\| \to 0, \qquad n \to \infty \tag{1}$$

where $\|\mu\| = \sup_{|g| \leq 1} |\mu(g)|$ denotes the usual total variation norm for a signed measure μ.

Here the rate function can be any sequence of positive numbers $r(n) \to \infty$: the most common examples include geometric sequences such as $r(n) =$

$r^n, r > 1$ as in [8] or subgeometric sequences such as $r(n) = n^\beta, \beta > 0$ as in [15].

Although the total variation norm is commonly used in convergence results, it is possible to consider stronger forms of convergence using the "f–norm" $\|\mu\|_f := \sup_{|g| \leq f} |\mu(g)|$ where $f \geq 1$ is a more general function ([8, Chapter 14]). We then consider rate functions such that

$$r(n)\|P^n(x, \cdot) - \pi\|_f \leq M(x) \tag{2}$$

for some bounding function $M(x) < \infty$. Note that (2) can be equivalently written in terms of convergence of moments: often the more appealing formulation is

$$\sup_{|g| \leq f} |\mathsf{E}_x[g(\Phi_n)] - \pi(g)| \leq M(x)[r(n)]^{-1}. \tag{3}$$

In all of these contexts we will be interested in this paper in the actual values of the rate function (for example, in the geometric case with $r(n) = r^n$ we are interested in the largest value that r can take) as well as the bounding function $M(x)$.

This appears to be a deep problem in general, although recently some very generally applicable (although far from tight) bounds on these quantities have been found in [9], [13] and [1]. When the chain is stochastically ordered, however, it is possible to get much more explicit and even tight results. In the geometric case, it is shown in [5] that the rate of convergence is bounded by the (common in x) radius of convergence r^* of the generating functions $\mathsf{E}_x[z^{\tau_0}]$ and under minor extra conditions is actually identical with r^*.

In Section 2 we generalise this result to find a reasonable (although not as tight) bound in terms of radii of convergence of the more general generating functions $\mathsf{E}_x[z^{\tau_C}]$ where $C = [0, c]$ for $c > 0$: in practice this then enables us to use versions of the drift conditions (or geometric Foster-Lyapunov conditions) $PV \leq \lambda V + b \mathbb{1}_C$ for such sets C to get bounds on the rate of convergence. These extensions widen the range of practical implementation of the methods introduced in [5] considerably, as we show in Section 4.

In Section 3 we then study subgeometric rate functions using sequences $r(n)$ in the class Λ discussed in detail in [15]: this includes for example all functions of the form

$$r(n) = n^\beta [\log n]^\gamma \vee 1, \qquad \beta > 0, \ \gamma > 0. \tag{4}$$

Again we show that if the distribution of τ_0 has suitably bounded tails, of "order r(n)", then the convergence in (2) can be described explicitly.

In Section 4 we apply these results to a number of stochastically monotone chains that arise in storage theory and in time series analysis, and calculate explicit bounds for rates of convergence in both these contexts.

2   Geometric Rates of Convergence

In [5] it is shown that, for the case of geometric rate functions $r(n) = r^n$, there is an almost complete characterisation of the best value of r in (2). Our first result is essentially deducible from Theorems 3.1 and 5.1 of [5], with a slightly different form of the bounding function M.

Theorem 2.1 *Suppose* Φ *is an irreducible stochastically ordered Markov chain. If there exists some* $x > 0$ *and* $r > 1$ *such that* $\mathsf{E}_x[r^{\tau_0}] < \infty$ *then there exists a stationary distribution* π *such that*

$$\sum r^n \|P^n(x, \cdot) - \pi\| \le M_0(x) \tag{5}$$

for every n, where

$$M_0(x) \le \frac{1}{r-1}\left[\mathsf{E}_x[r^{\tau_0}] + \frac{r}{r-1}\left[\mathsf{E}_0[r^{\tau_0}] - 1\right]\right] < \infty. \tag{6}$$

Proof

The existence of π and the geometric ergodicity follow from the finiteness of $\mathsf{E}_x[r^{\tau_0}]$ for some $r > 1$ as in [8, Chapter 15]. The result we focus on here is the fact that the same r gives a convergence rate. To prove this we run simultaneous copies Φ_n^1, Φ_n^2 starting from x and π and denote the coupling time at $\{0\}$ by

$$T = \min\{n \ge 1 : \Phi_n^1 = \Phi_n^2 = 0\}.$$

By the pathwise ordering we know that when the chain with the larger starting value is in $\{0\}$ then the chain with the smaller starting value is also in $\{0\}$; consequently we have

$$P_{x,\pi}(T > n) \le P_\nu(\tau_0 > n) \tag{7}$$

where ν is the distribution of $\max(x, X)$ and X has the distribution π. The stochastic monotonicity thus reduces the two-dimensional calculation of T to a one-dimensional calculation involving the tails of τ_0.

Using the stationarity of π in the normal coupling manner, and applying (7), for any $A \in \mathcal{B}(\mathsf{X})$

$$\sum_n r^n |P^n(x, A) - \pi(A)| \le \sum_n r^n P_\nu(\tau_0 > n). \tag{8}$$

Now the right side of (8) is bounded by

$$[r-1]^{-1} \sum_{n=1}^\infty r^n P_\nu(\tau_0 = n) \le [r-1]^{-1}\left[\mathsf{E}_x[r^{\tau_0}] + \mathsf{E}_\pi[r^{\tau_0}]\right].$$

The first term here is finite by assumption. Using the remark after Theorem 3.6 of [15] and rearranging terms we we also have

$$\mathsf{E}_\pi[r^{\tau_0}] = \pi(0)\mathsf{E}_0[\sum_1^{\tau_0} r^k] = \pi(0)\frac{r}{r-1}\left[\mathsf{E}_0[r^{\tau_0}] - 1\right].$$

This is clearly finite by stochastic monotonicity and we have the required result. □

Hence if r^* is the radius of convergence of $E_0[r^{\tau_0}]$ then for all $r < r^*$ we have (2) with $f = 1$. Geometric "drift conditions" now give us an approach to bounding r^*, and indeed getting even more. Suppose that $V \geq 1$ and that for some $\lambda < 1$ and $b < \infty$

$$PV(x) \leq \lambda V(x) + b\mathbb{1}_{\{0\}}(x). \tag{9}$$

Then as in [8, Theorem 15.2.5], the function $V(x)$ leads to an upper bound on $E_x[r^{\tau_0}]$. We will make use of the following lemma, which is a discrete time version of the result in [6], and which gives a slightly better and more explicit bound of this form.

Lemma 2.2 *Suppose that*

$$PV \leq \lambda_C V + b_C \mathbb{1}_C \tag{10}$$

is satisfied for some $V \geq 1$ and some set C. Then for any $r \leq \lambda_C^{-1}$,

$$V(x) \geq E_x[r^{\tau_C}] + (1 - r\lambda_C)E_x\Big[\sum_0^{\tau_C - 1} r^k V(\Phi_k)\Big]. \tag{11}$$

PROOF

If we define the function $Z_k(\Phi_k) = r^k V(\Phi_k)$ then from (10) we get

$$E_x[E[Z_k|\mathcal{F}_{k-1}] - Z_{k-1}] = E_x[r^k E[V(\Phi_k)|\mathcal{F}_{k-1}] - r^{k-1}V(\Phi_{k-1})]$$

$$\leq E_x[(r\lambda_C - 1)r^{k-1}V(\Phi_{k-1}) + b_C r^k \mathbb{1}_C(\Phi_{k-1})]. \tag{12}$$

Use the discrete form of Dynkin's formula (see [8, Theorem 11.3.1]) with the stopping time $\tau^n = \inf\{\tau_C, n, \sup(n : V(\Phi_n) \leq n)\}$ for fixed n to get

$$E_x[r^{\tau^n}V(\Phi_{\tau^n})] = V(x) + E_x\Big[\sum_1^{\tau^n}[E[Z_k|\mathcal{F}_{k-1}] - Z_{k-1}]\Big]$$

$$\leq V(x) + (r\lambda_C - 1)E_x\Big[\sum_1^{\tau^n} r^{k-1}V(\Phi_{k-1})\Big]. \tag{13}$$

Since $V \geq 1$ we thus have

$$V(x) \geq E_x[r^{\tau^n}] + (1 - r\lambda_C)E_x\Big[\sum_1^{\tau^n} r^{k-1}V(\Phi_{k-1})\Big].$$

Now from (10), it follows from [8, Chapter 15] that τ_C is almost surely finite from any initial point x. Hence $\tau^n \to \tau_C$ as $n \to \infty$ and the result follows by Fatou's Lemma. □

Suppose now that V satisfies (9): from Lemma 2.2 with $r = \lambda^{-1}$ we have $V(x) \geq E_x[r^{\tau_0}]$ and hence from Theorem 2.1 the rate of convergence is at least λ^{-1}, as was shown in [5]. However, we can get rather more out of (9). The next result is close to [5, Theorem 2.1], although the expressions for the bounding constants are not the same and the method of proof exploits the stochastic monotonicity more obviously.

Theorem 2.3 *Suppose that $V \geq 1$ is monotone increasing and satisfies (9). Then for any $r < \lambda^{-1}$,*

$$\sum_0^\infty \|P^n - \pi\|_V r^n \leq \frac{2}{1 - r\lambda}[V(x) + \frac{b}{1 - \lambda}] < \infty. \tag{14}$$

PROOF

We need to refine the coupling argument used in (8). Let us now run copies Φ_n^1, Φ_n^2 starting from x and 0, and other copies Φ_n^3, Φ_n^4 starting from π and 0, and denoting the coupling times at $\{0\}$ by T_1 and T_2 in the respective cases. The same monotonicity conditions as were utilised in (8) show that for any g such that $|g| \leq V$ we have (from the monotonicity of V) that

$$|\mathsf{E}_x[g(\Phi_n)] - \mathsf{E}_\pi[g(\Phi_n)]|$$

$$\leq \quad \mathsf{E}_{x,0}[|g(\Phi_n^1)|\mathbb{1}(T_1 > n)] + \mathsf{E}_{x,0}[|g(\Phi_n^2)|\mathbb{1}(T_1 > n)]$$

$$+ \mathsf{E}_{\pi,0}[|g(\Phi_n^3)|\mathbb{1}(T_2 > n)] + \mathsf{E}_{\pi,0}[|g(\Phi_n^4)|\mathbb{1}(T_2 > n)]$$

$$\leq \quad \mathsf{E}_{x,0}[V(\Phi_n^1)\mathbb{1}(T_1 > n)] + \mathsf{E}_{x,0}[V(\Phi_n^2)\mathbb{1}(T_1 > n)] \tag{15}$$

$$+ \mathsf{E}_{\pi,0}[V(\Phi_n^3)\mathbb{1}(T_2 > n)] + \mathsf{E}_{\pi,0}[V(\Phi_n^4)\mathbb{1}(T_2 > n)]$$

$$\leq \quad 2\mathsf{E}_x[V(\Phi_n)\mathbb{1}(\tau_0 > n)] + 2\mathsf{E}_\pi[V(\Phi_n)\mathbb{1}(\tau_0 > n)].$$

Now from Lemma 2.2 with $r\lambda < 1$ we have

$$\sum_0^\infty \mathsf{E}_x[V(\Phi_n)\mathbb{1}(\tau_0 > n)]r^n \leq (1 - r\lambda)^{-1}V(x).$$

Taking the supremum of $|g| \leq V$ on the left of (15), multiplying by r^n and summing, we have

$$\sum_0^\infty r^n\|P^n - \pi\|_V \leq 2(1 - r\lambda)^{-1}[V(x) + \pi(V)]. \tag{16}$$

Now from (9) and [8, Chapter 14] we have that $\pi(V) < \infty$, and then from (9) it further follows that $\pi(V) \leq b/(1 - \lambda)$; and thus we have (14) as required. $\quad\square$

In many cases it is possible to verify (9) directly, as in [5]. However, it is often the case that one can more easily establish drift to a larger set

182

$C = [0, c]$ in the sense that (10) holds. In this case we will assume that

$$0 < \delta := P(c, 0) = \inf_{x \in C} P(x, 0), \tag{17}$$

where the last equality is a consequence of stochastic monotonicity. The assumption that $\delta > 0$ can often be weakened, since for some n we have $P^n(c, 0) > 0$, and we could work with the n-skeleton in this case: but for ease of exposition we will assume $n = 1$.

Let us further define the bounding constants separately on and off $\{0\}$ in (10) and write

$$PV \leq \lambda_C V + b_C^* \mathbb{1}_{(0,c]} + b_0^* \mathbb{1}_0. \tag{18}$$

Theorem 2.4 *Under (17) and (18), there exists a function V' with $V \leq V' \leq V + b_C^*/\delta$ such that*

$$PV' \leq \lambda V' + b \mathbb{1}_0 \tag{19}$$

where

$$\lambda = [\lambda_C + b_C^*/\delta]/[1 + b_C^*/\delta] < 1 \tag{20}$$

and

$$b = b_0^* + b_C^*/\delta < \infty. \tag{21}$$

Thus Theorem 2.3 holds with this choice of λ, b and with V' in place of V.

PROOF

The construction of V' is given in Theorem 6.1 of [9]. It is shown there that if we choose

$$V'(x) = \begin{cases} V(x) & x = 0 \\ V(x) + b_C^*/\delta & x > 0 \end{cases} \tag{22}$$

then (19) holds with the values of λ and b in (20) and (21). Since this construction maintains the monotonicity of V' if V is monotone, the result follows. □

3 Subgeometric rates of convergence

In this section we use similar techniques to identify rate sequences for which (2) holds, and of course again to identify the bounding constants, when the sequences $r(n)$ are in the class Λ discussed in detail in [15]. We do not define Λ in detail here, since we do not need its specific properties, but note it does cover the class given in (4).

Following [15] we will say that the chain Φ is (f, r)–*ergodic*, where $f \geq 1$ and $r \in \Lambda$, if for some x

$$\mathsf{E}_x\left[\sum_{k=0}^{\tau_0-1} r(k) f(\Phi_k)\right] < \infty \tag{23}$$

and $(1, r)$–*ergodic* if (23) holds for $f \equiv 1$. We will write

$$r^0(n) := \sum_{k=0}^{n} r(k), \qquad \Delta r(n) := r(n) - r(n-1)$$

with the convention $r(-1) = 0$. A $(1, r)$-ergodic chain is then precisely one for which $E_x[r^0(\tau_0)] < \infty$ for some x. Note that by irreducibility, for a given sequence $r(n)$ this expectation is finite for all or for no $x \in \mathsf{X}$.

Using the Remark on p. 786 of [15], we have the relation

$$E_\pi[\sum_0^{\tau_0-1} r(n)f(\Phi_n)] = \pi(0)E_0[\sum_0^{\tau_0-1} r^0(n)f(\Phi_n)] \qquad (24)$$

and we shall use this duality in our next theorem. This parallels the results of Theorem 2.1, and gives conditions for subgeometric rates of convergence in terms of the hitting time probabilities on $\{0\}$.

Theorem 3.1 *Suppose* Φ *is an irreducible, stochastically-ordered Markov chain,* f *is a monotone increasing function on* X, *and* $r(n)$ *is monotone increasing on* $\mathbb{Z}_+$.

(i) If Φ *is* (f, r^0)–*ergodic then*

$$\sum_0^{\infty} r(n) \|P^n(x, \cdot) - \pi\|_f \leq M_0(x) \qquad (25)$$

for every n, *where*

$$M_0(x) \leq 2\Big[E_x[\sum_0^{\tau_0-1} r(n)f(\Phi_n)] + \pi(0)E_0[\sum_0^{\tau_0-1} r^0(n)f(\Phi_n)]\Big] < \infty. \qquad (26)$$

(ii) If Φ *is* (f, r)–*ergodic then for each* x

$$r(n)\|P^n(x, \cdot) - \pi\|_f \to 0, \qquad n \to \infty. \qquad (27)$$

PROOF

To see (26), we use the coupling in (15). From (24) we have

$$2\sum_0^{\infty} r(n)E_x[f(\Phi_n)\mathbb{1}(\tau_0 > n)] \quad + \quad 2\sum_0^{\infty} r(n)E_\pi[f(\Phi_n)\mathbb{1}(\tau_0 > n)]$$

$$= 2E_x[\sum_0^{\tau_0-1} r(n)f(\Phi_n)] + 2\pi(0)E_0[\sum_0^{\tau_0-1} r^0(n)f(\Phi_n)]$$

and the last term is finite as required from our assumption of (f, r^0)-ergodicity.

The second result is proved in two steps. Consider first the case with $f = 1$. Then from the simple coupling as in (8) we have from the monotonicity of $r(n)$ that

$$r(n)\|P^n(x,\cdot) - \pi\| \le r(n)\mathsf{P}_x(\tau_0 > n) + r(n)\mathsf{P}_\pi(\tau_0 > n)$$

$$\le \sum_{j=n+1}^{\infty} r(j-1)\mathsf{P}_x(\tau_0 = j) + \sum_{j=n+1}^{\infty} r(j-1)\mathsf{P}_\pi(\tau_0 = j) \qquad (28)$$

These two terms are respectively the upper tails of the sums

$$\sum_0^{\infty} r(j-1)\mathsf{P}_x(\tau_0 = j) = \mathsf{E}_x \sum_0^{\tau_0-1} \Delta r(j),$$

$$\sum_0^{\infty} r(j-1)\mathsf{P}_\pi(\tau_0 = j) = \mathsf{E}_\pi \sum_0^{\tau_0-1} \Delta r(j) = \pi(0)\mathsf{E}_0 \sum_0^{\tau_0-1} r(j).$$

The first of these is finite even under $(1, \Delta r)$–ergodicity, and the second requires $(1, r)$–ergodicity, which we have (more than) assumed. Thus the terms on the right in (28) tend to zero as $n \to \infty$.

This simple argument seems to fail for more general f–norm convergence. But now we can use the Regenerative Decomposition of [8, Theorem 13.2.5] to find that for general $f \ge 1$, as in (39) of [15],

$$r(n)\|P^n(x,\cdot) - \pi\|_f \le A_n + B_n + C_n \qquad (29)$$

where as in the proof of [15, Theorem 4.1] the terms A_n and C_n go to zero from the assumption of (f, r)–ergodicity, being terms or tails of the convergent series $\mathsf{E}_x[\sum_0^{\tau_0} f(\Phi_j)r(j)]$ and $\pi(0)\mathsf{E}_0[\sum_0^{\tau_0} f(\Phi_j)r(j)]$; and the middle term B_n goes to zero (from (42) of [15]) provided $r(n)|P^n(0,0) - \pi(0)| \to 0$. But this follows from the convergence of (28), and the result is proved. $\qquad \square$

We next give results on subgeometric rates that can be deduced from a drift function approach.

Theorem 3.2 *Suppose Φ is an irreducible, stochastically-ordered Markov chain and that (V_n) is a sequence of functions $\mathsf{X} \to [0, \infty)$ such that*

$$PV_{n+1}(x) \le V_n(x) - r^0(n)f(x), \qquad x > 0. \qquad (30)$$

where f is monotone increasing on $[0, \infty)$. Then Φ is (f, r^0)–ergodic, and (25) holds for all x with

$$M_0(x) \le 2(1 + \pi(0))V_0(x). \qquad (31)$$

PROOF

It follows from [8, Proposition 11.3.3] that when (30) holds we have, for $x > 0$,

$$\mathsf{E}_x[\sum_0^{\tau_0-1} f(\Phi_n)r^0(n)] \le V_0(x).$$

Now by stochastic ordering and the monotonicity of f we have

$$\mathsf{E}_0[\sum_0^{\tau_0-1} f(\Phi_n)r^0(n)] \le \mathsf{E}_x[\sum_0^{\tau_0-1} f(\Phi_n)r^0(n)] < V_0(x) \qquad (32)$$

and so the chain is (f, r^0)–ergodic. In fact, because V_0 is finite for all x, the chain is (f, r^0)–regular [15], so that the convergence in (25) holds for all x.

Since both terms in (26) are bounded as in (32), the inequality (31) follows and we are done. □

Note that here we have been able to utilise stochastic ordering so that we only require the condition (30) rather than extra conditions at $\{0\}$ as in Theorem 2.1(ii) or Proposition 2.2 of [15].

We conjecture that this result can be extended to the case where the drift is only to a set $C = [0, c]$ as in Theorem 2.4 but have so far been unable to prove this.

4 Examples

4.1 Random walk on a half–line: geometric case

We give a number of simple examples which illustrate the implementation of these explicit rate calculations. We first apply the results of Section 2 to the random walk on $\mathbb{R}_+ \cup 0$, defined by

$$\Phi_{n+1} = (\Phi_n + W_{n+1})^+, \quad n \in \mathbb{Z}_+, \qquad (33)$$

where W_n is a sequence of i.i.d. real-valued random variables, independent of Φ_0. We write W for a generic increment variable and denote the distribution function of W by Γ. We assume $\Gamma[0, \infty) > 0$. As is well known [5], this chain is stochastically ordered and irreducible when $\mathsf{E}[W] < 0$.

We can obtain bounds on the convergence rates for this random walk using a stochastic comparison approach combined with explicit construction of drift functions for truncated versions of the chain enabling the use of Theorem 3.1.

Theorem 4.1 *Suppose that* $\mathsf{E}[W] < 0$ *and* $\mathsf{E}[\gamma^W] < \infty$ *for some* $\gamma > 1$. *Choose any* x_0 *such that* $\Gamma((-\infty, -x_0]) = \delta > 0$ *and* $\Gamma([-x_0, 0]) = d > 0$, *and so that* $\mathsf{E}[W^*] < 0$ *where* $W^* = \max(-x_0, W)$. *Then if* $\lambda_0 =$

$\min_{\gamma > 1} \mathsf{E}[\gamma^W]$ *and γ_0 is the value for which this minimum is attained, and if*

$$\lambda = [\lambda_0 + d/\delta]/[1 + d/\delta], \qquad b = d[1 + 1/\delta] \tag{34}$$

we have that for any $r < \lambda^{-1}$

$$\sum_0^\infty \|P^n(x, \cdot) - \pi\|_V r^n \leq \frac{2}{1 - r\lambda}[V(x) + \frac{b}{1 - \lambda}] < \infty \tag{35}$$

where $\gamma_0^x \leq V(x) \leq \gamma_0^x + d/\delta$

PROOF

Clearly $\mathsf{E}[W^*] \to \mathsf{E}[W]$ as $x_0 \to \infty$ so that there is some x_0 for which $\mathsf{E}[W^*] < 0$. We will show below that the stated rates and bounds hold for the random walk on $[0, \infty)$ with "truncated" increment W^*.

But then the same rates and bounds hold for the original chain by stochastic comparison. To see this, note that the truncated chain has larger values of $\mathsf{E}_x[r^{\tau_0}]$ and $\mathsf{E}_0[r^{\tau_0}]$ than the original chain. Examination of the proofs of Theorem 2.3 and Theorem 2.4 reveals that bounds on $\mathsf{E}_x[r^{\tau_0}]$ and $\mathsf{E}_0[r^{\tau_0}]$ are all that is required to establish the results, and since the bounds on these quantities established for the truncated chain also hold for the original chain, the convergence results hold for the original chain also.

Now let us prove the theorem under the truncation condition. Write $\Psi(\gamma) = \mathsf{E}(\gamma^{W^*})$. Then $\Psi(1) = 1$ and $\Psi'(1) = \mathsf{E}(W^*) < 0$ which implies there exists a $\gamma > 1$ with $\mathsf{E}(\gamma^{W^*}) < 1$. Clearly $\mathsf{E}(\gamma^{W^*}) \geq \mathsf{E}(\gamma^{W^*} \mathbb{1}_{\{W^* \geq 0\}})$ which approaches ∞ as γ approaches $+\infty$. Hence there is at least a local minimum of $\mathsf{E}(\gamma^{W^*})$ in the range $\gamma > 1$. We can thus define λ_0 and γ_0 as in the statement of the theorem.

Now let $C = [0, x_0]$. We show that for $V(x) = \gamma_0^x$,

$$PV(x) \leq \lambda_0 V(x) + d\mathbb{1}_C. \tag{36}$$

First note that for $x > x_0$

$$PV(x) = \int_{-x_0}^\infty \gamma_0^{x+y} \Gamma(dy) = \lambda_0 \gamma_0^x = \lambda_0 V(x);$$

and next, for $0 \leq x \leq x_0$,

$$PV(x) = \int_{-x}^\infty \gamma_0^{x+y} \Gamma(dy) + \int_{-x_0}^x \Gamma(dy)$$

$$\leq \lambda_0 V(x) + \Gamma([-x_0, 0]).$$

This establishes (36), and applying Theorem 2.4 with the values (34) completes the proof. □

We may get some idea of the effect of the approximation (34) by considering this result for the simple random walk on the integer lattice $\mathbb{Z}_+$: here

$P(n, \{n+1\}) = q$ for $n \geq 0$, $P(n, \{n-1\}) = p$ for $n \geq 1$, and $P(0, \{0\}) = p$. Assume that $p + q = 1$ and $p < 1/2$ so that $\mathsf{E}(W) < 0$.

Take $x_0 = 1$, so $\delta = p = d$, leading to

$$V(x) \leq \gamma^x + 1, \qquad \lambda = [\lambda_0 + 1]/2, \quad b = p + 1.$$

We have $\mathsf{E}(\gamma^W) = p\gamma^{-1} + q\gamma$ which is minimized for $\gamma_0 = \sqrt{p/q}$. This minimum value is attained at $\lambda_0 = \sqrt{4pq}$. The convergence results in Theorem 2.4 thus hold for $r < 2/(\sqrt{4pq}+1)$. This differs from the optimal value $1/\sqrt{4pq}$ (see [5]) by a factor of $2\sqrt{4pq}/(\sqrt{4pq} + 1)$, which is approximately one for p near $1/2$ and approaches ∞ as p approaches 1. The first table below gives the values of the bound on r provided by (34), and the optimal rate provided by [5] for various values of p. Note that for practical purposes (34) does not lose much until the very light traffic case.

Bounds on the Rate of Convergence for Random Walk						
p	0.51	0.60	0.70	0.80	0.90	0.95
λ^{-1} from (34)	1.0001	1.0102	1.0436	1.1111	1.2500	1.3929
λ_0^{-1} from [5]	1.0002	1.0206	1.0911	1.2500	1.6667	2.2942
Ratio	1.0001	1.0103	1.0455	1.1250	1.3333	1.6471

It is also of interest to see how the constants in (14) behave in this case. In the next table, we give the values for the bound on convergence in (14) for $r = 1$ and $x = 0$ for different values of p.

Bounds on Geometric Convergence Rate						
p	0.51	0.60	0.70	0.80	0.90	0.95
Bound	301,979,593	31,753	2047	400	115	63

In the light traffic case these are not unreasonable but in the heavy traffic case they are clearly of limited practical use, even in this simple case.

4.2 Random walk on a half–line: subgeometric case

To illustrate Theorem 3.1 we again consider the random walk $\mathbf{\Phi}$ defined by (33) but we need some restrictions because of the absence of a result similar to Theorem 2.4 for the subgeometric case. We will assume that the random walk is restricted to the non-negative integers, and that $\mathbf{\Phi}$ is "left-continuous" in the sense that the increment distribution Γ has support restricted to $\{-1, 0, 1, \ldots\}$.

Proposition 4.2 *For a left continuous random walk on $\{0, 1, \ldots, \}$ suppose that $\mathsf{E}[W] < 0$ and $\mathsf{E}[(W^+)^m] < \infty$ for some $m \in \mathbb{Z}_+, m \geq 2$. Then there exist computable constants $a > 0$ and $c > 0$ such that (25) holds for the functions*

$$f(x) = (ax + c)^{k-1} \vee 1, \quad r(n) = (n + c)^{m-k} \vee 1 \quad (k = 1, \ldots, m),$$

with the bound in (26) given by $M_0(x) \leq 4(ax + c)^m$.

PROOF

We will show that condition (30) holds for $V_n(x) = (ax + n + c)^m$. For $x \geq 1$, we have

$$
\begin{aligned}
PV_{n+1}(x) &= \int_{-1}^{\infty} (ax + y + n + 1 + c)^m \Gamma(dy) \\
&= \int_{-1}^{\infty} \sum_{j=0}^{m} \binom{m}{j} (ax + n + c)^{m-j} (ay + 1)^j \Gamma(dy) \\
&= (ax + n + c)^m + \sum_{j=1}^{m} (ax + n + c)^{m-j} \binom{m}{j} \mathsf{E}[(aW + 1)^j] \\
&= V_n(x) + (ax + n + c)^{m-1} \\
&\qquad \left[ma\mathsf{E}(W) + m + \sum_{j=2}^{m} \frac{\mathsf{E}[(aW + 1)^j]}{(ax + n + c)^{j-1}} \right]
\end{aligned}
$$
(37)

Now choose $a = -(2 + m)/[m\mathsf{E}(W)]$ so that a satisfies $ma\mathsf{E}(W) + m = -2$; and select c so that

$$
S = \sum_{j=2}^{m} \frac{\mathsf{E}[(aW + 1)^j]}{(ax + n + c)^{j-1}} < 1 :
$$

note that this is possible since we have the crude bound

$$
S < \max_{2 \leq j \leq m} \mathsf{E}[(aW + 1)^j](2^m - m - 1)/c.
$$

Then from (37)

$$
\begin{aligned}
PV_{n+1}(x) &\leq V_n(x) - (ax + n + c)^{m-1} \\
&\leq V_n(x) - (ax + c)^{k-1}(n + x)^{m-k}.
\end{aligned}
$$

Thus (30) holds, and the result follows from Theorem 3.1 with

$$
M_0(x) \leq 2(1 + \pi(0))V_0(x) \leq 4(ax + c)^m
$$

as required. $\qquad\qquad\qquad\qquad\qquad\qquad\qquad\qquad\qquad\qquad\qquad\square$

4.3 A multiplicative autoregressive model

We conclude with an application to the multiplicative model

$$
Y_n = W_n Y_{n-1}^{\alpha}, \quad n = 1, 2, \ldots
$$
(38)

where W_n is again a sequence of i.i.d. random variables taking values in $(0, \infty)$, and W_n is independent of $Y_0, Y_1, \ldots$ for each n. We assume that the distribution Γ of W_n has a density with respect to Lebesgue measure and obeys the moment conditions

$$\mathsf{E}[\log |W_n|] < \infty, \quad \mathsf{E}[\log(W_n)] < 0. \tag{39}$$

We will assume also that $0 < \alpha < 1$: this will ensure that the chain Y_n is geometrically ergodic, since taking logarithms we obtain the chain $\log(Y_n) = \alpha \log(Y_{n-1}) + \log(W_n)$ and this AR(1) model is known to be geometrically ergodic for $|\alpha| < 1$ under our assumptions [2]. Moreover, if $\alpha > 0$, the Y_n chain is easily seen to be stochastically ordered.

Unfortunately, however, since $(0, \infty)$ does not have a minimal element, we need to modify the process given by (38) if we are to use the results of Section 2. One possibility is to take

$$Z_n = \max(W_n Z_{n-1}^{\alpha}, a) \tag{40}$$

for some a which we will take without loss of generality to satisfy $a < e$. This chain is still stochastically monotone and outside $(0, a)$ has the same behaviour as Y_n. In using drift conditions, it will be convenient to consider the shifted version $\Phi_n = Z_n - a$ on $[0, \infty)$: from (40), this satisfies

$$\Phi_n = \max(W_n(\Phi_{n-1} + a)^{\alpha}, a) - a. \tag{41}$$

For Φ_n we will take the drift function $V(x) = \log(x + e)$. Then since $a < e$,

$$
\begin{aligned}
PV(x) &= \int_0^{\infty} \log(\max(y(x+a)^{\alpha}, a) - a + e)\Gamma(dy) \\
&= \alpha V(x) + \int_0^{\infty} \log\left[\frac{\max(y(x+a)^{\alpha}, a) - a}{(x+e)^{\alpha}} + \frac{e}{(x+e)^{\alpha}}\right]\Gamma(dy) \quad (42) \\
&\leq \alpha V(x) + \int_0^{\infty} \log\left[y + \frac{e}{(x+e)^{\alpha}}\right]\Gamma(dy).
\end{aligned}
$$

Since Γ is assumed to have a density, $\mathsf{E}[\log(W+d)] \downarrow \mathsf{E}[\log(W)] < 0$ as $d \downarrow 0$, so that we can find $d_0 > 0$ such that $\mathsf{E}[\log(W + d_0)] < 0$. We now choose c such that $x > c$ implies that $e/(x + e)^{\alpha} < d_0$. Then in (18), we can take $C = [0, c]$ and $\lambda_C = \alpha$: for, from (42), for $x > c$ we have $PV(x) \leq \alpha V(x)$, and for $x \in [0, c]$ we similarly have $PV(x) \leq \alpha V(x) + b_0$ where

$$b_0 = \int_0^{\infty} \log(y + e^{1-\alpha})\Gamma(dy) < \infty. \tag{43}$$

Thus from Theorem 2.4, we have proved

Theorem 4.3 *The chain Z_n defined by (40) is geometrically ergodic with rate bounded by*

$$\lambda = \frac{\alpha + b_0/\delta_a}{1 + b_0/\delta_a}$$

where b_0 is given by (43) and

$$\delta_a = P(c,0) = \Gamma\left[(0, \frac{a}{(c+a)^\alpha})\right]$$

To try and obtain an explicit bound on the geometric rate of convergence of the original process, we might let $a \to 0$. But then δ_a converges to 0, and the value of $\lambda \to 1$ in Theorem 4.3. Thus we have an explicit rate of convergence for the modified chain Z_n, but not for the original chain.

Acknowledgements

We are grateful to Sean Meyn for discussion of these results, and in particular for showing us the coupling idea in (15).

5 References

[1] Peter H. Baxendale. Uniform estimates for geometric ergodicity of recurrent Markov processes. Unpublished report, University of Southern California, 1993.

[2] P. D. Feigin and R. L. Tweedie. Random coefficient autoregressive processes: a Markov chain analysis of stationarity and finiteness of moments. *J. Time Ser. Anal.*, 6:1–14, 1985.

[3] T. Lindvall. On coupling of discrete renewal processes. *Z. Wahrscheinlichkeitstheorie und Verw. Geb.*, 48:57–70, 1979.

[4] T. Lindvall. *Lectures on the Coupling Method*. John Wiley & Sons, New York, 1992.

[5] R.B. Lund and R.L. Tweedie. Geometric convergence rates for stochastically ordered Markov chains. *Maths. of O.R.* (accepted for publication).

[6] R.B. Lund, S.P. Meyn and R.L. Tweedie. Computable exponential convergence rates for stochastically ordered Markov processes. *Ann. Appl. Probab.* (accepted for publication).

[7] S. P. Meyn and R. L. Tweedie. Stability of Markovian processes I: Discrete time chains. *Adv. Appl. Probab.*, 24:542–574, 1992.

[8] S. P. Meyn and R. L. Tweedie. *Markov Chains and Stochastic Stability*. Springer-Verlag, London, 1993.

[9] S. P. Meyn and R. L. Tweedie. Computable bounds for convergence rates of Markov chains. *Ann. Appl. Probab.*, 4:981–1011, 1994.

[10] E. Nummelin and P. Tuominen. Geometric ergodicity of Harris recurrent Markov chains with applications to renewal theory. *Stoch. Proc. Applns.*, 12:187–202, 1982.

[11] E. Nummelin and P. Tuominen. The rate of convergence in Orey's theorem for Harris recurrent Markov chains with applications to renewal theory. *Stoch. Proc. Applns.*, 15:295–311, 1983.

[12] E. Nummelin and R. L. Tweedie. Geometric ergodicity and *R*-positivity for general Markov chains. *Ann. Probab.*, 6:404–420, 1978.

[13] J.S. Rosenthal. Minorization conditions and convergence rates for Markov chain Monte Carlo. *J. Amer. Statist. Assoc.*, 90:558–566, 1995.

[14] D. Stoyan. *Comparison Methods for Queues and Other Stochastic Models.* John Wiley and Sons, London, 1983.

[15] P. Tuominen and R.L. Tweedie. Subgeometric rates of convergence of *f*-ergodic Markov chains. *Adv. Appl. Probab.*, 26:775–798, 1994.

[16] R. L. Tweedie. Criteria for rates of convergence of Markov chains with application to queueing and storage theory. In J. F. C. Kingman and G. E. H. Reuter, editors, *Probability, Statistics and Analysis.* London Mathematics Society Lecture Note Series, Cambridge University Press, Cambridge, 1983.

MULTI-TYPE AGE-DEPENDENT BRANCHING PROCESSES WITH STATE-DEPENDENT IMMIGRATION

Maroussia N. Slavtchova-Bojkova*

Abstract

This work continues the study of the age-dependent branching processes allowing two types of immigration , i.e. one in the state zero and another one according to the i.i.d. times of an independent ergodic renewal process. The multidimensional case is considered and asymptotic properties and limit theorems are established. These results generalise both the results of the discrete theory and those for the one-dimensional continuous-time model.

multi-type age-dependent branching processes; state-dependent immigration; limit theorems
60J80

1   Introduction.

The theory of multi-type Bellman-Harris branching processes (BHBP), together with its discrete counterpart, the Bienaymé-Galton-Watson processes, has been treated by many authors. Excellent surveys are contained in Mode (1971), Athreya and Ney (1972), Sevastyanov (1971), etc. For the first time single-type branching processes with state-dependent immigration appeared in Foster's (1971) and Pakes' (1971, 1975) papers. In these works a Bienaymé-Galton-Watson process allowing immigration whenever the number of particles is zero was investigated. Foster (1971) studied the asymptotics of the probability of extinction and of the first two moments, and obtained the limit distribution of the processes under proper normalization in the critical case. Later the continuous-time Markov analog of this process was studied by Yamazato (1975).

BHBPIO (i.e. BHBP with immigration only in the state zero) were introduced and investigated in the critical case by Mitov and Yanev (1985, 1989). Their asymptotic results generalized those obtained by Foster (1971) and Yamazato (1975). In the non-critical cases limit results for the above-mentioned processes were obtained by Slavtchova

---

*Postal address: Department of Probability and Statistics, Institute of Mathematics and Informatics, Bulgarian Academy of Sciences, 1113 Sofia, Bulgaria.

Supported by the National Foundation for Scientific Investigations, grant MM-418/94.

and Yanev (1991). Mitov (1989, 1995) extended the results for the multidimensional model in the critical case, while the non-critical cases were studied in an author's paper (1991).

Weiner (1991) placed the last model in a new setting by allowing in addition a renewal immigration component. For BHBPIOR (i.e. BHBPIO which admit in addition a renewal immigration component) he proved that in the critical case the rate of convergence in probability is $Ct^2/\log t$ as $t \to \infty$, where C is a certain constant. In contrast to this, Slavtchova-Bojkova and Yanev (1995) obtained that the subcritical processes have a linear growth. In the supercritical case they generalized Athreya's result (1969) for the BHBPIOR refining and making more precise the estimates of the growth of the processes on the set of non-extinction. Under the classical $X \log X$ condition a convergence in distribution was proved and the properties of the limit were investigated. It is the purpose of this paper to carry on this investigation for the multidimensional model.

On the other hand, Kaplan and Pakes (1974) studied the supercritical BHBP allowing in addition an immigration at the event times of an ergodic renewal process (BHBPRI). It turns out that their approach could be applied to the multidimensional BHBPIOR to establish almost sure convergence (Theorem 5.3). The reason is that the convergence depends only on the asymptotics of the underlying age-dependent processes, so we can apply similar approach to study the more general model with two types of state-dependent immigration.

However, Kaplan (1974) obtained a sufficient condition for the existence of a proper limiting distribution of the subcritical multi-type BHBPRI and generalized many of the results of the discrete theory and those of the one-dimensional continuous time model. In comparison with his result the situation with BHBPIOR is rather different. In contrast to Kaplan's and the author's (1991) results, we here establish that the rate of convergence in probability of the subcritical processes is $\mathbf{D}t$, as $t \to \infty$, where $\mathbf{D}$ is a constant vector (Theorem 5.1). It would be desirable to have a proof of this result under only a second moment hypothesis.

2 Definitions and notation.

The prototype of the branching processes to be studied in this paper is the model of the BHBPIOR defined by Weiner (1991).

Let $p > 1$ be an integer constant. About p-dimensional vectors $\mathbf{x} = (x_1, x_2, \ldots, x_p)$, $\mathbf{y} = (y_1, y_2, \ldots, y_p)$, $\mathbf{1} = (1, 1, \ldots, 1)$, $\mathbf{0} = (0, 0, \ldots, 0)$ etc., we denote $\mathbf{xy} = \sum_{i=1}^{p} x_i y_i$, $\mathbf{x}^{\mathbf{y}} = (x_1^{y_1}, x_2^{y_2}, \ldots, x_p^{y_p})$ and $\mathbf{x} \geq \mathbf{y}$ or $\mathbf{x} > \mathbf{y}$ if $x_i \geq y_i$ or $x_i > y_i$ for $1 \leq i \leq p$ respectively.

Let $\{\mathbf{X}(t) = (X^{(1)}(t), \ldots, X^{(p)}(t))\}_{t \geq 0}$ be a p-dimensional population process, wherein the individuals reproduce according to a p-dimensional BHBPIO augmented by an independent immigration component $\{\nu_i\}_{i \geq 1}$ of the same processes at the event times $\{\tau_i\}_{i \geq 1}$ of a given renewal process.

$\mathbf{X}(t)$ counts the number of the particles of the various types alive at time t, $t > 0$, $\mathbf{X}(0) = \mathbf{0}$. The intervals $T_1 = \tau_1, T_2 = \tau_2 - \tau_1, \ldots$ between succesive immigrations and the sizes of the immigrants are assumed to be independent identicaly distributed random variables (i.i.d.r.v.) with a common distibution function (d.f.) $G_0(t)$.

The $\{\nu_i\}_{i \geq 1}$ are i.i.d. with common probability generating function (p.g.f.) $f_0(s)$.

The p-dimensional BHBPIO $\{\mathbf{Z}(t) = (Z^{(1)}(t), \ldots, Z^{(p)}(t))\}_{t \geq 0}$ is governed by a vector of the life-time distributions $\mathbf{G}(t) = (G^{(1)}(t), \ldots, G^{(p)}(t))$, a vector of the offspring p.g.f. $\mathbf{h}(\mathbf{s}) = (h^{(1)}(\mathbf{s}), \ldots, h^{(p)}(\mathbf{s}))$, a multidimensional p.g.f. $f(\mathbf{s})$ of the random vectors $\{\mathbf{Y}_i = (Y_i^{(1)}, \ldots, Y_i^{(p)})\}_{i \geq 1}$ of the immigrants in the state zero and the common d.f. $K(t)$ of the duration $\{X_i\}_{i \geq 1}$ of staying in the state zero, where $\mathbf{s} = (s_1, \ldots, s_p)$. It is assumed that $a = \int_0^\infty t\,dK(t) < \infty$.

Now we recall the definition of the p-dimensional BHBPIO given by Mitov (1989):

$$\mathbf{Z}(t) = \mathbf{Z}_{N(t)+1}(t - S_{N(t)} - X_{N(t)+1})\mathbb{I}_{\{S_{N(t)}+X_{N(t)+1}<t\}}, \quad \mathbf{Z}(0) = \mathbf{0}, \qquad (2.1)$$

where $\mathbf{Z}_i(t) = (Z_i^{(1)}(t), \ldots, Z_i^{(p)}(t))$, $t > 0$, $\mathbf{Z}_i(0) = \mathbf{Y}_i$, $i \geq 1$, is a p-dimensional BHBP starting with random vector of particles, with particle life d.f. $G^{(k)}(t)$, $G^{(k)}(0+) = 0$, and p.g.f. of the offsprings $h^{(k)}(\mathbf{s})$, $k = 1, \ldots, p$. As usual $N(t) = \max\{n \geq 0 : S_n \leq t\}$ is the number of renewal events for the renewal process $\{S_n\}_{n=0}^\infty$ with $S_0 = 0, S_n = \sum_{i=1}^n U_i, U_i = X_i + \sigma_i$, where $\sigma_i = \inf\{t : \mathbf{Z}_i(t) = \mathbf{0}\}$.

Let us mention that the process defined by (2.1) could be interpreted as follows: starting from the zero state, the process stays at that state random time X_i with d.f. $K(t)$ and after that a random vector $\mathbf{Y}_i$ of immigrants of different types enters the population, according to the p.g.f. $f(\mathbf{s})$. The further evolution of the particles is independent and in accordance with a vector $\mathbf{G}(t)$ of the life-time distributions and a vector $\mathbf{h}(\mathbf{s})$ of the p.g.f. of the offsprings. Then the process hits zero after a random period σ_i, depending of the evolution of the inner BHBP $\mathbf{Z}_i(t)$. The following evolution of the process could be presented as the replication of such i.i.d. cycles.

Slavtchova-Bojkova and Yanev (1991) analyzed the above model for the case $p = 1$ and the problem of determining necessary and sufficient conditions for the existence of a limiting distribution in the non-critical cases was investigated.

Using the defifition (2.1) the p- dimensional BHBPIOR $\mathbf{X}(t)$ admits the following representation:

$$\mathbf{X}(t) = \sum_{i=1}^{n(t)} \sum_{j=1}^{\nu_i} \mathbf{Z}_{ij}(t - \tau_i), \qquad (2.2)$$

where $\{\mathbf{Z}_{ij}(t)\}_{t \geq 0, i, j \geq 1}$ is the set of i.i.d. stochastic processes defined on a common probability space, each having the same distribution as the multidimensional BHBPIO $\mathbf{Z}(t)$ and

$$n(t) = \max\{n : \tau_n \leq t\}. \qquad (2.3)$$

Set:

$$m_{ij} = \frac{\partial h^{(i)}(\mathbf{1})}{\partial s_j}, \quad b_{ij}^k = \frac{\partial^2 h^{(k)}(\mathbf{1})}{\partial s_i \partial s_j},$$

$$\beta_i = \frac{\partial f(\mathbf{1})}{\partial s_i}, \quad n_{ij} = \frac{\partial^2 f(\mathbf{1})}{\partial s_i \partial s_j},$$

$c_0' = f_0'(1)$, $c_0'' = f_0''(1)$, $L(t) = \mathbb{P}\{X_i + \sigma_i \leq t\}$, $\nu_0 = \int_0^\infty t\,dL(t)$, $\mu_0 = \int_0^\infty x\,dG_0(x)$,

$\mu_i = \int_0^\infty x\,dG^{(i)}(x), i, j, k = 1, \ldots, p$, $\mathbf{M} = \{m_{ij}\}_{1 \leq i, j \leq p}$, $H_{ij}(t) \equiv m_{ij} G^{(i)}(t)$, $\mathbf{H}(t) = \{H_{ij}\}_{1 \leq i, j \leq p}$.

194

Let $H_{ij}^{(n)}(t)$ be the n-th fold convolution of $H_{ij}(t)$, where recursively, $H_{ij}^{(1)}(t) = H_{ij}(t)$,
$$H_{ij}^{(n)}(t) = \int_0^t \sum_{l=1}^m H_{il}^{(n-1)}(t-u)dH_{lj}(u), \text{ for } n \geq 1 \text{ and } H_{ij}^{(0)}(t) = U(t), \text{ where } U(t) = 1,$$
$t \geq 0$, $U(t) = 0$, $t < 0$.

In order to avoid technical difficulties, we make the following assumptions.

Assumption I.

(i) $f_0(0) < 1, 0 < m_{ij} < \infty, 1 \leq i, j \leq p$;

(ii) $G_0(0^+) = 0, G^{(i)}(0^+) = 0, 1 \leq i \leq p, \nu_0 < \infty, \mu_i < \infty, \mu_0 < \infty, \beta < \infty$;

(iii) $\mathbf{h}(\mathbf{s})$ is not singular;

(iv) $\mathbf{M}$ is positive regular;

(v) $G_0(t), G^{(i)}(t), 1 \leq i \leq p, K(t)$ and $L(t)$ are non-lattice distributions.

Assumption II.

The "Malthusian" parameter α_0 exists for the p-dimensional BHBP $\{\hat{\mathbf{Z}}(t)\}$. Let $\hat{\mathbf{M}}(t)$ be the matrix whose (i, j) entry is $m_{ij} \int_0^\infty e^{-\alpha_0 t} dG^{(i)}(t)$. The Malthusian parameter is that number α_0 (unique if it exists) such that the maximal eigenvalue of $\hat{\mathbf{M}}(t)$ is one.

It follows from Assumption I that the matrix $\mathbf{M}$ has a maximal eigenvalue ρ which is positive, simple, and has associated positive left and right eigenvectors $\mathbf{u}$ and $\mathbf{v}$; $\mathbf{u}$ and $\mathbf{v}$ are normalized so that $(\mathbf{u}, \mathbf{1}) = (\mathbf{v}, \mathbf{u}) = 1$. As it is done in the classical theory, we will call the process $\{\mathbf{X}(t)\}_{t \geq 0}$ supercritical, critical or subcritical depending on whether $\rho > 1, = 1$ or < 1.

3 Functional equations.

Let us denote $\Phi(t, \mathbf{s}) = \mathbf{E}\mathbf{s}^{\mathbf{Z}(t)}$, $\Phi(0, \mathbf{s}) = 1$, $\Phi_0(t, \mathbf{s}) = \mathbf{E}\mathbf{s}^{\mathbf{X}(t)}$, $\Phi_0(0, \mathbf{s}) = 1$, $\Phi_0(t, \tau, \mathbf{s}_1, \mathbf{s}_2) = \mathbf{E}\{\mathbf{s}_1^{\mathbf{X}(t)}\mathbf{s}_2^{\mathbf{X}(t+\tau)}\}$, $\tau \geq 0$, $\mathbf{F}(t, \mathbf{s}) = \mathbf{E}\mathbf{s}^{\hat{\mathbf{Z}}(t)} = (F_1(t, \mathbf{s}), \ldots, F_p(t, \mathbf{s}))$, $\mathbf{F}(0, \mathbf{s}) = \mathbf{s}$.

It is well-known (see e.g. Sevastyanov (1971)), that the functions $F_k(t, \mathbf{s}) = \mathbf{E}\{\mathbf{s}^{\hat{\mathbf{Z}}_k(t)}|\hat{\mathbf{Z}}_k(0) = \mathbf{e}_k\}$, ($\mathbf{e}_k$ is p-dimensional vector which k-th component is one and the others are equal to zero), satisfy the following system of integral equations:

$$F_k(t, \mathbf{s}) = \int_0^t h^{(k)}(\mathbf{F}(t-u, \mathbf{s}))dG^{(k)}(u) + s_k(1 - G^{(k)}(t)),$$
$$F_k(0, \mathbf{s}) = s_k, k = 1, \ldots, p. \tag{3.1}$$

It is not difficult to show that the p.g.f. $\Phi_0(t, \mathbf{s})$ admits the representation

$$\Phi_0(t, \mathbf{s}) = \int_0^t \Phi_0(t-u, \mathbf{s})f_0(\Phi(t-u, \mathbf{s}))dG_0(u) + 1 - G_0(t), \tag{3.2}$$

where the p.g.f. $\Phi(t, \mathbf{s})$, satisfies the renewal equation (see Mitov (1989))

$$\Phi(t, \mathbf{s}) = \int_0^t \Phi(t-u, \mathbf{s})dL(u) + 1 - K(t) - L(t) + \int_0^t f(\mathbf{F}(t-u, \mathbf{s}))dK(u). \tag{3.3}$$

The proof is quite similar to that for the single-type case and we omit it.

By the law of total probabilities it is not difficult to obtain the equation

$$
\begin{aligned}
\Phi_0(t,\tau,\mathbf{s}_1,\mathbf{s}_2) &= \int_0^t \Phi_0(t-u,\tau,\mathbf{s}_1,\mathbf{s}_2) f_0(\Phi(t-u,\tau,\mathbf{s}_1,\mathbf{s}_2)) dG_0(u) \\
&+ \int_0^{t+\tau} \Phi_0(t+\tau-u,\mathbf{s}_2) f_0(\Phi(t+\tau-u,\mathbf{s}_2)) dG_0(u) \\
&+ (1 - G_0(t+\tau)),
\end{aligned}
\tag{3.4}
$$

where $\Phi(t,\tau,\mathbf{s}_1,\mathbf{s}_2) = \mathbb{E}\mathbf{s}_1^{Z(t)}\mathbf{s}_2^{Z(t+\tau)}$.

Denote the moments

$$
M_{01}^{(k)}(t) = \left.\frac{\partial \Phi_0(t,\mathbf{s})}{\partial s_k}\right|_{\mathbf{s}=1} = \mathbb{E}X^{(k)}(t), \quad M_{02}^{(k,l)}(t) = \left.\frac{\partial^2 \Phi_0(t,\mathbf{s})}{\partial s_k \partial s_l}\right|_{\mathbf{s}=1} = \mathbb{E}X^{(k)}(t)X^{(l)}(t),
$$

$$
M_1^{(k)}(t) = \left.\frac{\partial \Phi(t,\mathbf{s})}{\partial s_k}\right|_{\mathbf{s}=1} = \mathbb{E}Z^{(k)}(t), \quad M_2^{(k,l)}(t) = \left.\frac{\partial^2 \Phi(t,\mathbf{s})}{\partial s_k \partial s_l}\right|_{\mathbf{s}=1} = \mathbb{E}Z^{(k)}(t)Z^{(l)}(t),
$$

$$
M_{0n}^{(k)}(t) = \left.\frac{\partial^n \Phi_0(t,\mathbf{s})}{\partial s_k^n}\right|_{\mathbf{s}=1} \equiv \mathbb{E}[X^{(k)}(t)]^n, \quad M_n^{(k)}(t) = \left.\frac{\partial^n \Phi(t,\mathbf{s})}{\partial s_k^n}\right|_{\mathbf{s}=1} \equiv \mathbb{E}[Z^{(k)}(t)]^n,
$$

$$
A_{kn}^{(l)}(t) = \left.\frac{\partial^n F_k(t,\mathbf{s})}{\partial s_l^n}\right|_{\mathbf{s}=1} \equiv \mathbb{E}[\hat{Z}_k^{(l)}(t)]^n, \quad B_m^{(k,l)}(t) = \left.\frac{\partial^2 F_m(t,\mathbf{s})}{\partial s_k \partial s_l}\right|_{\mathbf{s}=1} \equiv \mathbb{E}[\hat{Z}_m^{(k)}\hat{Z}_m^{(l)}],
$$

$$
N_0^{(k,l)}(t,\tau) = \left.\frac{\partial^2 \Phi_0(t,\tau,\mathbf{s}_1,\mathbf{s}_2)}{\partial s_{1,k} \partial s_{2,l}}\right|_{s_1=s_2=1} = \mathbb{E}X^{(k)}(t)X^{(l)}(t+\tau).
$$

Under the Assumption I by differentiating (3.1) - (3.4) and setting $\mathbf{s} = \mathbf{s}_1 = \mathbf{s}_2 = 1$ one obtains

$$
M_{01}^{(k)}(t) = \int_0^t M_{01}^{(k)}(t-u) dG_0(u) + c_0' \int_0^t M_1^{(k)}(t-u) dG_0(u),
\tag{3.5}
$$

$$
\begin{aligned}
M_{02}^{(k,l)}(t) &= \int_0^t M_{02}^{(k,l)}(t-u) dG_0(u) + c_0' \int_0^t M_2^{(k,l)}(t-u) dG_0(u) \\
&+ c_0' \int_0^t M_{01}^{(l)}(t-u) M_1^{(k)}(t-u) dG_0(u) \\
&+ c_0'' \int_0^t M_1^{(l)}(t-u) M_1^{(k)}(t-u) dG_0(u) \\
&+ c_0' \int_0^t M_1^{(l)}(t-u) M_{01}^{(k)}(t-u) dG_0(u),
\end{aligned}
\tag{3.6}
$$

$$
M_1^{(k)}(t) = \int_0^t M_1^{(k)}(t-u) dL(u) + \int_0^t \sum_{i=1}^P \beta_i A_{k1}^{(i)}(t-u) dK(u),
\tag{3.7}
$$

$$
\begin{aligned}
M_2^{(k,l)}(t) &= \int_0^t \sum_{i=1}^P \sum_{j=1}^P n_{ij} A_{l1}^{(j)}(t-u) A_{k1}^{(i)}(t-u) dK(u) \\
&+ \int_0^t \sum_{i=1}^P \nu_i B_i^{(kl)}(t-u) dK(u) + \int_0^t M_2^{(k,l)}(t-u) dL(u),
\end{aligned}
\tag{3.8}
$$

196

$$N_0^{(k,l)}(t, \tau) = \int_0^t N_0^{(k,l)}(t - u, \tau) dG_0(u)$$

$$+ c_0' \int_0^t M_{01}^{(k)}(t - u) M_1^{(l)}(t + \tau - u) dG_0(u)$$

$$+ c_0' \int_0^t M_{01}^{(l)}(t + \tau - u) M_1^{(k)}(t - u) dG_0(u) \qquad (3.9)$$

$$+ c_0'' \int_0^t M_1^{(k)}(t - u) M_1^{(l)}(t + \tau - u) dG_0(u)$$

$$+ c_0' \int_0^t N_2^{(k,l)}(t - u, \tau) dG_0(u),$$

where $N_2^{(k,l)}(t, \tau) = \mathbb{E}[Z^{(k)}(t) Z^{(l)}(t + \tau)], \quad \tau > 0.$

4   Preliminary results.

In addition to its own intrinsic interest, the asymptotics of the moments of the multidimensional BHBPIOR, plays a key role for establishing limit theorems. We concentrate our study on the non-critical cases.

I. Subcritical case.

Before stating the results about the moments of the processes of interest we need the following preliminary lemmas.

Lemma 4.1 *Let* $\hat{\mathbf{Z}}_k(t)$ *be the vector of the number of particles alive at time* t, $k = 1, \ldots, p$, *starting with one new particle of type* k *in a subcritical multi-type BHBP satisfying the Assumptions I and II. Then, if* $\int_0^\infty y e^{-n\alpha_0 y} dG^{(i)}(y) < \infty$, $\int_0^\infty y e^{-n\alpha_0 y} dG_0(y) < \infty$ *and all moments of* $h^{(k)}(\mathbf{s})$, $k = 1, \ldots, p$ *exist at* $\mathbf{s} = \mathbf{1}$, *for* $n \geq 1$, *as* $t \to \infty$,

$$A_{kn}^{(l)}(t) \sim \bar{A}_{kn}^{(l)} \exp^{\alpha_0 n t}, \qquad (4.1)$$

where $0 < \bar{A}_{kn}^{(l)} < \infty$.

Proof. We will establish the result using induction on n.

For $n = 1$ it is known (see Sevastyanov, Th.VIII.3, p.312) that for the subcritical multi-type BHBP, as $t \to \infty$

$$A_{k1}^{(l)}(t) \sim \bar{A}_{k1}^{(l)} \exp^{\alpha_0 t}, \qquad (4.2)$$

where

$$\bar{A}_{k1}^{(l)} = \frac{u_k v_l \int_0^\infty e^{-\alpha_0 u} [1 - G^{(l)}(u)] du}{\sum_{k,l=1}^p M_{\alpha_0 k}^l u_k v_l}, \qquad (4.3)$$

197

u_i and v_j are the i-th and j-th components of the right and left eigenvectors respectively, corresponding to the Perron root ρ of the matrix $E - \hat{M}$ and $M_{\alpha_0 k}^l = m_{kl} \int_0^\infty e^{-\alpha_0 u} dG^{(k)}(u)$. Denoting

$$Q_i(t, \mathbf{s}) = 1 - F_i(t, \mathbf{s}), \qquad (4.4)$$

after expanding the integrand on the right hand side of (3.1) in a Taylor series we obtain componentwise

$$
\begin{aligned}
Q_i(t, \mathbf{s}) = {} & (1 - s_i)(1 - G^{(i)}(t)) + \sum_{j=1}^{p} Q_j(t - u, \mathbf{s}) m_{ij} dG^{(i)}(u) \\
& - \sum_{k=2}^{\infty} \frac{1}{k!}(-1)^k \sum_{l_1,\dots,l_k=1}^{p} \frac{\partial^k h^{(i)}(\mathbf{1})}{\partial s_{l_1} \dots \partial s_{l_k}} \int_0^t \prod_{r=1}^{k} Q_{l_r}(t - u, \mathbf{s}) dG^{(i)}(u).
\end{aligned} \qquad (4.5)
$$

Writing (4.5) in vector form, taking Laplace transforms and re-inverting, it follows that

$$
\begin{aligned}
Q_i(t, \mathbf{s}) = {} & \sum_{j=1}^{p} (1 - s_j)(1 - G^{(j)}(t - u)) dK_{ij}(u) \\
& - \sum_{k=2}^{\infty} \frac{1}{k!}(-1)^k \int_0^t \sum_{l_1,\dots,l_k=1}^{p} \prod_{r=1}^{k} Q_{l_r}(t - u, \mathbf{s}) \sum_{j=1}^{p} \frac{\partial^k h^{(j)}(\mathbf{1})}{\partial s_{l_1} \dots \partial s_{l_k}} dR_{ij}(u),
\end{aligned} \qquad (4.6)
$$

where $K_{ij}(t) = \sum_{n=0}^{\infty} H_{ij}^{(n)}(t)$, $R_{ij} = \sum_{n=0}^{\infty} G^{(j)} * H_{ij}^{(n)}(t)$, and $F * G$ denotes convolution.

To compute $A_{k2}^{(l)}(t) \equiv \mathbb{E}[\hat{Z}_k^{(l)}(t)(\hat{Z}_k^{(l)}(t) - 1)] = \frac{\partial^2 Q_k(t,\mathbf{1})}{\partial s_l^2}$ for large t, using the fact that $Q_i(t, \mathbf{1}) \equiv 0$ for all $1 \le i \le p$, after differentiating (4.6) at $\mathbf{s} = \mathbf{1}$ we have

$$A_{k2}^{(l)}(t) = \frac{2}{2!} \sum_{l_1, l_2=1}^{p} \int_0^t A_{l_1 1}^{(l)}(t - u) A_{l_2 1}^{(l)}(t - u) \sum_{j=1}^{p} \frac{\partial^2 h^{(j)}(\mathbf{1})}{\partial s_{l_1} \partial s_{l_2}} dR_{kj}(u). \qquad (4.7)$$

Using (4.2) and (4.3) it is clear that $A_{k2}^{(l)}(t) \sim \bar{A}_{k2}^{(l)} e^{2\alpha_0 t}$, $t \to \infty$, where $0 < \bar{A}_{k2}^{(l)} < \infty$. Assume the result of the theorem for $n - 1$. Then, considering orders of magnitude of t, $(t \to \infty)$, it follows by the induction hypothesis that the asymptotic behaviour of $\mathbb{E}[Z_k^{(l)}(t)]^n$ is determined solely from the n-th derivatives with respect to s_l of the term

$$\frac{(-1)^2}{2!} \sum_{j=1}^{p} \sum_{l_1, l_2=1}^{p} \frac{\partial^2 h^{(j)}(\mathbf{1})}{\partial s_{l_1} \partial s_{l_2}} \int_0^t Q_{l_1}(t - u, \mathbf{s}) Q_{l_2}(t - u, \mathbf{s}) dR_{kj}(u)$$

in (4.6), evaluated at $\mathbf{s} = \mathbf{1}$.

By Leibnitz's rule for successive differentiation, since $Q_l(t, \mathbf{1}) \equiv 0$, $1 \le l \le p$,

$$\frac{\partial^n Q_r(t, \mathbf{1}) Q_m(t, \mathbf{1})}{\partial s_l^n} = \sum_{k=1}^{n} \binom{n}{k} \frac{\partial^k Q_r(t, \mathbf{1})}{\partial s_l^k} \frac{\partial^{n-k} Q_m(t, \mathbf{1})}{\partial s_l^{n-k}}. \qquad (4.8)$$

Then using the induction hypothesis one obtains $A_{kn}^{(l)}(t) \sim \bar{A}_{kn}^{(l)} e^{n\alpha_0 t}$ as $t \to \infty$, where $0 < \bar{A}_{kn}^{(l)} < \infty$, proving Lemma 4.1.

198

Lemma 4.2 *Let $\mathbf{Z}(t)$ be the vector of the number of particles alive at time t in a sub-critical case BHBPIO. Then, under the assumptions of Lemma 4.1 and if all moments of $f(\mathbf{s})$ exist at $\mathbf{s} = \mathbf{1}$, for $n \geq 1$*

$$M_n^{(k)}(t) \sim R_{kn} < \infty, \, as \quad t \to \infty.$$

Proof. Denoting $W(t, \mathbf{s}) = 1 - \Phi(t, \mathbf{s})$, then by (4.4) and (3.3), it follows that

$$W(t, \mathbf{s}) = \int_0^t W(t - u, \mathbf{s}) dL(u) + \int_0^t f(1 - \mathbf{Q}(t - u, \mathbf{s})) dK(u). \tag{4.9}$$

As $M_1^{(l)}(t) = -\frac{\partial W(t,1)}{\partial s_l} \equiv \mathbb{E}[Z^{(l)}(t)]$ after differentiating (4.9) and setting $\mathbf{s} = \mathbf{1}$ we have

$$M_1^{(l)}(t) = \int_0^t M_1^{(l)}(t - u) dL(u) + \int_0^t \sum_{i=1}^p \frac{\partial f(\mathbf{1})}{\partial s_i} \frac{\partial Q_i(t - u, \mathbf{s})}{\partial s_l} dK(u)$$

and via direct renewal methods, using Lemma 4.1 it is not difficult to obtain, that

$$M_1^{(l)}(t) \sim \frac{\int_0^t \sum_{i=1}^p \beta_i A_l^{(i)}(u) du}{\nu_0} \equiv R_{l1}, l = 1, \ldots, p.$$

After second differentiation of (4.9) at $\mathbf{s} = \mathbf{1}$ we obtain the equation

$$
\begin{aligned}
M_2^{(l)}(t) &= \int_0^t M_2^{(l)}(t - u) dL(u) \\
&+ \int_0^t \sum_{i,j=1}^p \frac{\partial^2 f(\mathbf{1})}{\partial s_i \partial s_j} \frac{\partial Q_i(t - u, \mathbf{1})}{\partial s_l} \frac{\partial Q_j(t - u, \mathbf{1})}{\partial s_l} dK(u) \\
&+ \int_0^t \sum_{i=1}^p \frac{\partial f(\mathbf{1})}{\partial s_i} \frac{\partial^2 Q_i(t - u, \mathbf{1})}{\partial s_l^2} dK(u).
\end{aligned}
$$

Hence applying Lemma 4.1 by similar renewal techniques (see Slavtchova (1991)) it follows that

$$M_2^{(l)}(t) \sim R_{l2} < \infty, \quad t \to \infty.$$

The rest of the proof is straightforward using the fact that the higher moments of $\mathbf{Z}(t)$ satisfy renewal type equations to which we can apply renewal methods and Lemma 4.1.

We can now present the asymptotics of the multi-type BHBPIOR.

Theorem 4.1 *Under the assumptions of Lemmas 4.1 and 4.2 if in addition all moments of $f_0(\mathbf{s})$ exist at $\mathbf{s} = \mathbf{1}$, then*

$$M_{0n}^{(k)}(t) \equiv \mathbb{E}[X^{(k)}(t)]^n = \frac{\partial^n \Phi_0(t, \mathbf{1})}{\partial s_k^n} \sim \{D_k t\}^n, \tag{4.10}$$

$$M_{0n}^{(kl)}(t) \equiv \mathbb{E}[X^{(k)}(t)]^{n_1} \mathbb{E}[X^{(l)}(t)]^{n_2} = \frac{\partial^n \Phi_0(t, \mathbf{1})}{\partial s_k^{n_1} \partial s_l^{n_2}} \sim D_k^{n_1} D_l^{n_2} t^n, \tag{4.11}$$

as $t \to \infty$, $n, n_1, n_2 \geq 1$, such that $n_1 + n_2 = n$, where $0 < D_k < \infty$, $k, l = 1, \ldots, p$ are explicitly computed.

199

Proof. Expanding the integrand on the right hand side of (3.2) in a Taylor series about **1** we have

$$
\begin{aligned}
\Phi_0(t,s) \;=\; & 1 - G_0(t) + \int_0^t \Phi_0(t-u,s)dG_0(u) \\
& - \int_0^t [1 - \Phi(t-u,s)]\Phi_0(t-u,s)dG_0(u) \\
& + \sum_{l=2}^{\infty} \int_0^t \frac{(-1)^l f_0^{(l)}(1)}{l!}(1 - \Phi(t-u,s))^l \Phi_0(t-u,s)dG_0(u).
\end{aligned}
$$

Taking Laplace transforms and re-inverting, it follows that

$$
\begin{aligned}
\Phi_0(t,s) \;=\; & 1 + c_0' \int_0^t [\Phi(t-u,s)-1]\Phi_0(t-u,s)dH_0(u) \hspace{2cm} (4.12) \\
& + \sum_{l=2}^{\infty} \int_0^t \frac{(-1)^l f_0^{(l)}(1)}{l!}(1 - \Phi(t-u,s))^l \Phi_0(t-u,s)dH_0(u),
\end{aligned}
$$

where $H_0(t) = \sum_{l=0}^{\infty} G_0^{*l}(t)$.

As $M_1^{(l)}(t) \equiv \mathbb{E}[Z^{(l)}(t)]$, differentiating (4.12) by s_l and setting $\mathbf{s} = \mathbf{1}$ yields

$$
M_{01}^{(l)}(t) = c_0' \int_0^t M_1^{(l)}(t-u)dH_0(u). \hspace{2cm} (4.13)
$$

Therefore from (4.13) and Lemma 4.1 it follows that as $t \to \infty$

$$
M_{01}^{(l)}(t) \sim c_0' \int_0^\infty \frac{R_{l1}}{\mu_0}du = D_l t,
$$

where $D_l \equiv c_0' R_{l1}/\mu_0$, $l = 1, \ldots, p$. After second differentiating of (4.12) by s_l and setting $\mathbf{s} = \mathbf{1}$ it is not difficult to obtain

$$
\begin{aligned}
M_{02}^{(l)}(t) \;=\; & c_0' \int_0^t M_2^{(l)}(t-u)dH_0(u) \\
& + 2c_0' \int_0^t M_1^{(l)}(t-u)M_{01}^{(l)}(t-u)dH_0(u) \hspace{2cm} (4.14) \\
& - c_0'' \int_0^t [M_1^{(l)}(t-u)]^2 dH_0(u),
\end{aligned}
$$

$l = 1, \ldots, p$. Then by Lemmas 4.1 and 4.2, applying similar renewal techniques to the equation (4.14) as in Theorem 3.1 (see Slavtchova (1991)), it follows that the asymptotics of the second moment $M_{02}^{(l)}(t)$ is determined by the second term on the right side, i.e.

$$
M_{02}^{(l)}(t) \sim 2c_0' \int_0^\infty R_{l1} D_l u du = D_l^2 t^2, \quad t \to \infty.
$$

Considering orders of magnitude of t, $(t \to \infty)$, it follows by induction, using Lemmas 4.1 and 4.2, that for $n \geq 2$,

$$
M_{0n}^{(l)}(t) \sim nc_0' \int_0^t M_1^{(l)}(t-u)M_{0n-1}^{(l)}(t-u)dH_0(u).
$$

200

Again using induction , assume (4.10) for $n - 1$. Then by the methods of Yanev and Mitov (1985, Theorem 2, p.761) one obtains, componentwise, for $n \geq 2$,

$$M_{0n}^{(l)}(t) \sim nc_0' \int_0^\infty \frac{R_l D_l^{n-1}}{\mu_0} du = [D_l t]^n,$$

completing the induction and establishing (4.10). The assertion (4.11) follows by the similar arguments and we omit it.

II. Supercritical case.

Theorem 4.2 *Under the Assumption I, if $\rho > 1$, then $\lim_{t\to\infty} M_{01}^{(k)}(t) \exp\{-\alpha_0 t\} = M_{01}^{(k)}$, where $\alpha_0 > 0$ is the Malthusian parameter,*

$$M_{01}^{(k)} = \frac{c_0' M_1^{(k)} \Delta(\alpha_0)}{1 - \Delta(\alpha_0)}, \quad \Delta(\lambda) = \mathbb{E} e^{-\lambda T_1}, \tag{4.15}$$

$$M_1^{(k)} = \frac{\int_0^\infty e^{-\alpha_0 u} dK(u) [\sum_{l=1}^p \beta_l \bar{A}_{k1}^{(l)}]}{1 - \int_0^\infty e^{-\alpha_0 u} dL(u)}, \tag{4.16}$$

and $\bar{A}_{k1}^{(l)}$ are defined by (4.3), $l, k = 1, \ldots, p$.

Theorem 4.3 *Assume the conditions of Theorem 4.2. Then, if $b_{ij}^k < \infty$ and $n_{ij}^k < \infty$, $i, j, k = 1, \ldots, p$, it follows that $\lim_{t\to\infty} M_{02}^{(k)}(t) e^{-2\alpha_0 t} = M_{02}^{(k)}$, where*

$$M_{02}^{(k)} = \frac{(c_0' M_2^{(k)} + c_0''[M_1^{(k)}]^2 + 2c_0' M_{01}^{(k)} M_1^{(k)}) \int_0^\infty e^{-2\alpha_0 u} dG_0(u)}{1 - \int_0^\infty e^{-2\alpha_0 u} dG_0(u)}, \tag{4.17}$$

$M_2^{(k)} = \lim_{t\to\infty} M_2^{(k)}(t)$ a.s., $M_{01}^{(k)}$, $M_1^{(k)}$ are defined by (4.15) and (4.16) respectively.

Theorem 4.4 *Let the conditions of Theorem 4.3 hold. Then componentwise $N_0^{(k)}(t, \tau) = e^{\alpha_0(2t+\tau)} M_{02}^{(k)}(1 + o(1))$, uniformly for $\tau \geq 0$, where $M_{02}^{(k)}$ is defined by (4.17).*

The proofs of Theorems 4.2, 4.3 and 4.4 follow by the quite similar renewal approach applied to the functional equations (3.5), (3.6) and (3.9).

5 Limit theorems.

Theorem 5.1 *Under the assumptions of Theorem 4.1, as $t \to \infty$, it is hold componentwise*

$$X^{(k)}(t)/t \xrightarrow{\mathbb{P}} D_k, k = 1, \ldots, p.$$

Proof. From (4.10) and (4.11) we get $\mathbb{E}\{X^{(k)}(t)/t\}^n \sim D_k$, $n \geq 1$. The method of the moments yields that $\mathbf{X}(t)/t$ converges in probability to a constant random vector, whose distribution is determined by the asymptotic moments of the process.

It is interesting to mention, that while for the multi-type subcritical both BHBPIO (see Slavtchova (1991)) and BHBPRI (see Kaplan (1974)) there exist a stationary limit distribution, here we obtain convergence in probability to a constant random vector.

Theorem 5.2 *Assume the conditions of Theorem 4.4.*

(i) Then the vector process $\mathbf{W}(t) = \mathbf{X}(t)/e^{\alpha_0 t}$ *converges in mean square to a positive vector random variable* $\mathbf{W} = (W_1, \ldots, W_p)$, *whose Laplace transform (L.T.)* $\varphi(\mathbf{y}) = \mathbb{E}e^{-\mathbf{y}\mathbf{W}}$, $\mathbf{y} \geq 0$ *satisfies the equation:*

$$\varphi(\mathbf{y}) = \int_0^\infty \varphi(\mathbf{y}e^{-\alpha_0 u}) f_0(\psi(\mathbf{y}e^{-\alpha_0 u})) dG_0(u), \tag{5.1}$$

and $\psi(\mathbf{y})$ *is the unique solution of the equation*

$$\psi(\mathbf{y}) = \int_0^\infty \psi(\mathbf{y}e^{-\alpha_0 u}) dL(u) + \int_0^\infty f(\theta(\mathbf{y}e^{-\alpha_0 u})) dK(u) - f(\mathbf{q}), \tag{5.2}$$

where $\mathbf{q} = (q_1, \ldots, q_p)$ *with* $q_k = \lim_{t \to \infty} \mathbb{P}\{\hat{\mathbf{Z}}_k(t) = \mathbf{0}|\hat{\mathbf{Z}}_k(0) = \mathbf{e}_k\}$ *and* $\theta(\mathbf{y}) = (\theta_1(\mathbf{y}), \ldots, \theta_p(\mathbf{y}))$ *have components* $\theta_i(u)$, $1 \leq i \leq p$, *satisfying the system of integral equations:*

$$\theta_i(u) = \int_0^\infty h^{(i)}(\theta_1(ue^{-\alpha_0 t}), \ldots, \theta_p(ue^{-\alpha_0 t})) dG^{(i)}(t), \quad i = 1, \ldots, p. \tag{5.3}$$

(ii) Furthermore,

$$\mathbb{E}W^{(l)} = M_{01}^{(l)} \quad and \quad Var[W^{(l)}] = M_{02}^{(l)} - [M_{01}^{(l)}]^2.$$

(iii) Moreover, there exists a scalar r.v. w *such that* $\mathbf{W} = w\mathbf{u}$ *a.s. and*

$$\mathbb{E}w = d = \frac{(\sum_{l=1}^p \beta_l v_l)(\int_0^\infty e^{-\alpha_0 u} dG_0(u))(\int_0^\infty e^{-\alpha_0 u} dK(u))}{(1 - \int_0^\infty e^{-\alpha_0 u} dL(u))(\sum_{k,l=1}^p M_{\alpha_0 k}^l u_k v_l)},$$

where $\mathbf{u}$ *is the left eigenvector of the matrix* $\hat{\mathbf{M}}$.

Proof. To prove (i) consider

$$\begin{aligned}
\mathbb{E}[W^{(k)}(t+\tau) - W^{(k)}(t)]^2 &= e^{-2\alpha_0 t} M_{02}^{(k)}(t) + e^{-2\alpha_0 t} M_{01}^{(k)}(t) \\
&+ e^{-2\alpha_0(t+\tau)} M_{02}^{(k)}(t+\tau) + e^{-2\alpha_0(t+\tau)} M_{01}^{(k)}(t) \\
&+ -2e^{-\alpha_0(2t+\tau)} N_0^{(k)}(t,\tau)
\end{aligned} \tag{5.4}$$

and observe that according to the Theorems 4.2, 4.3 and 4.4 the right-hand side of (5.4) approaches zero as $t \to \infty$, uniformly in $\tau \geq 0$ for all $k = 1, \ldots, p$. By completeness of the space $L_2(\Omega, \mathbf{F}, \mathbb{P})$ there exist random variables $W^{(k)}$ such that $W^{(k)}(t) \xrightarrow{L_2} W^{(k)}$ as $t \to \infty$.

The rest of the argument is a concequence of the results of Slavtchova (1991) and Mode (1971).

Denote $\lim_{t \to \infty} \hat{Z}_i^{(l)}(t)/e^{\alpha_0 t} = \hat{W}_i^{(l)}$ a.s., $i, l = 1, \ldots, p$.

Theorem 5.3 *Under the assumptions of Theorem 4.4. if in addition*

$$\int_0^\infty \mathbb{E}[\hat{\mathbf{Z}}_i(t)/e^{\alpha_0 t} - \hat{\mathbf{W}}_i]^2 dt < \infty, \tag{5.5}$$

then

$$\lim_{t \to \infty} \mathbf{W}(t) = \mathbf{W} \quad a.s.$$

Proof. We use the representation (2.2) of the process $\mathbf{X}(t)$.

It has been proven (see Slavtchova (1991), Theorem 4.3) that under the condition (5.5) $\mathbf{Z}_{ij}(t)/e^{\alpha_0 t} \to \tilde{\mathbf{W}}$ a.s. This implies that for each i there exists

$$\lim_{t \to \infty} \left\{ \sum_{j=1}^{\nu_i} \mathbf{Z}_{ij}(t)/e^{\alpha_0 t} \right\} = \mathbf{W}_i \quad a.s.$$

The random vector $\mathbf{W}_i$ has L.T. $f_0(\psi(\mathbf{u}))$. It follows from the assumptions of Section 1 that $\{\mathbf{W}_i\}$ are i.i.d. and independent of the $\{\tau_i\}$. Define $\mathbf{W} = \sum_{i=1}^{\infty} e^{-\alpha_0 \tau_i} \mathbf{W}_i$.

Proceeding as in Harris (1989, Ch. VI) assume that $h^{(i)}(0) = 0$ for each $i = 1, \ldots, p$, which forces the process $\mathbf{X}(t)$ to have nondecreasing sample paths with probability 1. Therefore, it is sufficient to show that $\int_0^\infty \mathbb{E}[W^{(l)}(t) - W^{(l)}]^2 dt < \infty$. Observe that

$$[W^{(l)}(t) - W^{(l)}]^2 \le 2 \left[(W^{(l)}(t) - \sum_{i=1}^{n(t)} e^{-\alpha_0 \tau_i} W_i^{(l)})^2 + (\sum_{i=n(t)+1}^{\infty} e^{-\alpha_0 \tau_i} W_i^{(l)})^2 \right] = 2[J_1(t) + J_2(t)].$$

From (2.2) by Schwarz's inequality,

$$J_1(t) \le \left[\sum_{i=1}^{n(t)} e^{-\alpha_0 \tau_i} \right] \left[\sum_{i=1}^{n(t)} e^{-\alpha_0 \tau_i} \left\{ \sum_{j=1}^{\nu_i} Z_{ij}^{(l)}(t - \tau_i)/e^{\alpha_0(t-\tau_i)} - W_i^{(l)} \right\}^2 \right].$$

It is not difficult to show that for the last sum we have

$$\int_0^\infty \left\{ \sum_{j=1}^{\nu_i} Z_{ij}^{(l)}(t - \tau_i)/e^{\alpha_0(t-\tau_i)} - W_i^{(l)} \right\}^2 dt$$

$$= \sum_{i=1}^{\infty} e^{-\alpha \tau_i} \int_{\tau_i}^\infty \left\{ \sum_{j=1}^{\nu_i} Z_{ij}^{(l)}(t - \tau_i)/e^{\alpha_0(t-\tau_i)} - W_i^{(l)} \right\}^2 dt$$

$$= \sum_{i=1}^{\infty} e^{-\alpha \tau_i} \int_0^\infty \left\{ \sum_{j=1}^{\nu_i} Z_{ij}^{(l)}(t)/e^{\alpha_0 t} - W_i^{(l)} \right\}^2 dt.$$

By independence we conclude that

$$\mathbb{E} \left[\int_0^\infty J_1(t) dt \right] \le \mathbb{E} \left\{ \left(\sum_{i=1}^{\infty} e^{-\alpha_0 \tau_i} \right)^2 \right\} \mathbb{E} \left\{ \int_0^\infty \left[\sum_{j=1}^{\nu_1} (Z_{1j}^{(l)}(t)/e^{\alpha_0 t} - W_1^{(l)}) \right]^2 dt \right\}.$$

Note that $\mathbb{E} \left[\sum_{i=1}^{\infty} e^{-\alpha_0 \tau_i} \right]^2 < \infty$. Also by Schwarz's inequality,

$$\mathbb{E} \left\{ \int_0^\infty \left[\sum_{j=1}^{\nu_1} (Z_{1j}^{(l)}(t)/e^{\alpha_0 t} - \sum_{j=1}^{\nu_i} \tilde{W}_j^{(l)}) \right]^2 dt \right\} \le \mathbb{E}\{\nu_1^2\} \int_0^\infty \mathbb{E}[Z_{11}^{(l)}(t)/e^{\alpha_0 t} - \tilde{W}_1^{(l)}]^2 dt.$$

203

However, Slavtchova and Yanev (1991) have shown that the last integral is finite under the condition (5.5). Similarly,

$$\mathbb{E}\left[\int_0^\infty J_2(t)dt\right] \leq \mathbb{E}\left\{\left(\sum_{i=0}^\infty e^{-\alpha_0 \tau_i}\right)\left(\sum_{i=1}^\infty \tau_i e^{-\alpha_0 \tau_i}(W_i^{(l)})^2\right)\right\}\mathbb{E}[W_1^{(l)}]^2$$

$$\leq \delta\mathbb{E}\left(\sum_{i=1}^\infty e^{(-\alpha_0/2)\tau_i}\right)^2 \mathbb{E}[W_1^{(l)}]^2,$$

for $\delta > 0$ such that $xe^{-\alpha_0 x} \leq \delta e^{-\alpha_0 x/2}$, $x > 0$. On the other hand $\mathbb{E}[W_1^{(l)}]^2 = \mathbb{E}\left(\sum_{j=1}^{\nu_1} W_{1j}^{(l)}\right)^2 \leq \mathbb{E}\{\nu_1^2\}\mathbb{E}\{[W_1^{(l)}]^2\}$. Second moment assumption implies $\mathbb{E}\{[W_1^{(l)}]^2\} < \infty$. Therefore $\mathbb{E}\int_0^\infty J_2(t)dt < \infty$, which completes the proof.

Finally, we would like to mention that it would be interesting to obtain limit results if the immigration component were not independent of the inner process.

Acknoledgements. The author is deeply grateful to Prof. Nickolay M. Yanev for his helpful comments and useful discussions.

References

Athreya, K. and Ney, P.(1972) *Branching Processes.* Springer Verlag, Berlin.

Feller W. (1971) *An Introduction to Probability Theory and Its Applications.*, 2nd ed., Vol. 2, John Wiley and Sons, New York.

Foster, J.H. (1971) A limit theorems for a branching process with state-dependent immigration. *Ann. Math. Stat.* **42**, 1773-1776.

Harris, T. (1989) *The Theory of Branching Processes.* Dover Publications Inc., New York.

Kaplan, N. and Pakes A. (1974) Supercritical age-dependent branching processes with immigration. *Stochastic Process Appl.* **2**, 371-389.

Kaplan, N. (1974) Multidimensional age-dependent branching processes allowing immigration: the limiting distribution. *J. Appl. Prob.* **11**, 225-236.

Mitov, K.V. (1995) Some results for multitype Bellman-Harris branching processes with state-dependent immigration. *Lecture Notes in Statistics.* **99**, C.C. Heyde (Ed.), Branching Processes, First World Congress, Springer Verlag, New York.

Mitov, K.V. (1989) Multitype Bellman-Harris branching processes with state-dependent immigration. *Proc. of the Eighteen Spring Conf. of UBM*, 423-428.

Mitov, K. V. and Yanev N. M. (1985) Bellman-Harris Branching processes with state-dependent immigration. *J. Appl. Prob.* **22**, 757-765.

Mitov, K. V. and Yanev N. M. (1989) Bellman-Harris Branching processes with a special type of state-dependent immigration. *Adv. Appl. Prob.* **21**, 270 - 283.

Mode C. (1971) *Multitype branching processes*, American Elsevier Publishing Company, Inc., New York.

Pakes, A. G. (1971) A branching processes with a state-dependent immigration component.*Adv. Appl. Prob.* **3**, 301-314.

Pakes, A. G. (1975) Some results for non-supercritical Galton-Watson processes with immigration. *Math. Biosci.* **24**, 71 - 92.

Sevastyanov, B. A. (1971) *Branching processes*, Mir, Moscow.

Slavtchova-Bojkova, M. and Yanev, N. (1995) Age-dependent branching proceses with state-dependent immigration. *Lecture Notes in Statistics.* **99**, C.C. Heyde (Ed.), Branching Processes, First World Congress, Springer Verlag, New York.

Slavtchova M. N. (1991) Limit theorems for multitype Bellman-Harris branching processes with state-dependent immigration. *Serdica* **17**, 144-156.

Slavtchova, M. and Yanev, N. (1991) Non-critical Bellman - Harris branching processes with state-dependent immigration. *Serdica* **17**, 67-79.

Weiner, H. (1991) Age-dependent Branching Processes with Two Types of Immigration. *Journal of Information and Optimization Sciences* **2**, 207-218.

Yamazato, M. (1975) Some results on continuous time branching processes with state-dependent immigration. *J. Math. Soc. Japan* **27** , 479 - 496.

The non homogeneous semi-Markov system in a stochastic environment.

P.-C. G. Vassiliou, Department of Mathematics, University of Thessaloniki, Thessaloniki, Greece

ABSTRACT In the present we introduce and define for the first time the concept of a non-homogeneous semi Markov system in a stochastic environment (S-NHSMS). We study the problem of finding the expected population structure as a function of the basic parameters of the system. Important properties are established among the basic parameters of a non-homogeneous semi Markov system in a stochastic environment.

1   Introduction

Semi-Markov models have found important applications in manpower systems, especially in the work by McClean [McC80], [McC86], [McC93], [MM95] , Mehlman [Meh79] and Bartholomew [Bar82]. A good account of applications can be found in Bartholomew, Forbes and McClean [DBM91]. The concept of a non-homogeneous semi Markov system (NHSMS) was firstly introduced and defined in Vassiliou and Papadopoulou [VP92]. This provided among others a general framework for a number of semi Markov models for population systems. The ergodic behaviour of NHSMS is provided in Papadopoulou and Vassiliou [PV94] and the study of the entrance probabilities and counting transitions in Papadopoulou [Pap95] .The concept of a non-homogeneous Markov system in a stochastic environment which introduces the idea of having a pool of transition probability matrices to choose from was firstly defined and studied in Tsantas and Vassiliou [TV93]. The idea is also included in a special case in Cohen [Coh76] [Coh77] for a case where the transition matrices are Leslie matrices. In the present paper we introduce and define for the first time the concept of a non-homogeneous semi Markov system in a stochastic environment (S-NHSMS) . The problem of finding the expected population structure in an S-NHSMS is studied, and important properties among the basic parameters of the S-NHSMS are established. For further work remains the question of whether the mean gives the real growth rate of the population. For population growth models based on products of stationary random (typically Leslie type) matrices it is well known that the real (stochastic) growth rate is strictly less than that suggested by the expected population structure. See for example Heyde and Cohen [HC85] ,and Tuljapurkar [Tul90, Page 28]

2 The non-homogeneous semi Markov system in a stochastic environment.

Consider a population which is stratified into classes according to various characteristics. The members of the population could be sections of human societies, animals, biological microorganisms, particles in a physical phenomenon, various types of machines e.t.c. Assume that $T(t)$, the total number of members of the system is known or that it is a known realization of a stochastic process. Let $S=\{1,2,...,k\}$ be the set of states that are assumed to be exclusive so that each member of the population may be in one and only one state at any given time. The state of the population at any time t=0,1,2,... is represented by the vector $\mathbf{N}(t) = [N_1(t), N_2(t), ..., N_k(t)]$ where $N_i(t)$ is the number of members in the system in the i-th state at time t.

Now let $f_{ij}(t) = prob\{$ a member of the system which entered state i at time t to move in state j at its next transition $\}$, while $p_{i,k+1}(t) = prob\{$a member of the system which entered state i at time t to leave the system at its next transition$\}$ and $p_{0j}(t) = prob\{$a member of the system to enter state j as a replacement of a member of the system given that the member entered his last state at time t $\}$.

We consider that initially there are $T(0)$ memberships in the system and a member entering the system holds a particular membership which moves within the states with the member. When a member leaves, the membership is taken by a new recruit and moves within the system with the replacement and so on. Then let

$p_{ij}(t) = prob\{$ a "membership" of the system which entered state i at time t to move to state j at its next transition $\}$

$$= f_{ij}(t) + p_{i,k+1}(t) p_{0j}(t).$$

When the system is expading($\Delta T(t) = T(t) - T(t-1) > 0$) then new memberships are created in the system and let that $r_{0i}(t) = prob\{$ a new "membership" to enter state i | a new membership is entering the system at time t$\}$. Let $\mathbf{F}(t) = \{f_{ij}(t)\}_{i,j \in S}$ and $\Im_I(t)$ be the set of all possible transition matrices $\mathbf{F}(t)$ and let $\Im_I(t) = \{\mathbf{F}_1(t), \mathbf{F}_2(t), ..., \mathbf{F}_\nu(t)\}$ be a finite subset such that $\{I - \mathbf{F}_h(t)\}\mathbf{1}' = \mathbf{p}'_{k+1}(t)$ for every $h \in I = 1, 2, ..., \nu$ and $t = 1, 2, ...$

Suppose that at every time point t=1,2,.. the system has the choice of selecting a transition matrix from the pool $\Im_I(t)$ while with no loss of generality $\mathbf{F}(0) = \mathbf{F}_1(0)$. Furthermore, assume that it makes its choice in a stochastic way and more specifically let

$$c_{hm}(t) = prob\{\mathbf{F}(t) = \mathbf{F}_m(t) \mid \mathbf{F}(t-1) = \mathbf{F}_h(t-1)\}, h, m \in I.$$

Then $\mathbf{C}(t) = \{c_{ij}(t)\}_{i,j \in S}$ uniquely defines a non homogeneous Markov chain. We call the sequence $\mathbf{C}(t) = \{c_{ij}(t)\}_{i,j \in S}$ the compromise non homogeneous Markov chain in the sense that it is the outcome of the choice of strategy under the various pressures in the environment. Also the idea of the existence of the compromise non homogeneous Markov chain seems to be an interesting generalization of the cyclic homogeneous Markov systems first introduced in Gani [Gan63], Bartholomew [Bar82] and the cyclic non-homogeneous Markov systems studied in Vassiliou [Vas84], Vassiliou [Vas86] and Georgiou and Tsantas [GT95]. In all of the above cases the matrix $\mathbf{C}(t)$ has the same special form.

By now it is evident that we can no longer speak of a specific transition matrix $\mathbf{P}(t) = \{p_{ij}(t)\}_{i,j \in S}$ of the "memberships" at time t. The transition matrix for the time interval $[t, t+1)$ will be selected from the pool $\Im_I(t) = \{\mathbf{F}(t), \mathbf{F}_2(t), ..., \mathbf{F}_\nu(t)\}$ by a stochastic mechanism which is described by the compromise non homogeneous Markov chain $\mathbf{C}(t)$. Thus it is logical to introduce $E[\mathbf{P}(t)]$ the (i, j) element of which is the expected value of the probability that whenever a "membership" enters state i at time t it determines the next state j to which it will move. However, after j has been selected but before making this transition from state i to state j, the membership "holds" for a time τ_{ij} in state i. The holding times τ_{ij} are positive integer valued random variables which are governed by a probability mass function defined as

$h_{ij}(x) = prob\{\tau_{ij}=$x | the "membership" entered state i at its last transition, state j has been selected$\} =$

$prob\{$ a "membership" of the system which entered state i at its last transition to hold x time units in state i before making its next transition given that state j has been selected$\}$

Now define the following probabilities which we call interval transition probabilities:

$q_{ij}(t, s) =$prob$\{$ a "membership" of the system which entered state i at time s to be in state j after t steps $\}$.

Also let $\mathbf{Q}\,(t,s) = \{q_{ij}\,(t,s)\}_{i,j \in S}$ and $\mathbf{H}\,(t) = \{h_{ij}\,(t)\}_{i,j \in S}$ .A population whose evolution is adequately described by a model as above is called a non homogeneous semi - Markov system in a stochastic environment (S-NHSMS).

3   The expected population structure.

One of the basic questions in population system theory is to find the expected population structure $E[\mathbf{N}(t)]$ as a function of the basic parameters of the sequences of the S-NHSMS i.e. $\{\mathbf{H}\,(t)\}_{t=0}^{\infty}$, $\{\mathbf{P}\,(t)\}_{t=0}^{\infty}$, $\{\mathbf{C}\,(t)\}_{t=0}^{\infty}$, $\{\mathbf{r}_o\,(t)\}_{t=0}^{\infty}$ and $\{T\,(t)\}_{t=0}^{\infty}$. It is important for prediction purposes and for the solution of other basic problems that such a relationship should be found in closed analytic form. With careful analysis of the probabilistic meaning of each part it is possible to verify that

$$E[N_j\,(t)] \;\; = \;\; \sum_{i=1}^{k} N_i\,(0)\,E[q_{ij}\,(0,t)] +$$

$$+ \sum_{m=1}^{t}\sum_{i=1}^{k} \Delta T\,(m)\,r_{0i}\,(m)\,E[q_{ij}\,(t-m,m)]. \quad (3.1)$$

The above equation in matrix notation could be written as

$$E\,[\mathbf{N}\,(t)] = \mathbf{N}(0)E\,[\mathbf{Q}\,(t,0)] + \sum_{m=1}^{t} \Delta T\,(m)\,\mathbf{r}_0\,(m)\,E[\mathbf{Q}\,(t-m,m)]. \quad (3.2)$$

Now from the above it is evident that we need to find $E[\mathbf{Q}(n,s)]$ as a function of the basic parameters of the system. In this respect define
$w_i\,(m,t) = prob\{$ a "membership" of the system which entered state i at time t to stay m time units in state i before its next transition $\}$.

Let $\mathbf{W}\,(n,s)$ the k.k matrix which has zeros everywhere apart from the diagonal which has in position i the element

$$\sum_{m=n+1}^{\infty} w_i\,(m,s) = 1 - \sum_{m=0}^{n} w_i\,(m,s)\;.$$

Using the fact that $E[E(X \mid Y)] = E[X]$ we could prove that

$$E[\mathbf{W}(n,s)] \;\; = \;\; \sum_{m=n+1}^{\infty} \mathbf{I} \diamond \{[E[\mathbf{P}(s)] \diamond \mathbf{H}(m)]\mathbf{U}\} =$$

$$\mathbf{I} \diamond \{[E[\mathbf{P}(s)] \diamond \sum_{m=n+1}^{\infty} \mathbf{H}(m)]\mathbf{U}\}. \quad (3.3)$$

where $\mathbf{A}\Diamond\mathbf{B}$ is the Hadamard product of the two matrices and $\mathbf{U} = \{u_{ij}\}_{i,j\in S}$ with $u_{ij} = 1$ for every i.j. From Tsantas and Vassiliou [TV93] we get from theorem 2.1 that

$$E[\mathbf{P}(s)] = \sum_{h\in I} \mathbf{e}_h [\mathbf{e}_1 \prod_{r=1}^{s} \mathbf{C}(r)]^\top \mathbf{P}_h(s). \qquad (3.4)$$

Consider the following analysis of events

(i) The (i,j) element of the matrix $E[\mathbf{P}(s)]\Diamond\mathbf{H}(m)$ is the expected value of the probability

$prob\{$ a membership of the system, which entered state i, makes its next transition to state j and makes that transition after n time units $\}$

(ii)The (i,i) element of the matrix $\mathbf{W}(n,s)$ is the expected value of the probability

$prob\{$ a membership of the system, which entered state i at time s, to remain at state i at least n time units$\}$

(iii)The (i,l) element of the matrix

$$\sum_{j=2}^{n} E\{[\mathbf{P}(s)\Diamond\mathbf{H}(j-1)] \times$$

$$\times [\mathbf{W}(n-j+1, s+j-1) + \mathbf{P}(s+j-1)\Diamond\mathbf{H}(n-j+1)]\}$$

could be proved that it is the expected value of the probabilities

$$\sum_{j=2}^{n} prob\{ \text{ a membership of the system, which entered state i at time s,}$$

to make one transition in the time interval (s,s+n) at time s+j-1 and be in state l at time s+n$\}$.

(iv) Consider the matrix

$$\mathbf{S}_j(k, s, m_k) = \sum_{m_k=2}^{j-k} \left\{ \sum_{m_{k-1}=1+m_k}^{j-k+1} \cdots \right.$$

$$\cdots \left\{ \sum_{m_1=1+m_2}^{j-1} \prod_{r=-1}^{k-1} \mathbf{P}(s+m_{k-r}-1)\Diamond\mathbf{H}(m_{k-r-1}-m_{k-r}) \right\} \right\}. \qquad (3.5)$$

for every $j \geq k+2$, while for every $j < k+2$ we have $\mathbf{S}_j(k,s,m_k) = 0$ and $m_o = j$ and for every $k = 1, 2, ..., j-2$ we have $m_{k+1} = 1$.

Then following a similar methology as in remark 3.1 in Papadopoulou and Vassiliou [PV94] we get that the (i,l) element of the matrix (3.5) is the probability

$prob\{$ a membership of the system, which entered state i at time s, makes k transitions in the time interval $(s, s+j-1)$ at times $s+m_k-1$, $s+m_{k-1}$, $s+m_{k-1}-1, ..., s+m_1-1$ and enters state l at time $s+j-1\}$

Now consider the matrix

$$\sum_{j=2}^{n}\sum_{k=1}^{j-2} E\left\{S_j\left(k,s,m_k\right)\times\left[W\left(n-j+1,s+j-1\right)+\right.\right.$$

$$\left.\left.+P\left(s+j-1\right)\Diamond H\left(n-j+1\right)\right]\right\} \tag{3.6}$$

the (i,l) element of which is the expected value of the probabilities

$$\sum_{j=2}^{n}\sum_{k=1}^{j-2} prob\{$$ a membership of the system, which entered state i at time s, makes k+1 transitions in the time interval $(s,s+n)$ and the (k+1)-th transition at time $s+j-1$ and be at time s+n in state l$\}$.

Now from (i), (ii), (iii) and (iv) we get that

$$
\begin{aligned}
E[Q(n,s)] &= E[W(n,s)] + E[P(s)]\Diamond H(m) + \sum_{j=2}^{n} E\{[P(s)\Diamond H(j-1)]\\
&\times\ [W(n-j+1,s+j-1)+P(s+j-1)\Diamond H(n-j+1)]\}\\
&+ \sum_{j=2}^{n}\sum_{k=1}^{j-2} E\{S_j(k,s,m_k)\times \\
&\times\ [W(n-j+1,s+j-1)+P(s+j-1)\Diamond H(n-j+1)]\}
\end{aligned}
\tag{3.7}
$$

Now in equation (3.7) in the right hand side the first two parts have been found in equation (3.3) and (3.4). For the third part we have

$$\sum_{j=2}^{n} E\{[P(s)\Diamond H(j-1)]\times[W(n-j+1,s+j-1)+$$

$$+P(s+j-1)\Diamond H(n-j+1)]\}=$$

$$=\sum_{j=2}^{n}\sum_{h_s}\sum_{h_{s+j-1}} prob\left\{P(s)=P_{h_s}(s),P(s+j-1)=P_{h_{s+j-1}}(s+j-1)\right\}\cdot$$

$$\cdot\{P_{h_s}(s)\Diamond H(j-1)\}\left[W_{h_{s+j-1}}(n-j+1,s+j-1)+\right.$$

$$\left.P_{h_{s+j-1}}(s+j-1)\Diamond H(n-j+1)\right]=$$

$$=...\text{(using recursively properties of conditional probabilities)}...=$$

$$\sum_{j=2}^{n}[\sum_{h_s}\sum_{h_{s+1}}\cdots\sum_{h_{s+j-1}} P_{h_s}(s)\Diamond H(j-1)\times$$

$$\times\left[W_{h_{s+j-1}}(n-j+1,s+j-1)+P_{h_{s+j-1}}(s+j-1)\Diamond H(n-j+1)\right]\times$$

$$\times c_{h_{s+j-2},h_{s+j-1}}(s+j-1)\prod_{r=0}^{j-2} prob[P(s+r)=P_{h_{s+r}}(s+r)]$$

$$= \sum_{j=2}^{n} SC_j^0 (s, h_s) \qquad (3.8)$$

Now what remains from equation (3.7) is to find

$$\sum_{j=2}^{n} \sum_{k=1}^{j-2} E[\mathbf{S}_j (k, s, m_k)] \times$$

$$\times [\mathbf{W} (n - j + 1, s + j - 1) + \mathbf{P} (s + j - 1) \Diamond \mathbf{H} (n - j + 1)] =$$

$$\sum_{j=2}^{n} \sum_{k=1}^{j-2} \sum_{m_k=2}^{j-k} * \sum_{m_{k-1}=1+m_k}^{j-k+1} * \cdots * \sum_{m_1=1+m_2}^{j-1} \sum_{h_s} \sum_{h_{s+1}} \cdots \sum_{h_{s+j-1}}$$

$$* \prod_{r=-1}^{k-1} P_{h_s+m_{k-r}-1} (s + m_{k-r} - 1) \Diamond H (m_{k-r-1} - m_{k-r}) \times$$

$$\times \left[\mathbf{W}_{h_s+j-1} (n - j + 1, s + j - 1) + \mathbf{P}_{h_s+j-1} (s + j - 1) \Diamond \mathbf{H} (n - j + 1) \right]$$

$$prob \left\{ \mathbf{P}(s) = \mathbf{P}_{h_s}(s), \mathbf{P}(s + m_k - 1) = \mathbf{P}_{h_s+m_k-1} (s + m_k - 1), \cdots, \right.$$

$$\mathbf{P} (s + m_1 - 1) = \mathbf{P}_{h_s+m_1-1} (s + m_1 - 1), \mathbf{P} (s + j - 1) = \mathbf{P}_{h_s+j-1} (s + j - 1) \right\}$$

(where the asterisk (*) means appropriate parenthesis)
=...(using recursively properties of conditional probabilities)...=

$$\sum_{j=2}^{n} \sum_{k=1}^{j-2} E \left\{ \mathbf{S}_j (k, s, m_k) \times \right.$$

$$\times [\mathbf{W} (n - j + 1, s + j - 1) + \mathbf{P} (s + j - 1) \Diamond \mathbf{H} (n - j + 1)] \right\} =$$

$$= \sum_{j=2}^{n} \sum_{k=1}^{j-2} \sum_{m_k=2}^{j-k} * \sum_{m_{k-1}=1+m_k}^{j-k+1} * \ldots * \sum_{m_1=1+m_2}^{j-1} \sum_{h_s} \sum_{h_{s+1}} \cdots \sum_{h_{s+j-1}}$$

$$* \prod_{r=-1}^{k-1} c_{h_s+m_{k-r-1}, h_s+m_{k-r-1}-1} (s + m_{k-r} - 1) \, \mathbf{P}_{h_s+m_{k-r}-1} (s + m_{k-r} - 1) \Diamond$$

$$\diamond \mathbf{H}\left(m_{k-r-1} - m_{k-r}\right)) \times \prod_{q=s+m_{k-r}}^{s+m_{k-r}-2} prob[\mathbf{P}(q) = \mathbf{P}_{h_q}(q)]\times$$

$$\times \left[\mathbf{W}_{h_{s+j-1}}(n-j+1, s+j-1) + \mathbf{P}_{h_{s+j-1}}(s+j-1)\diamond\mathbf{H}(n-j+1)\right] =$$

$$= \sum_{j=2}^{n}\sum_{k=1}^{j-2} SC_j\left(s, h_s, k, m_k\right). \tag{3.9}$$

Now from equation (3.3), (3.4), (3.7), (3.8) and (3.9) we get

$$E[\mathbf{Q}(n,s)] = E[\mathbf{P}(s)]\diamond\mathbf{H}(n) + \mathbf{I}\diamond\left\{\left[E[\mathbf{P}(s)]\diamond\sum_{m=n+1}^{\infty}\mathbf{H}(m)\right]\mathbf{U}\right\} +$$

$$+ \sum_{j=2}^{n}\left[SC_j^0(s, h_s) + \sum_{k=1}^{j-2} SC_j(s, h_s, k, m_k)\right]. \tag{3.10}$$

Combining equation (3.10) with equation (3.2) we have the expected population structure.

* The author wish to thank Dr. A. Papadopoulou for her valuable help with some of the difficulties of the present work.

References

[Bar82]   D.J. Bartholomew. Stochastic models for social processes. Wiley, Chichester 3rd edn, 1982.

[Coh76]   J. E. Cohen. Ergodicity of age structure in populations with markovian vital rates i : countable states. J. Amer. Statist. Assoc., 71, 335-339, 1976.

[Coh77]   J.E. Cohen. Ergodicity of age structure in populations with markovian vital rates ii : general states. J. Appl. Prob., 9, 18-37, 1977.

[DBM91] A.F. Forbes D.J. Bartholomew and S.I. McClean. Statistical teqniques for manpower planning. Wiley, Chichester, 1991.

[Gan63]   J. Gani. Formulae for projecting enrolments and degrees awarded in universities. J. R. Statist Soc.,126, 400-409., 1963.

[GT95]   A. Georgiou and N. Tsantas. Non stationary cyclic behaviour in markov systems. Linear Algebra and its Applications (to appear), 1995.

[HC85]   C.C. Heyde and J.E. Cohen. Confidence intervals for demographic projections based on products of random matrices. Theor. Popul. Biol., 27, 120-153, 1985.

[McC80]   S.I. McClean. A semi-markov model for a multigrade population with poisson recruitment. J. Appl. Prob., 17, 846-852, 1980.

[McC86] S. I. McClean. Semi-markov models for manpower planning. In *Semi-Markov models: Theory and Applications*. ed. J. Janssen. Plenum Press, New York, 1986.

[McC93] S.I. McClean. Semi-markov models for human resource modelling. IMA Journal of mathematics Applied in Business and Industry, 4, 307-315, 1993.

[Meh79] A. Mehlmann. Semi markovian manpower models in continuous time. J. Appl. Prob., 6, 416-422, 1979.

[MM95] S. I. McClean and E. Montgomery. Estimation for continuous time non homogeneous markov and semi markov models. In *The Ins and Outs of Solving Real Problems*. ed. J. Janssen and S. McClean. Jonh Wiley., 1995.

[Pap95] A. A. Papadopoulou. Counting transitions and entrance probabilities in non-homogeneous semi-markov systems. In *The Ins and Outs of Solving Real Problems*. ed. J. Janssen and S. McClean. Jonh Wiley, 1995.

[PV94] A.A. Papadopoulou and P. C.G. Vassiliou. Asymptotic behavior of non- homogeneous semi-markov systems. Linear Alg. and its Appl., 10, 153-198, 1994.

[Tul90] S. Tuljapurkar. In *Population Dynamics in Variable Environments*. Springer Lecture Notes in Biomathematics 85, 1990.

[TV93] N. Tsantas and P.-C.G. Vassiliou. The nonhomogeneous markov system in a stochastic environment. J. Appl. Prob.,30, 285-301, 1993.

[Vas84] P.-C.G. Vassiliou. Cyclic behaviour and asymptotic stability of non-homogeneous markov systems. J. Appl. Prob., 21, 315-325, 1984.

[Vas86] P.-C.G. Vassiliou. Asymptotic variability of non-homogeneous markov systems under cyclic behaviour. Eur. J. of Oper. Res., 27, 215-228, 1986.

[VP92] P.-C.G. Vassiliou and A.A Papadopoulou. Non- homogeneous semi-markov systems and maintainability of the state sizes. J. Appl. Prob. ,29, 519-534, 1992.

Branching Processes with Two Types Emigration and State-Dependent Immigration

George P. Yanev[*]   Nickolay M. Yanev[*]

Abstract

We consider branching processes allowing a random migration component. In each generation the following three situations are possible: (i) with probability p - family emigration and individual emigration (possibly dependent); (ii) with probability q - no migration, i.e. the reproduction is as in the classical BGW process; (iii) with probability r - state dependent immigration; $p + q + r = 1$. In the critical case an additional parameter of recurrence is obtained. The asymptotic behaviour of the hitting zero probability and of the first two moments is investigated. Limiting distributions are also obtained depending on the range of the recurrence parameter.

random migration; hitting zero; moments; limiting distributions

60J80

1   Introduction

The classical Bienaymé-Galton-Watson (BGW) branching process can be interpreted as a model of an isolated population which individuals evolve independently of each other. The following definition is well-known

$$\mu_t = S_t(\mu_{t-1}) , \qquad t = 1, 2, \ldots; \quad \mu_0 = 1 , \tag{1.1}$$

where $S_t(n) = X_{1,t} + \ldots + X_{n,t}$, $n \geq 1$, $S_t(0) = 0$ and $\{X_{i,t}\}$ are i.i.d.r.v.

In this paper we consider a generalization of the BGW process when the evolution is not isolated and admits a random migration component. Let us have on some probability space three independent sets of integer-valued r.v, i.i.d. in each set: $X = \{X_{i,t}\}$, $\eta = \{(\eta_{1,t}, \eta_{2,t})\}$ and $I = \{(I_t, I_t^0)\}$. Then the process is defined as follows

$$Y_t = (S_t(Y_{t-1}) + M_t)^+, \ t = 1, 2, \ldots; \quad Y_0 \geq 0, \tag{1.2}$$

where $p + q + r = 1$, $P\{M_t = -(S_t(\eta_{1,t}) + \eta_{2,t})\} = p$, $P\{M_t = 0\} = q$,
$P\left\{M_t = I_t 1_{\{Y_{t-1} > 0\}} + I_t^0 1_{\{Y_{t-1} = 0\}}\right\} = r$, and Y_0 is independent of X, η and I. As usual $a^+ = \max(0, a)$.

---

[*]Postal address for both authors: Department of Probability and Statistics, Institute of Mathematics, Bulgarian Academy of Sciences, 1113 Sofia, Bulgaria.

Supported by the National Foundation for Scientific Investigations, grant MM-418/94.

The definition (1.2) admits the following interpretation. As usual, $X_{i,t}$ is the offspring in the t-th generation of the i-th individual which exists in the $(t-1)$-th generation. Then in the t-th generation the following three situations are possible: (i) emigration with probability p, i.e. $\eta_{1,t}$ families emigrate which give $S_t(\eta_{1,t}) = \sum_{i=1}^{\eta_{1,t}} X_{i,t}$ emigrants (family emigration) and, additionally, $\eta_{2,t}$ individuals randomly chosen from different families are eliminated (individual emigration); (ii) no migration with probability q, i.e. the reproduction is as in the classical BGW process; (iii) state dependent immigration with probability r, i.e. I_t new individuals immigrate in the non-zero states or I_t^0 in the state zero.

If $q = 1$ then (1.1) follows from (1.2), i.e. in this case $\{Y_t\}$ is a BGW process. Let $F(s) = Es^{X_{i,t}}$ be the offspring p.g.f. Then (1.1) is equivalent to the functional equation

$$F_{t+1}(s) = F_t(F(s)), \qquad t = 0,1,2\ldots; \qquad F_0(s) = s. \tag{1.3}$$

It follows from (1.1) and (1.3) that the lines of the descendants are independent, i.e. $E(s^{\mu_t}|\mu_0 = n) = F_t^n(s)$. Note that this basic property is not valid for the general case (1.2) (see (2.3) and (2.4) in Section 2).

If $r = 1$ then $\{Y_t\}$ is a branching process with two (possible dependent) immigration components depending on the states of the process. The case $r = 1$ and $I_t = I_t^0$ a.s. is the classical BGW process with immigration. The process with $r = 1$ and $I_t \equiv 0$ a.s. was investigated firstly by Foster(1971) and Pakes(1971).

The process (1.2) with $p = 1$ (i.e. with pure emigration) was studied in the case $\eta_{1,t} \equiv 1$ and $\eta_{2,t} \equiv 0$ a.s. by Vatutin (1977); Kaverin (1990) considered family emigration ($\eta_{2,t} \equiv 0$ a.s.) and Grey (1988) - individual emigration ($\eta_{1,t} \equiv 0$ a.s.).

Some particular cases of (1.2) were investigated by Yanev and Mitov (1980,1981) when $\eta_{1,t} \equiv 1$, $\eta_{2,t} \equiv 0$ and $I_t \equiv I_t^0$ a.s.; Nagaev and Han(1980), Han(1980) and Dyakonova(1992) studied the case $0 \leq \eta_{1,t} \leq N_1$, $\eta_{2,t} \equiv 0$ and $I_t \equiv I_t^0 = \sum_{i=1}^{\eta_{3,t}} X_{i,t}$ a.s, where $\eta_{3,t}$ is a r.v. Some results for the processes (1.2) with $\eta_{1,t} \equiv 0$ and $I_t \equiv I_t^0$ a.s. were announced by Yanev and Yanev (1991). Models with time non-homogeneous migration (i.e. $p = p_t$, $q = q_t$ and $r = r_t$)) were investigated by Yanev and Mitov (1985) (see also Rahimov(1995), Chapter 3).

Note that the state zero is a reflecting barrier for the regenerative process $\{Y_t\}$. We will say that $\tau = \tau(T)$ is a life-period started at a moment $T \geq 0$ if $Y_{T-1} = 0$, $Y_{T+n} > 0$ for $0 \leq n < \tau$ and $Y_{T+\tau} = 0$.

We also consider the process stopped at zero defined as follows

$$Z_t \overset{d}{=} Y_{T+t} 1_{\{Z_{t-1} > 0\}}, \quad t = 1,2,\ldots; \quad Z_0 \overset{d}{=} Y_T > 0, \tag{1.4}$$

where T is the begginning of a life-period, i.e. $P(Y_T = n) = P(I_1^0 = n|I_1^0 > 0)$, $n \geq 1$. In other words, the process $\{Z_t\}$ is the positive part of the process $\{Y_t\}$ between two succesive points of regeneration. Then

$$P(Z_t > 0) = P(\tau > t) = u_t, \qquad \text{say.} \tag{1.5}$$

Yanev, Vatutin and Mitov (1986) investigated the process $\{Z_t\}$ in the case $\eta_{1,t} \equiv 1$ and $\eta_{2,t} \equiv 0$ a.s. The general model (1.4) was considered by Yanev and Yanev (1995b).

Note that the asymptotics of u_t plays an important role in investigation of the limit behaviour of the process $\{Y_t\}$ which is the aim of the paper. The basic results are given

217

in Section 2 (Theorems 2.1 - 2.3). In Section 3 three lemmas are proved. The probability of hitting zero is investigated in Section 4. The asymptotic results for the mathematical expectation and the variance are proved in Section 5. Finally, in Section 6 the limiting distributions are obtained.

2   Equations and Basic results.

In the sequal $F(s) = Es^{X_{i,t}}$ is the offspring p.g.f. $H(s_1, s_2) = Es_1^{\eta_{1,t}} s_2^{\eta_{2,t}}$ is the p.g.f of the emigration components, $G(s_1, s_2) = Es_1^{I_t} s_2^{I_t^0}$ is the p.g.f. of the immigration components where $g(s) = G(s, 1)$ is the p.g.f. of immigration in the positive states and $g_0(s) = G(1, s)$ is the p.g.f of immigration in the state zero. Without any restriction we can suppose that $Y_0 = 0$ a.s. The case of random $Y_0 \geq 0$ and $EY_0 < \infty$ is similar. We use the following basic notations

$$
\left.
\begin{aligned}
&\delta(s) = pH(F^{-1}(s), s^{-1}) + q + rg(s), \qquad \Delta(s) = 1 - \delta(s) - r(1 - g_0(s)), \\
&\gamma_t(s) = \prod_{j=0}^{t-1} \delta(F_j(s)), \ \gamma_t = \gamma_t(0), \ \gamma_0 = 1, \qquad F_t = F_t(0), \\
&a = P(Y_{t+1} > 0 \mid Y_t = 0) = r(1 - g_0(0)).
\end{aligned}
\right\} \tag{2.1}
$$

It is convenient to introduce the functionals

$$
\left.
\begin{aligned}
W(\xi_{t-1}, s) =\ & E\left\{ \left(1 - F^{-(\eta_{1,t} - \xi_{t-1})}(s) s^{-\eta_{2,t}}\right) 1_{\{\eta_{1,t} \geq \xi_{t-1} > 0\}} \right\} \\
&+ E\left\{ \left(1 - s^{-(\eta_{2,t} - S_t(\xi_{t-1} - \eta_{1,t}))}\right) 1_{\{\eta_{2,t} \geq S_t(\xi_{t-1} - \eta_{1,t}), \xi_{t-1} > \eta_{1,t}\}} \right\}, \\
W(\xi_{t-1}, 0) =\ & E\left\{ 1_{\{\eta_{1,t} \geq \xi_{t-1} > 0\}} \right\} + E\left\{ 1_{\{\eta_{2,t} \geq S_t(\xi_{t-1} - \eta_{1,t}), \xi_{t-1} > \eta_{1,t}\}} \right\},
\end{aligned}
\right\} \tag{2.2}
$$

where $s \neq 0$ and ξ_{t-1} is a r.v., $t = 1, 2, \ldots$.

The p.g.f. $\Phi(t, s) = Es^{Y_t}$ of the process (1.2) satisfies the equation (see Yanev and Yanev (1995a), Theorem 2.1)

$$
\Phi(t+1, s) = \delta(s)\Phi(t, F(s)) + \Delta(s)\Phi(t, 0) + pW(Y_{t-1}, s) \tag{2.3}
$$

and by iterating

$$
\Phi(t+1, s) = \Phi(0, F_{t+1}(s))\gamma_{t+1}(s) + \sum_{k=0}^{t} \Phi(t-k, 0)\Delta(F_k(s))\gamma_k(s) + p\sum_{k=0}^{t} W(Y_{t-k}, F_k(s))\gamma_k(s). \tag{2.4}
$$

The generating function of $\{u_t\}$ admits the following expression (see Yanev and Yanev(1995a), Lemma 3.1)

$$
U(s) = \sum_{k=0}^{\infty} u_k s^k = B(s)/((1-s)\gamma(s)), \tag{2.5}
$$

where $\gamma(s) = \sum_{k=0}^{\infty} \gamma_k s_k$, $B(s) = \sum_{k=0}^{\infty}[1 - Q(F_k) - p\sum_{i=1}^{\infty} W(Z_{i-1}, F_t)s^i]\gamma_k s^k$ and $Q(s) = Es^{Z_0} = (g_0(s) - g_0(0))/(1 - g_0(0))$.

Further on we investigate the critical case under the conditions:

$$
\left.
\begin{aligned}
&F'(1) = 1, \ 0 < F''(1) = 2b < \infty, \ 0 < g'(1) < \infty, \ 0 < \lambda_0 = g_0'(1) < \infty \\
&0 \leq \eta_{1,t} \leq N_1 < \infty, \ 0 \leq \eta_{2,t} \leq N_2 < \infty.
\end{aligned}
\right\} \tag{2.6}
$$

We obtain an additional parameter of criticality which plays an important role for the asymptotic behaviour of the processes:

$$\theta = E(M_t|Y_{t-1} > 0)/(\frac{1}{2}VarX_{1,1}) = (rEI_1 - pE\{\eta_{1,1} + \eta_{2,1}\})b^{-1} . \qquad (2.7)$$

Sometimes we need the stronger moment conditions

$$\begin{aligned}
EI_1 \log(1 + I_1) < \infty, \quad EX_{1,1}^2 \log(1 + X_{1,1}) < \infty \quad &\text{for} \quad 0 < \theta \le 1 , \\
EI_1 \log(1 + I_1) < \infty \quad &\text{for} \quad \theta = 0 .
\end{aligned} \right\} \qquad (2.8)$$

We have (Yanev and Yanev (1995a), Lemma 3.3) that under the conditions (2.6),

$$\begin{aligned}
W &= \sum_{k=1}^{\infty} \{E(\eta_{1,k} + \eta_{2,k} - Z_{k-1}) \\
&\quad + E(\eta_{2,k} - S_k(Z_{k-1} - \eta_{1,k})1_{\{\eta_{2,k} \ge S_k(Z_{k-1} - \eta_{1,k}), Z_{k-1} > \eta_{2,k}\}}\} \qquad (2.9) \\
&= \sum_{k=1}^{\infty} W'_s(Z_{k-1}, 1) < \infty.
\end{aligned}$$

Further on we use also the next basic results (see Yanev and Yanev (1995a), Theorem 2.2 and (1995b), Theorem 2.1). Remember that by Lemma 3.2 of Yanev and Yanev (1995a) one has $0 < A(\theta) = \lim_{t \to \infty} t^\theta \gamma_t < \infty$.

Theorem C. *Assume (2.6) and (2.8). Then as $t \to \infty$,*

$$u_t \sim \begin{cases} C(\theta) & , \quad \theta > 1, \\ C(1)/log\ t & , \quad \theta = 1, \\ C(\theta)/t^{1-\theta} & , \quad 0 \le \theta < 1, \end{cases} \qquad (2.10)$$

where $C(\theta)$ are positive constants and

$$C(\theta) = \begin{cases} B(1)/\gamma(1) & , \quad \theta > 1, \\ B(1)/(\Gamma(\theta)\Gamma(1-\theta)A(\theta)) & , \quad 0 < \theta \le 1, \\ (r\lambda_0 + apW)/(ab) & , \quad \theta = 0. \end{cases} \qquad (2.11)$$

(i) If $0 < \theta \le 1$ and only (2.6) holds then $C(\theta) = C(\theta, t)$ is a s.v.f. as $t \to \infty$.
(ii) If $\theta < 0$ then $E\tau = U(1) = (r\lambda_0 + apW)/(-ab\theta) < \infty$.
Note that limiting distributions of $\{Z_t\}$ were studied by Yanev and Yanev (1995b). Now we will give the results for $\{Y_t\}$ obtained by investigating the equation (2.3).

Theorem 2.1 *Assume conditions (2.6) and (2.8).*
(i) If $\theta \le 1/2$ then as $t \to \infty$,

$$P(Y_t = 0) \sim \begin{cases} D(\theta) & , \quad \theta < 0, \\ D(0)/log\ t & , \quad \theta = 0, \\ D(\theta)/t^\theta & , \quad 0 < \theta \le 1/2, \end{cases} \qquad (2.12)$$

where $D(\theta)$ are positive constants and

$$D(\theta) = \begin{cases} (-b\theta)/(r\lambda_0 - b\theta + paW) & , \quad \theta < 0, \\ b/(r\lambda_0 + paW) & , \quad \theta = 0, \\ A(\theta)\Gamma(1-\theta)\Gamma(1+\theta)/(aB(1)) & , \quad 0 < \theta \le 1/2. \end{cases} \qquad (2.13)$$

219

(ii)If $\theta > 1/2$ then as $t \to \infty$,

$$\sum_{k=0}^{t} P(Y_k = 0) \sim \begin{cases} K(\theta)t^{1-\theta} & , \quad 1/2 < \theta < 1, \\ K(1)\log t & , \qquad \theta = 1, \\ K(\theta) & , \qquad \theta > 1, \end{cases} \tag{2.14}$$

where $K(\theta)$ are positive constants and

$$K(\theta) = \begin{cases} A(\theta)/(aB(1)(1-\theta)) & , \quad 1/2 < \theta < 1, \\ A(1)/(aB(1)) & , \qquad \theta = 1, \\ \gamma(1)/(aB(1)) & , \qquad \theta > 1. \end{cases} \tag{2.15}$$

(iii) If $0 < \theta \le 1$ and only (2.6) holds then $D(\theta) = D(\theta,t)$ and $K(\theta) = K(\theta,t)$ are s.v.f. as $t \to \infty$.

Corollary 2.1 *The branching migration process $\{Y_t\}$ is an aperiodic and irredicible Markov chain which is: (i) non-recurrent if $\theta > 1$; (ii) null-recurrent if $0 \le \theta \le 1$; (iii) positive-recurrent if $\theta < 0$.*

Hence θ is the recurrence parameter of the critical process $\{Y_t\}$.

Theorem 2.2 *Under the conditions (2.6) and (2.8) as $t \to \infty$,*

$$EY_t \sim \begin{cases} b\theta t & , \quad \theta > 0 \\ bt/\log t & , \quad \theta = 0, \\ o(t) & , \quad \theta < 0, \end{cases} \tag{2.16}$$

$$VarY_t \sim \begin{cases} b^2\theta(1+\theta)\, t^2 & , \quad \theta > 0, \\ b^2t^2/\log t & , \quad \theta = 0. \\ o(t^2) & , \quad \theta < 0, \end{cases} \tag{2.17}$$

Theorem 2.3 *Assume condition (2.6).*
   (i) If $\theta < 0$ then for $\{Y_t\}$ there exists a stationary distribution $\{v_k\}$, $\sum_{k=0}^{\infty} v_k = 1$, and its p.g.f. $V(s) = \sum_{k=0}^{\infty} v_k s^k$, $|s| \le 1$, is the unique solution of the functional equation

$$V(s) = V(F(s))\delta(s) + V(0)\Delta(s) + p\sum_{k=1}^{\infty} W(k,s) , \tag{2.18}$$

where $V(0) = (-b\theta)/(-b\theta + r\lambda_0 + paW)$.
   (ii) If $\theta = 0$ and $g''(1) < \infty$, then

$$\lim_{t \to \infty} P(\frac{\log Y_t}{\log t} \le x) = x \in [0,1]. \tag{2.19}$$

(iii) If $\theta > 0$, then

$$\lim_{t \to \infty} P(Y_t/(bt) \le x) = \frac{1}{\Gamma(\theta)} \int_0^x y^{\theta-1}e^{-y}dy , \quad x \ge 0. \tag{2.20}$$

220

Comment. To precise the statements of Theorem 2.2 and Theorem 2.3 in the case $\theta < 0$ one needs on additional efforts. In the particular case $\eta_{1,t} \equiv 1$ and $\eta_{2,t} \equiv 0$ it was obtained by Yanev and Yanev (1993), under slightly stronger moment assumptions, that $EY_t \sim ct^{1+\theta}$ for $-1 < \theta < 0$; $\sim C \log t$ for $\theta = -1$; $\sim C$ for $\theta < -1$. For the p.g.f. of the stationary distribution as $s \uparrow 1$ we got : $1 - V(s) \sim C(1-s)^{-\theta}$ for $-1 < \theta < 0$; $\sim C(1-s) \log(1/(1-s))$ for $\theta = -1$; $\sim C(1-s)$ for $\theta < -1$. The constants C depend on θ.

3 Preliminaries

Lemma 3.1 *Assume (2.6).*
   (i) For every $\alpha \geq 0$,

$$\lim_{t \to \infty} \gamma_t(\exp\{-\alpha/(bt)\}) = (1+\alpha)^{-\theta}. \tag{3.1}$$

(ii) If $\theta = 0$ and additionally $g''(1) < \infty$ then uniformly for $0 \leq s < 1$

$$\gamma_t(s) \sim \gamma(s) \left(1 - \frac{\delta''(1)/(2b)}{(1-s)^{-1}+bt} \right) , \qquad t \to \infty , \tag{3.2}$$

where $0 < \gamma(s) = \prod_{k=0}^{\infty} \delta(F_k(s)) < \infty$.

Proof.   (i) This part follows directly by Lemma 4.1 of Yanev and Yanev (1995b).
   (ii) Since for $0 \leq s < 1$,

$$F_k(s) = 1 - \frac{1+\varepsilon_k(s)}{(1-s)^{-1}+bk} , \qquad \lim_{k \to \infty} \sup_{s \in [0,1)} \varepsilon_k(s) = 0 , \tag{3.3}$$

and $\delta'(1) = b\theta = 0$ then

$$\delta(F_k(s)) = 1 + \frac{\delta''(1)}{2}(1-F_k(s))^2 + o((1-F_k(s))^2) = 1 + \frac{\delta''(1)/2}{((1-s)^{-1}+bk)^2} + o(1/k^2). \tag{3.4}$$

Therefore $0 < \gamma(s) = \prod_{k=0}^{\infty} \delta(F_k(s)) < \infty$. Now it is not difficult to obtain that

$$
\begin{aligned}
\gamma_t(s) &= \prod_{k=0}^{t-1} \delta(F_k(s)) = \gamma(s) \prod_{k=t}^{\infty} \frac{1}{\delta(F_k(s))} \\
&= \gamma(s) \exp\left\{ -\sum_{k=t}^{\infty} \log\left(1 + \frac{\delta''(1)/2}{((1-s)^{-1}+bk)^2} + o\left(\frac{1}{k^2}\right)\right) \right\} \\
&= \gamma(s) \exp\left\{ -\sum_{k=t}^{\infty} \frac{\delta''(1)/2}{((1-s)^{-1}+bk)^2} + o\left(\frac{1}{t}\right) \right\} \\
&= \gamma(s)(1 - \frac{\delta''(1)/(2b)}{((1-s)^{-1}+bt)}) + o\left(\frac{1}{t}\right) ,
\end{aligned}
$$

uniformly for $0 \leq s < 1$.

221

Lemma 3.2 *If* $a = P(Y_1 > 0|Y_0 = 0) = r(1 - g_0(0)) > 0$ *then*

$$\sum_{n=0}^{t} W(Y_n, s) = a \sum_{k=0}^{t-1} P(Y_k = 0) \sum_{j=0}^{t-k-1} W(Z_j, s). \tag{3.5}$$

Proof. Note that

$$
\begin{aligned}
P(Y_n = m) &= \sum_{k=0}^{n-1} P(Y_k = 0) P(Y_{n-k} = m, \min_{1 \le i \le n-k} Y_i > 0) \\
&= a \sum_{k=0}^{n-1} P(Y_k = 0) P(Z_{n-k-1} = m).
\end{aligned}
$$

Hence from (2.2) it follows that

$$
\begin{aligned}
\sum_{n=0}^{t} W(Y_n, s) &= \sum_{n=0}^{t} \sum_{m=1}^{\infty} P(Y_n = m) W(m, s) \\
&= a \sum_{n=0}^{t} \sum_{k=0}^{n-1} P(Y_k = 0) \sum_{m=1}^{\infty} P(Z_{n-k-1} = m) W(m, s) \\
&= a \sum_{k=0}^{t-1} P(Y_k = 0) \sum_{j=0}^{t-k-1} W(Z_j, s),
\end{aligned}
$$

which proves (3.5).

Lemma 3.3 *Let* $\alpha > 0$, $c > 0$ *and* $0 < x < 1$. *Then as* $t \to \infty$,

$$S(t, x) = \sum_{k=1}^{t-1} \left\{ \left(k + \frac{1}{(1 - \exp\{-\alpha/t^x\})c}\right) \log(t + 1 - k) \right\}^{-1} \to 1 - x . \tag{3.6}$$

Proof. For each $0 < \varepsilon < 1$ one has

$$S(t, x) = \sum_{k=1}^{[\varepsilon(t-1)]} (\cdot) + \sum_{k=[\varepsilon(t-1)]+1}^{t-1} (\cdot) = S_1(t, x) + S_2(t, x). \tag{3.7}$$

Now it is not difficult to see that for $t \to \infty$,

$$S_1(t, x) \sim \sum_{k=1}^{[\varepsilon(t-1)]} \{(k + t^x/(\alpha c)) \log(t + 1 - k)\}^{-1} \sim \frac{1}{\log t} \sum_{k=1}^{[\varepsilon(t-1)]} (k + t^x/(\alpha c))^{-1}. \tag{3.8}$$

On the other hand, as $t \to \infty$,

$$
\begin{aligned}
\sum_{k=1}^{[\varepsilon(t-1)]} (k + t^x/\alpha c)^{-1} &\sim \sum_{1 + t^x/\alpha c \le j \le \varepsilon(t-1) + t^x/(\alpha c)} j^{-1} \\
&\sim \log(\varepsilon(t - 1) + t^x/(\alpha c)) - \log(1 + t^x/\alpha c) \sim (1 - x) \log t.
\end{aligned}
\tag{3.9}
$$

222

By (3.7) it follows (similarly to (3.9)) that

$$S_2(t,x) \le \frac{1}{\varepsilon(t-1) + t^x/(\alpha c)} \sum_{j=2}^{[t(1-\varepsilon)]} \frac{1}{\log t} = O\left(\frac{1}{\log t}\right). \tag{3.10}$$

Now from (3.8)-(3.10) one has for every $0 < \varepsilon < 1$,

$$\lim_{t\to\infty} S_1(t,x) = 1 - x, \quad \lim_{t\to\infty} S_2(t,x) = 0, \tag{3.11}$$

which proves (3.6).

4   Probability of hitting zero

Let us recall the definition of "taboo probabilities" for $\{Y_t\}$, when the "taboo" is zero

$$_0p_{0n}(t) = P(Y_{T+t} = n, Y_{T+i} > 0, 1 \le i \le t, Y_T = 0), {}_0p_{0n}(0) = 0, \ n = 0, 1, 2, \ldots. \tag{4.1}$$

Proof of Theorem 2.1.   By (4.1) and the definition of the life-period it follows that

$$P(\tau = t) = u_{t-1} - u_t = {}_0p_{00}(t+1)/(1 - {}_0p_{00}(1)) = {}_0p_{00}(t+1)/a.$$

Clearly $P(Y_t = 0) = \sum_{k=1}^{t} {}_0p_{00}(k)P(Y_{t-k} = 0)$ and

$$\sum_{k=0}^{\infty} P(Y_k = 0)s^k = (1 - \sum_{k=0}^{\infty} {}_0p_{00}(k)s^k)^{-1} = \frac{1}{a(1-s)U(s)}, \tag{4.2}$$

where $U(s)$ is from (2.5).

If $U(1) < \infty$, then applying to (4.2) the well-known Erdös-Feller-Pollard renewal theorem (see Feller(1957), XIII, 3, p.286), one obtains

$$\lim_{t\to\infty} P(Y_t = 0) = (1 + aU(1))^{-1}. \tag{4.3}$$

If $U(1) = \infty$ and $u_t \sim ct^{-\beta}$, $1/2 \le \beta \le 1$, then applying Theorem 3 of Erickson (1970) to (4.2), one obtains

$$P(Y_t = 0) \sim (a \sum_{k=0}^{t} u_k)^{-1}, \quad t \to \infty. \tag{4.4}$$

Now (2.12) follows directly from (4.3) and (4.4) applying (2.10) and Theorem C(ii).

If $1/2 < \theta < 1$ then by Theorem C it follows that

$$U(s) \sim B(1)/(\Gamma(1-\theta)A(\theta)(1-s)^\theta), \quad s \uparrow 1,$$

and by (4.2) one has

$$\sum_{k=0}^{\infty} P(Y_t = 0)s^k \sim \Gamma(1-\theta)A(\theta)/(aB(1)(1-s)^{1-\theta}), \quad s \uparrow 1.$$

223

Therefore, applying Theorem 5 (Feller(1971), Ch.XII, 5) one obtains that

$$\sum_{k=0}^{t} P(Y_k = 0) \sim \frac{A(\theta)}{aB(1)(1 - \theta)} t^{1-\theta}, \quad t \to \infty,$$

which proves (2.14) in this case. The cases $\theta = 1$ and $\theta > 1$ can be proved similarly.

Comment The results generalize those obtained by Yanev and Mitov (1981) when a.s. $\eta_{1,t} \equiv 1$, $\eta_{2,t} \equiv 0$ and $I_t \equiv I_t^0$.

Proof of Corrolary 2.1 Let $p_{ij}(t) = P(Y_{t+k} = j | Y_k = i)$. Since $p_{00}(1) = 1 - r(1 - g_0(0)) > 0$ then the state zero has period 1. On the other hand, the set of the states of the process $\{Y_t\}$ consists of all states which can be reached by the state zero, i.e. for each j there exists t_j such that $p_{0j}(t_j) > 0$. Since $p_{i0}(1) \geq pF^{(i-N_1)}(0) + qF^i(0) + rF^i(0)g_0(0) > 0$, then for all states i and j one has $p_{ij}(t_j+1) \geq p_{i0}(1)p_{0j}(t_j) > 0$. Thus $\{Y_t\}$ is an aperiodic and irreducible Markov chain. The rest follows by Theorem 2.1.

5 Moments

The differentiation of (2.3) for $s = 1$ implies

$$\Phi_s'(t + 1, 1) = \Phi_s'(t, 1) + b\theta + (r\lambda_0 - b\theta)\Phi(t, 0) + pW_s'(Y_{t-1}, 1). \tag{5.1}$$

Now iterating (5.1) and applying Lemma 3.2 one obtains

$$\begin{aligned}
EY_{t+1} &= EY_0 + b\theta t + (r\lambda_0 - bt)\sum_{k=0}^{t} P(Y_k = 0) + p\sum_{k=1}^{t} W_s'(Y_{k-1}, 1) \tag{5.2} \\
&= EY_0 + b\theta t + (r\lambda_0 - b\theta)\sum_{k=0}^{t} P(Y_k = 0) + pa\sum_{k=0}^{t-2} P(Y_k = 0)\sum_{j=0}^{t-k-2} W_s'(Z_j, 1).
\end{aligned}$$

Proof of Theorem 2.2 If $\theta < 0$ then from (5.2), Theorem 2.1 and (2.9) one has

$$\lim_{t\to\infty} \frac{EY_{t+1}}{t} = b\theta - \frac{b\theta(r\lambda_0 - b\theta)}{r\lambda_0 - b\theta + paW} - \frac{b\theta paW}{r\lambda_0 - b\theta + paW} = 0$$

If $\theta = 0$ then from (5.2) applying Theorem 2.1 and (2.9) one obtains as $t \to \infty$,

$$EY_{t+1} \sim \frac{r\lambda_0 b}{r\lambda_0 + paW}\frac{t}{\log t} + \frac{paWb}{r\lambda_0 + paW}\frac{t}{\log t} = \frac{bt}{\log t}.$$

Similarly, if $\theta > 0$, applying Theorem 2.1 and (2.9) one gets from (4.2) as $t \to \infty$,

$$EY_{t+1} = b\theta t + O(t^{(1-\theta)^+}).$$

Hence (2.16) is proved. Now differatiating (2.3) twice for $s = 1$ one can obtain that

$$\begin{aligned}
\Phi_s''(t + 1, 1) &= EY_0(Y_0 - 1) + \delta''(1)t + 2b(1 + \theta)\sum_{n=0}^{t} EY_n \tag{5.3} \\
&+ (rg_0''(1) - \delta''(1))\sum_{n=0}^{t} P(Y_n = 0) + p\sum_{n=1}^{t} W_s''(Y_{n-1}, 1).
\end{aligned}$$

Similarly to Lemma 3.3 of Yanev and Yanev (1995b), it is not difficult to show that $\sum_{k=1}^{\infty} W_s''(Z_{k-1}, 1) < \infty$. Therefore, applying Theorem 2.1 and Lemma 3.2 to (5.3) one obtains for $\theta \geq 0$,

$$\Phi_s''(t+1, 1) \sim 2b(1+\theta) \sum_{n=0}^{t} EY_n, \quad t \to \infty.$$

Now (2.17) follows from (2.16).

6 Limiting distributions

Now we will shall prove Theorem 2.3, which is the main result of the paper.

(i) If $\theta < 0$, then by Corollary 2.1 there exists a stationary distribution $v_j = \lim_{t \to \infty} p_{ij}(t)$, $\sum_{k=0}^{\infty} v_k = 1$, such that $v_j = \sum_{k=0}^{\infty} v_k p_{kj}(1)$. Hence for the p.g.f. $V(s) = \sum_{j=0}^{\infty} v_j s^j$ one has

$$
\begin{aligned}
V(s) &= \sum_{k=0}^{\infty} v_k \sum_{j=0}^{\infty} P(Y_1 = j | Y_0 = k) s^j = \sum_{k=0}^{\infty} v_k E(s^{Y_1} | Y_0 = k) \\
&= \sum_{k=1}^{\infty} v_k F^k(s)(\delta(s) + pW(k,s)) + v_0(p + q + rg_0(s))) \\
&= V(F(s))\delta(s) - V(0)(1 - \delta(s) - r(1 - g_0(s)) + p\sum_{k=1}^{\infty} v_k W(k,s),
\end{aligned}
$$

where $V(0) = (-b\theta)/(r\lambda_0 - b\theta + paW)$ by Theorem 2.1.

(ii) Let now $\theta = 0$ and $s_t = \exp\{-\alpha/t^x\}$, $\alpha \geq 0,$, $0 < x < 1$. Then by (2.4) and (3.5) one gets

$$
\begin{aligned}
\Phi(t+1, s_t) &= \Phi(0, F_{t+1}(s_t))\gamma_{t+1}(s_t) + \sum_{k=0}^{t} P(Y_k = 0)\Delta(F_{t-k}(s_t))\gamma_{t-k}(s_t) \quad (6.1) \\
&\quad + pa\sum_{k=0}^{t-1} P(Y_k = 0) \sum_{j=0}^{t-k-1} W(Z_j, F_{t-k-j}(s_t))\gamma_{t-k-j}(s_t) \\
&= \Sigma_1(t) + \Sigma_2(t) + \Sigma_3(t), \quad \text{say.}
\end{aligned}
$$

By (3.2) and (3.3) it follows that as $t \to \infty$,

$$\Sigma_1(t) \to 1 . \tag{6.2}$$

On the other hand

$$
\begin{aligned}
\Sigma_2(t) &= \sum_{k=0}^{t} P(Y_k = 0)(1 - \delta(F_{t-k}(s_t)))\gamma_{t-k}(s_t) \tag{6.3} \\
&\quad - r\sum_{k=0}^{t} P(Y_k = 0)(1 - g_0(F_{t-k}(s_t)))\gamma_{t-k}(s_t) = \Sigma_{21}(t) - \Sigma_{22}(t).
\end{aligned}
$$

225

Applying Theorem 2.1 and Lemma 3.1 (see (3.2)-(3.4)) it is not difficult to show that as $t \to \infty$,

$$\Sigma_{21}(t) = o(\frac{1}{\log t}). \tag{6.4}$$

Similarly, by Theorem 2.1, Lemma 3.1 (see (3.2) and (3.3)) and Lemma 3.3 one obtains as $t \to \infty$,

$$\Sigma_{22}(t) \to \frac{r\lambda_0}{r\lambda_0 + paW}(1 - x). \tag{6.5}$$

By (3.3) it follows as k and t tend to ∞,

$$1 - F_k(s_t) \quad \sim \quad \frac{1}{(1 - s_t)^{-1} + bk} \sim \frac{1}{t^x/\alpha + bk}. \tag{6.6}$$

Therefore as k and t tend to ∞,

$$1 - F_k^{-n}(s_t)F_k^{-m}(s_t) \sim -(n+m)/((1-s_t)^{-1} + bk) \sim -\frac{n+m}{t^x/\alpha + bk}. \tag{6.7}$$

Now applying (6.7) and (6.6) it is not difficult to obtain (see (2.2)) that for $0 \le j \le t$ and $t \ge k \to \infty$,

$$W(j, F_k(s_t)) \sim -\frac{W_s'(j,1)}{(1-s_t)^{-1} + bk} \sim -\frac{W_s'(j,1)}{t^x/\alpha + bk}. \tag{6.8}$$

Hence using (6.8) and (2.9) and applying Theorem 2.1 and Lemma 3.3 one obtains

$$\Sigma_3(t) \to -\frac{paW}{r\lambda_0 + paW}(1 - x), \quad t \to \infty. \tag{6.9}$$

Finally, from (6.1)-(6.5) and (6.9) one has

$$\lim_{t \to \infty} \Phi(t+1, s_t) = 1 - (\frac{r\lambda_0}{r\lambda_0 + paW} + \frac{paW}{r\lambda_0 + paW})(1 - x) = x. \tag{6.10}$$

Note that (6.10) is equivalent to the following

$$\lim_{t \to \infty} P(\frac{Y_t}{t^x} \le y) = x \in (0,1), \quad 0 < y < \infty.$$

Now putting $y = 1$ it is not difficult to obtain (2.19).

(iii)Consider the case $\theta > 0$. We shall study the equation (6.1) with $s_t = \exp\{-\alpha/(bt)\}$, $\alpha \ge 0$. Note that by Lemma 3.1 (see (3.1)) one has

$$\Sigma_1(t) = \Phi(0, F_{t+1}(s_t))\gamma_{t+1}(s_t) \to (1+\alpha)^{-\theta}, \quad t \to \infty. \tag{6.11}$$

On the other hand, for every $\alpha > 0$ there exists $t_0 > 0$ such that for $t \ge t_0$ one has $s_t \in [s_0, 1)$ for some $s_0 < 1$. Since $\delta(1) = 1$ and $\delta'(1) = b\theta > 0$ one can choose $s_0 \in [0,1)$ such that $\delta(s)$ be increasing in $[s_0, 1)$. Therefore $\delta(F_k(s_t)) \le 1$ for $t \ge t_0$ and

$$\gamma_k(s_t) = \prod_{j=0}^{k-1} \delta(F_j(s_t)) \le 1. \tag{6.12}$$

226

By (2.1) and (2.6) it follows that $\delta(s) = R(s)F^{-N_1}(s)s^{-N_2}$, where $R(s) = pE\left\{F^{N_1-\eta_{1,t}}(s)s^{N_2-\eta_{2,t}}\right\} + qF^{N_1}(s)s^{N_2} + rF^{N_1}(s)s^{N_2}g(s)$ is a p.g.f. Hence for large enough t

$$
\begin{aligned}
1 - \delta(F_k(s_t)) &= \frac{1 - R(F_k(s_t)) - (1 - F_{k+1}^{N_1}(s_t)) - F_{k+1}^{N_1}(s_t)(1 - F_k^{N_2}(s_t))}{F_{k+1}^{N_1}(s_t)F_k^{N_2}(s_t)} \\
&\leq F^{-(N_1+N_2)}(0)(R'(1) + N_1 + N_2)(1 - s_t)(1 + o(1)), \\
1 - g_0(F_k(s_t)) &= \lambda_0(1 - s_t)(1 + o(1)).
\end{aligned}
$$

Therefore (see (2.1)) as $t \to \infty$,

$$
|\Delta(F_k(s_t))| \leq F^{-(N_1+N_2)}(0)(R'(1) + N_1 + N_2 + r\lambda_0)(1 - s_t)(1 + o(1)). \qquad (6.13)
$$

Now by Theorem 1.1, using (6.12) and (6.13), one can show that

$$
\Sigma_2(t) \leq \frac{C}{t}\sum_{k=0}^{t}P(Y_k = 0) \to 0, \quad t \to \infty. \qquad (6.14)
$$

Similarly to (6.6)-(6.8) one gets as k and t tend to ∞,

$$
\begin{aligned}
1 - F_k(s_t) &\sim \alpha/(b(t + \alpha k)), \\
1 - F_k^{-n}(s_t)F_k^{-m}(s_t) &\sim -(n + m)\alpha/(b(t + \alpha k)), \\
W(j, F_k(s_t)) &\sim -\alpha W_s'(j, 1)/(b(t + \alpha k)), \quad 0 \leq j \leq t, \quad t \geq k \to \infty.
\end{aligned}
$$

Therefore, by Theorem 1.1, using (2.9) and (6.12), one can prove that

$$
\begin{aligned}
|\Sigma_3(t)| &\leq pa\sum_{k=0}^{t}P(Y_k = 0)\sum_{j=0}^{t-k}\frac{\alpha W_s'(j, 1)}{b(t + \alpha(t - k - j))} \qquad (6.15) \\
&\leq \frac{paW}{bt}\sum_{k=0}^{t}P(Y_k = 0) \to 0, \quad t \to \infty.
\end{aligned}
$$

Finally from (6.1), (6.11), (6.14) and (6.15) it follows that $\lim_{t\to\infty}\Phi(t, s_t) = (1+\alpha)^{-\theta}$, which is equivalent to (2.20).

Acknowledgement. The authors are very grateful to the referee for useful comments.

References

1. Dyakonova E.E. (1993) On transient phenomena for branching migration processes. In: *Probabilistic Methods in Discrete Mathematics*, ed. V. Kolchin. *Prog. Pure and Appl. Discret. Math.*, **1**, 148-154, TVP, Moscow.
2. Erickson, K.B. (1970) Strong renewal theorem with infinite mean. *Trans. Amer. Math. Soc.*, **151**, 263-291.

3. Feller W. (1957) *An Introduction to Probability Theory and Its Applications*, **1**, 2nd edn., Wiley, New York.

4. Feller W. (1971) *An Introduction to Probability Theory and Its Applications*, **2**, 2nd edn., Wiley, New York.

5. Foster, J.H. (1971) A limit theorem for a branching process with state-dependent immigration. *Ann. Math. Statist.*, **42**, 1773-1776.

6. Grey, D.R. (1988) Supercritical branching processes with density independent catastrophes. *Math. Proc. Camb. Phil. Soc.* **104,** 413-416.

7. Han, L.V. (1980) Limit theorems for Galton-Watson branching processes with migration. *Siberian Math. J.,* **21**, 283-293.

8. Kaverin, S.V. (1990) A refinement of limit theorems for critical branching processes with an emigration. *Theory Prob. Appl.,* **35**, 574-580.

9. Nagaev, S.V. and Han, L.V. (1980) Limit theorems for critical Galton-Watson branching process with migration. *Theory Prob. Appl.* **25**, 514-525.

10. Pakes, A.G. (1975) Some results for non-supercritical Galton-Watson processes with immigration. Mathematical Biosciences, **24**. 71-92.

11. Rahimov, I. (1995) *Random Sums and Branching Stochastic Processes.* Lecture Notes in Statistics, **96**, Springer-Verlag, New York.

12. Vatutin, V.A. (1977) A critical Galton-Watson branching process with emigration. *Theory Prob. Appl.* **22**, 465-481.

13. Yanev, G.P. and Yanev, N.M. (1991) On a new model of branching migration processes.*C. R. Acad. Bulg. Sci.,* **44**, 19-22.

14. Yanev, G.P. and Yanev, N.M. (1993) On critical branching migration processes with predominating emigration. *Institute of Mathematics*, Sofia, Preprint no. 1, pp.37.

15. Yanev G.P. and Yanev N.M. (1995a) Critical branching processes with random migration. In: *Branching Processes,* ed. C. C. Heyde. *Lecture Notes in Statistics*, **99**, 36-46, Springer-Verlag, New York .

16. Yanev G.P. and Yanev N.M. (1995b) Limit theorems for branching processes with random migration stopped at zero. In: *Classical and Modern Branching Processes,* eds. K. B. Athreya and P. Jagers. *IMA Volumes in Mathematics and its Applications,* **84**, 323-336, Springer-Verlag, New York.

17. Yanev, N.M. and Mitov, K.V. (1980) Controlled branching processes: The case of random migration. *C. R. Acad. Bulg. Sci.* **33**, 433-435.

18. Yanev, N.M. and Mitov, K.V. (1981) Critical branching migration processes. In: *Mathematics and Mathematical Education,* Proc. 10th Spring Conf. of UBM, Publ. House of Bulg. Acad. Sci., Sofia, 321-328 (In Russian).

19. Yanev, N.M. and Mitov, K.V. (1985) Critical branching processes with non-homogeneous migration. *Ann. Prob.,* **13**, 923-933.

20. Yanev, N.M., Vatutin, V.A. and Mitov, K.V. (1986) Critical branching processes with random migration stopped at zero. *Mathematics and Mathematical Education,* Proc. 15th Spring Conf. of UBM, Publ. House of Bulg. Acad. Sci., Sofia, 511-517 (In Russian).

21. Zubkov, A.M. (1972) Life-periods of a branching process with immigration. *Theory Prob. Appl.,* **17**, 174-183.

Remarks on the Law of Succession

W. J. Ewens

Department of Mathematics
Monash University

Abstract

Applied probability and time series, the respective areas in which Joe Gani and Ted Hannan made their names, meet in the classical problem of the law of succession. Much of the work on this problem assumes a multinomial distribution for the observations. In this paper assumptions are made which imply a non-multinomial distribution, which in turn implies a rather different approach to that used in the multinomial case to estimating probabilities of novel events.

1 Introduction

We consider a sequence of trials, on each one of which one or other of the events $\{E_1, E_2, E_3 \ldots\}$ will occur. The "Law of Succession" asks the question: if an event has occurred n_i times in the first n trials, $(n_i > 0)$, what is the probability that it occurs again on trial $n+1$? A closely associated question is: what is the probability that a previously unobserved event occurs on trial $n + 1$? We focus on the latter question, but note also the implications of our analysis for the former.

The vast literature on this subject can be classified in various ways. First, some authors consider the *conditional* probability that a previously unobserved event occurs on trial $n+1$, given the outcomes of the first n trials, others the unconditional probability. In our model the two probabilities are the same, a matter of independent interest in itself. Second, some approaches are Bayesian, others non-Bayesian. Ours is non-Bayesian, although connections with a Bayesian approach are noted. Finally, some authors assume a finite number of possible events, others an infinite number. This distinction is important and is discussed further below.

229

Within the non-Bayesian literature, by far the most commonly made assumption is the multinomial, that successive trials are independent, with the event E_i having unknown probability p_i on each trial (see, for example, Good (1965), Robbins (1968), Starr (1979) and Clayton and Frees (1987)). Under this assumption the unconditional probability that trial $n + 1$ yields a previously unobserved event is, clearly,

$$\sum_i p_i (1 - p_i)^n. \tag{1}$$

Next, it is necessary to distinguish the case where the set of possible events is finite from the case where it is infinite. In the infinite case the appropriate objects of study are the random *partitions* generated by the events observed. This implies that the only probabilities which we consider are those relating to (unlabelled) partitions rather than labelled events. (In the multinomial case, this requires that the probabilities of interest are those, such as (1), which involve an infinite symmetric summation (or summations) over events.) We assume the infinite case throughout. It is convenient to use the "sampling of species" paradigm: entering a new territory, we observe a sequence of animals (trials), each being of one or other species (event), and we assume, first, that there are infinitely many species, and second that *a priori* we do not even have a list of the possible species in the territory. The (unlabelled) probabilities of interest are thus of outcomes such as: "in a sample of ten animals, six are of one species, three of another and one of a third", it being meaningless to ask for probabilities involving labelled species. The probabilities described in the opening paragraph refer to outcomes of this type. Thus the only outcomes to which we assign probabilities are the species partition vectors $a(n) = (a_1, a_2, \ldots a_n)$, indicating that in a sample of n animals, a_1 species have only one animal represented (i.e. are "singleton" species, discussed frequently later), a_2 species each having two animals represented, and so on.

2 Assumptions

The basic assumption we make is that the probabilities of the partitions of animals into species form a *partition structure* (Kingman, 1878a,b, 1980). In effect this assumption requires consistency and exchangeability for partitions analogous to the corresponding requirements for random variables. Thus it is not surprising that there is a representation for partition structures analagous to the (de Finetti) representation for random variables. Specifically, if $a(n)$ is any species partition of a sample of n animals, with partition structure probability $P(a(n))$, and if $\sum P_m(a(n))$ is the sum of all multinomial probabilities for that partition for some multinomial probability vector $\boldsymbol{p} = p(p_1, p_2, \ldots)$, then (Kingman, 1978a,b) $P(a(n))$ can be expressed in the form

$$P(a(n)) = \int \sum P_m(a(n)) d\mu(\boldsymbol{p}) \tag{2}$$

for some probability measure μ. Multinomial probabilities satisfy the partition structure requirements, assigning the entire probability μ to some fixed vector. However there are many non-multinomial partition structure probabilities, as we note by introducing separately two further requirements.

Requirement 1 (subset deletion, or non-interference). Suppose, from a sample of n animals, we choose one at random, and then remove all (r) animals of the same species as the one chosen. Then we require that the partition probability of the remaining $n - r$ animals is the same as that of an original sample of $n - r$ animals.

Requirement 2 (fixed probability of new species and sufficientness). We require, first, that the conditional and unconditional probabilities that animal $n + 1$ be of a new species be equal and further that, if animal $n+1$ is of one or other species already seen, the probability that it is of a given species seen n_i times among the first n animals depends only on n_i and n. (This latter requirement is the sufficientness postulate of Johnson (1932).)

Taking Requirement 2 first, Donnelly (1986) has shown that if conditional and unconditional probabilities are equal, then they must be of the form $\theta/(\theta + n)$, for some positive parameter θ. It follows that, for a partition structure, the probability that animal $n + 1$ is of a given species seen n_i times among the first n animals is $n_i/(n + \theta)$ (see Appendix 1). These conclusions imply that the distribution μ is Kingman's (1975) Poisson-Dirichlet distribution and that the distribution for the species partition in the sample is given by the *Ewens Sampling Formula*, (or ESF). This prescribes for the partition $(a(n))$ the probability

$$P(a(n)) = \frac{n!}{S_n(\theta)} \prod_j \frac{(\theta/j)^{a_j}}{a_j!} , \tag{3}$$

where $S_n(\theta) = \theta(1 + \theta)(2 + \theta) \ldots (n - 1 + \theta)$ and θ is a finite positive parameter (Ewens, 1972).

Kingman (1978a,b) showed that the ESF also follows from Requirement 1, so that Requirements 1 and 2, perhaps surprisingly, are equivalent (for partition

231

structures).

We now consider properties of the ESF relevant to the law of succession: further general properties of the ESF are given by Tavaré and Ewens (1995).

3   Properties of the Ewens sampling formula

From (3), the (random) total number K of species among the first n animals seen has probability distribution

$$P(K = k) = \frac{S(k,n)\theta^k}{S_n(\theta)} , \quad (k = 1, 2, ..., n) , \tag{4}$$

where $S(k,n)$ is the coefficient of θ^k in $S_n(\theta)$, i.e. is the absolute value of a Stirling number (of the first kind). The mean of K is $\xi(\theta)$, defined by

$$\xi(\theta) = \frac{\theta}{\theta} + \frac{\theta}{1+\theta} + \cdots + \frac{\theta}{n-1+\theta} . \tag{5}$$

From (3) and (4), K is a complete sufficient statistic for θ. Thus minimum variance unbiased (MVU) estimation of any estimable function of θ is obtained by using some (unique) function of K only. Those functions of θ admitting unbiased estimation are linear combinations of expressions of the form

$$[(i + \theta)(j + \theta) \cdots (m + \theta)]^{-1}, \tag{6}$$

where $i, j, \ldots, m$ are integers with $1 \leq i < j < \ldots < m \leq n - 1$. Thus there is an unbiased estimator of the function $\alpha_j(\theta)$, defined by

$$\alpha_j(\theta) = \frac{(j - 1)!\theta}{(n - 1 + \theta)(n - 2 + \theta) \cdots (n - j + \theta)]} , \tag{7}$$

for $j = 1, 2, ..., n - K$, namely $h_j(K)$, defined by

$$h_j(K) = \frac{(j - 1)!S(K - 1, n - j)}{S(K,n)} . \tag{8}$$

In particular, the unbiased estimator of $\theta/(n - 1 + \theta)$ is $h_1(K)$.

Next, the maximum likelihood estimator $\hat{\theta}$ of θ is found implicitly as the solution of the equation

$$\xi(\hat{\theta}) = K. \tag{9}$$

Turning to probabilities relevant to the law of succession, we note again Donnelly's result that, after n animals have been observed,

$$\Pr(\text{animal } n + 1 \text{ is of a new, unobserved, species}) = \frac{\theta}{(n + \theta)} , \tag{10}$$

while the discussion above shows that the probability that animal $n+1$ is of a species seen n_i (> 0) times among the first n animals is $n_i/(n + \theta)$. Since this is a linear function of the probability (10), it is sufficient to focus attention on estimation of (10).

4   Estimating the probability that the next animal observed will be of a new species

The expression (6) shows that, for the ESF, there is no unbiased estimator, given the species composition of the first n animals, of the probability $\theta/(n + \theta)$ that animal $n + 1$ will be of a previously unobserved species. On the other hand, there exists an unbiased estimator of the closely related function $\theta/(n - 1 + \theta)$, namely $h_1(k)$. It is clear why this is so. Given the observed partition vector $a(n)$, an empirical estimator of the latter probability is possible by considering all permutations of the order in which the animals were observed, leading to the unbiased estimator a_1/n, where a_1 is the number of singleton species. The MVU estimator of $\theta/(n-1+\theta)$ is thus, from the sufficiency of K for θ, the expected value of a_1/n, given K, and this is easily shown to be $h_1(K)$, as expected directly from the comment below equation (8). Given an observed sample of n animals, this form of empirical procedure is not available, however, to estimate the probability that animal $n + 1$ is of a new species, so it is not surprising that there is no unbiased estimator of $\theta/(n + \theta)$ in our model.

A parallel conclusion holds in the multinomial case. Robbins (1968) was unable to find an unbiased estimator for the unconditional probability (1) (corresponding to $\theta/(n + \theta)$) and, as noted by Clayton and Frees (1987), there is none. Focusing perhaps unduly on unbiased estimation, Robbins (and many subsequent authors) thus imagined observing a further $((n+1)^{\text{th}})$ animal and found the unbiased estimator a_1'/n for (1), where a_1' is the number of "singleton" species in the $n + 1$ animals. Clayton and Frees (1987) showed that, in the multinomial case, this is the MVU estimator of (1). However use of this estimator can be objected to on two grounds, first that it is unduly motivated by the aim of unbiased estimation, and secondly that if the sample size n is fixed in advance, it is not even possible to observe an $((n + 1)^{\text{th}})$ animal. For purposes of comparison, however, we assume in this section that we also can observe a further animal. (In the next section we drop this (surely unrealistic) assumption.) Thus from the comment below equation (8), (and replacing n by $n+1$), our MVU estimator of $\theta/(n+\theta)$ is $S(K'-1,n)/S(K',n+1)$, where K' is the number of species seen among the $n+1$ animals observed.

This leads to the first interesting difference between the ESF estimation procedure and that for the multinomial case, namely that the MVU estimator of the probability that animal $n+1$ is of a new species is, for the ESF, a function of the total number of species observed, while in the multinomial case it is a function only of the number of singleton species.

The Kingman representation (2) shows that the distribution (3) can be thought of as being derived from the Bayesian viewpoint, with Poisson-Dirichlet prior for a multinomial probability distribution $\mu(\boldsymbol{p})$. It would be interesting to find those priors for the multinomial distribution for which the MVU estimator of the probability that animal $n+1$ is of a new species is a function only of the total number of species K'.

In the multinomial literature several authors have gone further and have imagined sampling a total of $n+m$ animals, for arbitrary m, and then using the observations obtained to estimate (1). Clayton and Frees (1987) show that, defining $a_j(m)$ as the number of species each represented by j animals, the MVU estimator V_m of (1) is a linear combination of $a_1(m), \ldots, a_m(m)$. However, for the ESF, V_m, while being an unbiased estimator of the corresponding probability $\theta/(n+\theta)$, is not the MVU estimator: the sufficiency of K for θ shows that this is found by taking the conditional expectation of V_m, given $K(m)$, the total number of species seen among the first $n+m$ animals. This estimator is

$$\frac{(n+m)!}{j(n+m-j)!} \frac{S(K(m)-1,n+m-j)}{S(K(m),n+m)}.$$

We note that this is a non-linear function of the total number of species observed rather than, as in the multinomial case, a linear combination of the number of singleton, doubleton,..., m-ton species.

A further estimator of $\theta/(n+\theta)$ is the maximum likelihood estimator $\hat{\theta}/(n+\hat{\theta})$, where $\hat{\theta}$ is found from (5) and (9) (replacing n by $n+m$). This estimator is biased and, as noted below, has poor estimation properties.

We now return to the case $m=1$, and consider the variance of the estimator $V_1 \ (= a_1'/(n+1))$. For the ESF, the distribution of a_1' is obtained by using standard inclusion and exclusion formulae (Feller 1968), from which

$$\Pr(a_1' = a) = \frac{\theta^a}{a!} \left[\sum_{j=0}^{n-a} (-1)^j \frac{\theta^j}{j!} \frac{\binom{n}{a+j}}{\binom{n+\theta-1}{a+j}} \right] \tag{11}$$

234

(This generalizes the matching probability distribution applying in the random permutation case $\theta = 1$.) From (11),

$$\text{Var}(V_1) = \frac{n(n-1+2\theta)\theta}{(n+1)(n+\theta)^2(n-1+\theta)} \ . \tag{12}$$

This variance is of order n^{-2}, and is small for both small and large θ. The variance of the MVU estimator $V_1(K)$ is

$$\frac{\sum_k [h_1(k)]^2 \theta^k S(k, n+1)}{S_{n+1}(\theta)} - \theta^2/(n+\theta)^2 \ . \tag{13}$$

This does not simplify to a more easily handled form, and numerical calculation of (13) is necessary. On the other hand, standard theory shows that the variance of $V_1(K)$ approaches the Cramér-Rao bound as n increases. This bound is

$$\frac{(\theta n)^2}{(n+\theta)^4 \psi(\theta)} \ , \tag{14}$$

where

$$\psi(\theta) = \left(\frac{\theta}{1+\theta}\right) + \cdots + \left(\frac{\theta}{n+\theta}\right) - \left[\left(\frac{\theta}{1+\theta}\right)^2 + \cdots + \left(\frac{\theta}{n+\theta}\right)^2\right] \ .$$

For large n, this is of order $\theta/(n^2 \log n)$, less than the variance of the "singletons" estimator V_1 by a factor of order $\log n$. Extensive numerical calculations (not reported here) suggest that there exists an absolute upper bound, over all (θ, n) combinations, to the ratio of (14) to (13), and that this bound is 1.06649..., arising for $n = 3$, $\theta = 0.879$. These calculations also suggest that (13) and (14) are very close when n exceeds about 20. We therefore use (14) as a close approximation to the variance of $h_1(K)$.

Clayton and Frees (1987) give the asymptotic variance of the (MVU) estimator V_1 in the multinomial case (from their (2.4)). It is interesting to compare this asymptotic variance with (14). We do this by using "matched means": for a given n and a given multinomial probability vector p, we compute the multinomial probability (1) and then, from (10), find the value of θ which makes the means of the two MVU estimators agree. We then compare the asymptotic variances of the two estimators, using the Clayton and Frees expression for the multinomial estimator and the Cramér-Rao expression (14) for the ESF.

This comparison is of most interest if specific forms for p are assumed. Table 1 provides numerical values for two simple multinomial cases, the geometric (where

$p_i = p(1-p)^{i-1})$ and the "inverse power" (where $p_i = \text{const}/i^m$), together with the "matched" values for our model. For the values of n and p considered, the ESF asymptotic variance (14) is always less than that in the corresponding geometric multinomial (although for very small n the inequality is reversed), and as n increases this effect appears to become more marked. We conjecture that the ESF variance will always be asymptotically ($n \to \infty$) less than that for the corresponding geometric multinomial. The comparison with the "inverse power" distribution is more interesting. While (14) is generally less than the asymptotic multinomial variance for the values of n and m listed, for any given m there is a range of values of n, increasing with m, for which this not so. The values in Table 2 suggest that both lower and upper limits of this range approximately double when m increases by unity. This matter deserves further investigation.

5   Observing only n animals

We turn now to the much more realistic case where only n animals have been observed and we wish to estimate, in the ESF model, the probability $\theta/(n + \theta)$ that the next animal will be of a new species. This question is largely ignored in the multinomial literature, perhaps because of a focus, in our view misguided, on unbiased estimation.

Simple estimation bounds follows from the inequalities

$$\frac{n-1}{n}\frac{\theta}{n-1+\theta} < \frac{\theta}{n+\theta} < \frac{\theta}{n-1+\theta} . \tag{15}$$

Since $h_1(K)$ is an unbiased estimator of $\theta/(n-1+\theta)$, we easily obtain simple bounding estimates for $\theta/(n+\theta)$. We can, however, do better than use these bounds. The terms in the sequence

$$\alpha_1(\theta), \ \alpha_1(\theta) - \alpha_2(\theta), \ \alpha_1(\theta) - \alpha_2(\theta) + \alpha_3(\theta), \ \ldots \tag{16}$$

where $\alpha_j(\theta)$ is defined in (7), differ from $\theta/(n+\theta)$ by amounts of order $n^{-2}, n^{-3}, \ldots$, the difference being alternately positive and negative. Using (8), we see that the terms in the sequence of estimators

$$h_1(K), \ h_1(K) - h_2(K), \ h_1(K) - h_2(K) + h_3(K), \ldots \tag{17}$$

have respective biases of order $n^{-2}, \ n^{-3}, \ n^{-4}, \ldots$ as estimators of $\theta/(n+\theta)$, with the bias being alternately positive and negative. While the factorial term in these

236

estimators imply that the later terms in the sequence become unsatisfactory, this is not a problem in practice since, unless n is quite small, very tight bounds are obtained before this problem arises.

This is illustrated by the case $n = 300$, $k = 10$. The numerical values of the first four estimators in the sequence (17) are 0.006086362, 0.006066046, 0.006066182 and 0.006066181. Recalling that the successive terms in the sequence over- and underestimate $\theta/(n + \theta)$, it is clear that to eight decimal place accuracy, we can take 0.00606618 as an essentially unbiased estimate of the probability that, having seen 10 species represented in the first 300 animals, animal 301 will be of a new species. This value lies within the bounds derived from (15), namely 0.00606607 and 0.00608636. Table 3 gives results for further (n, k) combinations.

The maximum likelihood estimate 0.00615628 does not lie within these simple bounds, a property also occurring for a wide range of cases not reported here. We will note later further undesirable properties of the maximum likelihood estimator.

Extensive numerical calculations suggest that the mean square errors of the estimators in the sequence (17) have two remarkable and unexpected properties. First, for fixed n and θ, the mean square errors are not monotonic: the mean square error of any even-numbered estimator in the sequence is less than those of the adjacent odd-numbered estimators. Of the even-numbered estimators, that with the smallest mean square error is the first one, i.e. $h_1(K) - h_2(K)$. (This is despite a monotonic decrease in bias of the early terms in the sequence, indicating a second reason why an emphasis on unbiased estimation is inappropriate.) Second, although $h_1(K) - h_2(K)$ is a biased estimator of the probability $\theta/(n + \theta)$ and has no obvious optimality properties, while $h_1(K)$ is the MVU estimator of the closely related probability $\theta/(n - 1 + \theta)$, nevertheless its mean square error (as an estimator of $\theta/(n + \theta)$) appears, from extensive numerical calculations, to be less than the variance of $h_1(K)$. This is a third reason for de-emphasizing unbiased estimation. It also suggests that, in the multinomial case, a more fruitful approach than imagining (possibly unobservable) new animals is to consider, in a sample of n, the properties of the multinomial analogue of $h_1(K) - h_2(K)$, namely

$$\frac{a_1}{n} - (n - 2)^{-1} \left(\frac{a_1}{n} - \frac{a_1(a_1 - 1)}{n(n - 1)} \right) .$$

237

6   Further calculations

There are many possible generalizations of the above calculations. Suppose first we wish to estimate the probability that all of the next m animals will be of a new species. In the case $m = 2$, this probability is

$$\frac{\theta^2}{(n+\theta)(n+1+\theta)} .$$

(18)

There is of course no unbiased estimate of this probability. However the successive terms in the sequence

$$\alpha_2(\theta) - 2\alpha_3(\theta) + \cdots + (-1)^i(i-1)\alpha_i(\theta), \quad (i = 2, 3, 4, \ldots)$$

exceed, and are less than, (18), with successive terms differing from (18) by terms of order $n^{-3}, n^{-4}, \ldots$ Using (18), we again obtain a "bounding" sequence of estimators which successively over- and underestimate (18), with biases decreasing as $n^{-3}, n^{-4}, \ldots$

For the case $k = 10$, $n = 300$, the first four estimates in this sequence are 0.000032346 ($i = 2$), 0.000031912 ($i = 3$) and 0.000031918 ($i = 4, 5$). Thus to eight decimal place accuracy, an unbiased estimate of the probability that animals 301 and 302 will both be of new species types, given that we observed 10 species among the first 300 animals, is 0.00003192. This estimate is approximately the square of the estimate of $\theta/(n+\theta)$, as we expect. The maximum likelihood estimate of (18) is 0.00003776. In proportional terms, the bias in the maximum likelihood estimator is some twelve times larger than in the estimation of (10), a matter discussed further below.

A similar "bounding" estimation procedure can be carried out for further values of m. However, the method does not work for indefinitely large m. Estimation of the probability that the next m animals observed will all be of new species involves functions having numerators of the form $S(k-m, n-p)$, and thus the procedure will work in principle only as far as $m = k - 1$. Even when k is so large that this does not cause difficulties, there is a second limitation involving the value of n. The success of the bounding process relies on the rapid convergence of the estimators in the bounding sequences. For large m convergence does not occur, and asymptotic formulae for the Stirling numbers in (8) show that the bounding procedure does not work effectively when m is of order $n^{\frac{1}{2}}$ or more. For example, if the first 30 animals reveal 10 species, and we wish to estimate the probability that the next 6 animals

238

are all of new (and different) species, the initial (over-)estimate of this probability, using the bounding procedure, is 0.00002965, while the second (under-)estimate is -0.00001477. This latter value is of course useless. These observations indicate interesting limits, one involving k and the other n, to our ability to estimate probabilities of events sufficiently far in the future.

Of course, maximum likelihood estimate of probabilities of this type can still be calculated, but our calculations show that their biases will be considerable. We have noted that the maximum likelihood estimate of (10) exceeds the (effectively unbiased) estimate from the bounding procedure by about 1.5% and the maximum likelihood estimator of (18) exceeds the (effectively unbiased) bounding procedure estimate by 18%. For $m = 3$ the excess is approximately 100%. Maximum likelihood estimation does not, therefore, overcome this problem.

On the other hand, maximum likelihood estimation is useful, after n animals have been observed, for estimating the mean number of species that will be observed after $n + m$ animals have been seen. The details of this estimation procedure are straightforward.

The above may be called prospective estimation procedures. Suppose now that $n + m$ animals have been observed, and we wish to estimate the probability $\theta^m/[(\theta + n)(\theta + n + 1)\ldots(\theta + n + m - 1)]$ that all of the last m animals are of new species types. (This retrospective probability might be used as the basis of a test, having observed n+m animals, that new species arose significantly frequently towards the end of the sequence.) From (4), with n replaced by $n + m$, the MVU estimator $g(K(m), n + m)$ of this probability satisfies

$$\sum_k g(k, n + m)S(k, n + m)\theta^{k-m} = \theta(\theta + 1)\ldots(\theta + n - 1).$$

Estimation makes sense only when $m < K(m)$, and in this case the MVU estimator is, clearly, $S(K(m) - m, n)/S(K(m), n + m)$. Many further MVU estimators are derived with equal simplicity.

Similar estimation procedures, both prospective and retrospective, arise when the species are sequentially labeled as they enter the sample. For example, we can estimate the prospective probability, when n animals have been observed, that the ith species seen is observed j times in the next m animals. Having observed $n + m$ animals, we can also find the MVU estimator of the retrospective probability, again possibly used as the basis of a test of hypothesis, that the ith species arises j times

239

in the last m animals. These and similar procedures are carried out by simple generalizations of the above methods.

A Appendix

The aim of this Appendix is to prove that the probability that animal $n+1$ be of a given species seen n_i times among the first n animals is $n_i/(n+\theta)$ (see Requirement 2).

Suppose the sample of n animals yields n_1 of one species, n_2 of another, ... , n_k of a kth species. Then since animal $n+1$ must be of one or other species seen so far, or of a new species,

$$p(n_1, n) + p(n_2, n) + \cdots + p(n_k, n) + \theta/(n+\theta) = 1.$$

The case $k = n$, so that $n_1 = \ldots = n_k = 1$, gives $p(1, n) = 1/(n+\theta)$. Suppose then that at least one n_i value, say n_1, exceeds 1, and consider a sample of n animals consisting of $n_1 - 1$ of one species, n_2 of a second, ... , n_k of a kth species, and 1 of a $(k+1)^{\text{th}}$ species. Then arguing as above,

$$p(n_1 - 1, n) + p(n_2, n) + \cdots + p(n_k, n) + p(1, n) + \theta/(n+\theta) = 1 .$$

These equations show that $p(n_1, n) - p(n_1 - 1, n) = p(1, n)$ for all n_1, and hence by induction that $p(n_1, n) = n_1/(n+\theta)$ for all n_1. This is the required result.

Note that the proof does not rely essentially on the probability $\theta/(n+\theta)$ that animal $n+1$ is of a new species, and that the conditional probability n_i/n is free of this assumption. Note also that the result is a particular case of a more general result due to Pitman (1995) and Zabell (1995), and the above proof is included for its brevity compared to that of the more general case.

References

1. Clayton, M.K. and Frees, E.W. (1987), "Non-parametric Estimation of the Probability of Discovering a New Species," *J. Amer. Stat. Assocn.* 82, 305-311.

2. Donnelly, P.J. (1986), "Partition Structures, Polya Urns, the Ewens Sampling Formula and the Ages of Alleles," *Theoret. Pop. Biol.* 30,271-288.

3. Ewens, W.J. (1972), "The Sampling Theory of Selectively Neutral Alleles," *Theoret. Pop. Biol.* 3, 87-112.

4. Feller, W. (1968), *An Introduction to Probability Theory and its Applications.* Wiley, New York.

5. Good, I.J. (1965), *The Estimation of Probabilities.* M.I.T.Press, Cambridge, MA.

6. Johnson, W.E. (1932), "Probability: the Deductive and Inductive Problems," *Mind* 49, 409-423.

7. Kingman, J.F.C. (1975), "Random Discrete Distributions," *J. Roy. Stat. Soc. Ser. B* 37, 1-22.

8. Kingman, J.F.C. (1978a), "Random Partitions in Population Genetics," *Proc. Roy. Soc. London Series A*, 361, 1-20.

9. Kingman, J.F.C. (1978b), "The Representation of Partition Structures," *J. Lond. Math. Soc.* 18, 374-380.

10. Kingman, J.F.C. (1980), *The Mathematics of Genetic Diversity*, SIAM, Philadelphia.

11. Pitman, J. (1995), "Exchangeable and partially exchangeable random partitions," *Probability Theory and Related Fields* (to appear).

12. Robbins, H.E. (1968), "Estimating the Total Probability of the Unobserved Outcomes of an Experiment," *Ann. Math. Stat.* 39,256-257.

13. Starr, N. (1979), "Linear Estimation of the Probability of Discovering a New Species," *Ann. Stat.* 7, 644-652.

14. Tavaré, S. and Ewens, W.J. (1995), "The Ewens Multivariate Distribution," in *Discrete Multivariate Disributions*, (ed Johnson, N.L., Kotz S. and Kemp, A.W.), to appear.

15. Zabell, S.L. (1995), "The continuum of inductive methods revisited," *Pittsburgh-Konstanz series in the History and Philosophy of Science*, (to appear).

241

TABLE 1

Asymptotic variances of "matched means" estimators of the probability that animal n is of a new species using (a) multinomial and (b) the present approaches. Cases (i)-(iv) refer to geometric multinomial probabilities $p_i = p(1-p)^{i-1}$, with $p = .2, .4, .6, .8$. Cases (v)-(x) refer to multinomial probabilities of the form $p_i = \text{const}/i^m$, for $m = 2, 3, 4, 5, 6, 7$. See text for details. All variances are $\times 10^3$.

		$n = 60$	$n = 80$	$n = 100$
(i)	(a)	0.3061	0.1729	0.1109
	(b)	0.2910	0.1600	0.1000
(ii)	(a)	0.1338	0.0755.	0.0485
	(b)	0.1221	0.0659	0.0408
(iii)	(a)	0.0751	0.0420	0.0269
	(b)	0.0659	0.0354	0.0218
(iv)	(a)	0.0488	0.0266	0.0151
	(b)	0.0376	0.0197	0.0119
(v)	(a)	0.6758	0.4434	0.3193
	(b)	0.3513	0.2252	0.1593
(vi)	(a)	0.1737	0.1086	0.0747
	(b)	0.1033	0.0618	0.0413
(vii)	(a)	0.0772	0.0444	0.0306
	(b)	0.0487	0.0281	0.0186
(viii)	(a)	0.0539	0.0333	0.0208
	(b)	0.0377	0.0177	0.0108
(ix)	(a)	0.0213	0.0156	0.0129
	(b)	0.0274	0.0153	0.0093
(x)	(a)	0.0289	0.0962	0.0526
	(b)	0.0192	0.1183	0.0792

242

TABLE 2

Range of values of n for which the ESF asymptotic variance (14) exceeds the corresponding multinomial asymptotic variance, when multinomial probabilities are of the form $p_i = c(m)/i^m$, for selected values of m.

m	lower limit	upper limit
3	4	7
4	7	17
5	17	36
6	34	78
7	68	168
8	136	357
9	275	752
10	558	1567

TABLE 3

Estimation of the probability that the $(n+1)^{th}$ animal will
be of a previously unobservered species, for various n

and k. Values of (i) the estimator $h_1(k) - h_2(k)$, (ii) the
maximum likelihood estimator, and the upper (iii) and lower

(iv) bounds (from (15)).

n		$k = 5$	$k = 10$
100	(i)	0.00907806	0.02456437
	(ii)	0.00938698	0.02500769
	(iii)	0.00916916	0.02479618
	(iv)	0.00907747	0.02454389
200	(i)	0.00386603	0.01005127
	(ii)	0.00397079	0.01022055
	(iii)	0.00388540	0.01010134
	(iv)	0.00386597	0.01005083
300	(i)	0.00237392	0.00606618
	(ii)	0.00243087	0.00615628
	(iii)	0.00238184	0.00608636
	(iv)	0.00237390	0.00606607
400	(i)	0.00168662	0.00426545
	(ii)	0.00172392	0.00432372
	(iii)	0.00169084	0.00427610
	(iv)	0.00168661	0.00426541
500	(i)	0.00129655	0.00325561
	(ii)	0.00132353	0.00329740
	(iii)	0.00129946	0.00326212
	(iv)	0.00129655	0.00325559

Large deviations of the Wright-Fisher Process

F. Papangelou
University of Manchester

1 Introduction

In the present paper we will be concerned with the Wright-Fisher process of mathematical genetics in one of its simplest forms: the univariate case with no selection and with either no mutation or only one-way mutation. Consider a large biological population consisting of a single species in which a particular gene appears in two possible forms (alleles) A and a, say. Assume that we are dealing with a one-sex population (as with some plants) and that we are in the haploid case (in which chromosomes occur singly). The Wright-Fisher process models the way in which the proportion of individuals carrying the A form of the gene changes from generation to generation, as the population reproduces itself ([1], [4]). To be specific, let us first consider the case where the proportion of the A allele changes only through the effect of random sampling, with no mutation or selection. If the population consists of N individuals, i of whom are of type A and $N - i$ of type a, then the state of the process is the proportion $y = \dfrac{i}{N}$. To produce an offspring generation from this population of "genes" we sample N times with replacement from it, thus keeping the size of the population constant. The probability that the proportion of A-alleles will make a transition from state $y = \dfrac{i}{N}$ in the "current" generation to state $\tilde{y} = \dfrac{j}{N}$, say, in the following generation is then

$$P(y, \tilde{y}) = \binom{N}{j} y^j (1 - y)^{N-j}. \tag{1.1}$$

The process is repeated to produce the granddaughter generation and so on. This is trivially a Markov process on the state space $\{0, \dfrac{1}{N}, \dfrac{2}{N}, \ldots, \dfrac{N-1}{N}, 1\}$.

We will show in section 3 that if the state of this process is known to have changed over a large number of generations from y_0 to y_1 say, then we can trace with near certainty the history of the change provided the size of the population is much larger than the number of generations involved. This is obtained as a consequence of Wentzell's results on large deviations of discrete time Markov chains ([9], [10]), and the main purpose of the present paper is to bring out the scaling properties of the sample paths of the Wright-Fisher process and demonstrate the significance of these properties for mathematical genetics. The underlying analytic reason for the phenomenon observed here is similar to that which lies at the root of [8], although in that case the authors were concerned with the exit problem rather than the scaling properties of their processes. (See also [5] and [7]).

The simple case of no mutation amply illustrates the potential of large deviations theory to yield explicit results in more complex cases. In the brief final section of the paper we give a summary presentation, without discussion, of the fascinating picture that emerges if we allow one-way mutation.

2   The problem

To describe the nature of the problem, we suppose that we begin with an "initial" generation, in which the proportion of the A-allele is y_0, and follow the process through successive generations. It is a classical fact (cf. [1], [4]) that the discrete stochastic process so obtained approaches, when suitably scaled, a diffusion as $N \to \infty$. Specifically, if $y(0) = y_0$, $y(1), y(2), \ldots$ are the states of consecutive generations and we define $Y_t^{(N)} = y([Nt])$, $t \geq 0$ where $[Nt]$ is the integer part of Nt, then for fixed $T > 0$ the process $Y_t^{(N)}, 0 \leq t \leq T$ converges weakly as $N \to \infty$ to a diffusion on $[0, T]$, with diffusion parameter $\sigma^2(y) = y(1-y)$, drift parameter 0 and absorption at 0 and 1. The large deviations of such a diffusion, as the diffusion parameter is gradually suppressed ($\epsilon\sigma^2(y)$ with $\epsilon \to 0$), have been studied by Freidlin and Wentzell ([2]) but are not directly relevant for the Wright-Fisher process here. There is however another scaling of the Wright-Fisher process which leads to large deviations results indicative of the manner in which this process evolves in time.

Suppose that in the above Wright-Fisher process with population size N we adopt a time scaling under which the time coordinates of the generations are $0, \frac{1}{n}, \frac{2}{n}, \ldots$, and define $Y_t^{(n)} = y([nt])$, $t \geq 0$. If $\frac{N}{n} \to \infty$ as $n \to \infty$ then, on any fixed time interval $[0, T]$, this scaled process converges in distribution to the constant "process" $Z_t = y_0$, $0 \leq t \leq T$ and Wentzell's large deviations results imply that if $\phi(t), 0 \leq t \leq T$ is a "path" in $(0, 1)$ ($0 < \phi(t) < 1$ for $t \in [0, T]$), then the logarithm of the probability that $Y_t^{(n)}, 0 \leq t \leq T$ "follows closely the path $\phi(t)$, $0 \leq t \leq T$" is of order

$$-\frac{N}{n} \cdot \frac{1}{2} \int_0^T \frac{\phi'(t)^2}{\phi(t)(1 - \phi(t))} dt$$

where ϕ' is the derivative of ϕ. To make this more precise we state the following.

Proposition *Assume that $N = N(n)$ depends on n in such a way that $\frac{N}{n} \to \infty$ as $n \to \infty$. If $\phi(t), 0 \leq t \leq T$ is an absolutely continuous function such that $0 < \phi(t) < 1$ for all $t \in [0, T]$, then*

$$\frac{1}{2} \int_0^T \frac{\phi'(t)^2}{\phi(t)(1 - \phi(t))} dt = \qquad (2.1)$$

$$= -\lim_{\delta \to 0} \lim_{n \to \infty} nN^{-1} \log P(|Y_t^{(n)} - \phi(t)| < \delta \text{ for all } t \in [0, T]).$$

Since we have the explicit integral (2.1) for the "action functional" of the process, we can answer an obvious variational question which we state here somewhat imprecisely.

Question *If the scaled process was at y_0 at time 0 and is at y_1 ($y_1 \neq y_0, 0, 1$) at time T, what approximate path has it followed on $0 \leq t \leq T$?*

Asymptotically the process follows the path $\phi(t)$ which minimizes the integral (2.1) subject to $\phi(0) = y_0$, $\phi(T) = y_1$. This is a standard variational problem and the answer can be obtained by the equally standard methods of the Euler-Lagrange calculus of variations ([3]) or, in this simple case, directly. It will be seen in the next section that the path followed is an arc of

$$\phi(t) = \frac{1}{2} - \frac{1}{2}\cos(ct + k), \quad 0 \le t \le T$$

where the constants c and k are such that $\phi(0) = y_0$, $\phi(T) = y_1$ and ϕ is monotone on $0 \le t \le T$. This should be no surprise in view of Fisher's "angular transformation" ([6]).

Let us next introduce *one-way mutation*: allele A mutates to allele a at rate $\frac{\gamma}{n}$ per generation (γ per "unit time"), where $\gamma > 0$. The probability of transition from state $y = \frac{i}{N}$ to state $\tilde{y} = \frac{j}{N}$ is now $P(y, \tilde{y}) = \binom{N}{j}p_i^j(1 - p_i)^{N-j}$, where $p_i = y(1 - \frac{\gamma}{n})$. This introduces a "drift" equal to $-\gamma y$. The scaled process $Y_t^{(n)}$, $t \ge 0$, introduced as above, converges weakly as $n \to \infty$ to the fixed curve $y_0 e^{-\gamma t}$, $t \ge 0$, over any finite time interval $[0, T]$, where again y_0 is the starting state. This means that if n and $\frac{N}{n}$ are large then, with high probability, the scaled process follows closely the path $y_0 e^{-\gamma t}$, $0 \le t \le T$. Moreover, if $\phi(t), 0 \le t \le T$ is an absolutely continuous path in $(0, 1)$ then the logarithm of the probability that $Y_t^{(n)}, 0 \le t \le T$ follows closely the path $\phi(t), 0 \le t \le T$ is of order

$$-\frac{N}{n} \cdot \frac{1}{2} \int_0^T \frac{(\phi'(t) + \gamma\phi(t))^2}{\phi(t)(1 - \phi(t))} dt. \tag{2.2}$$

One may then pose the same variational problem as in the case of no mutation, and we will say more on this in section 4.

3 The case of no mutation.

As already mentioned, the key to the variational problem posed above is provided by the results of [9] and [10]. There is a slight difference in terminology and notation between [9] and [10] and to avoid ambiguity we adopt here the terminology of [10].

The scaled Wright-Fisher process $Y_t^{(n)}, t \ge 0$, with no mutation or selection, changes its state at times t which are multiples of $\tau = \frac{1}{n}$ and remains constant between any two consecutive multiples of τ. It was defined so as to be right-continuous and in the terminology of [10] (p.20) is a τ-process. We will assume for the sake of convenience that the starting state of the process can be any number in $(0, 1)$, although of course the state after one transition can only be a multiple of $\frac{1}{N}$. The transition probabilities at moments of transition are given by (1.1). The *cumulant* of $Y_t^{(n)}, t \ge 0$ is (cf. [10], p.27)

$$G^n(y, z) = n \log E_y \exp\{z(Y_\tau^{(n)} - y)\}, \quad -\infty < z < \infty$$

where E_y denotes conditional expectation given $Y_0^{(n)} = y$. An easy calculation shows that

$$G^n(y, z) = -nyz + nN \log[1 + y(e^{zN^{-1}} - 1)].$$

247

Our parameter n plays the role of θ in [10] and in order to determine the large deviation asymptotics of $Y_t^{(n)}, t \geq 0$ we set $k(n) = \dfrac{N(n)}{n} = \dfrac{N}{n}$, where $\dfrac{N}{n} \to \infty$ as $n \to \infty$, and proceed to calculate ([10], p.61)

$$k(n)^{-1}G^n(y, k(n)z) = nN^{-1}G^n(y, Nn^{-1}z)$$

$$= -nyz + n^2 \log[1 + y(e^{zn^{-1}} - 1)].$$

Notice that this is independent of N. Using first the expansion $\log(1 + s) = s - \dfrac{s^2}{2} + O(s^3)$ for $s \to 0$ and then the expansion $e^s - 1 = s + \dfrac{s^2}{2} + O(s^3)$, $s \to 0$, one can easily prove that

$$\lim_{n \to \infty} nN^{-1}G^n(y, Nn^{-1}z) = G(y, z)$$

where $G(y, z) = \dfrac{1}{2}y(1 - y)z^2$. As a function of z, $G(y, z)$ is the cumulant generating function of the Gaussian distribution with mean 0 and variance $y(1 - y)$ and its Legendre transform is

$$H(y, u) = \dfrac{1}{2}\dfrac{u^2}{y(1 - y)}, \quad -\infty < u < \infty.$$

Since $y(1-y)$ becomes arbitrarily small for values of y close to 0 or 1 and, moreover, 0 and 1 are absorbing barriers for our process, some of the hypotheses under which Wentzell's results were obtained are not satisfied. For instance the "uniformity" required in some hypotheses breaks down. One way to remedy this is to modify the process $Y_t^{(n)}$ when it gets close to 0 and 1. We take $(-\infty, \infty)$ as the state space of the modified process and retain the transition probabilities of the original process as long as the latter does not exit from $[\epsilon, 1 - \epsilon]$ for an appropriately small $\epsilon > 0$. If the process visits a state y with either $y > 1 - \epsilon$ or $y < \epsilon$, then the next jump is given a Gaussian conditional distribution with mean 0 and variance $\dfrac{v(y)}{N}$, where $v(y)$ is a smooth extension of the function $y(1 - y)$, $\epsilon \leq y \leq 1 - \epsilon$ such that $\dfrac{1}{2}\epsilon(1 - \epsilon) \leq v(y) \leq \epsilon(1 - \epsilon)$ for $y \notin [\epsilon, 1 - \epsilon]$. The cumulant $\tilde{G}^n(y, z)$ of the modified process satisfies $nN^{-1}\tilde{G}^n(y, Nn^{-1}z) = \tilde{G}(y, z) = \dfrac{1}{2}v(y)z$ if either $y < \epsilon$ or $y > 1 - \epsilon$. The original and the modified process have the same distribution up to the time of first exit from $[\epsilon, 1 - \epsilon]$ and, as will be seen from the following discussion, the probability of an exit from $[\epsilon, 1 - \epsilon]$ by time T is, for large n, of smaller order than the probability that the process follows closely a minimizing curve joining $(0, y_0)$ with (T, y_1) where $\epsilon < y_1 < 1 - \epsilon$. This enables us to handle things as if Wentzell's hypotheses were satisfied.

With the above modification it can be shown that conditions A, B, C, D and E on pp.56 and 61 of [10] are satisfied and (setting $\tilde{G}(y, z) = G(y, z)$ if $\epsilon \leq y \leq 1 - \epsilon$) that

$$\lim_{n \to \infty} nN^{-1}\tilde{G}^n(y, Nn^{-1}z) = \tilde{G}(y, z)$$

and

$$\lim_{n \to \infty} \dfrac{\partial}{\partial z}(nN^{-1}\tilde{G}^n(y, Nn^{-1}z)) = \dfrac{\partial}{\partial z}\tilde{G}(y, z)$$

uniformly with respect to y and z, provided the latter is restricted to a bounded set. Also

$$\left| \frac{\partial^2}{\partial z^2} (nN^{-1} \tilde{G}^n(y, Nn^{-1}z)) \right| \leq \text{const} < \infty$$

for all y, all z in a bounded set and all sufficiently large n. All this means that Theorem 3.2.3' on p.68 of [10] applies and yields the action functional of the sequence of processes $Y_t^{(n)}, 0 \leq t \leq T$ for curves $\phi(t), 0 \leq t \leq T$ with $\epsilon < \phi(t) < 1 - \epsilon$ for all $t \in [0,T]$: this is $Nn^{-1}S_{0,T}(\phi)$, where $S_{0,T}(\phi)$ is equal to $\int_0^T H(\phi(t), \phi'(t))dt = \frac{1}{2} \int_0^T \frac{\phi'(t)^2}{\phi(t)(1 - \phi(t))} dt$ if ϕ is absolutely continuous, and to ∞ otherwise.

Turning now to the question of minimizing $S_{0,T}(\phi)$ subject to $\phi(0) = y_0$, $\phi(T) = y_1$, where $0 < y_0 < y_1 < 1$ say, we may without loss restrict attention to non-decreasing ϕ since if $\phi(t_1) = \phi(t_2)$ $(t_1 < t_2)$ $S_{0,T}(\phi)$ does not increase if we change ϕ to the constant $\phi(t_1)$ on $[t_1, t_2]$. If ϕ is in addition absolutely continuous, then

$$S_{0,T}(\phi) \geq \frac{1}{2T} \left(\int_0^T \frac{\phi'(t)}{[\phi(t)(1 - \phi(t))]^{1/2}} dt \right)^2 = \frac{1}{2} c^2 T$$

where $c = T^{-1} \left(\int_{y_0}^{y_1} [y(1-y)]^{-1/2} dy \right)$, and this inequality becomes an equality if and only if $\phi'(t)[\phi(t)(1 - \phi(t))]^{-1/2} = c$ almost surely, i.e. if ϕ is the curve

$$\phi_0(t) = \frac{1}{2} - \frac{1}{2} \cos(ct + k) \tag{3.1}$$

where $k \in (0, \pi)$ satisfies $y_0 = \frac{1}{2} - \frac{1}{2} \cos k$. Note that $\frac{1}{2} c^2 T$ is, for given y_0, y_1, a decreasing function of T and this is sufficient to justify our modification of the process.

The probabilistic theorem that follows from the above minimization can be formulated in a number of ways and we choose the following. Let y_0, y_1 and T be given $(0 < y_0 < y_1 < 1, T > 0)$. Then the conditional distribution of $Y_t^{(n)}, 0 \leq t \leq T$, given $Y_T^{(n)} \geq y_1$, converges weakly to the distribution assigning mass 1 to $\{\phi_0\}$. More precisely

Theorem *If $0 < y_0 < y_1 < 1$ and $T > 0$, then for every $\delta > 0$*

$$\lim_{n \to \infty} P(\phi_0(t) - \delta < Y_t^{(n)} < \phi_0(t) + \delta \text{ for all } t \in [0,T] \mid Y_T^{(n)} \geq y_1) = 1$$

where $\phi_0(t)$ is given by (3.1).

Proof Denote by V the set of functions ϕ on $[0,T]$ that are right-continuous with left-hand limits ("cadlag") and satisfy $\phi_0(t) - \delta < \phi(t) < \phi_0(t) + \delta$ for all $t \in [0,T]$. Likewise let U be the set of cadlag functions ϕ with $\phi(T) \geq y_1$. The conditional probability in the statement of the theorem is no less than

$$P(Y^{(n)} \in V^0 \cap U^0)[P(Y^{(n)} \in V^0 \cap U^0) + P(Y^{(n)} \in \overline{V^c} \cap U)]^{-1}$$

where V^c denotes the complement of V and $^0, ^-$ denote interior and closure with respect to the uniform topology. Now $\liminf_{n \to \infty} nN^{-1} \log P(Y^{(n)} \in V^0 \cap U^0)$ is $\geq -\inf\{S_{0,T}(\phi) :$

$\phi \in V^0 \cap U^0\}$ and hence also $\geq -S_{0,T}(\phi_0)$ since one can easily construct a sequence $\phi_n \in V^0 \cap U^0$ such that $\phi_n \to \phi_0$ uniformly and $S_{0,T}(\phi_n) \to S_{0,T}(\phi_0)$. On the other hand

$$\limsup_{n\to\infty} nN^{-1} \log P(\overline{V^c} \cap U) \leq -\inf\{S_{0,T}(\phi) : \phi \in \overline{V^c} \cap U\} < -S_{0,T}(\phi_0)$$

since if we had $\inf\{S_{0,T}(\phi) : \phi \in \overline{V^c} \cap U\} \leq S_{0,T}(\phi_0)$, there would exist $\phi \in \overline{V^c} \cap U$ such that $S_{0,T}(\phi) \leq S_{0,T}(\phi_0)$, contradicting the extremality and uniqueness of ϕ_0. The modification of our process does not undermine this argument, in view of the comment following (3.1). This completes the proof.

4 The case of one-way mutation

We mentioned at the end of section 2 that, in the case of one-way mutation, the action functional for paths $\phi(t), 0 \leq t \leq T$ staying clear of 0 and 1 is given by (2.2). Since the behaviour of our process is more delicate in this case, a full discussion cannot be given here. We confine ourselves to merely describing the extremals of the functional (2.2), i.e. the solutions of the Euler-Lagrange equation, without making any probability statements.

Suppose we seek a path $\phi(t), 0 \leq t \leq T$ that minimizes (2.2) subject to $\phi(0) = y_0$, $\phi(T) = y_1$. Such a path satisfies the Euler-Lagrange equation and hence is an extremal. It turns out that such an extremal can be an exponential, a hyperbolic cosine, a trigonometric cosine or a parabola, depending on the values of y_0, y_1 and T. As is well known from variational calculus, along each extremal of a time homogeneous functional the so-called "energy" (a term borrowed from Hamiltonian mechanics) is constant. In our particular case this translates into the fact that for each extremal ϕ there is a constant c such that

$$\frac{\phi'(t)^2 - \gamma^2 \phi(t)^2}{\phi(t)(1 - \phi(t))} = c \tag{4.1}$$

for all t in the appropriate domain. This constant c is obviously determined by $y_0 = \phi(0)$ and $\phi'(0)$ and to each admissible value of c there correspond two values of $\phi'(0)$, unless $\phi'(0) = 0$. The extremals can be obtained as non-singular solutions of (4.1). Let us fix $\phi(0) = y_0$. In what follows, the different types of extremals arising are described in terms of the possible values of $\phi'(0)$; naturally, each extremal is assumed terminated at the point where it exits from the strip $\{(t, y) : t \geq 0, 0 < y < 1\}$. The reader should note that each point (T, y_1) $(0 < y_1 < 1)$ is joined to $(0, y_0)$ by one of the following extremals.

(i) If $\phi'(0) = 0$ (in which case c has its smallest possible value $-\dfrac{\gamma^2 y_0}{1 - y_0}$), the extremal is the hyperbolic cosine

$$\phi(t) = \frac{1}{2} y_0 \left[\cosh \frac{\gamma t}{\sqrt{1 - y_0}} + 1\right].$$

(ii) If $0 < \phi'(0) < \gamma y_0$ or $-\gamma y_0 < \phi'(0) < 0$ (i.e. if $-\dfrac{\gamma^2 y_0}{1 - y_0} < c < 0$), then

$$\phi(t) = \frac{-c}{2(\gamma^2 - c)} \left[\cosh((\sqrt{\gamma^2 - c})(t - k)) + 1\right]$$

where the constant k is negative in the first case and positive in the second.

250

(iii) If $\phi'(0) = \gamma y_0$ or $-\gamma y_0$ ($c = 0$), then $\phi(t) = y_0 e^{\gamma t}$ or $y_0 e^{-\gamma t}$ respectively, the latter representing "mean behaviour" for our process.

(iv) If $\gamma y_0 < \phi'(0) < \gamma \sqrt{y_0}$ or $-\gamma \sqrt{y_0} < \phi'(0) < -\gamma y_0$ ($0 < c < \gamma^2$), then

$$\phi(t) = \frac{c}{2(\gamma^2 - c)} \left[\cosh((\sqrt{\gamma^2 - c})(t - k)) - 1 \right]$$

where k is negative in the first case and positive in the second.

(v) If $\phi'(0) = \pm \gamma \sqrt{y_0}$ ($c = \gamma^2$), then

$$\phi(t) = \frac{1}{4}(\gamma t \mp 2\sqrt{y_0})^2.$$

(vi) If $\phi'(0) > \gamma \sqrt{y_0}$ or $\phi'(0) < -\gamma \sqrt{y_0}$ ($c > \gamma^2$), then the extremal is a monotone arc of the trigonometric cosine

$$\phi(t) = \frac{c}{2(c - \gamma^2)} \left[1 - \cos((\sqrt{c - \gamma^2})(t - k)) \right]$$

with $k < 0$ in the first case and $k > 0$ in the second.

References

1. Ewens, W.J.: Mathematical Population Genetics. Biomathematics Vol. 9. Springer-Verlag, New York, 1979.

2. Freidlin, M.I. and Wentzell, A.D.: Random Perturbations of Dynamical Systems. Springer-Verlag, New York, 1984.

3. Gelfand, I.M. and Fomin, S.V.: Calculus of Variations. Prentice-Hall, Englewood Cliffs, 1963.

4. Karlin, S. and Taylor, H.M.: A Second Course in Stochastic Processes. Academic Press, New York, 1981.

5. Kifer, Yu.: A discrete-time version of the Wentzell-Freidlin theory, Ann. Probab. **18** (1990), 1676-1692.

6. Kimura, M.: Diffusion Models in Population Genetics. Methuen's Review Series in Applied Probability, Methuen, London, 1964.

7. Klebaner, F.C. and Zeitouni, O.: The exit problem for a class of density-dependent branching systems, Ann. Appl. Probab. **4** (1994), 1188-1205.

8. Morrow, G.J. and Sawyer, S.: Large deviation results for a class of Markov chains arising from population genetics, Ann. Probab. **17** (1989), 1124-1146.

9. Wentzell, A.D.: Rough limit theorems on large deviations for Markov processes, Th. Probab. Appl. **21** (1976), 227-242 and 499-512; **24** (1979), 675-692; **27** (1982), 215-234.

10. Wentzell, A.D.: Limit Theorems on Large Deviations for Markov Stochastic Processes. Kluwer Academic Publishers, Dordrecht, 1990.

Department of Mathematics
University of Manchester
Manchester M13 9PL
U.K.

Threshold behaviour in stochastic epidemics among households

Frank Ball

ABSTRACT A very general model for the spread of an epidemic among a population consisting of m households, each of size n, is presented. The asymptotic situation in which the number of households m tends to infinity, whilst the household size n remains fixed, is analysed. For large m the process of infected households can be approximated by a branching process. A coupling argument is used to make this approximation precise as m tends to ∞, thus enabling a threshold theorem to be developed. Specialisation to the case where infective individuals during their infectious period make infectious contacts within their own group with rate λ_W and outside their own group with rate λ_B is briefly considered. Generalisations to populations with unequal household sizes and different types of individuals are outlined.

1 Introduction

In a recent paper, Daley and Gani [DG94] considered the threshold behaviour of a deterministic model for the spread of an epidemic among a population consisting of m interacting communities. This paper is concerned with the threshold analysis of a stochastic analogue of a special case of that model. Specifically, the spread of an epidemic among a population divided into a large number of relatively small groups or households, with different infection rates for within-household and between-household infections, is considered. To put our analysis into perspective, it is instructive to give first a brief overview of epidemic threshold theorems.

To focus attention consider the general deterministic epidemic, which is defined by the following system of differential equations:

$$\frac{dx}{dt} = -\beta xy, \qquad \frac{dy}{dt} = \beta xy - \gamma y, \qquad \frac{dz}{dt} = \gamma y \qquad (\beta, \gamma > 0). \qquad (1.1)$$

Here, $x(t)$, $y(t)$ and $z(t)$ denote, respectively, the numbers of susceptible, infectious and removed individuals at time t, and β and γ are known as the infection and removal rates. Suppose that initially $(x(0), y(0), z(0)) = (n, a, 0)$. Then it is immediate from the second equation in (1.1) that a build up of infection will occur if and only if $n > \gamma/\beta$. This is part of the celebrated Kermack and McKendrick threshold theorem [KM27]; they also derived, amongst other things, a non-linear equation for the final size of the epidemic and an approximation to that final size. [DG94] extended all three results to the situation where there are m distinct subpopulations, labelled $1, 2, \ldots, m$, with associated infection rates β_{ij} $(i, j = 1, 2, \ldots, m)$ and removal rates γ_i $(i = 1, 2, \ldots, m)$; see also Watson [Wat72].

The general stochastic epidemic (see eg Bailey [Bai75], Chapter 6) is obtained by replacing the infinitesimal transition rates governing (1.1) by infinitesimal transition probabilities. Thus, if $X(t)$, $Y(t)$ and $Z(t)$ denote, respectively, the numbers of susceptibles, infectives and removed cases at time t, then $\{(X(t), Y(t), Z(t)); t \geq 0\}$ follows a continuous time Markov chain with transition probabilities

$$P\big(X(t + \Delta t), Y(t + \Delta t), Z(t + \Delta t)\big) = (i - 1, j + 1, k)|$$
$$(X(t), Y(t), Z(t)) = (i, j, k)\big) = \beta ij \Delta t + o(\Delta t)$$

for an infection, and

$$P\big(X(t + \Delta t), Y(t + \Delta t), Z(t + \Delta t)\big) = (i, j - 1, k + 1)|$$
$$(X(t), Y(t), Z(t)) = (i, j, k)\big) = \gamma j \Delta t + o(\Delta t),$$

for a removal. Note that the process of infectives $\{Y(t); t \geq 0\}$ follows a birth-and-death process, with random birth-rate $\beta X(t)$ and death-rate γ. Whittle [Whi55] obtained a threshold theorem for the general stochastic epidemic by noting that for epidemics of intensity $\leq i$, ie that infect less than a fixed proportion i of susceptibles, the process of infectives can be sandwiched between two birth-and-death processes, each having the same death-rate γ, but with birth-rates βn and $\beta n(1 - i)$. By analysing the bounding birth-and-death processes, [Whi55] showed that for large n, if $\beta n \leq \gamma$ there is zero probability of an epidemic exceeding any fixed intensity $i > 0$, while if $\beta n > \gamma$ the probability of an epidemic is approximately $1 - (\gamma/\beta n)^a$, for small i. The general stochastic epidemic implicitly assumes that the infectious period of infectives follows a negative exponential distribution with mean γ^{-1}. Becker [Bec77] and Kendall [Ken94] extended Whittle's threshold theorem to models with non-exponential infectious periods.

A second threshold theorem for the general stochastic epidemic was given by Williams [Wil71]. He effectively considered a sequence of epidemics, indexed by the initial number of susceptibles n, in which $\beta = \alpha/n$, and showed that as $n \to \infty$ the total size of the epidemic converges in distribution to that of a birth-and-death process with birth-rate α and death-rate γ. A *major* epidemic is said to occur if this limiting birth-and-death process does not go extinct. Thus, from standard results for birth-and-death processes, a major epidemic cannot occur if $\alpha \leq \gamma$ (ie if $\beta n \leq \gamma$), whilst if $\alpha > \gamma$ the probability of a major epidemic is $1 - (\gamma/\alpha)^a$. Ball [Bal83] and

Ball and Donnelly [BD95a] used a coupling argument to prove an almost sure version of Williams' threshold theorem, and also considered convergence of sample paths of the epidemic process, see also Metz [Met78]. This coupling argument can be applied to a very wide class of single population epidemics, and it also extends to the stochastic version of the model studied by [DG94], see [Bal83].

The threshold theorem is the practically most useful result of mathematical epidemic theory (see eg Becker [Bec79]). A fundamental aim of any vaccination programme is to bring the initial number of susceptibles down to below its threshold value. Note that the threshold condition for epidemics to occur, $n > \gamma/\beta$, is the same for both the deterministic and stochastic general epidemics, though, of course, the interpretation of the threshold behaviour is quite different. It is useful to express the threshold condition as $R_0 > 1$, where $R_0 = n\beta/\gamma$ is the *reproductive ratio* for the epidemic, ie the expected number of infectious contacts made by a typical infective in an otherwise susceptible population (see eg Heesterbeek and Dietz [HD96]). This concept of R_0 is very general and it extends to a broad class of deterministic epidemic models, which includes the model of [DG94]. It also extends to multi-population stochastic epidemic models, where R_0 is the maximal eigenvalue of the multi-type branching process corresponding to the limiting (upper bounding) birth-and-death process of Williams' (Whittle's) threshold theorem, see eg Ball [Bal94]. Again, the deterministic and stochastic threshold conditions coincide. However, the multipopulation version of Williams' threshold theorem requires that the initial numbers of susceptibles in each of the subpopulations is large, which is clearly inappropriate for a population of households. The aim of this paper is to prove a Williams' threshold theorem for a very general household epidemic model.

Stochastic models for the spread of an epidemic among a community of households have been considered by Bartoszyński [Bar72], Ball, Mollison and Scalia-Tomba [BMST95] , Becker and Dietz [BD95b] and Becker and Hall [BH95] . [BMST95] consider a model in which a typical infective has two infection rates, one for susceptibles in its own household and another, much smaller, for susceptibles in other households. [BD95b] consider a model for highly infectious diseases in which it is assumed that once one member of a household becomes infected the whole household becomes infected. [BH95] consider the case when there are p types of individuals. [BD95b] and [BH95] derive threshold parameters for their models, which they use to evaluate various vaccination strategies. None of the above papers provides a formal proof of the branching process approximation, on which the threshold parameters are based. [Bar72] essentially models the household epidemic as a branching process.

2 Threshold theorem for stochastic household epidemics

2.1 Household epidemic model

Consider the following model for the spread of an epidemic among a population of m groups (households), each containing n individuals. Infectious individuals have independent and identically distributed *life histories*, $\mathcal{H} = (T^I, \eta^W, \eta^B)$, where T^I is the time elapsing between an individual's infection and eventual removal or death, and η^W and η^B are point processes of times, relative to an individual's infection, at which *within-group* and *between-group* contacts are made, respectively. Each within-group contact of a specific infective is with an individual chosen independently and uniformly from the n initial individuals in the same group. Each between-group contact of a specific infective is with an individual chosen independently and uniformly from the mn initial individuals in the population. After death or removal, an individual is immune to further infection. If the individual contacted by the infective is susceptible, then it becomes infected, otherwise nothing happens. Initially there are $mn - 1$ susceptibles and one infective, who has just been infected. The epidemic ceases as soon as there are no infectives present in the population.

The above model, which we denote by $E_{m,n}$, is a generalisation of the single population epidemic model of [BD95a]. Note that the components T^I, η^W and η^B of $\mathcal{H}$ need not be independent and further the intensities of the point processes governing an individual's within-group and between-group contacts need not be constant throughout the infectious period. Thus a latent period, during which a recently infected individual is unable to infect others, is incorporated in our model. Note also that the possibilities of within-group contacts with oneself and between-group contacts with one's own group involve no real loss of generality since, for any realistic model, the point process η^W and η^B can be modified accordingly. (The basic model could be defined so that the within-group contacts of a specific infective are chosen independently and uniformly from the remaining $n - 1$ individuals in the same group, whilst the between-group contacts of a specific infective in a particular group are chosen independently and uniformly from the remaining $(m - 1)n$ individuals in the other $m - 1$ groups. However, this turns out to be inconvenient for the coupling argument used to prove our threshold theorem.) The idea of a life history corresponds to the reproduction and dispersal kernel of deterministic spatial epidemic models, see eg van den Bosch, Metz and Diekmann [vdBMD90] or Mollison [Mol91].

2.2   Construction of epidemic model and approximating branching process

In order to study the asymptotic behaviour of $E_{m,n}$, as the numbers of groups $m \to \infty$, it is fruitful to give a more detailed construction of our epidemic.

Let $(\Omega, \mathcal{F}, P)$ be a probability space on which are defined the following independent sets of random quantities:

(i) life histories $\mathcal{H}_{ij} = (T_{ij}^I, \eta_{ij}^W, \eta_{ij}^B)$ $(i = 1, 2, \dots, j = 1, 2, \dots, n)$, independent and identically distributed according to $\mathcal{H}$,

(ii) C_{ij} $(i = 1, 2, \dots, j = 1, 2, \dots)$, independent and uniformly distributed on $\{1, 2, \dots, n\}$,

(iii) $\chi_k^{(m)}$ $(m = 1, 2, \dots, k = 1, 2, \dots)$, where for each $m = 1, 2, \dots, \chi_1^{(m)}$, $\chi_2^{(m)}, \dots$ are independent and uniformly distributed on $\{1, 2, \dots, m\}$.

The epidemic $E_{m,n}$ is constructed as follows. Label the groups $1, 2, \dots, m$ and the individuals in each group $1, 2, \dots, n$. For $i = 1, 2, \dots, m$, let $r_i^{(m)}$ denote the ith group to be infected in $E_{m,n}$, and assume without loss of generality that $r_1^{(m)} = 1$ $(m = 1, 2, \dots)$. The initial infective in group 1 adopts the life history $\mathcal{H}_{11}$, making contacts governed by the point processes η_{11}^W and η_{11}^B. For $i = 1, 2, \dots, m$ and $j = 1, 2, \dots, C_{ij}$ is the individual contacted at the jth contact occurring in group $r_i^{(m)}$. If that individual is still susceptible then it becomes infected and adopts the life history $\mathcal{H}_{i,j'}$, where $j' - 1$ is the number of individuals that have previously been infected in group $r_i^{(m)}$. Thus the life histories $\mathcal{H}_{i1}, \mathcal{H}_{i2}, \dots, \mathcal{H}_{in}$ are assigned sequentially to the infectives in group $r_i^{(m)}$, in the order that they become infected. Finally, for $k = 1, 2, \dots, \chi_k^{(m)}$ is the group contacted at the kth between-group contact made in $E_{m,n}$, the individual contacted being chosen using the appropriate C_{ij}.

For $m = 1, 2, \dots$, let $\chi_0^{(m)} = 1$ and $M_m = \min\{k \in \mathbb{N} : \chi_k^{(m)} \in \{\chi_0^{(m)}, \chi_1^{(m)}, \dots, \chi_{k-1}^{(m)}\}\}$. This is the birthday problem and $m^{-\frac{1}{2}} M_m \overset{D}{\to} M$ as $m \to \infty$, where M has density $f(x) = x \exp(-x^2/2)$ $(x > 0)$, see eg Aldous [Ald85], p96. Thus, by the Skorokhod representation theorem, we can assume that $(\Omega, \mathcal{F}, P)$ is also endowed with a random variable M having the above density and $m^{-\frac{1}{2}} M_m \overset{a.s.}{\to} M$ as $m \to \infty$. This can be achieved by first constructing $(\Omega, \mathcal{F}, P)$ on which are defined appropriately distributed random variables $M_1, M_2, \dots$ and M such that $m^{-\frac{1}{2}} M_m \overset{a.s.}{\to} M$ as $m \to \infty$, and then augmenting $(\Omega, \mathcal{F}, P)$ to carry for each m, independent and identically distributed random variables $\chi_1^{(m)}, \chi_2^{(m)}, \dots$ that are consistent with M_m (see [BD95a]).

We can also define the following general (Crump–Mode–Jagers) branching process on $(\Omega, \mathcal{F}, P)$, which will approximate $E_{m,n}$ for large m. For fixed $i = 1, 2, \ldots$ the life histories $\mathcal{H}_{i1}, \mathcal{H}_{i2}, \ldots, \mathcal{H}_{in}$ and the contact numbers $C_{i1}, C_{i2}, \ldots$ can be used to define a single group epidemic initiated by one infective among $n-1$ susceptibles, by simply ignoring the point processes $\eta_{i1}^B, \eta_{i2}^B, \ldots, \eta_{in}^B$. Let D_i be the duration of this single population epidemic, ie the time from infection of the initial infective until removal of the last infective, and η_i^G be the point process, obtained using $\eta_{i1}^B, \eta_{i2}^B, \ldots, \eta_{in}^B$, of times (relative to infection of the initial infective) at which infectives in this single group epidemic would make between-group contacts if they were allowed to. Then (D_i, η_i^G) $(i = 1, 2, \ldots)$ are independent and identically distributed, so they can be used to define a general branching process in which a typical individual lives until age D and reproduces at ages according to η^G (see eg Jagers [Jag75]).

2.3   Main convergence theorem

We need to impose some mild technical conditions on the above branching process to guarantee convergence of certain of its functionals. Specifically we assume that

(i) there exists a Malthusian parameter $\alpha \in (0, \infty)$ such that

$$\int_0^\infty e^{-\alpha t} \mu(dt) = 1,$$

where $\mu(t) = E[\eta^G[0, t]]$ $(t \geq 0)$;

(ii) there exists a non-increasing positive integrable function g such that

$$\int_0^\infty \frac{1}{g(t)} e^{-\alpha t} \mu(dt) < \infty.$$

For $t \geq 0$, let $Y_m(t)$ be the number of groups in $E_{m,n}$ containing infectives at time t $(m = 1, 2, \ldots)$ and let $Y(t)$ be the number of individuals that are alive in the branching process at time t.

Theorem 2.1 Let $A = \{\omega \in \Omega : \lim_{t \to \infty} Y(t, \omega) = 0\}$ denote the set on which the branching process becomes extinct. Then

(a) $\lim_{m \to \infty} \sup_{0 \leq t < \infty} |Y_m(t, \omega) - Y(t, \omega)| = 0$ for P-almost all $\omega \in A$;

(b) for any $c_1 < (2\alpha)^{-1}$ and any $c_2 > (2\alpha)^{-1}$,

$$(i) \qquad \lim_{m \to \infty} \sup_{0 \leq t \leq c_1 \log m} |Y_m(t, \omega) - Y(t, \omega)| = 0$$

and

$$(ii) \qquad \lim_{m \to \infty} \sup_{0 \leq t \leq c_2 \log m} |Y_m(t, \omega) - Y(t, \omega)| = \infty,$$

for P-almost all $\omega \in A^c$.

Proof: The proof is similar to that given in [BD95a] for single population epidemics. The key observation is that $Y_m(t,\omega) = Y(t,\omega)$ $(0 \le t < \tau_m(\omega))$, where $\tau_m(\omega)$ is the time of the first between-group contact in $E_{m,n}$ with a previously infected group. Since $m^{-\frac{1}{2}} M_m \overset{a.s.}{\to} M$ as $m \to \infty$, there exists $B \in \mathcal{F}$, with $P(B) = 1$, such that

$$\lim_{m \to \infty} m^{-\frac{1}{2}} M_m(\omega) = M(\omega) \qquad (\omega \in B). \tag{1.2}$$

For $t > 0$, let $T(t)$ be the total number of births occurring during $(0, t]$ in the branching process and let $T(\infty) = \lim_{t \to \infty} T(t)$. Fix $\omega \in A \cap B$. Then $T(\infty, \omega) < \infty$, so by (1.2) $T(\infty, \omega) < M_m(\omega)$ for all sufficiently large m. For such m every birth in the branching process will correspond to a between-group infection in $E_{m,n}$, so part (a) follows since $P(A \cap B) = P(A)$.

To prove part (b)(i), first note that Nerman [Ner81], Theorem 5.4, implies

$$\lim_{t \to \infty} e^{-\alpha t} T(t) = cW \ a.s.,$$

where $c > 0$ is an appropriate constant and the (finite) non-negative random variable $W(\omega) = 0$ if and only if $\omega \in A$. Thus there exists $C \in \mathcal{F}$, with $C \subseteq A^c$ and $P(C) = P(A^c)$, such that

$$\lim_{t \to \infty} e^{-\alpha t} T(t, \omega) = cW(\omega) \qquad (\omega \in C). \tag{1.3}$$

Fix $\omega \in B \cap C$ and $c_1 < (2\alpha)^{-1}$. Then it follows from (1.3) that there exists $\beta < \frac{1}{2}$ and $m_1(\omega) > 0$ such that $T(c_1 \log m, \omega) < m^\beta$ for all $m \ge m_1(\omega)$. From (1.2), there exists $m_2(\omega) > 0$ such that $M_m(\omega) > m^\beta$ for all $m \ge m_2(\omega)$. Thus $\chi_k^{(m)}$ $(k = 0, 1, \dots, T(c_1 \log m, \omega))$ are distinct for all $m \ge \max(m, (\omega), m_2(\omega))$ and part (b)(i) follows since $P(B \cap C) = P(A^c)$.

To prove part (b)(ii), first fix $c_2 > (2\alpha)^{-1}$ and choose $\beta \in (\frac{1}{2}, \alpha c_2)$ and $\gamma \in (0, \frac{1}{2})$ such that $\beta + \gamma > 1$. For $m = 1, 2, \dots$, let L_m be the set consisting of the first $[m^\gamma]$ distinct elements of the sequence $(\chi_k^{(m)})$ and $l_m = \min\{k \in \mathbb{N} : L_m \subseteq \{\chi_0^{(m)}, \chi_1^{(m)}, \dots, \chi_k^{(m)}\}\}$. (For $x \in \mathbb{R}$, $[x]$ denotes the greatest integer $\le x$.) For $m = 1, 2, \dots$, let A_m be the set consisting of those individuals alive in the branching process at time $c_2 \log m$ that were not among the first l_m births in the branching process. If $|A_m| \ge [m^\beta]$ let D_m be the number of the $[m^\beta]$ youngest individuals in A_m whose associated $\chi_k^{(m)}$ belong to L_m. If $|A_m| < [m^\beta]$ let $D_m = \sum_{k=1}^{[m^\beta]} 1_{\{\chi_{l_m+k}^{(m)} \in L_m\}}$. Then D_m follows the binomial distribution $\text{Bin}([m^\beta], [m^\gamma]/m)$ and, since $\beta + \gamma > 1$, it follows using Leibniz's integral test, as in [BD95a], that $\sum_{m=1}^{\infty} P(D_m \le k) < \infty$ $(k = 1, 2, \dots)$. Thus by the first Borel–Cantelli lemma, $P(D_m \ge k$ infinitely often $) = 0$ $(k = 1, 2, \dots)$ and hence $P(D) = 1$, where $D = \{\omega \in \Omega : \lim_{m \to \infty} D_m(\omega) = \infty\}$.

Now fix $\Delta > 0$ and let $Y_\Delta(t)$ be the number of individuals who are alive in the branching process at time t and are of age less than Δ. Then from

[Ner81], Theorem 5.4,

$$\lim_{t \to \infty} e^{-\alpha t} Y_\Delta(t, \omega) = c' W(\omega) \qquad (\omega \in C), \tag{1.4}$$

where $c' > 0$ is an appropriate constant. Fix $\omega \in B \cap C \cap D$. Then from (1.4), $Y_\Delta(c_2 \log m, \omega) \geq m^\beta$ for all sufficiently large m, so using part (b)(i), $|A_m(\omega)| \geq m^\beta$ for all sufficiently large m. For such m,

$$\sup_{0 \leq t \leq c_2 \log m} |Y_m(t, \omega) - Y(t, \omega)| \geq D_m(\omega), \tag{1.5}$$

since the $D_m(\omega)$ individuals corresponding to the right hand side of (1.5) do not give rise to new groups being infected in $E_{m,n}$. Now $\lim_{m \to \infty} D_m(\omega) = \infty$ and $P(B \cap C \cap D) = P(A^c)$, so

$$P(\lim_{m \to \infty} \sup_{0 \leq t \leq c_2 \log m} |Y_m(t) - Y(t)| = \infty | A^c) = 1 \qquad (c_2 > (2\alpha)^{-1}).$$

Hence,

$$P(\lim_{m \to \infty} \sup_{0 \leq t \leq c_2 \log m} |Y_m(t) - Y(t)| = \infty, \ c_2 \in ((2\alpha)^{-1}, \infty) \cap \mathbf{Q} | A^c) = 1$$

and part (b)(ii) follows since $\sup_{0 \leq t \leq c_2 \log m} |Y_m(t) - Y(t)|$ is increasing with c_2.

2.4   Final outcome and threshold theorem

For $m = 1, 2, \ldots$, let $T_m(\infty)$ be the total number of groups ultimately infected by the epidemic $E_{m,n}$. Recall that $T(\infty)$ is the total number of individuals born into the branching process.

Theorem 2.2

$$\lim_{m \to \infty} T_m(\infty) = 1 + T(\infty) \ a.s.$$

Proof. If $T(\infty) < \infty$ then the result follows immediately from Theorem 2.1(a), since $Y_m(\cdot)$ and $Y(\cdot)$ coincide for all sufficiently large m. If $T(\infty) = \infty$, fix $\omega \in B \cap C$ and for $k = 1, 2, \ldots$ let $\tilde{\tau}_k(\omega) = \inf\{t : T(t) \geq k\}$. Then by Theorem 2.1(b)(i), $Y_m(t, \omega) = Y(t, \omega)$ $(0 \leq t \leq \tilde{\tau}_k(\omega))$ for all sufficiently large m $(k = 1, 2, \ldots)$. Thus $\lim_{m \to \infty} T_m(\infty) \geq k + 1$ $(k = 1, 2, \ldots)$ and the result follows.

We are now in a position to prove a threshold theorem for the household epidemic model $E_{m,n}$. We say that a *global epidemic* occurs if in the limit as $m \to \infty$ the epidemic infects infinitely many groups.

Corollary 2.1 *Let $R = \eta^G([0, \infty))$ be the number of between-group contacts resulting from a typical single group epidemic, $R_T = E[R]$ and $f(s) = E[s^R]$ be the probability generating function of R. Then, as $m \to \infty$,*

(i) a global epidemic occurs with non-zero probability if and only if $R_T > 1$,

(ii) the probability of a global epidemic is $1 - p$, where p is the smallest solution of $f(s) = s$ in $[0, 1]$,

(iii) the probability generating function of the limiting total number of groups infected $T(\infty)+1$, $h(s)$ say, satisfies $h(s) = sf(h(s))$. Further,

$$P(T(\infty) + 1 = k) = k^{-1}P(R_1 + R_2 + \ldots + R_k = k) \quad (k = 1, 2, \ldots),$$

where $R_1, R_2, \ldots$ are independent and identically distributed copies of R.

Proof The distribution of $T(\infty)$ is the same as that of the total size of the embedded Galton–Watson process with offspring distribution the distribution of $R = \eta^G([0, \infty))$. Corollary 2.1 follows immediately from Theorem 2.2 and standard results (see eg [Jag75]) for Galton–Watson processes.

Note that Corollary 2.1 (i) still holds if there are $a < \infty$ initial infectives. If these a initial infectives are in different groups then the probability of a global epidemic is $1 - p^a$ and (iii) can also be modified appropriately. If there are more than one initial infectives in some groups then the situation is more complicated since the offspring distributions of the corresponding ancestors in the branching process no longer have the same distribution as R. In such cases, the probability of a global epidemic is most easily found by conditioning on the size of the first generation in the embedded Galton–Watson process, cf [BMST95]. Note also that Theorem 2.2 (and hence Corollary 2.1) do not require conditions (i) and (ii), given just prior to the statement of Theorem 2.1.

3 Specialisation to model of Ball, Mollison and Scalia-Tomba

Suppose now that, as in [BMST95], T^I, η^W and η^B are mutually independent, with T^I following an arbitrary but specified distribution, and η^W and η^B following homogeneous Poisson processes with rates $n\lambda_W$ and λ_B, respectively. Recall that n denotes group size. Thus a given infective makes contact with a given susceptible in its own group at rate $\lambda_W + (mn)^{-1}\lambda_B$ and with a given susceptible in another group at rate $(mn)^{-1}\lambda_B$. Note that the lack of a latent period involves no loss of generality as far as the final outcome of the epidemic is concerned, since, as is well known, the distribution of the final outcome of a closed population epidemic is invariant to very general assumptions concerning a latent period (see eg Ball [Bal86]).

Consider the single population epidemic, with initially 1 infective and $n - 1$ susceptibles, in which any given infective makes contacts with any given susceptible at the points of a Poisson process with rate λ_W during an infectious period distributed according to T^I. Let N^* be the total number of initial susceptibles that are ultimately infected by the epidemic and T_A be the severity of the epidemic, ie the sum of the infectious periods of the N^*+1 infectives in the epidemic. In the epidemic $E_{m,n}$, each of these N^*+1 infectives would make between-group contacts at rate λ_B. Thus the total number of between-group contacts emanating from the above single group epidemic, $R = \eta^G([0,\infty))$, follows a Poisson distribution with random mean $\lambda_B T_A$. Hence,

$$R_T = E[R] = \lambda_B E[T_A] = \lambda_B E[T^I](1 + E[N^*]), \qquad (1.6)$$

where the final equality comes from a Wald's identity for epidemic models, see [Bal86].

Let $\phi(\theta) = E[\exp(-\theta T^I)]$ ($\theta \geq 0$) be the moment generating function of T^I and write $\mu_n = E[N^*]$, to show the dependence on the single group population size. Then it is shown in [Bal86] that

$$\mu_n = n - 1 - \sum_{k=0}^{n-1} \binom{n-1}{k} \alpha_k \phi(\lambda_W k)^{n-k} \qquad (n = 0, 1, \ldots), \qquad (1.7)$$

where $\alpha_0, \alpha_1, \ldots$ are defined recursively by

$$\sum_{i=0}^{k} \binom{k}{i} \alpha_i \phi(\lambda_W k)^{k-i} = k \qquad (k = 0, 1, \ldots). \qquad (1.8)$$

Equations (1.6), (1.7) and (1.8) enable R_T to be calculated for any choice of λ_W, λ_B and T^I.

Suppose now that T^I follows a negative exponential distribution with mean γ^{-1}, so that the deterministic version of our model is a special case of the model of [DG94]. Rescale the time axis so that $\gamma = 1$. Then the reproductive ratio is given by $R_0 = (n - 1)\lambda_W + \frac{mn-1}{mn}\lambda_B$, accounting for potential contacts with oneself. Thus, if we now assume that η^W has rate $n\lambda_W/(n - 1)$ then, as $m \to \infty$, $R_0 = \lambda_W + \lambda_B$ and is independent of group size n. (It was not convenient to make this assumption earlier as it causes problems with (1.7) and (1.8).) Figure 1 shows graphs of critical values of (λ_W, λ_B) so that $R_T = 1$ for various group sizes n, together with the corresponding graph for $R_0 = 1$. Note that, unless n is large, R_0 is not a good indicator as to whether global epidemics can occur. Note also that if the within-group epidemic is above its internal threshold, ie $\lambda_W \geq 1$, then, whatever the value of λ_B, global epidemics can always occur provided that the group size n is large enough, whilst if $\lambda_W < 1$, global epidemics can only occur for sufficiently large n if $\lambda_B > 1 - \lambda_W$. Thus the household epidemic model displays a phase transition as the within-group epidemic passes through its internal threshold. The above comments

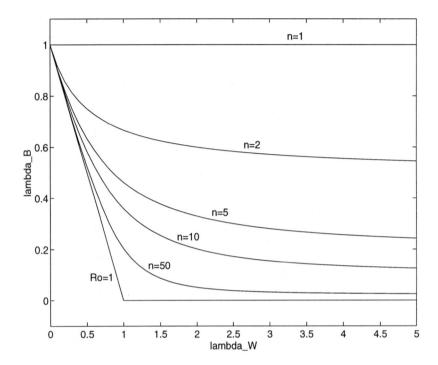

FIGURE 1. Critical values of (λ_W, λ_B) so that $R_T = 1$.

are discussed further in [BMST95], where other properties of the limiting branching process and the distribution of the number of susceptibles in a typical household that are infected by a global epidemic are determined. [BMST95] extend their results to the situation of unequal group sizes and briefly discuss the estimation of R_T from household final size data. Implications of their results for vaccination strategies are also considered.

4   Extensions

We briefly outline generalisation of our results to two extensions of the household epidemic model of Section 2.1. First suppose that the households are not all the same size. Specifically, for $k = 1, 2 \ldots$, let θ_k be the proportion of households that are of size k and suppose that $\theta = \sum_{k=1}^{\infty} k\theta_k < \infty$. Then the limiting branching process of Theorem 2.1 becomes a multitype branching process, with type corresponding to household size. The mean matrix of the associated embedded multitype Galton–Watson process has multiplicative structure, so its maximal eigenvalue, the threshold parameter R_T, is easily determined. This reflects the fact that the limiting process can also be described by a single-type branching process, in which the probability that an individual contacted by a between-group contact is in a household of size k is given by the size-biased distribution $\theta^{-1}\theta_k$ ($k = 1, 2, \dots$). See [BMST95] for details when the model is the special case of Section 3.

The second generalisation is to a model in which there are p types of individuals, labelled $\alpha_1, \alpha_2, \dots, \alpha_p$. The model of Section 2.1 is then extended in the obvious fashion, replacing $\mathcal{H} = (T^I, \eta^W, \eta^B)$ by $\mathcal{H}_{\alpha_i} = (T^I_{\alpha_i}, (\eta^W_{\alpha_i,\alpha_j}, \eta^B_{\alpha_i,\alpha_j}) (j = 1, 2, \dots, p)) (i = 1, 2, \dots, p)$. The limiting branching process, in which the unit is a single-group epidemic, is again multitype, with type corresponding to the initial composition of susceptibles in a group augmented by the type of the initial infective in that group. This can be reduced, utilising appropriate size-biased sampling, to a p-type branching process, with type corresponding to the type of the initial infective in a group. The model can be specialised along the lines of Section 3, and then the mean matrix of the associated embedded p-type Galton–Watson process, and hence the threshold parameter R_T, can be determined using the analogue of (1.7) and (1.8) for multipopulation epidemics given in [Bal86]. Note that in this latter model, the spread of infection within a household follows a stochastic version of the model in [DG94]. Becker and Hall [BH95] consider a household epidemic model with p types of individuals, in which household outbreaks are classified according to the eventual number of cases of each type and the model is defined in terms of the mean number of household outbreaks of different types generated by individuals of different types. They also analyse the special case of highly infectious diseases (see Section 1). The model outlined above is defined at a more basic level than Becker and Hall's model. Moreover, it provides one approach to determining the mean numbers that parameterise their model.

Acknowledgments: This work follows on from collaborations with Peter Donnelly, Denis Mollison and Gianpaolo Scalia-Tomba. In this paper, the method of proving strong approximation of single population epidemics by branching processes, developed jointly with Peter Donnelly, is extended to a generalisation of the household epidemic model analysed in collaboration with Denis Mollison and Gianpaolo Scalia-Tomba.

5 References

[Ald85]   D.J. Aldous. Exchangeability and related topics. In P.L. Hennequin, editor, *Ecole dÉté de Probabilités de Saint-Flour XIII, 1983, Lecture Notes in Mathematics, Vol. 1117*, pages 1–198, Berlin, 1985. Springer-Verlag.

[Bai75]   N.T.J. Bailey. *The Mathematical Theory of Infectious Diseases and its Applications*. Griffin, London, 1975.

[Bal83]   F.G. Ball. The threshold behaviour of epidemic models. *J. Appl. Prob.*, 20:227–241, 1983.

[Bal86]   F.G. Ball. A unified approach to the distribution of total size and total area under the trajectory of infectives in epidemic models. *Adv. Appl. Prob.*, 18:289–310, 1986.

[Bal94]   F.G. Ball. *J. Roy. Statist. Soc. A*, 157:129–131, 1994. Contribution to the discussion of "Epidemics: models and data." by D. Mollison, V. Isham and B.T. Grenfell.

[Bar72]   R. Bartoszyński. On a certain model of an epidemic. *Applicationes Mathematicae*, 13:139–151, 1972.

[BD95a]   F.G. Ball and P.J. Donnelly. Strong approximations for epidemic models. *Stoch. Proc. Appl.*, 55:1–21, 1995.

[BD95b]   N.G. Becker and K. Dietz. The effect of the household distribution on transmission and control of highly infectious diseases. *Math. Biosci.*, 127:207–219, 1995.

[Bec77]   N.G. Becker. On a general epidemic model. *Theor. Popul. Biol.*, 11:23–36, 1977.

[Bec79]   N.G. Becker. The uses of epidemic models. *Biometrics*, 35:295–305, 1979.

[BH95]   N.G. Becker and R. Hall. Immunization levels for preventing epidemics in a community of households made up of individuals of different types. 1995. Preprint.

[BMST95]   F.G. Ball, D. Mollison, and G. Scalia-Tomba. Epidemics in populations with two levels of mixing. 1995. Submitted to *Ann. Appl. Probab.*

[DG94]   D.J. Daley and J. Gani. A deterministic general epidemic model in a stratified population. In F.P. Kelly, editor, *Probability, Statistics and Optimisation – A Tribute to Peter Whittle*, pages 117–132, Chichester, 1994. Wiley.

[HD96]   J.A.P. Heesterbeek and K. Dietz. The concept of r_0 in epidemic theory. *Staistica Neerlandica*, 50, 1996. To appear.

[Jag75]   P. Jagers. *Branching Processes with Biological Applications*. Wiley, Chichester, 1975.

[Ken94]   W.S. Kendall. *J. Roy. Statist. Soc. A*, 157:143, 1994. Contribution to the discussion of "Epidemics: models and data." by D. Mollison, V. Isham and B.T. Grenfell.

[KM27]   W.O. Kermack and A.G. McKendrick. A contribution to the mathematical theory of epidemics. *Proc. Roy. Soc. London, Ser. A*, 115:700–721, 1927.

[Met78]   J.A.J. Metz. The epidemic in a closed population with all susceptibles equally vulnerable; some results for large susceptible populationsand small initial infections. *Acta Biotheoretica*, 27:75–123, 1978.

[Mol91]   D. Mollison. Dependence of epidemic and population velocities on basic parameters. *Math Biosci*, 107:255–287, 1991.

[Ner81]   O. Nerman. On the convergence of supercritical general (c-m-j) branching processes. *Z. Wahrscheinlichkeitsth*, 57:365–395, 1981.

[vdBMD90]   F. van den Bosch, J.A.J. Metz, and O. Diekmann. The velocity of spatial epidemic expansion. *J Math Biol*, 28:529–565, 1990.

[Wat72]   R.K. Watson. On an epidemic in a stratified population. *J. Appl. Prob.*, 9:659–666, 1972.

[Whi55]   P. Whittle. The outcome of a stochastic epidemic – a note on bailey's paper. *Biometrika*, 42:116–122, 1955.

[Wil71]   T. Williams. An algebraic proof of the threshold theorem for the general stochastic epidemic (abstract). *Adv. Appl. Prob.*, 3:223, 1971.

Reproduction Numbers and Critical Immunity Levels for Epidemics in a Community of Households

NIELS G. BECKER

School of Statistics, La Trobe University, Bundoora VIC 3083, Australia

KLAUS DIETZ

Institut für Medizinische Biometrie, Universität Tübingen, Westbahnhofstraße 55, D 72070 Tübingen 1, Germany

ABSTRACT

Epidemic threshold parameters, also called reproduction numbers, play a central role in computing the vaccination coverage required to prevent epidemics. It is possible to define several different reproduction numbers for infectives in a community of households. To illustrate this we compute four distinct reproduction numbers for infectives for a community consisting of a large number of households of size three, using assumptions similar to the so-called general epidemic model. It is found that when individuals are selected independently for immunization the proportion that needs to be immunized so as to prevent epidemics is $v_I^* = 1 - 1/R_{30}$, where R_{30} is one of these reproduction numbers. When a proportion of households is selected and every member of each selected household is immunized, then the proportion of households that needs to be immunized is $v_H^* = 1 - 1/R_{40}$, where R_{40} is another of the basic reproduction numbers. The result for v_H^* applies for an arbitrary household distribution and a disease with an arbitrary infectivity function. However, the result for v_I^* is more complicated for a community containing larger households.

1. INTRODUCTION

Epidemic threshold parameters, or reproduction numbers, are central to discussions on the prevention of epidemics in a community. A basic reproduction number R_0 is a parameter of an epidemic model, for a community in which no one is immune, with the property that major epidemics and persistence of endemic infection are possible in the community only when $R_0 > 1$. Here we consider reproduction numbers for epidemics in a community of households, where a household is interpreted as a group of individuals with a considerably higher contact rate within the group. Such reproduction numbers have been discussed previously by Bartoszyński [3] and, more recently, by Becker and Dietz [6], Becker and Hall [7] and Ball et al. [2].

In the so-called general epidemic model for a closed community of n homogeneously mixing individuals, see Bailey [1, Chapter 6], disease transmission occurs with a rate $\beta I_t S_t / n$, where S_t and I_t denote the number of susceptibles and infectives at time t, respectively.

The infectious period is assumed to have an exponential distribution with mean $1/\gamma$. For this model $R_0 = \beta/\gamma$ is the basic reproduction number, interpreted as the mean number of infectives generated by a single initial infective during the entire infectious period of the latter, in a large population consisting entirely of susceptibles. Immunizing a fraction v of the susceptibles changes the reproduction number for infectives in this partially immunized community to $R_v = R_0(1 - v)$. The level of immunity required to prevent major epidemics, denoted by v^* and called the critical immunity level, is calculated by setting $R_v = 1$. This gives the well known result $v^* = 1 - 1/R_0$, see Smith [9], Becker [4] and Dietz [8]

Our main objective is to determine whether the critical immunity level for the prevention of epidemics retains such a direct, and simple, relationship to a basic reproduction number when the community is made up of households. In other words, we are seeking to generalize the above result for the critical immunity level v^*. As will be seen, this problem is complicated by the fact that for a community of households several meaningful reproduction numbers for infectives can be defined and different immunization strategies can differ substantially in their effectiveness. However, there is a strong a $priori$ impression that the result $R_v = R_0(1 - v)$, and therefore the result $v^* = 1 - 1/R_0$, should hold quite generally, because under random vaccination of individuals a fraction $1 - v$ of the individuals are initially susceptible.

We compute reproduction numbers for infectives by approximating the size of the population of infectives by a branching process and identifying the event of a minor epidemic with extinction of the branching process. Different types of infectives need to be introduced, because infectives differ in their potential to infect others, and so a multi-type branching process is used. For a multi-type branching process the largest eigenvalue of the mean matrix is a reproduction number.

In Section 2 four basic reproduction numbers are derived for the simplified setting of a community consisting entirely of households of size three, using assumptions akin to the Markovian general epidemic model. The magnitudes of these basic reproduction numbers are compared in Section 2.5. The computation of critical immunity levels and their relationships to certain basic reproduction numbers are considered in Section 3 for a community of households of size three. In Section 4 we show how critical immunity levels are computed for communities with arbitrary household size distributions, and point out that their relationship to basic reproduction numbers is not so simple when some households are larger than four.

Throughout, the community is assumed to consists of a large number of households. We refer to an infective as he with the understanding that this means he or she.

2. DIFFERENT REPRODUCTION NUMBERS

We choose a simple setting to show that there are several plausible candidates for the role of basic reproduction number. Consider a community consisting of a large number of households of size three and suppose that every individual is susceptible to the infectious disease. To make our discussion more specific, we adopt, initially, assumptions similar to those of the so-called general epidemic model. That is, there is no latent period and the duration of the infectious period has an exponential distribution with mean $1/\gamma$. During the infectious period each infective generates primary cases in other households according to a Poisson process with rate β_C, where the subscript C refers to community, and exerts a force of infection β_H on $each$ susceptible member of his own household.

Four basic reproduction numbers for infectives are now introduced for this setting. Reproduction numbers are determined by the embedded jump process of the epidemic which has a probability model depending only on $b_c = \beta_c/\gamma$ and $b_H = \beta_H/\gamma$. For notational convenience we introduce the terms A_1, A_2, B_1 and B_2 given by

$$A_r = \frac{rb_H}{1 + rb_H} = 1 - B_r \qquad r = 1, 2.$$

2.1 REPRODUCTION BASED ON ACTUAL CASES

We begin with the situation in which each infection is attributed to the infective whose contact was directly responsible for the infection. The primary case is responsible for the infection of the second infective in a household, if any. However, since there is no latent period and the duration of the infectious period has an exponential distribution, which lacks memory, the last remaining susceptible has an equal chance of being infected by either the primary infective or the second infective of the household. In other words, when there are two infectives and one susceptible in the household the infectives are essentially 'competing', equally, to infect the remaining susceptible.

We need to introduce types of infectives, because infectives differ in their potential to infect others. Infectives are classified according to how many susceptibles remain in the household at the time of their infection. Primary infectives of the household are labelled type 2, the first secondary case is type 1 and the last remaining susceptible, if infected, becomes a type 0 infective. In other words, the label given to an infective indicates the number of susceptibles remaining in the household at the time of their infection.

The mean matrix for these types of infectives is calculated to be

$$(m_{ij}) = \begin{array}{c} \\ 2 \\ 1 \\ 0 \end{array} \begin{array}{ccc} 2 & 1 & 0 \\ \left(\begin{array}{ccc} b_c & A_2 & A_1 A_2 (1 + B_1)/2 \\ b_c & 0 & A_1 (1 + B_1)/2 \\ b_c & 0 & 0 \end{array} \right), \end{array}$$

where m_{ij} denotes the mean number of type j infectives generated by one type i infective during the infectious period of the latter. The characteristic equation associated with this mean matrix is

$$f_2(\lambda) = -\lambda^3 + b_c \lambda^2 + b_c (2A_1 - A_1^2 A_2/2)\lambda + b_c A_1 A_2 B_1 + b_c A_1^2 A_2/2 = 0. \qquad (1)$$

The largest root of this equation gives the basic reproduction number R_{20} corresponding to these types of infectives and this way of attributing infections to infectives. The 0 in the subscript indicates that this is a *basic* reproduction number, while the 2 distinguishes it from reproduction numbers given below and also indicates its ranking, as discussed in Section 2.5.

2.2 INFECTIVES OF EQUIVALENT CURRENT STATUS

Above we classified infectives according to the number of susceptibles remaining in the household at the *times of their infection*. This classification does not reflect the fact that, in this Markovian model, upon infection of a second household member, both infectives

269

have the same transmission potential thereafter. To reflect this explicitly we now reclassify the primary case as type 1 if, and when, he infects a second household member. It means that upon infection of the second household member we have two infectives of type 1.

A consequence of this reclassification is that the primary infective remains of type 2 for a mean duration of $(\gamma + 2\beta_H)^{-1}$, rather than γ^{-1}. This adjusts the first element of the mean matrix, which is now computed to be

$$
(m_{ij}) = \begin{matrix} 2 \\ 1 \\ 0 \end{matrix} \begin{pmatrix} \overset{2}{b_c B_2} & \overset{1}{2A_2} & \overset{0}{0} \\ b_c & 0 & A_1(1+B_1)/2 \\ b_c & 0 & 0 \end{pmatrix},
$$

The characteristic equation of this mean matrix is

$$
f_1(\lambda) = -\lambda^3 + b_c B_2 \lambda^2 + 2b_c A_2 \lambda + b_c A_1 A_2(1 + B_1) = 0. \tag{2}
$$

The largest root of this equation gives the basic reproduction number R_{10} corresponding to these types of infectives and this way of attributing infections to infectives.

2.3 REPRODUCTION BASED ON POTENTIAL CASES

Above we attributed infections to infectives only if their contact was the direct cause of the infection. This does not give the introductory infective his full potential to infect the other two household members, because when the second individual is infected the two infectives are 'competing' with each other to infect the remaining susceptible. We now calculate a basic reproduction number in a way that gives the introductory infective his full potential to infect the other two household members. We do this by considering all the infectious contacts an infective has with other individuals, irrespective of whether the contacted individual has already been infected. An *infectious* contact is one that is close enough to lead to infection if the contact is with a susceptible.

Infectives are now classified according to the number of infectives sharing their 'generation' and the number of susceptibles exposed to this generation of infectives. The primary infective makes up the first generation and is labelled 12, to indicate that there is 1 infective in the generation and there are 2 susceptible household members exposed to this infective. We attribute any infection in the household to the primary case if at least one infectious contact occurred with the primary case during the latter's infectious period, *irrespective* of whether the actual infection was acquired earlier from a contact with a secondary case of the household. Consider now a household in which the first individual has just been infected from outside. If exactly one of the remaining two susceptibles makes at least one infectious contact with the primary infective, then we label this secondary infective 11, to indicate that there is one infective in the second generation and there remains one susceptible household member who is exposed to this infective. Any infectives belonging to a generation that has no susceptible household members exposed to them are labelled •0.

The idea of constructing generations of epidemic chains in terms of avoidance, or non-avoidance, of infectious contacts is the basis of the epidemic chain probability models given by Becker [5, Chapter 3] and we use methods from there to compute the mean matrix. The mean matrix for these types of infectives is calculated to be

$$
\begin{array}{c}
\quad\quad 12 \quad\quad 11 \quad\quad \bullet 0 \\
\begin{array}{c} 12 \\ 11 \\ \bullet 0 \end{array}
\left(
\begin{array}{ccc}
b_c & A_2 B_1 & 2 A_1 A_2 \\
b_c & 0 & A_1 \\
b_c & 0 & 0
\end{array}
\right).
\end{array}
$$

The characteristic equation associated with this mean matrix is

$$
f_3(\lambda) = -\lambda^3 + b_c \lambda^2 + 2 b_c A_1 \lambda + b_c A_1 A_2 B_1 = 0. \tag{3}
$$

The largest root of this equation gives the basic reproduction number R_{30} corresponding to these types of infectives and this way of attributing infections to infectives.

2.4 REPRODUCTION OF INFECTED HOUSEHOLDS

A basic reproduction number for the propagation of household outbreaks was introduced by Bartoszyński [3]. Here this reproduction number is interpreted as a reproduction number for infectives, by attributing infections to infectives in a certain way.

We now attribute to an infective all of the primary infectives he generates in other households as well as any secondary cases arising in those households. In other words, the secondary cases in a household are not attributed to the infectives within the household, whose contacts actually caused their infections, but instead to the infective who generated the primary case of the household.

In this setting there is only one type of infective and the basic reproduction number R_{40} is simply the mean number of infections attributed to an infective. Therefore $R_{40} = b_c \nu$, where ν is the mean size of a household outbreak assuming that everybody is susceptible. Under the present assumptions, with households of size three, we simply have

$$
R_{40} = b_c (1 + 2 A_1 + A_1 A_2 B_1). \tag{4}
$$

For the purpose of comparisons, made in the next section, it is useful to view this as the largest root of the characteristic equation

$$
f_4(\lambda) = -\lambda^3 + b_c (1 + 2 A_1 + A_1 A_2 B_1) \lambda^2 = 0. \tag{5}
$$

2.5 COMPARISON OF REPRODUCTION NUMBERS

The basic reproduction numbers R_{10}, R_{20}, R_{30} and R_{40} are all epidemic threshold parameters. That is, epidemics can occur only when their value is greater than unity. The critical values for transmission rates, i.e. the bound on transmission rates so that epidemics cannot occur, is given by setting the basic reproduction number to 1. We do not have explicit expressions for R_{10}, R_{20} and R_{30}, but can find the criticality conditions for them by setting $\lambda = 1$ in (2), (1) and (3), respectively. We find, as we should, that all four threshold parameters lead to the same criticality condition, namely

$$
b_c (1 + 2 A_1 + A_1 A_2 B_1) = 1. \tag{6}
$$

Any control strategy that reduces the values of b_c and b_H sufficiently to make the left hand side of (6) less than unity will prevent epidemics and persistent endemic infection.

271

The basic reproduction numbers can be ranked. Specifically, we show that R_{10}, R_{20}, R_{30} and R_{40} are equal to 1 simultaneously, while $R_{10} < R_{20} < R_{30} < R_{40}$ when they exceed 1 and $R_{10} > R_{20} > R_{30} > R_{40}$ when their values are less than 1.

First note that $f_i(0) \geq 0$ and $f_i(\lambda) \to -\infty$ as $\lambda \to \infty$, so that R_{i0}, the largest root of $f_i(\lambda) = 0$, is non-negative. From the differences between the characteristic functions, expressed as

$$f_2(\lambda) - f_1(\lambda) = b_c A_2(\lambda - 1)[\lambda + A_1(1 + B_1)/2],$$

$$f_3(\lambda) - f_2(\lambda) = b_c A_1^2 A_2(\lambda - 1)/2,$$

$$f_4(\lambda) - f_3(\lambda) = b_c A_1(\lambda - 1)[(2 + A_2 B_1)\lambda + A_2 B_1].$$

it follows immediately that

$$f_1(\lambda) > f_2(\lambda) > f_3(\lambda) > f_4(\lambda), \quad \text{when } 0 \leq \lambda < 1, \tag{7}$$

$$f_1(\lambda) = f_2(\lambda) = f_3(\lambda) = f_4(\lambda), \quad \text{when } \lambda = 1, \tag{8}$$

and

$$f_1(\lambda) < f_2(\lambda) < f_3(\lambda) < f_4(\lambda), \quad \text{when } \lambda > 1. \tag{9}$$

Therefore, if the common $f_i(1)$ value is positive then R_{10}, R_{20}, R_{30} and R_{40} exceed 1 and from (9) we have $R_{10} < R_{20} < R_{30} < R_{40}$. If, on the other hand, the common $f_i(1)$ value is 1 then $R_{40} = 1$, and by (8) and (9) all of the basic reproduction numbers are 1. Finally, if the common $f_i(1)$ value is negative then $R_{40} < 1$, so that (7) and (9) imply $1 > R_{10} > R_{20} > R_{30} > R_{40}$. This completes the proof.

3. CRITICAL IMMUNITY COVERAGE

We are interested in the proportion of individuals that need be immunized in a community of households in order to prevent epidemics. That is, we want to compute v^*, the critical immunity coverage. Immunization changes the number of susceptibles in households, which changes each of the reproduction numbers. To compute v^* we need to work with one of the reproduction numbers. Any of the reproduction numbers may be used for this purpose, but it is easiest to work with the reproduction number for infected households considered in Section 2.4, since we have an explicit expression for it.

It emerges that the critical immunity coverage depends on the way individuals are selected for immunization, that is, the immunization strategy. We comsider two different strategies.

3.1 IMMUNIZING A PROPORTION OF HOUSEHOLDS

Suppose that a proportion v_H of households are selected and every member of each selected household is immunized. The only effect this has on the calculations made above is that b_c changes to $b_c(1 - v_H)$, because only a proportion $1 - v_H$ of the infectious contacts made with individuals from other households are with susceptibles. As a consequence the reproduction number for the partially immunized community is given by

$$R_{4V} = (1 - v_H)R_{40}.$$

272

The critical immunity level for households is obtained by setting $R_{4V} = 1$, giving

$$v_{\mathrm{H}}^{*} = 1 - 1/R_{40}. \qquad (10)$$

The same result can be obtained from the characteristic equations (1) (2) and (3), by substituting $b_{\mathrm{C}} = b_{\mathrm{C}}(1 - v_{\mathrm{H}})$ and $\lambda = 1$ into these equations.

We have therefore found that for this immunization strategy the critical immunity coverage has the same simple relationship to a basic reproduction number. That the basic reproduction number in this relationship is R_{40} seems appropriate since it is the reproduction number for infected households and this strategy focusses on immunizing entire households.

The results (10) can be shown to hold for a community in which the household sizes have an arbitrary distribution and for a completely general model for disease transmission within households.

3.2 IMMUNIZING A RANDOM SELECTION OF INDIVIDUALS

Now suppose instead that every individual, independently, is immunized with probability v_{I}. Then the proportion of households with s susceptibles is $\binom{3}{s}(1 - v_{\mathrm{I}})^{s} v_{\mathrm{I}}^{3-s}$, $s = 0, 1, 2, 3$. Let ν_{s} denote the mean size of an outbreak in a household with s non-immunized individuals when one of them is infected from outside. Then $\nu_{1} = 1$ and $\nu_{2} = 1 + A_{1}$, while $\nu_{3} = 1 + 2A_{1} + A_{1}A_{2}B_{1}$ as before. The reproduction number for infected households for this partially immunized community is

$$R_{4V} = b_{\mathrm{C}} \sum_{s=1}^{3} \binom{3}{s} v_{\mathrm{I}}^{s}(1 - v_{\mathrm{I}})^{3-s} \cdot \frac{s}{3} \cdot \nu_{s}, \qquad (11)$$

where the term $s/3$ is the probability that the contact is with one of the s susceptibles of the household.

To determine the critical immunity level we need to substitute $R_{4V} = 1$ into equation (11) and solve for v_{I}. If we substitute $R_{4V} = 1$ and $v_{\mathrm{I}} = 1 - 1/\lambda$ into this equation we find, after some algebra, that it leads to the equation (3). In other words, the solution of (3), namely R_{30}, determines the critical immunity level by the relation

$$v_{\mathrm{I}}^{*} = 1 - 1/R_{30}. \qquad (12)$$

It seems reasonable that R_{30}, rather than R_{10} or R_{20}, should determine v_{I}^{*}, because R_{30} is calculated in a way that gives the primary infective his full potential to infect others and a control strategy must overcome the full infection potential.

3.3 GENERAL ASSUMPTIONS ABOUT TRANSMISSION

We now ask how sensitive the relationship $v_{\mathrm{I}}^{*} = 1 - 1/R_{30}$ is to assumptions we made about disease transmission. Consider therefore a general model based on avoidance and non-avoidance of infection. The infectives are classified into types just as in Section 2.3. The mean matrix is now written as

$$
\begin{array}{c}
\quad\quad 12 \quad\quad 11 \quad\quad \bullet 0 \\
\begin{array}{c} 12 \\ 11 \\ \bullet 0 \end{array}
\begin{pmatrix}
b_{\mathrm{C}} & p_{12,1} & 2p_{12,2} \\
b_{\mathrm{C}} & 0 & p_{11,1} \\
b_{\mathrm{C}} & 0 & 0
\end{pmatrix},
\end{array}
$$

273

where $p_{i,j,k}$ denotes the probability that k cases result when j susceptibles are exposed to i infectives of the same household for the duration of their infectious periods. The characteristic equation associated with this mean matrix is

$$-\lambda^3 + b_c\lambda^2 + b_c(p_{12,1} + 2p_{12,2})\lambda + b_c p_{12,1}p_{11,1} = 0. \tag{13}$$

Now suppose that each individual is immunized with probability v_{I}. The expression for the reproduction number R_{4V} is given by (11), but now $\nu_1 = 1$, $\nu_2 = 1 + p_{11,1}$ and $\nu_3 = 1 + p_{12,1} + 2p_{12,2} + p_{12,1}p_{11,1}$. With these expressions for the ν_s, as well as $R_{4V} = 1$ and $v_{\text{I}} = 1 - 1/\lambda$ in (11), we find after some algebra that

$$-\lambda^3 + b_c\lambda^2 + 2b_c p_{11,1}\lambda + b_c(p_{12,1} + 2p_{12,2} + p_{12,1}p_{11,1} - 2p_{11,1}) = 0.$$

This equation agrees with (13) provided that

$$p_{12,1} + 2p_{12,2} = 2p_{11,1}. \tag{14}$$

This means that we need the mean number infected when two susceptibles are exposed to one infective for the entire infectious period to be twice the mean number infected when one susceptible is exposed to one infective for the entire infectious period. This is indeed true for virtually all transmission models considered, because the event of escaping infection is always assumed to be independent for different susceptibles.

We have therefore shown that the relationship $v_{\text{I}}^* = 1 - 1/R_{03}$ for the critical immunity coverage is not sensitive to the form of the transmission model for within household disease transmission, when households are of size 3.

We now ask whether the result is sensitive to household size. It turns out that the result remains true for any mixture of households of sizes 1, 2 and 3, but it is not true when larger household sizes are included.

4. HOUSEHOLDS OF LARGER SIZES

Consider a community consisting of a large number of households with an arbitrary household size distribution. We need to keep track of both the household size and the number of susceptibles in the household. The proportion of households consisting of n members of whom s are susceptibles is denoted by h_{ns}, where $s = 0, 1, 2, \ldots, n$ and $n = 1, 2, \ldots N$. The largest household size is N. Our transmission assumption is based on individuals meeting individuals and that disease is transmitted from person to person. With this assumption it is useful to introduce g_{ns}, the proportion of individuals belonging to households with n members of whom exactly s are initially susceptible. The relationship between the g_{ns} and the h_{ns} is given by

$$g_{ns} = \frac{nh_{ns}}{\mu_{\text{H}}} \quad \text{and} \quad h_{ns} = \frac{g_{ns}/n}{\sum_{j=1}^{N} g_{j+}/j},$$

where $\mu_{\text{H}} = \sum_{n=1}^{N} nh_{n+}$ is the mean household size, and a '+' in place of a subscript denotes summation over that subscript.

Suppose now that each individual is immunized, independently of others, with probability v_{I} and all unimmunized individuals are susceptible. The reproduction number for infected households for such a partially immunized community is

$$R_{4V} = b_c \sum_{n=1}^{N} \sum_{s=1}^{n} g_{ns}\frac{s}{n}\nu_s = \frac{b_c}{\mu_{\text{H}}} \sum_{n=1}^{N} \sum_{s=1}^{n} s\binom{n}{s}(1 - v_{\text{I}})^s v_{\text{I}}^{n-s} h_{n+}\nu_s.$$

The critical immunity level v_I^* is obtained by setting $R_{4V} = 1$ and solving for v_I. It is of interest to see if v_I^* can, in general, be related to the basic reproduction number R_{03}, as was the case for households of size three. Accordingly we make the substitution $v_I = 1 - 1/R$, and see that we need to solve

$$F_N(R) = -R^N + \frac{b_c}{\mu_H} \sum_{n=1}^{N} \sum_{s=1}^{n} s \binom{n}{s} R^{N-n}(R-1)^{n-s} h_{n+} \nu_s = 0. \tag{15}$$

We now construct the characteristic equation for R_{03} using a more general household size distribution than that used in sections 2 and 3. As in Section 2.3, we label infectives by is when there are i infectives in their generation and s susceptible household members are exposed to them. For convenience, we continue our discussion for a maximum household size $N = 4$. The mean matrix for these types of infectives is

$$
\begin{array}{cccccc}
 & 13 & 21 & 12 & 11 & \bullet 0 \\
13 & \begin{pmatrix} b_4 & 2p_{13,2} & b_3 + p_{13,1} & b_2 & b_1 + 3p_{13,3} \\
21 & b_4 & 0 & b_3 & b_2 & b_1 + \frac{1}{2}p_{21,1} \\
12 & b_4 & 0 & b_3 & b_2 + p_{12,1} & b_1 + 2p_{12,2} \\
11 & b_4 & 0 & b_3 & b_2 & b_1 + p_{11,1} \\
\bullet 0 & b_4 & 0 & b_3 & b_2 & b_1 \end{pmatrix} ,
\end{array}
$$

where

$$b_s = b_c \sum_{n=s}^{N} g_{ns} \frac{s}{n}, \qquad s = 1, 2, \ldots .$$

The characteristic equation for this mean matrix is computed to be

$$\lambda G(\lambda) = -\lambda^5 + \sum_{i=1}^{4} \xi_i \lambda^{5-i} = 0, \tag{16}$$

where ξ_i is the mean number of generation i infectives generated by an infective during the early stages of the epidemic. If we write ν_{is} for the mean number of infections resulting when s susceptibles are exposed, for the entire duration of their infectious periods, to i infectives of a given generation, then

$$\xi_1 = b_1 + b_2 + b_3 + b_4, \qquad \xi_2 = b_2 \nu_{11} + b_3 \nu_{12} + b_4 \nu_{13},$$

$$\xi_3 = b_3 p_{12,1} \nu_{11} + b_4 (p_{13,1} \nu_{12} + p_{13,2} \nu_{21}) \quad \text{and} \quad \xi_4 = b_4 p_{13,1} p_{12,1} \nu_{11} ,$$

where $\nu_{is} = \sum_{j=0}^{s} j p_{is,j}$.

The basic reproduction number R_{03} is the largest root of the equation (13). We should therefore have $G(\lambda) = F_4(\lambda)$, where F_4 is given by (15), if $v_I^* = 1 - 1/R_{03}$ is true in this setting. The requirements for this equation to be satisfied, in addition to those required for households of size 3, are

$$p_{31,1} + 2p_{31,2} + 3p_{31,3} = 3p_{11,1} \tag{17}$$

and

$$p_{13,1}(p_{12,1} + 2p_{12,2}) + p_{13,2} p_{21,1} = 3p_{12,1} p_{11,1} . \tag{18}$$

The requirement (17) is that the mean number infected when three susceptibles are exposed to one infective for the entire infectious period is three times the mean number infected when one susceptibles is exposed to one infective for the entire infectious period. This is true for virtually all transmission models considered, because susceptibles are usually assumed to avoid infection independently of each other. The requirement (18) is that the mean number of cases in the second generation for an outbreak in a household with 4 initial susceptibles, when one of them is infected from outside the household, is 3 times the mean number of cases in the second generation for an outbreak in a household with 3 initial susceptibles. Virtually all transmission models fail to satisfy this requirement.

The conclusion is that the relationship $v_I^* = 1 - 1/R_{30}$, which one might expect to be true in general, is not true for a community of households when households of sizes larger than three are present.

Support by a grant from the Victorian Health Promotion Foundation is gratefully acknowledged.

References

[1] N. T. J. Bailey. *The Mathematical Theory of Infectious Diseases and its Applications*, Griffin, London (1975).

[2] F. Ball, D. Mollison and G. Scalia-Tomba. Epidemics with two levels of mixing. Submitted (1995).

[3] R. Bartoszyński. On a certain model of an epidemic. *Applicationes Mathematicae* XIII(2):139-151 (1972).

[4] N. G. Becker. The use of mathematical models in determining vaccination policies. *Bull Int Statist Inst* **46**, Book 1, 478-490 (1975).

[5] N. G. Becker. *Analysis of Infectious Disease Data*. Chapman and Hall, London (1989).

[6] N. G. Becker and K. Dietz. The effect of the household distribution on transmission and control of highly infectious diseases. *Mathematical Biosciences* **127**, 207-219 (1995).

[7] N. G. Becker and R. Hall. Immunisation levels for preventing epidemics in a community of households made up of individuals of different types. *Mathematical Biosciences*, in press.

[8] K. Dietz. Transmission and control of arboviruses. In: D. Ludwig and K. L. Cooke (eds) *Proceedings of the SIMS Conference on Epidemiology*, Society for Industrial and Applied Mathematics, 122-131 (1975).

[9] C. E. G. Smith. Factors in the transmission of virus infections from animals to man. *Scientific Basis of Medicine. Ann. Rev.* 125-150 (1964).

Modelling the spread of HIV in prisons

J. Gani

Stochastic Analysis Group, CMA

The Australian National University

Canberra   ACT 0200,   Australia

Dedicated to the memory of Ted Hannan, 1921–1994

Abstract

This paper is concerned with models for the spread of HIV in prisons. We first consider a single prison of size N in which there is homogeneous mixing, and where there is an inflow and outflow of $n < N$ prisoners at times $t = 0, 1, 2, \ldots$. Conditions are derived for the stability of the system. The interaction of such a prison with the outside world is then studied, and stability conditions obtained for this case. Finally, a quarantine policy is examined and its cost analyzed, both in the non-stable and stable cases.

Key words: HIV/AIDS, homogeneous mixing, epidemic model, stability, quarantine, cost effectiveness.

Stochastic Analysis Group

Centre for Mathematics and its Applications

The Australian National University

Canberra   A.C.T. 0200

Australia

July 1995

1. Introduction

When the International Conference in Applied Probability and Time Series Analysis, held in Athens between 22 and 26 March 1995, was first planned in 1993, the organizers kindly invited both Ted Hannan and me to present papers on our current research. The Conference was partly designed to review progress in these active fields, and partly to honour Ted's 74th and my 70th birthdays. Sadly, Ted died on 7 January 1994, just 2 weeks short of turning 73. What was to have been a joyful celebration of his contributions to Time Series Analysis became a memorial meeting in his honour. His sudden death cut short a friendship of 41 years: I dedicate this review to his memory, in gratitude for the many happy hours we spent together.

It may be relevant to provide a little background to my interest in epidemic modelling, particularly as my early research under the guidance of Pat Moran between 1953 and 1956 had been on dam theory and inference on Markov chains. A Nuffield Fellowship in 1956-7 enabled me to join Maurice Bartlett at the University of Manchester, and to acquire some familiarity with his epidemic studies (Bartlett, 1949, 1956, 1957). I soon became fascinated by the possibility of using stochastic processes to model biological populations and study the spread of epidemics; in the mid 1960s I began to work actively in epidemic modelling (Gani, 1965, 1967). Thereafter, right up to this last decade, my research efforts have been divided between dam and queueing theory (as for example in Gani, Todorovic and Yakowitz, 1987 and Gani, 1992), the theory of Markov processes (as in Gani and Tin, 1986), biological population theory (as in Brockwell, Ewens, Gani and Resnick, 1987), statistical methods in literature (as in Daley, Gani and Ratkowsky, 1990), and epidemic models (increasingly in recent years, as in Gani, 1991, Gani, Yakowitz and Hayes, 1992, Daley and Gani, 1993, 1994).

In this last area, my current interest lies in qualitative results associated with the HIV/AIDS epidemic. These can be used to compare alternative strategies for the reduction of levels in HIV infection. In this respect, prisons provide a convenient laboratory for testing a range of policies which can be used to curtail the spread of HIV among the prison population. Some of the results presented in this paper are available in fuller detail in Gani, Yakowitz and Blount (GYB 1995), and Yakowitz, Gani and Blount (YGB 1995); our purpose here is to concentrate on the basic ideas, which we shall outline simply, in a deterministic context, with a minimum of mathematical technicalities.

HIV is spread in prisons by sexual contacts and needle sharing among intravenous drug users (Brewer *et al.*, 1988). In its *1992 Update: HIV/AIDS in Correctional Facilities*, the US Department of Justice (1993) reports that there were 11,500 known cases of AIDS in Federal prisons for the five months from November 1992 to May 1993, with an average growth of over 50% annually during the past several years (see Table 1). Thomas and Moerings' book (1994) has documented worldwide concern about the spread of HIV/AIDS in prisons. The question of mandatory HIV testing of prisoners on entry into jail has been raised by Blumberg and Langston (1991), but such a procedure is not currently acceptable. However, voluntary reporting or testing for HIV is possible (Stevens, 1993), and both medical and educational help then becomes available to HIV+ prisoners; Siegal *et al.* (1993) state that HIV+ prisoners who do not report their status do not receive medical help. A model of HIV screening which incorporates the quarantine of infectives has already been considered by Hsieh (1991); we shall later outline a quarantine model designed to minimize overall medical costs.

Table 1

CUMULATIVE TOTAL AIDS CASES AMONG
US CORRECTIONAL INMATES 1985–1992

	No. of cases	Per cent increase from preceding survey
November 1985	766	–
October 1986	1,232	61%
October 1987	1,964	59%
October 1988	3,136	60%
October 1989	5,411	72%
October 1990	6,985	29%
November 1992	11,565	66%

From Thomas and Moerings (1994) p.139.

Although we shall deal only with deterministic models, their stochastic equivalents can also be analyzed in detail, as they have in the two previously mentioned papers (GYB 1995, YGB 1995). Since these have yet to appear, it may be useful to give a brief summary of their contents.

The first paper (GYB 1995) compares a deterministic single prison model, as given in Section 2 below, with its stochastic analogue in which infection is spread binomially. The stochastic mean for the number of infectives is shown to be identical with the deterministic result. The model is generalized to a two-prison system, also with an exchange of prisoners. Finally, the effect and cost of a screening and quarantine policy carried out over a fixed horizon of T time units are analyzed; the results are summarized in Example 4.1 below.

The second paper (YGB 1995) is concerned with the case of $m > 2$ prisons, with exchanges of prisoners. Both deterministic and stochastic models are considered, and a simplified method for computing the marginal probabilities of the numbers of infectives in the prisons is outlined.

The present paper differs from these in the following ways:

(a) it considers only deterministic models for a single isolated prison, and for a prison interacting with the outside world;

(b) it obtains stable solutions to the equations for the number of prison infectives, and gives the conditions under which they are valid;

(c) it discusses a screening and quarantine procedure under such conditions of stability; for this, the overall medical costs at any time t. including those of treating undiagnosed HIV+ prisoners, are minimized.

2. The single prison model

Suppose that a prison consisting of N inmates, allows a simultaneous inflow and outflow of $n < N$ prisoners at time t ($t = 0, 1, 2, \ldots$), after which there are $y(t)$ prisoners who are HIV+ and $N - y(t)$ susceptibles. During the interval $(t, t+1)$, which we take to be relatively short so that $y(t)$ does not vary greatly in it, we assume that homogeneous mixing occurs with an infection rate β, so that $\beta y(t)(N - y(t))$ new infectives are produced. This is a discrete time approximation based on the continuous time simple epidemic; for details, the reader is referred to Bailey (1975), Chapter 5. Thus at time $t + 1$, there are

$$z(t + 1) = y(t) + \beta y(t)(N - y(t)) \tag{2.1}$$

infectives.

An outflow of n prisoners leaves the prison at time $t + 1$, among them

$$v(t + 1) = \frac{n}{N} z(t + 1) = \frac{n}{N} y(t)(1 + \beta N - \beta y(t)) \tag{2.2}$$

infectives, leaving behind $(1 - n/N)\, y(t)(1 + \beta N - \beta y(t))$ infectives.

At the same time, an inflow of n new prisoners enters the prison, with a proportion of HIV+ individuals $0 \le \mu(t + 1) \le 1$, so that

$$u(t + 1) = n\mu(t + 1) \tag{2.3}$$

infectives are added. Thus

$$y(t + 1) = n\mu(t + 1) + y(t)(1 + \beta N - \beta y(t))\left(1 - \frac{n}{N}\right) \tag{2.4}$$

expresses the number of HIV+ prisoners $y(t + 1)$ at time $t + 1$ in terms of $y(t)$ at time t and $\mu(t + 1)$.

Assuming that we start at $t = 0$ with $y(0) > 0$ infectives, one can ask under what conditions stability of the number of HIV+ prisoners can be achieved. Let us first consider the case $\mu(t) = 0$ where the only infectives are in prison; then

$$y(t + 1) = y(t)(1 + \beta N - 3y(t))\left(1 - \frac{n}{N}\right) \tag{2.5}$$

and if y_s is the stable value, then

$$y_s = y_s(1 + \beta N - \beta y_s)\left(1 - \frac{n}{N}\right) = f(y_s). \tag{2.6}$$

Note that this is equivalent to stating that $\beta y_s(N - y_s) = n(y_s + \beta y_s(N - y_s))/N$, or the increase in prison infectives equals the number of departing infectives. We

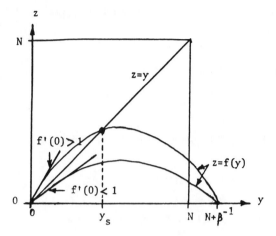

Figure 1. The graphs $z = y$ and $z = f(y)$ for $f'(0) \gtrless 1$.

see from the graphs in Figure 1 that two cases are possible, depending on the slope $f'(0)$ at $y = 0$.

Now $f'(0) = (1 + \beta N)\left(1 - \frac{n}{N}\right)$ can be greater than, equal to or smaller than 1 depending on whether

$$(1 + \beta N)\left(1 - \frac{n}{N}\right) = 1 - \frac{n}{N} + \beta(N - n) \gtrless 1$$

or

$$\beta \gtrless \frac{n}{N(N - n)}. \tag{2.7}$$

If $f'(0) \leq 1$, the stable value is $y_s = 0$, while if $f'(0) > 1$, it is readily seen from (2.6) that $y_s = N - \beta^{-1} n(N - n)^{-1} > 0$, a positive value smaller than N. These cases are illustrated in Figure 1.

If $\mu(t)$ is a positive constant $\mu > 0$, then the slope of $f(y)$ at $y = 0$ is no longer relevant as Figure 2 shows. For the stable value y_s is now given by the positive solution of

$$y_s = n\mu + y_s(1 + \beta N - \beta y_s)\left(1 - \frac{n}{N}\right) = n\mu + f(y_s),$$

or

$$n\mu + y_s\left(\beta(N - n) - \frac{n}{N}\right) - \beta\left(1 - \frac{n}{N}\right)y_s^2 = 0. \tag{2.8}$$

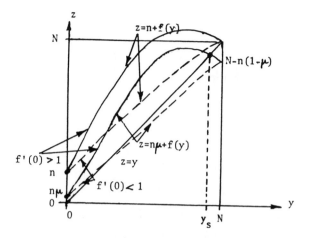

Figure 2. The graphs $z = y$ and $z = n\mu + f(y)$ for $\mu > 0$ when $f'(0) > 1$ $(-)$, and $f'(0) < 1$ $(--)$.

This is

$$y_s = \tfrac{1}{2} \left\{ N - \beta^{-1} n(N-n)^{-1} + \sqrt{\left[N - \beta^{-1} n(N-n)^{-1}\right]^2 + 4nN\mu\beta^{-1}(N-n)^{-1}} \right\}$$
(2.9)

which is clearly smaller than N when $\mu < 1$, as can be seen from Figure 2, since $n\mu + f(N) = N - n(1 - \mu) < N$. When $\mu = 1$, $y_s = N$.

Example 2.1: We illustrate these results with an example in which $N = 500$, $n = 25$ and $\beta = 0.01$ and 0.0001 respectively. For the case $\mu = 0$, the slope at $y = 0$ is

$$f'(0) = (1 + \beta N)\left(1 - \frac{n}{N}\right) = 5.7 > 1, \qquad \text{when } \beta = 0.01$$
$$= 0.9975 < 1, \text{ when } \beta = 0.0001.$$

Thus for $\beta = 0.01$ the stable value for the number of HIV+ prisoners will be

$$y_s = N - \beta^{-1} n(N - n)^{-1} = 494.74$$

while for $\beta = 0.0001$, it will be $y_s = 0$.

For the case $0 < \mu < 1$, setting $\mu = 0.01$, from (2.9) we obtain that for $\beta = 0.01$

$$y_s = 494.79$$

while for $\beta = 0.0001$

$$y_s = 39.8\,.$$

Taking $\mu = 0.01$, it can readily be shown that for $\beta = 0.1, y_s = 499.48$, while for $\beta = 0.00001, y_s = 5.52$. In general, as β decreases, we find that y_s ranges from a value close to N, to one close to 0.

3. Interaction of a prison with the outside world

Let us consider a city of fixed population size M with $Y(t)$ infectives at time $t = 0, 1, 2, \ldots$, in which there is a single prison of size N with $y(t)$ infectives. For the prison, we shall make the same assumptions as before; there is homogeneous mixing in the relatively short interval $(t, t + 1)$ so that

$$\beta y(t)(N - y(t))$$

new infectives are produced. At $t + 1$, $n < N$ prisoners leave the prison of whom

$$\frac{n}{N}\, y(t)(1 + \beta(N - y(t))) \tag{3.1}$$

are infective, leaving $\left(1 - \frac{n}{N}\right) y(t)(1 + \beta(N - y(t)))$ infectives behind. The departing prisoners are replaced by n new inmates from the city with infection rate $\mu(t + 1)$, depending on the number of city infectives.

For the city, in the interval $(t, t + 1)$, homogeneous mixing produces

$$\beta Y(t)(M - Y(t))$$

new infectives. There will, however, be some deaths at the rate $0 < \gamma < 1$ so that at time $t + 1$, assuming that deaths are replaced by healthy new births, the number of infectives will be

$$Y(t)(1 - \gamma + \beta(M - Y(t)))\,, \tag{3.2}$$

and the infection rate

$$\mu(t + 1) = \frac{Y(t)}{M}\,(1 - \gamma + \beta(M - Y(t)))\,. \tag{3.3}$$

We can now write the following equations for $Y(t+1)$ and $y(t+1)$:

$$Y(t+1) = \left(1 - \frac{n}{M}\right) Y(t)(1 - \gamma + \beta(M - Y(t))) + \frac{n}{N} y(t)(1 + \beta(N - y(t)))$$

$$y(t+1) = \frac{n}{M} Y(t)(1 - \gamma + \beta(M - Y(t))) + \left(1 - \frac{n}{N}\right) y(t)(1 + \beta(N - y(t))).$$

$$(3.4)$$

Such a system can also achieve stability. Suppose the stable values of $Y(t)$, $y(t)$ are $Y_s > 0$, $y_s > 0$; these will then satisfy the matrix

$$\begin{bmatrix} Y_s \\ y_s \end{bmatrix} = \begin{bmatrix} \left(1 - \frac{n}{M}\right)(1 - \gamma + \beta(M - Y_s)) & \frac{n}{N}(1 + \beta(N - y_s)) \\ \frac{n}{M}(1 - \gamma + \beta(M - Y_s)) & \left(1 - \frac{n}{N}\right)(1 + \beta(N - y_s)) \end{bmatrix} \begin{bmatrix} Y_s \\ y_s \end{bmatrix}.$$

$$(3.5)$$

Clearly, since $0 \leq y_s \leq N$, then if $N < \beta^{-1}$ as it usually is, $n(1 + \beta N - \beta y_s) y_s / N$ is monotonic increasing in the range, and

$$M - \rho - \eta \leq Y_s \leq \frac{1}{2}\left\{(M - \rho - \eta) + \sqrt{(M - \rho - \eta)^2 + 4M\eta}\right\},$$

$$\rho = \frac{\gamma}{\beta}, \qquad \eta = \frac{n}{\beta(M - n)},$$

depending on whether the prison output of infectives into the general population is 0 or n. We assume for simplicity that $M > \rho$.

Example 3.1: If, for example, we take

$M = 50,000$	$\gamma = 0.03$	
	$\frac{n}{M} = 0.002$	$\eta = 2004.008$
$N = 1,000$	$\beta = 10^{-6}$	
	$\frac{n}{N} = 0.1$	$M - \rho - \eta = 17,995.992$
$n = 100$	$\rho = 30.000$	

then, it follows that $17,996 \leq Y_s \leq 22,458$, avoiding decimals. Let us take as a first guess for Y_s the value $Y_0 = 20,000$, roughly intermediate between the two bounds. Then, since from (3.5)

$$\begin{bmatrix} \left(1 - \frac{n}{M}\right)(1 - \gamma + \beta(M - Y_s)) - 1 & \frac{n}{N}(1 + \beta(N - y_s)) \\ \frac{n}{M}(1 - \gamma + \beta(M - Y_s)) & \left(1 - \frac{n}{N}\right)(1 + \beta(N - y_s)) - 1 \end{bmatrix} \begin{bmatrix} Y_s \\ y_s \end{bmatrix} = 0,$$

we have

$$\begin{vmatrix} \left(1 - \frac{n}{M}\right)(1 - \gamma + \beta(M - Y_s)) - 1 & \frac{n}{N}(1 + \beta(N - y_s)) \\ \frac{n}{M}(1 - \gamma + \beta(M - Y_s)) & \left(1 - \frac{n}{N}\right)(1 + \beta(N - y_s)) - 1 \end{vmatrix} = 0, \quad (3.6)$$

and also

$$\left[\left(1 - \frac{n}{M}\right)(1 - \gamma + \beta(M - Y_s)) - 1\right]Y_s + \left[\frac{n}{N}(1 + \beta(N - y_s))\right]y_s = 0. \quad (3.7)$$

Thus, for our example, (3.6) gives us on writing $X_s = M - Y_s$, $x_s = N - y_s$, the relation

$$\left(-.03194 + 998 \times 10^{-9} X_s\right)\left(-.1 + 9 \times 10^{-7} x_s\right) - .0002\left(.97 + 10^{-6} X_s\right)\left(1 + 10^{-6} x_s\right) = 0 \quad (3.8)$$

and (3.7) yields

$$\left(-.03194 + 998 \times 10^{-9} X_s\right)Y_s + .1\left(1 + 10^{-6} x_s\right)y_s = 0, \quad (3.9)$$

so that one can in theory solve these equations for Y_s, y_s exactly.

It is easier, however, to adopt a recursive method of solution based on (3.4) and (3.8). Thus if we start with some initial number of infectives in the outside world, say $Y_0 = 20,000$, then from (3.9) we obtain for x_0, the relation

$$(-.03194 + .02994)\,20,000 + .1\left(1 + 10^{-6} x_0\right)\left(10^3 - x_0\right) = 0,$$

or $x_0 = 600.2399$ (600 omitting the decimals), so that $y_0 = 400$ in the prison. Using (3.4), we find after several recursions (284 to be precise) that

$$Y_s = 20,012, \qquad y_s = 402.399 \quad (3.10)$$

the convergence being rather slow.

Clearly, the parameters γ and β must satisfy certain conditions for stability, and for $Y_s > 0$, $y_s > 0$, it is readily seen from (3.6) that

$$\begin{vmatrix} \left(1 - \frac{n}{M}\right)(1 - \gamma + \beta X_s) - 1 & \frac{n}{N}(1 + \beta x_s) \\ \frac{n}{M}(1 - \gamma + \beta X_s) & \left(1 - \frac{n}{N}\right)(1 + \beta x_s) - 1 \end{vmatrix} = 0, \quad (3.11)$$

which, after some algebra yields

$$X_s = \rho - \frac{x_s \frac{n}{M}}{\frac{n}{N} - \beta x_s\left(1 - \frac{n}{M} - \frac{n}{N}\right)}. \quad (3.12)$$

We may assume what is usually the case, namely that n/M, n/N are small and $(1 - n/M - n/N) > 0$. In theory, two possibilities may arise:

(a) $n/N > \beta x_s(1 - n/M - n/N)$,

(b) $n/N < \beta x_s(1 - n/M - n/N)$.

We dismiss the case where $n/N = \beta x_s(1 - n/M - n/N)$.

In the case (a), since $x_s \leq N$, the inequality will certainly hold if

$$\frac{n}{N} > \beta N\left(1 - \frac{n}{M} - \frac{n}{N}\right)$$

or

$$\beta < \frac{nM}{N[MN - n(M + N)]}.\qquad(3.13)$$

In our example (3.1), this would require

$$\beta < \frac{100 \times 50,000}{1000[50,000,000 - 100(51,000)]} = 1.1136 \times 10^{-4}.$$

Thus our $\beta = 10^{-6}$ satisfies this condition.

If the case (b) were possible, then from (3.12) we have

$$Y_s = M - \rho - \frac{x_s \frac{n}{M}}{\beta x_s\left(1 - \frac{n}{M} - \frac{n}{N}\right) - \frac{n}{N}}\qquad(3.14)$$

and from the inequality below (3.5), we see that this would imply

$$\eta = \frac{n}{\beta(M - n)} \geq \frac{x_s \frac{n}{M}}{\beta x_s\left(1 - \frac{n}{M} - \frac{n}{N}\right) - \frac{n}{N}}.$$

But this is equivalent to

$$\beta x_s\left(1 - \frac{n}{M} - \frac{n}{N}\right) - \frac{n}{N} \geq 3 x_s\left(1 - \frac{n}{M}\right),$$

which is not possible.

4. A quarantine policy in prison

Gani, Yakowitz and Blount (1995) have studied a cost effective model for a prison in which prisoners may be screened regularly, and quarantined if found to be HIV+. We follow the single prison model of Section 2 with the difference that at time $t = 0, 1, 2, \ldots$, the prison population of size N is now subdivided into three groups: susceptibles $x(t)$, infectives not in quarantine $y(t)$, and quarantined prisoners $q(t)$, where

$$x(t) + y(t) + q(t) = N.\qquad(4.1)$$

There is homogeneous mixing between susceptibles and infectives not in quarantine between times t and $t + 1$, so that at $t + 1$ there are

$$z(t + 1) = y(t) + \beta x(t)\, y(t) = y(t)(1 + \beta x(t)) \tag{4.2}$$

infectives.

Suppose that a new inflow n of prisoners joins the prison at time $t + 1$, with a proportion $0 \le \mu(t + 1) \le 1$ of HIV+ individuals, so that

$$u(t + 1) = n\mu(t + 1) \tag{4.3}$$

infectives are added. At the same time, n prisoners leave, among them

$$v(t + 1) = \frac{n}{N}\, z(t + 1), \qquad w(t + 1) = \frac{n}{N}\, q(t) \tag{4.4}$$

non-quarantined and quarantined infectives. Thus, before a new screening and quarantine of the prisoners, the total number of non-quarantined infectives at $t + 1$ is

$$z(t + 1) + u(t + 1) - v(t + 1) = y(t)(1 + \beta x(t))\left(1 - \frac{n}{N}\right) + n\mu(t + 1) \tag{4.5}$$

while there are

$$q(t) - w(t + 1) = \left(1 - \frac{n}{N}\right) q(t) \tag{4.6}$$

quarantined infectives.

Let us now screen a proportion $0 \le \tau \le 1$ of the non-quarantined prison population, and quarantine those who test HIV+, leaving only

$$y(t + 1) = (1 - \tau)\left(n\mu(t + 1) + y(t)(1 + \beta x(t))\left(1 - \frac{n}{N}\right)\right) \tag{4.7}$$

non-quarantined infectives,

$$q(t + 1) = q(t)\left(1 - \frac{n}{N}\right) + \tau\left(n\mu(t + 1) + y(t)(1 + \beta x(t))\left(1 - \frac{n}{N}\right)\right) \tag{4.8}$$

quarantined infectives, and

$$x(t + 1) = N - y(t + 1) - q(t + 1). \tag{4.9}$$

susceptibles.

Suppose the costs involved per prisoner in such screening, quarantining and medical treatment of prisoners are:

a for the HIV test ,

b for quarantining ,

c for treating an infective, even if HIV is not confirmed .

Then the cost $C(t)$ at time t will be

$$C(t) = a\tau(N - q(t)) + bq(t) + cy(t).$$ (4.10)

The total cost over a time horizon of T units will be

$$J(\tau, T) = \sum_{t=1}^{T} C(t).$$ (4.11)

The following example was given in Gani, Yakowitz and Blount (1995).

Example 4.1: If we assume $b > c > a$, we may expect $\tau < 1$ for minimum $J(\tau, T)$, and the following simple example confirms this. Suppose we set the costs at

$$a = 0.1, \qquad b = 1.1, \qquad c = 1$$

and let $\mu = 0.1$ a constant, $\beta = 3 \times 10^{-4}$, $n = 10$, $N = 500$ with the initial number of infectives $y_0 = 100$. We allow τ to take different values between 0 and 1, and calculate $J(\tau, T)$ for $T = 50$. The graph in Figure 3 indicates the values of this total cost, which has a minimum at $\tau = 0.525$; we note that beyond this level, additional screening does not yield savings, possibly because further infection is now relatively small.

We can also find the cost of screening when stability is reached, that is when (for a fixed μ) the stable values y_s and q_s are achieved, satisfying

$$y_s = (1 - \tau)\left(n\mu + y_s(1 + \beta x_s)\left(1 - \frac{n}{N}\right)\right),$$

$$q_s = q_s\left(1 - \frac{n}{N}\right) + \tau\left(n\mu + y_s(1 + \beta x_s)\left(1 - \frac{n}{N}\right)\right),$$ (4.12)

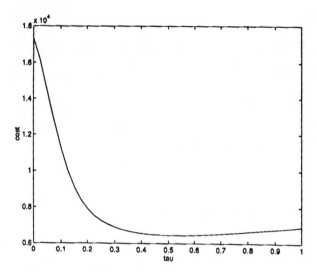

Figure 3. Cost $J(\tau, 50)$ as a function of τ.

where $x_s = N - y_s - q_s$. We can readily see that

$$q_s = \frac{N}{n}\tau\left(n\mu + y_s(1 + \beta x_s)\left(1 - \frac{n}{N}\right)\right)$$

$$= \begin{cases} \frac{N\tau}{n(1-\tau)}\,y_s & , \quad 0 < \tau < 1 , \\ N\mu & , \quad \tau = 1 , \end{cases} \tag{4.13}$$

so that the first equation of (4.12) yields for $0 < \tau < 1$,

$$y_s^2 + \frac{1 - (1-\tau)\left(1 - \frac{n}{N}\right)(1 + \beta N)}{\beta\left(1 - \frac{n}{N}\right)\left(\frac{(N-n)\tau+n}{n}\right)}\,y_s - \frac{(1-\tau)n\mu}{\beta\left(1 - \frac{n}{N}\right)\left(\frac{(N-n)\tau+n}{n}\right)} = 0 , \tag{4.14}$$

and $y_s = 0$ for $\tau = 1$, since complete screening means that all infectives are quarantined. Here $(1-\tau)\left(1 - \frac{n}{N}\right)(1 + \beta N)$ may be greater, less than or equal to 1.

The non-negative solution of (4.14) is

$$y_s = \frac{1}{2}\left\{\left[\frac{(1-\tau)\left(1 - \frac{n}{N}\right)(1 + \beta N) - 1}{\beta\left(1 - \frac{n}{N}\right)\left(\frac{(N-n)\tau+n}{n}\right)}\right] \right.$$

$$\left. + \sqrt{\left[\frac{(1-\tau)\left(1 - \frac{n}{N}\right)(1 + \beta N) - 1}{\beta\left(1 - \frac{n}{N}\right)\left(\frac{(N-n)\tau+n}{n}\right)}\right]^2 + \frac{4n\mu(1-\tau)}{\beta\left(1 - \frac{n}{N}\right)\left(\frac{(N-n)\tau+n}{n}\right)}}\right\} \tag{4.15}$$

with

$$q_s = \frac{N\tau}{n(1-\tau)}\,y_s , \qquad x_s = N - y_s\left(\frac{(N-n)\tau+n}{n(1-\tau)}\right) , \qquad 0 < \tau < 1 , \tag{4.16}$$

and $y_s = 0$, $q_s = N\mu$, $x_s = N(1 - \mu)$ when $\tau = 1$.

Example 4.2: We reconsider the case of Example 4.1, with $\mu = 0.1$, $\beta = 3 \times 10^{-4}$, $n = 10$ and $N = 500$. Then for $0 < \tau < 1$,

$$y_s = \frac{1}{2} \left\{ \left[\frac{(1 - \tau)(1.127) - 1}{2.94 \times 10^{-4}(49\tau + 1)} \right] \right.$$

$$\left. + \sqrt{\left[\frac{(1 - \tau)(1.127) - 1}{2.94 \times 10^{-4}(49\tau + 1)} \right]^2 + \frac{4(1 - \tau)}{2.94 \times 10^{-4}(49\tau + 1)}} \right\}$$

$$q_s = \frac{50\tau}{1 - \tau} y_s .$$

Calculating y_s and q_s for different τ, we find

τ	0	.125	.25	.4	.425	.45	.5	.525	.75	1
y_s	439.71	17.39	4.37	1.79	1.58	1.41	1.12	1.01	0.346	0
q_s	0	124.21	72.83	59.76	58.39	57.68	56.19	55.56	51.94	50

Note the rapid decline of y_s, even for a screening proportion of $\tau = 0.125$.

For these values, the cost $C(t)$ of (4.10) can be calculated for $a = 0.1$, $b = 1.1$ and $c = 1$ as before to give

$$C(t) = .1\,\tau(500 - q_s) + 1.1\,q_s + y_s = 50\,\tau + (1.1 - .1\,\tau)\,q_s + y_s$$

which yields the following values for different values of τ:

τ	0	.125	.25	.4	.425	.45	.5	.525	.75	1
$C(t)$	439.71	158.71	95.16	85.14	84.58	84.76	85.12	85.46	91.08	100

The time horizon T of (4.11) is no longer relevant, since $C(t)$ remains constant for all t. We see that in the case where stability is reached, a screening rate of $\tau = 0.425$ leads to optimal cost effectiveness, rather than $\tau = 0.525$ as in example 4.1.

Such elementary calculations enable one to formulate a cost effective screening and quarantine policy on the basis of real data. While the values of μ, β, n, and N and the costs a, b, c may vary, the principles of cost calculation and its optimization will always result in an estimate of τ which will keep costs to a minimum. Thus, a realistic policy can be developed to reduce the spread of AIDS in prisons.

Acknowledgement. This research was carried out with the partial support of NIH Grant R01–AI29426–01. My thanks are due to Dr. Z. Leyk for his calculation of Y_s and y_s in (3.10).

References

Bailey, N.T.J. (1975) *The Mathematical Theory of Infectious Diseases and its Applications.* Griffin, London.

Bartlett, M.S. (1949) Some evolutionary stochastic processes. *J. Roy. Statist. Soc.* B11, 211–229.

Bartlett, M.S. (1956) Deterministic and stochastic models for recurrent epidemics. *Proc. 3rd Berkeley Symp. Math. Statist. Prob.* 4, 81–109. U. of California Press, Berkeley.

Bartlett, M.S. (1957) Measles periodicity and community size. *J. Roy. Statist. Soc.* A120, 48–70.

Blumberg, M. and Langston, D. (1991) Mandatory HIV testing in criminal justice settings. *Crime and Delinquency* 37, 5–18.

Brewer, T.F., Vlahov, D., Taylor, E., Hall, D., Munoz, A. and Polk, F. (1988) Transmission of HIV1 within a statewide prison system. *AIDS* 2, 363–367.

Brockwell, P.J., Ewens, W.J., Gani, J. and Resnick, S.I. (1987) Minimum viable population size in the presence of catastrophes. In *Viable Populations for Conservation,* M.E. Soule (Editor), 59–68. Cambridge U. Press.

Daley, D.J., Gani, J. and Ratkowsky, D.A. (1990) Markov chain models for type-token relationships. In *Statistical Inference in Stochastic Processes,* N.U. Prabhu and I.V. Basawa (Editors), 209–232. Marcel Dekker, New York.

Daley, D.J. and Gani, J. (1993) A random allocation model for carrier-borne epidemics. *J. Appl. Prob.* 30, 751–765.

Daley, D.J. and Gani, J. (1994) A deterministic general epidemic model in a stratified population. In *Probability, Statistics and Optimization: A Tribute to Peter Whittle*, F.P. Kelly (Editor) 117–132. John Wiley, Chichester.

Gani, J. (1965) On a partial differential equation of epidemic theory I. *Biometrika* 52, 617–622.

Gani, J. (1967) On the general stochastic epidemic. *Proc. 5th Berkeley Symp. Meth. Statist. Prob.* 4, 271–279. U. of California Press, Berkeley.

Gani, J. (1991) A carrier-borne epidemic with two stages of infection. *Stoch. Models* 7, 83–105.

Gani, J. (1992) Random allocation in a waiting room problem. In *Queueing and Related Models*, I.V. Basawa and U.N. Bhat (Editors), 34–46. Oxford U. Press.

Gani, J. and Tin, P. (1986) A class of transformed Markov processes. *Stoch. Anal. Applns* 4, 77–86.

Gani, J., Todorovic, P. and Yakowitz, S. (1987) Silting of dams by sedimentary particles. *Math. Scientist* 12, 81–90.

Gani, J., Yakowitz, S. and Hayes, R. (1992) Automatic learning for dynamic Markov fields with application to epidemiology. *Operat. Res.* 40, 867–876.

Gani, J., Yakowitz, S. and Blount, M. (1995) The spread and quarantine of HIV infection in a prison system. *Submitted for publication.*

Hsieh, Y.-H. (1991) Modelling the effects of screening in HIV transmission dynamics. In *Differential Equations Models in Biology. Epidemiology and Ecology.* Lecture Notes in Biomathematics 92, Busenberg. S. and Martelli, M. (Editors). Springer–Verlag, Berlin.

Siegal, H.A., Carlson, R.G., Falck, R., Reece, R.D. and Perlin, T. (1993) Conduction HIV outreach and research among incarcerated drug abusers. *J. Substance Abuse Treatment* 10, 71–75.

Stevens, S. (1993) HIV prevention programs in a jail setting: educational strategies.

The Prison Journal 73, 379–390.

Thomas, P.A. and Moerings, M. (Editors) (1994) *AIDS in Prison*. Dartmouth, Brookfield.

U.S. Department of Justice (1993) *1992 Update: HIV/AIDS in Correctional Facilities*. US Department of Justice, Washington DC.

Yakowitz, S., Gani, J. and Blount, M. (1995) Computing marginal probabilities for large compartmentalized models, with application to AIDS evolution in a prison system. *Submitted for publication.*

An Algorithmic Study of S-I-R Stochastic Epidemic Models

Marcel F. NEUTS
Jian-Min LI

ABSTRACT Algorithms for the *S-I-R* epidemic with an initial population with m infectives and n susceptibles are examined. We propose efficient algorithms for the distributions of the total and the maximum size of the epidemic, and for the joint distribution of the maximum and the time of its occurrence. We also discuss the joint distribution of the sizes of the epidemic at two epochs. By studying the Markov chain describing an indefinite replication of the epidemic, we obtain new descriptors of the process of infections.

Key Words: Algorithmic probability, stochastic epidemics, maximum size of epidemic, total size of epidemic.

1 Introduction

This paper deals with algorithmic aspects of some stochastic epidemic models to which Professor Gani's research has extensively contributed, see, for example, Gani and Purdue [GANI84]. These models are known as *S-I-R* (susceptible $\rightarrow$ infected $\rightarrow$ removed) epidemics. We show how a stochastic model can be examined by interactive use of algorithms of varying degrees of complexity.

Our overall approach is to explore the epidemic model, first by simpler, faster algorithms and to draw qualitative conclusions from their numerical results. Subsequently, more time consuming algorithms are implemented to yield more detailed information about the epidemic.

Initially the population consists of m infectives and n susceptibles. For a generic time t, the numbers of infected and susceptible individuals are denoted by i and j respectively. When in state (i, j), $i > 0$, $j > 0$, the process either moves to $(i + 1, j - 1)$ (an infection) at rate $\lambda_{i,j}$, or to $(i - 1, j)$ (removal of an infective) at rate μ_i. In the states $(i, 0)$, $i > 0$, only a removal can occur, and $(0, j)$, for $j \geq 0$ are absorbing states. For the familiar *general stochastic epidemic* $\lambda_{i,j} = \beta i j$, $\mu_i = \eta i$. Our examples deal with two parameter choices, designated as *Model-1* and *Model-2*. These are:

1. *Model-1*: $\lambda_{i,j} = \beta i^\alpha j, \mu_i = \eta i$. The parameter α, $0 < \alpha \leq 1$, models

the degree of interaction between infectives and susceptibles. With a smaller α, there is less exposure of susceptibles to the infectives.

2. *Model-2*: $\lambda_{i,j} = \beta i \min\{j, \epsilon n\}, \mu_i = \eta i$. The quantity ϵ is the fraction of the susceptible population that is exposed to each infective.

Clearly $\lambda_{i,j}$ and μ_i can be specified in infinitely many ways. For example Saunders [SAUN80] showed that $\lambda_{i,j} = ij/(i+j)^{\frac{1}{2}}$ is better suited for modeling the transmission of myxomatosis among rabbits. We propose flexible algorithms valid for arbitrary specifications, and Models 1 and 2 serve to illustrate the algorithms. The continuous time Markov chain describing the epidemic has the structured generator Q:

$$\begin{pmatrix} A(n) & B(n) & & & & & & T(n) & & & \\ & A(n-1) & B(n-1) & & & & & & T(n-1) & & \\ & & \ddots & \ddots & & & & & & \ddots & \\ & & & \ddots & B(1) & & & & & & \\ & & & & A(0) & & & & & T(0) \\ 0 & 0 & \cdots & \cdots & 0 & 0 & 0 & \cdots & 0 \end{pmatrix}.$$

The blocks $A(k)$ and $B(k)$ correspond to the states with k *susceptibles* present. In the matrix $A(k)$ of order $m + n - k$, all elements not on the diagonal or the superdiagonal are zero. The elements on the superdiagonal are $\mu_{m+n-k}, \mu_{m+n-k-1}, \cdots, \mu_2$. The diagonal elements are such that the row sums of Q vanish; they are:

$$-\lambda_{m+n-j,j} - \mu_{m+n-j}, \cdots, -\lambda_{1,j} - \mu_1.$$

The $(m+n-k) \times (m+n-k+1)$ matrices $B(k)$ have diagonal elements $\lambda_{m+n-k,k}, \cdots, \lambda_{1,k}$. All other elements are zero. The $T(k)$ are column vectors of dimension $m+n-k$. Only their last component differs from zero and equals μ_1. With reference to the discussion in Section 1.2, we call that Markov chain the *ordinary epidemic*.

2   Visualization and the Replicated Epidemic

2.1   *Visualizing the Ordinary Epidemic*

Computer visualization of the behavior of a stochastic model gives insight into its most informative mathematical descriptors. It helps in formulating conjectures and in identifying parameter ranges for which interesting phenomena occur. Its educational value is well-recognized. For some materials on visualization of stochastic models see Chapter 6 of Neuts [NEUT95].

To see the effect of the parameters α and ϵ, we ran the simulation for various parameter values. For each choice, many paths are superimposed in

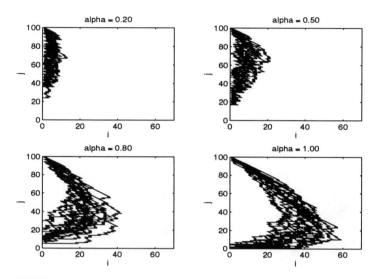

FIGURE 1. Model-1: Visualization, with $m = 1$, $n = 100$, $\eta/\beta = 25$, sample size $= 25$

the same graph. With the numbers of infectives and susceptibles as abscissa and ordinate, a path starts at the point (m, n) and ends at a point $(0, j)$ with $0 \leq j \leq n$. Steps to the southeast and the west respectively correspond to infections and removals. Examples of simulated path functions are shown in Figures 1 and 2. These suggest the following observations:

1. In general, the larger α or ϵ, the more susceptibles eventually get infected. That corresponds to our original motivation for introducing these parameters.

2. With α small, there is much variability in the number of susceptibles who get infected eventually. That is confirmed by the flatness of the computed total size distributions.

3. If ϵ is gradually increased, the total size increases dramatically in a narrow range of ϵ−values.

2.2 The Replicated Epidemic

Some descriptors of the epidemic, notably of the point process of the *times of infections*, can be defined by considering an irreducible Markov chain constructed from the ordinary epidemic. We describe that construction in

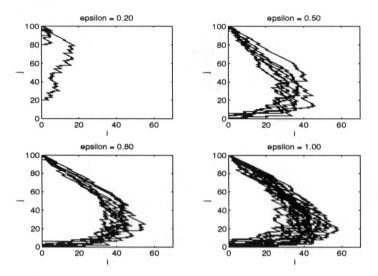

FIGURE 2. Model-2: Visualization, with $m = 1$, $n = 100$, $\eta/\beta = 25$, sample size $= 25$

general for absorbing Markov chains with one absorbing state and finitely many transient states. Consider a Markov chain with generator

$$Q_1 \;=\; \begin{pmatrix} L & \mathbf{L}^0 \\ \mathbf{0} & 0 \end{pmatrix}$$

and the initial probability vector $(\boldsymbol{\gamma}, 0)$. The matrix L is nonsingular. The time until absorption τ has the $PH-$distribution $F(\cdot)$ with representation $(\boldsymbol{\gamma}, L)$, see Chapter 2 of Neuts [NEUT81]. The $PH-$renewal process with distribution $F(\cdot)$ is obtained by instantaneously restarting the absorbing Markov chain upon each absorption, according to the same initial vector $\boldsymbol{\gamma}$. We assume, without loss of generality, that the generator $Q_1^0 \;=\; L + \mathbf{L}^0 \boldsymbol{\gamma}$ is irreducible.

For the epidemic model, that construction lumps all absorbing states into one and at each extinction restarts the epidemic in the same state (m, n). The matrix Q_1^0 is obtained by adding the vector $[\mathbf{T}(n), \; ..., \; \mathbf{T}(0)]$ to the first column of Q and deleting all rows and columns corresponding to absorbing states. We call that Markov chain the *replicated epidemic*.

The stationary probability vector $\boldsymbol{\theta}^0$ of Q_1^0 is $\boldsymbol{\theta}^0 \;=\; (\mu_1')^{-1}\boldsymbol{\gamma}(-L)^{-1}$, where μ_1' is the mean of $F(\cdot)$. The vector $\boldsymbol{\theta}^0$ partitioned as $[\boldsymbol{\theta}^0(n), \; ..., \boldsymbol{\theta}^0(1)]$, is computed by solving the equations

$$\boldsymbol{\theta}^0(n)A(n) \;=\; [-\mu', 0, ..., 0],$$

$$\theta^0(j+1)B(j+1) + \theta^0(j)A(j) = 0, \quad \text{for } 0 \leq j \leq n-1.$$

With θ^0 as the initial probability vector, we obtain the stationary version of the replicated epidemic. If we mark the epochs of new infections in the replicated epidemic, that point process is a stationary MAP with coefficient matrices D_0 and D_1, see Lucantoni [LUCA93] and Neuts [NEUT92]. D_0 is obtained by replacing all matrices $B(j)$ in Q_1^0 by zero matrices. D_1 has the $B(j)$ in the same locations as in Q_1^0. The rate λ_{infect} of infections in the replicated epidemic is given by

$$\lambda_{infect} = \theta^0 D_1 \mathbf{e} = \sum_{j=1}^{n} \theta^0(j)B(j)\mathbf{e}.$$

By using the special structure of the matrices $B(j)$, the quantity λ_{infect} is easily computed. Its significance is clear from the following theorem.

Theorem 2.1 *The expected number of infections up to extinction in the ordinary epidemic is equal to $\lambda_{infect}\,\mu_1'$.*

Proof. The elementary probability of an infection in $(u, \ u+du)$ is $\gamma exp(Lu)D_1\mathbf{e}du$. Hence, the expected number of infections during the ordinary epidemic is

$$\int_0^{\infty} \gamma exp(Lu)D_1 \mathbf{e}du = \gamma(-L)^{-1}D_1\mathbf{e} = \mu_1'\theta^0 D_1\mathbf{e} = \lambda_{infect}\,\mu_1'.$$

∎

The Laplace-Stieltjes transform of the time between two successive infections in the stationary replicated epidemic is

$$\Psi_{infect}(s) = [\lambda_{infect}]^{-1}\theta^0 D_1(sI - D_0)^{-1}D_1\mathbf{e}.$$

The time between an arbitrary infection and the next, therefore has a PH−distribution with representation $\{[\lambda_{infect}]^{-1}\theta^0 D_1, D_0\}$. Its mean is given by

$$[\lambda_{infect}]^{-1}\theta^0 D_1(-D_0)^{-1}\mathbf{e} = [\lambda_{infect}]^{-1}, \quad \text{since } \theta^0(D_0 + D_1) = 0.$$

The *relative infection rate* at time t in the ordinary epidemic is defined by

$$\lambda(t) = (\mu_1')^{-1}\gamma \ exp(Lt)D_1\mathbf{e}, \quad \text{for } t \geq 0.$$

Since clearly, $\int_0^{\infty} \lambda(t)dt = \lambda_{infect}$, the increasing function

$$\psi(t) = \lambda_{infect}^{-1}\int_0^t \lambda(u)du = \lambda_{infect}^{-1}\theta^0[I - exp(Lt)]D_1\mathbf{e}$$

$$= 1 - \lambda_{infect}^{-1}\theta^0 exp(Lt)D_1\mathbf{e},$$

is the expected fraction of infections occurring up to time t in the ordinary epidemic. A graph of its derivative, the ratio of $\lambda(t)/\lambda_{infect}$, gives an informative picture of the evolution of the ordinary epidemic.

Computational experience shows that $\lambda(t)$ is usually decreasing. Since $\mathbf{p}(t)_i = [\gamma exp(Lt)]_i$ is the probability that at time t the process is in a non-absorbing state i, and $\mathbf{b} = D_1\mathbf{e}$ is a positive column vector, independent of time,

$$\gamma exp(Lt)D_1\mathbf{e} = \sum b_i\mathbf{p}(t)_i.$$

3   The Total Size Distribution

The total size is the number of susceptibles who contract the disease. Its probability distribution has received much attention, see, for example, Bailey [BAIL53], Whittle [WHIT55], Bailey [BAIL64], Gani [GANI66], Williams [WILL71], and Bailey [BAIL75].

The following algorithms for the total size distribution are suitable for general single-population S-I-R stochastic epidemic models.

3.1   Solving an Upper-Triangular System of Linear Equations

In matrix notation, the probability P_w can be concisely written

$$P_w = \gamma(-L)^{-1}\hat{\mathbf{T}}(n - w),$$

where $\gamma = (\, 1, \ 0, \ \cdots \ ,0\,)$ is the initial probability vector, and

$$\hat{\mathbf{T}}(n - w) = (\, 0 \quad \cdots \quad 0 \quad \mathbf{T}(n - w) \quad 0 \quad \cdots \quad 0\,)^{T}.$$

Setting $\mathbf{z}(w) = (-L)^{-1}\hat{\mathbf{T}}(n - w)$, we have that $(-L)\,\mathbf{z}(w) = \hat{\mathbf{T}}(n - w)$. As L is upper-triangular and $\hat{\mathbf{T}}(n - w)$ has only one non-zero element, that system of linear equations is easily solved. P_w is just the first element of $\mathbf{z}(w)$.

3.2   A Recursive Algorithm

Considering the Markov chain embedded at the epochs when infection or removals occur, we obtain a recursive algorithm which is similar to that of Bailey in [BAIL53]. An array $\{a_{i,j}\}$ is initialized by setting all its elements to 0, but for $a_{m,n} = 1$. We then perform the following recursive calculations:

1. For $i = m - 1, m - 2, \cdots, 0$,

$$a_{i,n} \ \leftarrow \ a_{i+1,n}\frac{\mu_{i+1}}{\mu_{i+1} + \lambda_{i+1,n}}.$$

2. For $k = 1, \cdots, n,$

(a)

$$a_{m+k,n-k} \leftarrow a_{m+k-1,n-k+1} \frac{\lambda_{m+k-1,n-k+1}}{\mu_{m+k-1} + \lambda_{m+k-1,n-k+1}}.$$

(b) For $i = m + k - 1, m + k - 2, \cdots, 2,$

$$a_{i,n-k} \leftarrow a_{i+1,n-k} \frac{\mu_{i+1}}{\mu_{i+1} + \lambda_{i+1,n-k}} + a_{i-1,n-k+1} \frac{\lambda_{i-1,n-k+1}}{\mu_{i-1} + \lambda_{i-1,n-k+1}}.$$

(c) For $i = 1, 0,$

$$a_{i,n-k} \leftarrow a_{i+1,n-k} \frac{\mu_{i+1}}{\mu_{i+1} + \lambda_{i+1,n-k}}.$$

For $0 \leq w \leq n$, the terms of the total size density are then given by $P_w = a_{0,n-w}$.

In our computational experience, implementation of this recursion was very fast and numerically stable. For instance, computation of the P_w for a *Model-1* example with $m = 1$, $n = 1000$, $\rho = 25$, and $\alpha = 0.20$, took only 31 CPU-seconds on a Sun Sparc-10 workstation. The Markov chain for that example has 501501 transient and 1001 absorbing states.

4   The Maximum Size Distribution

The maximum size I^* of an epidemic is the largest number of infectives ever present during its course. By Brownian motion techniques, Daniels [DANI74] obtained a normal approximation to the maximum size distribution for large n. Daniels [DANI88] also derived a formula for $E(I^*|t^*)$, in which some parameters need to be estimated from simulations. Here, t^* is the time at which the maximum is first attained.

We shall calculate the probabilities $P\{I^* = i_0\}$, $m \leq i_0 \leq m + n$ by considering the Markov chain embedded at the epochs which infections or removals occur. Setting $P\{I^* \geq m + n + 1\} = 0$, it is obvious that for $m \leq i_0 \leq m + n$, $P\{I^* = i_0\} = P\{I^* \geq i_0\} - P\{I^* \geq i_0 + 1\}$. To calculate $P\{I^* \geq i_0\}$ we add a second absorbing boundary consisting of the states for which $i = i_0$, and we sum the probabilities of the paths leading to that boundary. Daniels [DANI74] also describes that approach without the implementational details.

For special values of i_0, $P\{I^* \geq i_0\}$ is explicitly available:

$$P\{I^* \geq m\} = 1, \qquad P\{I^* \geq m + n\} = \prod_{i=m}^{m+n-1} \frac{\lambda_{i,m+n-i}}{\mu_i + \lambda_{i,m+n-i}},$$

but the other probabilities $P\{I^* \geq i_0\}$ must be evaluated recursively. Let $\{a_{i,j}\}$ be our working array, initialized by setting $a_{m,n} = 1$, as for the total size density. To determine $P\{I^* \geq i_0\}$, for $m + 1 \leq i_0 \leq m + n - 1$, we proceed as follows:

1. Set $P\{I^* \geq i_0\} = 0$,

2. For $i = m - 1, m - 2, \cdots, 0$,

$$a_{i,n} \leftarrow a_{i+1,n} \frac{\mu_{i+1}}{\mu_{i+1} + \lambda_{i+1,n}}.$$

3. For $k = 1, \cdots, n$,

   (a)

   $$a_{m+k,n-k} \leftarrow a_{m+k-1,n-k+1} \frac{\lambda_{m+k-1,n-k+1}}{\mu_{m+k-1} + \lambda_{m+k-1,n-k+1}}.$$

   (b) For $i = m + k - 1, m + k - 2, \cdots, 2$,

   $$a_{i,n-k} \leftarrow a_{i+1,n-k} \frac{\mu_{i+1}}{\mu_{i+1} + \lambda_{i+1,n-k}} + a_{i-1,n-k+1} \frac{\lambda_{i-1,n-k+1}}{\mu_{i-1} + \lambda_{i-1,n-k+1}}.$$

   (c)

   $$a_{1,n-k} \leftarrow a_{2,n-k} \frac{\mu_2}{\mu_2 + \lambda_{2,n-k}},$$

While calculating $a_{i,n-k}$, for $i = m + k, m + k - 1, \cdots, 1$, if $i = i_0$ then we do the substitution

$$P\{I^* \geq i_0\} \leftarrow P\{I^* \geq i_0\} + a_{i_0,n-k},$$

and set $a_{i_0,n-k} = 0$.

Based on our computational experience, this algorithm is also numerically stable and reasonably fast. For example, 98 CPU-seconds of computation on a Sun Sparc-10 workstation were required to obtain the $P\{I^* = i_0\}$ for a *Model-2* example with $m = 1$, $n = 200$, $\rho = 20$, and $\alpha = 0.80$. That Markov chain has 20301 transient and 201 absorbing states. Similarly, computation of the $P\{I^* = i_0\}$ for a *Model-1* example with $m = 1$, $n = 1000$, $\rho = 5$, and $\alpha = 0.50$, involving a Markov chain with 501501 transient and 1001 absorption states, requires 375 CPU-minutes on the same computer.

5   The Maximum Size and its Occurrence Times

The maximum size I^* and the epochs t^* and t^*_{end} at which the maximum size attained for the first and last times are interesting features of stochastic epidemics. To undertake their study, we rearrange the generator as a block tridiagonal matrix, which is used henceforth.

$$
Q = \begin{pmatrix}
D_1 & B_1 & & & & T_n & T_{n-1} & \cdots & T_1 & T_0 \\
A_2 & D_2 & B_2 & & & 0 & 0 & \cdots & 0 & 0 \\
& & \ddots & \ddots & & \vdots & \vdots & \vdots & \vdots & \vdots \\
& & & \ddots & B_{m+n-1} & & & & & \\
& & & & D_{m+n} & & & & & \\
0 & 0 & \cdots & \cdots & 0 & 0 & 0 & \cdots & 0 & 0
\end{pmatrix}.
$$

The kth block corresponds to states with k infectives present. With d_k denoting the number of rows of the kth block, $d_k = n + 1$, if $k \leq m$; $d_k = m+n+1-k$, if $m+1 \leq k \leq m+n$. If $k \leq m$, A_k is a diagonal matrix of dimension $d_k \times d_k$ with diagonal elements equal to μ_k. If $m+1 \leq k \leq m+n$, A_k is a matrix of dimension $d_k \times (d_k + 1)$ with superdiagonal elements equal to μ_k and other elements zero. B_k is a matrix of dimension $d_k \times d_k$ if $k \leq m - 1$, $d_k \times (d_k - 1)$ if $m \leq k \leq m + n$. In B_k, the superdiagonal elements are $\lambda_{k,h}, \lambda_{k,h-1}, \cdots, \lambda_{k,1}$, where $h = \min\{n, d_k - 1\}$; all elements not on the superdiagonal are zero. D_k is a diagonal matrix of dimension $d_k \times d_k$. Its diagonal elements are such that the row sums of Q vanish. T_k is a column vector with $(n - k + 1)$-th component equal to μ_1 and others equal to zero.

From the original Markov chain Q, we construct the following Markov chain with two absorbing states

$$
Q_r = \begin{pmatrix}
T_r & T_1^0 & T_2^0 \\
0 & 0 & 0 \\
0 & 0 & 0
\end{pmatrix},
$$

where $T_1^0 = (T_n + \cdots + T_0 \ 0 \ \cdots \ 0)^T$, corresponding to the absorbing state with no infectives left, and $T_2^0 = (0 \ \cdots \ 0 \ B_r)^T$, corresponding to the absorbing state with $r+1$ infectives present. T_r is the upper left submatrix of Q up to and including the rth block.

We define $c(r, u)$ as a row vector consisting of the components of the rth block of $\gamma exp(T_r u)$. v_r is a column vector given the rth block by $(-T_r)^{-1} T_1^0$.

Let $\phi_1(r, u) \, du$ be the elementary probability that the maximum size of the epidemic is r and that it is first attained in $(u, u + du)$. By a standard argument we have that

$$
\phi_1(r, u) = c(r - 1, u) B_{r-1} v_r, \qquad m + 1 \leq r \leq m + n.
$$

For $r = m$, a special definition is needed. It is readily seen that when $u = 0$, $\phi_1(m,0) = P\{I^* = m\}$, which is probability mass. For $u > 0$, $\phi(m, u) = 0$.

If $\phi_2(r, u)\,du$ is the elementary probability that the maximum epidemic is r and that it is last seen in $(u, u+du)$, then $\phi_2(r, u) = c(r, u)A_r\mathbf{v}_{r-1}$, $\quad m \leq r \leq m + n$.

If $m = 1$, some care is needed for $r = 1$. Then,

$$\phi_2(1, u) = \frac{\mu_1}{\mu_1 + \lambda_{1,n}}e^{-\mu_1 u}.$$

Clearly,

$$P\{I^* = r\} = \int_0^\infty \phi_1(r, u)du = \int_0^\infty \phi_2(r, u)du.$$

6   The Epidemic at Two Time Instants

Let $Y(t)$ be the number of infectives at time t. For $t_1 < t_2$, $P\{Y(t_1) = i_1, Y(t_2) = i_2\}$, $0 \leq i_1, i_2 \leq m + n$, can be written as

$$P\{Y(t_1) = i_1, Y(t_2) = i_2\} = \sum_{k_1:\ i=i_1} [\gamma\,exp(Qt_1)]_{k_1} \sum_{k_2:\ i=i_2} [exp(Q(t_2 - t_1))]_{k_1,k_2},$$

where i is the number of infectives, $\sum_{k_1:\ i=i_1}$ means that the summation is over those indices k_1 with $i = i_1$, similar for other summations. If we let $t_2 \to \infty$, since $P\{Y(\infty) = 0\} = 1$, the marginal distribution of $Y(t_1)$ is given by

$$P\{Y(t_1) = i_1\} = \sum_{k_1:\ i=i_1} [\gamma\,exp(Qt_1)]_{k_1}.$$

Consequently, we obtain the conditional distribution

$$P\{Y(t_2) = i_2|Y(t_1) = i_1\} = \frac{P\{Y(t_1) = i_1, Y(t_2) = i_2\}}{P\{Y(t_1) = i_1\}}.$$

Let $X(t)$ denote the number of susceptibles at time t. For $t_1 < t_2$. $P\{X(t_1) = j_1, X(t_2) = j_2\}$, $0 \leq j_1, j_2 \leq n$, can be written as

$$P\{X(t_1) = j_1, X(t_2) = j_2\} = \sum_{k_1:\ j=j_1} [\gamma\,exp(Qt_1)]_{k_1} \sum_{k_2:\ j=j_2} [exp(Q(t_2 - t_1))]_{k_1,k_2},$$

where j is the number of susceptibles.

The marginal distribution of $X(t_1)$ is given by

$$P\{X(t_1) = j_1\} = \sum_{j_2=0}^n P\{X(t_1) = j_1, X(t_2) = j_2\}.$$

Let $P_{j_2}(k_1)$ denote the total size probability if the epidemic starts at state k_1 and ends up with j_2 susceptibles. Another way to obtain the marginal distribution of $X(t_1)$ is then

$$P\{X(t_1) = j_1\} = \sum_{j_2=0}^{n} \sum_{k_1: \, j=j_1} [\gamma \, exp(Qt_1)]_{k_1} P_{j_2}(k_1),$$

where $P_{j_2}(k_1)$ can be easily calculated by the algorithm proposed for the total size distribution.

Consequently we obtain the conditional distribution

$$P\{X(t_2) = j_2 | X(t_1) = j_1\} = \frac{P\{X(t_1) = j_1, X(t_2) = j_2\}}{P\{X(t_1) = j_1\}}.$$

As $t_2 \to \infty$,

$$P\{X(t_1) = j_1, X(\infty) = j_2\} = \sum_{k_1: \, j=j_1} [\gamma \, exp(Qt_1)]_{k_1} P_{j_2}(k_1).$$

An informative graph would plot $P\{X(\infty) = j_2 | X(t_1) = j_1\}$ versus j_2 for different values of j_1. It would show how bad the epidemic will ultimately be if there already are $n - j_1$ susceptibles who have got infected by time t_1.

7   Acknowledgements

Research of this paper was supported in part by NSF Grant Nr. DMI-9306828. We thank Fangyuan Cui for developing computer codes to visualize path functions of epidemic models. We are grateful to a referee who brought the model of Saunders [SAUN80] to our attention. We have generated some numerical examples also for that model and we shall gladly provide them upon request.

8   References

[BAIL53] N. T. J. Bailey The total size of a general stochastic epidemic. *Biometrika*, 40:177–185, 1953.

[BAIL64] N. T. J. Bailey *The Elements of Stochastic Processes with Applications to the Natural Sciences*. John Wiley & Sons, Inc., 1964.

[BAIL75] N. T. J. Bailey *The Mathematical Theory of Infectious Diseases and its Applications*. Charles Griffin & Company Ltd., 1975.

[DANI74] H. E. Daniels The maximum size of a closed epidemic. *Advances in Applied Probability*, 6(4):607–621, 1974.

[DANI88] H. E. Daniels The time of occurrence of the maximum of a closed epidemic. In J. P. Gabriel, C. Lefèvre, and P. Picard, editors, *Stochastic Processes in Epidemic Theory*, pages 129–136, Luminy, France, October 1988. Springer-Verlag 1990.

[GANI66] J. Gani On the general stochastic epidemic. In *Proc. Fifth Berkeley Symp. Math. Statist. & Prob.*, pages 271–279. University of California Press, 1966.

[GANI84] J. Gani and P. Purdue Matrix-geometric methods for the general stochastic epidemic. *IMA Journal of Mathematics Applied in Medicine & Biology*, 1(4):333–342, 1984.

[LUCA93] D. M. Lucantoni The BMAP/G/1 queue: a tutorial. In L. Donatiello and R. Nelson, editors, *Performance Evaluation of Computer and Communication Systems, Joint Totorial Papers of Performance '93 and Sigmetrics '93*, pages 330–358. Springer-Verlag Berlin Heidelberg, 1993.

[NEUT81] M. F. Neuts *Matrix-geometric Solutions in Stochastic Models: An Algoritmic Approach*. The Johns Hopkins University Press, 1981; Dover Publications, Inc., New York, 1995.

[NEUT92] M. F. Neuts Models based on the Markovian arrival processes. *IEICE Transactions On Communications*, E75-B(12):1255–65, 1992.

[NEUT95] M. F. Neuts *Algorithmic Probability: A Collection of Problems*. Chapman & Hall, New York, New York, 1995.

[SAUN80] I. W. Saunders A model for myxomatosis. *Mathematical Biosciences*, 48(1):1–15, 1980.

[WHIT55] P. Whittle The outcome of a stochastic epidemic – a note on Bailey's paper. *Biometrika*, 42:116–122, 1955.

[WILL71] T. Williams An algebraic proof of the threshold theorem for the general stochastic epidemic. *Advances in Applied Probability*, 3(2):223–224, 1971.

Testing the Validity of Value-at-Risk Measures

Bertrand Gamrowski
Svetlozar Rachev

ABSTRACT   Value at Risk ($< VaR >$) is a notion meant to measure the risk linked to the holding of an asset or a portfolio of diverse assets. It is defined as the amount of money that one might lose with a certain confidence level and within a given time horizon. It is used by bankers because it allows them to measure, compare and consolidate "risk" that is linked to their trading activities. As it has recently been recommended to bankers by many financial institutions, it may increasingly become more and more widespread, with the possibility of becoming a market standard. Within a given time horizon, the portfolio return should not drop below a stated VaR number more often than predicted by the confidence interval. This paper aims to test this assertion under different methods of estimating VaR. Much of the current literature is devoted to the prediction of volatility, another measure of risk commonly used by bankers. However, the predictive power of Value-at-Risk has not been tested so far.
We show that "classical" ways of estimating VaR are valid only in reasonable ranges of confidence intervals. If bankers are looking at very low quantiles, then they might be advised to employ more sophisticated models.

1   Introduction

Value-at-Risk (VaR) is a notion meant to measure market risk linked to the holding of an asset or of a portfolio of diverse assets. Market risk could be defined as the amount of money that one may lose within a given time horizon and with a fixed confidence interval, because of market movements. It corresponds in fact to the statistical notion of quantile. VaR could be seen as just another measure of risk. Volatility, i.e. standard deviation, lower partial moments, i.e. the probability for a portfolio to get under a certain value, semi-variance, moments of different orders are already known as indicators of the risk of a portfolio. Sensitivity analysis of a portfolio (the so-called "greeks" in financial jargon) also show what is the responsiveness of holdings to market changes. Nonetheless, VaR may become more and more widespread, and perhaps even a market standard. In recent years, it was recommended by many professional institutions as well as reserve banks as a convenient and meaningful way to manage, control, reduce and disclose risks about their derivative activities. In 1993, the Group of Thirty, including leading banks, gave the following recommendation: *"Market risk is best measured as 'Value-at-Risk' using probability analysis based upon a common confidence interval (e.g. two standard deviations) and time horizon (e.g. a one-day exposure)."* The International Swap and Derivatives Association (ISDA) also stated that *"a popular and commonly accepted measure that could be disclosed is the average Value-at-Risk during the reporting period using a one-day holding period and a stated confidence interval."* Many other organizations like the Security Exchange Commission (SEC) and reserve banks gave similar advice. And, in April 1995, the importance of VaR was emphasized even more when the Bank of International Settlements (BIS) said that VaR figures produced by banks' internal models might be used in order to calculate capital requirements for market risks, as soon as the end of 1996, under some reasonable qualitative and quantitative conditions.

VaR not only has the advantage of being recommended by financial institutions; it also is a very intuitive concept which emphasizes what really matters in the volatility of a portfolio, i.e. heavy losses. VaR tries to answer the easily understable question: "How much could I lose because of my various positions on the financial markets? As a 'maximum' loss has no meaning, what would be a 'major' loss?" This question makes it a very intuitive concept. Finally, VaR is a uniform measure that could be adapted to portfolios of any size, positions on any instruments and market risks of any nature. As such, it can easily be used to consolidate risks through diverse activities and it allows comparisons between these activities.

Nonetheless, the recent popularity of VaR, along with the seeming consensus on its usefulness, should not hide theoretical difficulties that are raised. If everyone agrees that it should be measured, we are far from a single answer to the question of *how* to measure it. This paper will therefore start by introducing different methods to estimate VaR and will describe for each method the underlying hypothesis as well as the pros and cons of the methods.

The predictive power of volatility measures has often been tested. For example, Harvey and Whaley [HW91, HW92], Day and Lewis [DL92], Canina and Figlewski [CF93] and Jorion [Jor95] used options' implied volatilities on diverse markets. Boudoukh et al. [Bou95] compared the methods based on historical observations.

Our paper is organized as follows. Section 2 describes several VaR models, based on Gaussian, Paretian hypotheses or on order statistics. The third section will discuss the assertion of VaR models and present the testing procedures that will be used for this. In Section 4, we will give the results of simulations and statistical tests for VaR at different confidence intervals, based on the daily returns of CAC240 French Stock Index between 1976 and 1992. The last section will give comments and prospects.

2   Value-at-Risk: how to measure?

2.1   General framework

Most VaR models can be applied either to a single asset or to a multi-asset portfolio. Our tests will be based on a single asset, i.e. CAC240 French Stock Index. Therefore, we will confine ourselves to present VaR estimation in the case of a single asset. We will just give a brief explanation of what the extension is to several assets. We start first with the assumption that we have $N + n$ observations of returns of a given asset within a given time horizon:

$$R_{-n+1}, R_{-n+2}, \ldots R_0, R_1, \ldots R_N,$$

the return R_t being defined as the relative change of price of the asset between t and $t+1$. VaR at date t with confidence interval p, with the same time horizon as the above observations and a unit amount of numeraire invested in the considered asset, is defined as the quantity $VaR_t{}^1$ such that:

$$Pr(R_t < -VaR_t) = p.$$

We will assume that at each date t between 1 and N, VaR is estimated according to the latest n observations of the asset returns. That is:

$$\overline{VaR_t} = f(R_{t-1}, \ldots, R_{t-n}).$$

Note that here R_{t-i} should be seen as an observation of the random change in value of the portfolio containing one unit of the asset. If one had to manage a multi-position portfolio, one could extend the single asset method in the following way. Suppose that P is the pricing function of a portfolio, depending on factors $\#1, \ldots, F$, for which we have return observations:

$$R^1_{-n+1}, R^1_{-n+2}, \ldots, R^1_0, R^1_1, \ldots R^1_N, \ldots,$$

$$R^F_{-n+1}, R^F_{-n+2}, \ldots, R^F_0, R^F_1, \ldots R^F_N.$$

as well as current time t levels:

$$X^1_t, \ldots, X^F_t.$$

If we denote the current value of the portfolio by:

$$P_t = P(X^1_t, \ldots, X^F_t),$$

then VaR at confidence interval p is defined as:

$$Pr(P(X^1_t(1 + R^1_t), \ldots, X^F_t(1 + R^F_t)) - P_t < -VaR_t) = p$$

---

[1]In fact, $VaR_t = VarR_t(p)$, but usually for shortness, one omits the index p.

and the estimator of VaR as:

$$\overline{VaR} = f(\Delta P^t_{t-1}, \ldots, \Delta P^t_{t-n}),$$

where:[2]

$$\Delta P^t_{t-i} = P(X^1_t(1 + R^1_{t-i}), \ldots, X^F_t(1 + R^F_{t-i})) - P_t.$$

For a more detailed presentation of possible extensions of the most common methods (i.e. based on Gaussian hypothesis or order statistics), we refer the reader to RiskMetrics Technical Documentation [Mor95]. The methods we will introduce are the following.

- Non-parametric estimation: based on order statistics, it is commonly referred to as the "historical simulation method" by practitioners.

- Gaussian estimation: the normal law is widely used in traditional mathematical finance.

- RiskMetrics^{TM} method: The Global Research Division of JP Morgan & Co. introduced that method. It is also based on the Gaussian hypothesis but is similar to GARCH models in that it tries to capture natural heterskedasticity of financial data. This method and the two above are the most widely used in banks.

- Tail estimation using Hill method: The same exponential tail model as above will be used.

We will now go through each of these methods.

2.2   Non-parametric estimation ("historical simulation" method)

Consider the last n observations of the asset returns $R_{t-1}, \ldots, R_{t-n}$ on which $\overline{VaR}_t$ is based and sort them by ascending order:

$$R_{(1)}, \ldots, R_{(n)}.$$

Assume that they are independent and identically distributed (i.i.d.) with density function f and distribution function F. Then, as $n \to \infty$, and if the ratio $p = k/n$ is kept constant, $R_{(k)}$ tends asymptotically to a normal law with mean $\xi = F^{-1}(p)$ and standard deviation:

$$\frac{1}{f(\xi)} \sqrt{\frac{p(1-p)}{n}}.$$

---

[2]The purpose of calculating ΔP^t_{t-i} is to generate observations of random changes of the portfolio.

In this framework, $R_{(k)}$ can be considered as an approximation of the $100^*p\%$ quantile.

The advantage of this method is that asset returns are only assumed to be i.i.d. The estimator is asymptotically normal for almost any distribution one uses in finance. The first main drawback is that the range of quantiles that one can estimate are limited: one can not expect to calculate properly a 0.5% quantile with a hundred observations! The second significant drawback stems from a more practical point of view; simulating the returns of a large and complex portfolio can be extremely computer-intensive.

2.3   The Gaussian Method

This procedure is simpler. Recent historical observations allow one to calculate the volatility of asset returns with the well-known least-square estimator:

$$\hat{\sigma}_t^2 = \frac{1}{n-1} \sum_{i=1}^{n} (R_{t-i} - \overline{R_t})^2$$

where

$$\overline{R_t} = \frac{1}{n} \sum_{i=1}^{n} R_{t-i}.$$

Quantiles can be found easily by a numerical approximation of the distribution function. The typical assumption is the normality of the returns R_{t-i} and from there the name of the method (The Gaussian method). In this case, $\overline{VaR_t}$ is indeed equal to $\hat{\sigma}_t$ multiplied by the appropriate scalar determined by the quantile level.

2.4   RiskMetrics^{TM} method

JP Morgan & Co. also proposed an alternative way of estimating volatility with an exponentially weighted average of square returns:

$$\hat{\sigma}_t^2 = \sum_{i=1}^{\infty} w_i (R_{t-i} - \overline{R_t})^2,$$

where

$$w_i = (1 - \lambda)\lambda^i \qquad (0 < \lambda < 1).$$

This model, which gives all the more weight to returns in volatility estimation as they are recent, is similar to ARCH models (see [Bol86], [Bol87]). RiskMetrics is in fact an Integrated GARCH (IGARCH) model. The JP Morgan model assumes that the mean return of the asset is negligible and sets it to zero. With a short-term horizon, (e.g. one day), the volatility effect is indeed more important than the trend effect. The estimation formula of volatility leads us then to an easy recursive formula:

$$\hat{\sigma}_t^2 = \lambda\hat{\sigma}_{t-1}^2 + (1-\lambda)R_{t-1}^2.$$

[BR95] showed that the method proposed by the JP Morgan model gives good results in terms of volatility prediction. Under the assumption for normal returns VaR_t is determined by $\hat{\sigma}_t$ in the same way as in the Gaussian Method.

2.5   Tail Estimation Using Hill Method

We now abandon the Gaussian hypothesis for a Pareto tail hypothesis:

$$F(x) = \frac{C}{|x|^a}$$

or

$$lnF(x) \approx lnC - \alpha ln\,|\,x\,|\,.$$

Let us take the ordered latest n observations of returns $R_{(1)}, \ldots, R_{(m)}$. [Hil75] provides an estimator of α for distributions with Pareto tails. The formula of the estimator is the following:

$$\hat{\alpha}_{Hill} = \frac{1}{\frac{1}{m}\sum_{i=1}^{m} ln\,|R_{(i)}| - ln\,|R_{(m+1)}|}.$$

In this formula $m = n/10$ empirically seems to be a proper value. We define the sample distribution function as $\hat{F}(x) = \frac{1}{n}\sum_{i=1}^{n} 1_{\{R_{(i)}-x\}}$, $ln\hat{C}$ can then be estimated as:

$$ln\hat{C} = \frac{1}{m}\sum_{i=1}^{m} ln\hat{F}(R_{(i)}) + \hat{\alpha}\frac{1}{m}\sum_{i=1}^{m} ln\,|R_{(i)}|\,.$$

The estimator of VaR at $p\%$ confidence interval could then be derived from:[3]

$$\overline{VaR} = exp\{\frac{ln\hat{C} - lnp}{\hat{\alpha}}\}.$$

Hill estimator is widely used for tail index estimation (see e.g. [LP94], [BS95]). Even though it proves strongly unrobust for non exactly Pareto tails (see [MR95a]), we will use it in the first step because it is computationally simple.

---

[3]Under some regularity conditions similar to these for the Hill estimator, $\overline{VaR}$ is strongly consistent and asymptotically normal. Unfortunately, the use of tail estimator for $\overline{VaR}$ inherits the drawbacks of the Hill estimator, see [MR93b], [MR95b].

3   Value-at-Risk: how to test?

Our goal is to test the null hypothesis that for any t:

$$Pr(R_t < -VaR_t) = p.$$

In order to derive a testing procedure, we first have to make the approximation that $Pr(R_t < \overline{VaR_t}) = \bar{p}$ is not too different from p. We have in fact:

$$\bar{p} = \int_{-\infty}^{\infty} Pr(R_t < -v)\varphi(v)dv$$

where φ is the density function of the estimator $\overline{VaR}$. We will have strict equality between p and $\bar{p}$ when $\varphi(v) = \delta_{v=VaR}$. Nonetheless, we will suppose that the precision of the estimator is "good enough" to confirm the approximation. We did some simple calculations in order to quantify the quality of the approximation. Let us assume again a Pareto-tail for the returns, i.e.:

$$Pr(R_t < -v) \approx \frac{C}{|v|^\alpha},$$

as well as a normal distribution of the Var estimator:

$$\varphi(v) = \frac{1}{\sigma\sqrt{2\pi}} exp\left(-\frac{(v-VaR)^2}{\sigma^2}\right).$$

The following table gives us the ratio $\frac{\bar{p}-p}{p}$ according to the standard deviation of the VaR estimator and the value of the characteristic exponent α.

Characteristic exponent	1.5	1.5	3	3
Standard deviation	10%	20%	10%	20%
$\frac{\bar{p}-p}{p}$	1.6%	8.3%	5.96%	32.31%

If we accept this approximation, the hypothesis we want to test is now equivalent to the following:

$$\xi_t = 1_{\{R_t - VaR_t\}} = 1 \text{ with probability } p, \text{ and } 0 \text{ with probability } 1 - p.$$

$\frac{1}{\sqrt{n}}\left(\sum_{t=1}^{n}(\xi_t - p)\right)$ converges thus to a centered normal law with standard deviation $\sqrt{p(1-p)}$. If n is large enough, the fundamental hypothesis of VaR should therefore be rejected at $x\%$ confidence interval if $\sum_{t=1}^{n}\xi_t$ happens to be below $np - q_x\sqrt{p(1-p)}$ or above $np + q_x\sqrt{p(1-p)}$, where q_x is the corresponding normal quantile. Note that for very low quantiles (i.e. low values of p), convergence to the normal law happens to be slow. Thus, we considered $p = 0.5\%$ or $p = 1\%$ in the sequel and used the table for the binomial distribution of $\sum_{t=1}^{n} \xi_t$. Values S of this sum have been considered as contradictory with the hypothesis provided that they were such that:

$$1 - x \leq \sum_{s=0}^{S} C_N^s p^s (1 - p)^{N-s} \leq x.$$

Table 1 gives the range of VaR exceedings as well as the range of exceeding frequencies $\frac{S}{N}$ in which the null hypothesis can not be rejected. These ranges are provided for $p = 0.5\%, 1\%, 5\%$ and $N = 500, 3000$.

0.5% quantile

	500 obs.		3000 obs.	
5% c.i.	0	4	9	21
1% c.i.	0	6	7	24

0.5% quantile

	500 obs.		3000 obs.	
5% c.i.	0,0%	0,8%	0,3%	0,7%
1% c.i.	0,0%	1,2%	0,2%	0,8%

1% quantile

	500 obs.		3000 obs.	
5% c.i.	2	8	21	38
1% c.i.	1	10	18	42

1% quantile

	500 obs.		3000 obs.	
5% c.i.	0,4%	1,6%	0,7%	1,3%
1% c.i.	0,2%	2,0%	0,6%	1,4%

5% quantile

	500 obs.		3000 obs.	
5% c.i.	17	32	131	169
1% c.i.	14	35	123	177

5% quantile

	500 obs.		3000 obs.	
5% c.i.	3,4%	6,4%	4,4%	5,6%
1% c.i.	2,8%	7,0%	4,1%	5,9%

Table 1: Number and frequency of exceedings.

4 Value-at-Risk: should we trust?

VaR estimations at quantile levels $p = 0.5\%, 1\%, 5\%$ have been performed on a daily basis according to the diverse methodologies on CAC240 French Stock Index between 1976 and 1992. The testing procedure described above has been implemented on the whole sample (i.e. $N = 3000$) as well as for 6 consecutive sub-windows (i.e. $N = 500$). The number S of exceedings is displayed in Table 2; the frequency S/N of exceedings is given in Table 3.

Method: Hill method / Gaussian law

# obs.	100 obs.			250 obs.			50 obs			100 obs.			1000 obs		
c.i. of VaR	0.5%	1%	5%	0.5%	1%	5%	0.5%	1%	5%	0.5%	1%	5%	0.5%	1%	5%
pass 3000	13	23	160	12	30	162	37	53	148	35	42	127	38	46	109
pass 500	2	3	22	1	1	17	4	4	21	3	4	20	2	3	4
(6 sub-per.)	2	4	29	1	3	30	4	11	24	4	5	25	0	1	0
	4	5	34	6	14	47	12	15	33	14	15	30	24	27	48
	2	4	23	2	3	18	5	6	15	6	6	13	5	6	22
	2	4	28	2	4	26	11	14	33	5	7	23	4	4	14
	1	3	24	0	5	24	1	3	22	3	5	16	3	5	12

Method: Historical simulation / RiskMetrics

# obs.	100 obs.			1000 obs.			lambda = 0.85			lambda = 0.9			lambda = 0.95		
c.i. of VaR	0.5%	1%	5%	0.5%	1%	5%	0.5%	1%	5%	0.5%	1%	5%	0.5%	1%	5%
pass 3000	x	29	153	17	35	143	44	60	160	35	50	150	35	43	134
pass 500	x	4	22	1	2	12	5	9	25	3	7	22	3	5	20
(6 sub-per.)	x	5	28	0	0	22	8	12	27	6	11	26	4	7	23
	x	8	35	15	25	59	7	7	27	6	8	27	7	10	28
	x	4	20	0	1	24	8	8	22	6	6	19	4	6	13
	x	4	23	1	4	14	12	17	37	10	14	33	6	11	30
	x	4	25	0	3	12	4	7	22	4	4	23	1	4	20

Bold italics : rejected at the 1% c.i.   Bold : rejected at the 5% c.i.

Table 2: number of VaR exceedings on CAC240.

Method: Hill method / Gaussian law

# obs.	100 obs.			250 obs.			50 obs			100 obs.			1000 obs		
c.i. of VaR	0.5%	1%	5%	0.5%	1%	5%	0.5%	1%	5%	0.5%	1%	5%	0.5%	1%	5%
pass 3000	0,43%	0,77%	5,33%	0,40%	1,00%	5,40%	1,23%	1,77%	4,93%	1,17%	1,40%	4,23%	1,27%	1,53%	3,63%
pass 500	0,4%	0,6%	4,4%	0,2%	0,2%	3,4%	0,8%	0,8%	4,2%	0,6%	0,8%	4,0%	0,4%	0,6%	0,8%
(6 sub-per.)	0,4%	0,8%	5,8%	0,2%	0,6%	6,0%	0,8%	2,2%	4,8%	0,8%	1,0%	5,0%	0,0%	0,2%	1,8%
	0,8%	1,0%	6,8%	1,2%	2,8%	9,4%	2,4%	3,0%	6,6%	2,8%	3,0%	6,0%	4,8%	5,4%	9,6%
	0,4%	0,8%	4,6%	0,4%	0,6%	3,6%	1,0%	1,2%	3,0%	1,2%	1,2%	2,6%	1,0%	1,2%	4,4%
	0,4%	0,8%	5,6%	0,4%	0,8%	5,2%	2,2%	2,8%	6,6%	1,0%	1,4%	4,6%	0,8%	0,8%	2,8%
	0,2%	0,6%	4,8%	0,0%	1,0%	4,8%	0,2%	0,6%	4,4%	0,6%	1,0%	3,2%	0,6%	1,0%	2,4%

Method: Historical simulation / RiskMetrics

# obs.	100 obs.			1000 obs.			lambda = 0.85			lambda = 0.9			lambda = 0.95		
c.i. of VaR	0.5%	1%	5%	0.5%	1%	5%	0.5%	1%	5%	0.5%	1%	5%	0.5%	1%	5%
pass 3000	x	0,97%	5,10%	0,57%	1,17%	4,77%	1,47%	2,00%	5,33%	1,17%	1,67%	5,00%	0,83%	1,43%	4,47%
pass 500	x	0,8%	4,4%	0,2%	0,4%	2,4%	1,0%	1,8%	5,0%	0,6%	1,4%	4,4%	0,6%	1,0%	4,0%
(6 sub-per.)	x	1,0%	5,6%	0,0%	0,0%	4,4%	1,6%	2,4%	5,4%	1,2%	2,2%	5,2%	0,8%	1,4%	4,6%
	x	1,6%	7,0%	3,0%	5,0%	11,8%	1,4%	1,4%	5,4%	1,2%	1,6%	5,4%	1,4%	2,0%	5,6%
	x	0,8%	4,0%	0,0%	0,2%	4,8%	1,6%	1,6%	4,4%	1,2%	1,2%	3,8%	0,8%	1,2%	2,8%
	x	0,8%	4,6%	0,2%	0,8%	2,8%	2,4%	3,4%	7,4%	2,0%	2,8%	6,6%	1,2%	2,2%	6,0%
	x	0,8%	5,0%	0,0%	0,6%	2,4%	0,8%	1,4%	4,4%	0,8%	0,8%	4,6%	0,2%	0,8%	4,0%

Bold italics : rejected at the 1% c.i.   Bold : rejected at the 5% c.i.

Table 3: frequency of VaR exceedings on CAC240.

The shadings in both tables indicate the cases when we reject the null hypothesis.

The conclusions of this study are as follows:

- The use of too many data depreciates the quality of the prediction. This is simply due to natural heteroskedasticity of financial data. Note that our test is not a test of the precision of the estimator. Further comments will therefore concentrate on methods based on 100 observations.

- For 5% quantile level, RiskMetrics and historical simulation provide good results. They both pass the test on the whole period. One rejection out of 6 sub-periods can still be considered "acceptable".

- For 1% quantile level, only historical simulation and the Hill method provide good results. We are not surprised to see that these methods are more accurate since market representation is more realistic in them.

- For 0.5% quantile level, historical simulation can hardly be applied. A quantile estimator based on order statistics would require a lot of data, which would diminish the predictive power. Only the Hill method provides satisfactory results.

- 5% quantile levels are most easily handled than 1% or 0.5% quantiles. Precision of estimation is much better at the 5% level. The fact that extreme events are not properly taken into account under Gaussian hypothesis is also less sensitive at 5% than at 1%.

- RiskMetrics' exponential weighting really seems to improve calculations compared with traditional Gaussian estimators.

VaR over time is represented on Graph 1 (for 5% c.i.).

Daily VaR at the 5% confidence interval on CAC240

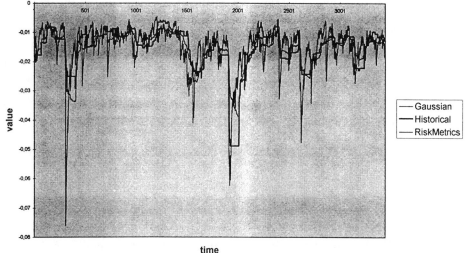

Graph 1

Additional remarks can be made.

- At the 5% confidence interval, figures obtained with the three methods are not too different. At the 1% c.i., they would be more divergent. Historical simulation and Hill methods allow variations of the ratio between 1% and 5% $VaR's$ whereas it is fixed to 2.33/1.64 in other methods.

- RiskMetrics data are not as reliable. A sharp change in the factor has more impact on the measure. On the other hand, this impact will fade away smoothly as time goes by. With the Gaussian or historical simulation method, the sharp change will have another impact when it goes out of the observation window.

5   Conclusion

This test shows clearly that non-traditional hypothesis for financial returns, i.e. not based on the Gaussian Law, dramatically increase the quality of VaR estimations. Whereas, the gain in quality is not always detectable in the case of "average" relations (see e.g. [GR95]), this gain is undeniable when one tries to model extreme events like in a VaR approach.

In further work, we will integrate other estimators of VaR in the above statistical test, namely:

- Pareto-stable distributed returns: Like the two above, this method assumes fat tails but contrarily, the characteristic exponent is supposed to be included between 1 and 2. Moreover, for this case, stable distribution estimators (see [MR95a]), are based on the whole sample.

- GARCH models: The GARCH model we plan to use assumes a Pareto-stable driver (see [MP95]). RiskMetrics^{TM} is a method which already mixes heteroskedasticity and Gaussian hypothesis. Moreover, our conviction is that Pareto-stable assumptions will better capture the features of financial data.

- Geo-stable hypothesis (see [MR95a],[MR95b], [Koz95].

- A more robust estimator of α than the Hill estimator (see [MR95b]).

- Predictions based on implied volatility.

6   REFERENCES

[Bac90] L. Bachelier, Théorie de la speculation. *Annales de l'Ecole Normale Superieure 3, 17, 21-86*, 1990.

[Bol87] J. Bollerslev, A conditional heteroskedastic time series model for speculative prices and rates of return. *Review of Economics and Statistics, 69, 542-547*, 1987.

[Bol86] T. Bollerslev, Generalized autoregressive conditional heterskedasticity. *Journal of Econometrics, 31, 307-327*, 1986.

[BC92] J. Bollerslev, T. Chou, S.A. and Kroner, K.F. Arch modelling in finance. *Journal of Econometrics, 7, 229-264*, 1992.

[BR95] J. Boudoukh, M. Richardson, and R. Whitelaw. The stochastic behavior of interest rates. *Working paper, Stern School of Business, New York University*, 1995.

[BS95] J.P. Bouchaud, D. Sornette, C. Walter, and J.P. Aguilar. Taming large events; optimal portfolio theory for strongly fluctuating assets. *Working paper, Stern School of Business, New York University*, 1995.

[CF93] L. Canina and S. Figlewski. The information content of implied volatility. *Review of Financial Studies, 6, 659-681*, 1993.

[DL92] T. Day and C. Lewis. Stock market volatility and the information content of stock index options. *Journal of Econometrics, 52, 267-287*, 1992.

[EK95] E. Eberlein and H. Keller. Hyperbolic distributions in finance. *Technical Report, Inst. für Math. Stochastik, Univ. Freiburg, 1995.*

[Eng95] R.F. Engle. Autoregressive conditional heterskedasticity with estimates of the variance of U.K. inflation. *Econometrica, 50, 987-1008, 1995.*

[Fel65] W. Feller. An introduction to probability theory and its applications. *Vol. 2,* Wiley Eastern Ltd., New York, 1965.

[GR95] B. Gamrowski and S.T. Rachev. A testable version of the Pareto-stable CAPM. *Working paper, Department of Statistics & Applied Probability, UCSB, Santa Barbara.*

[HW91] C. Harvey and R. Whaley. S & P 100 index option volatility. *Journal of Finance, 46, 1551-1561,* 1991.

[HW92] C. Harvey and R. Whaley. Market volatility prediction and the efficiency of the S & P 100 index option market. *Journal of Financial Economics, 31, 43-74,* 1992.

[Hil75] B.M. Hill. A simple general approach to inference about the tail of a distribution. *Annals of Statistics, 3, 1163-1174,* 1975.

[Jor95] P. Jorion. Predicting volatility in the foreign exchange market. *Journal of Finance, 50/2, 507-528,* 1995.

[KM95] L.B. Klebanov, J.A. Melamed, S. Mittnik and S.T. Rachev. Integral and asymptotic representations of geo-stable densities. *To appear in Applied Mathematics Letters,* 1995.

[Koz95] T.J. Kozubowski. Geometric stable laws: estimation and applications. *Working paper, Department of Mathematics, UTC, Chattanooga,* 1995.

[LP94] M. Loretan and P.C.B. Phillips. Testing the variance stationarity of heavy-tailed time series. *Journal of Empirical Finance, 1, 211-248,* 1994.

[Man77] B.B. Mandelbrot. The fractal geometry of nature. *W.M. Freeman & Co., New York,* 1977.

[Mcc86] J.H. McCulloch. Simple consistent estimators of stable distribution parameters. *Communications in Statistics, Computation and Simulation, 15, 1109-1136,* 1986.

[Mor95] J.P. Morgan & Co. *RiskMetrics–Technical Document,* New York, 1995.

[MR93a] S. Mittnik and S.T. Rachev. Modelling asset returns with alternative stable distributions. *Econometric Reviews, 12/3, 261-330,* 1993a.

[MR93b] S. Mittnik and S.T. Rachev. Reply to comments on modeling asset returns with alternative stable distributions. *Econometric Reviews, 12/3 347-389,* 1993b.

[MR95a] S. Mittnik and S.T. Rachev, S.T. Modeling financial assets with alternative stable models. *To appear in Wiley series in Financial Economics and Quantitative analysis, Wiley,* 1995a.

[MR95b] S. Mittnik and S.T. Rachev. Tail estimation of the stable index α. *Working paper, Department of Statistics & Applied Probability,* 1995b.

[MP95] S. Mittnik, A. Panorska and S.T. Rachev. Stable GARCH models for financial time series. *Applied Mathematics Letters, 8/5, 33-37,* 1995.

[Pet94] E.E. Peters. Fractal market analysis, *Wiley & Sons,* 1994.

[ST94] G. Samorodnitsky and M.S. Taqqu. Stable non-gaussian random processes. *Chapman and Hall, London,* 1994.

[Zol83] V.M. Zolotarev. One dimensional stable distributions. *Translation of Mathematical monographs, American Mathematical Society, Vol. 65,* 1983.

Option pricing for hyperbolic CRR model

Aleksander Rejman
Aleksander Weron

ABSTRACT Employing the Rachev and Rüschendorf (1994) idea to generalize Cox–Ross–Rubinstein (1979) binomial model we find the option price formula for the limiting model with hyperbolic distributions. It turns out that the results in the continuous and discrete models lead to markedly different fair prices.

1   Introduction

An option is a security which gives its owner the right to trade a fixed number of shares of a specified common stock at a fixed price K at an expiration date T (European option). A call option which is considered here gives the buyer the right to purchase the stock.

The current literature considers two main categories of option models: models that presume trading at discrete time intervals and those presuming continuous trading. The first one introduced by Cox, Ross and Rubinstein [CRR79] is often used as an approximation of the second model, particularly when it is not possible to find an algebraic solution for the continuous trading model.

A key role in option pricing play distributional assumptions concerning security returns. Beginning with the work of Mandelbrot [Man63], evidence indicates that an empirical distribution of the daily stock returns differs significantly from the Gaussian distribution. Researchers proposed many alternative hypotheses and among them the hyperbolic law. Eberlein and Keller [EK94] show that the hyperbolic law fits data from the German stock market much better than the Gaussian one and apply a hyperbolic Lévy motion process as the model of the stock returns. We shall use hyperbolic distributions in the model proposed by Rachev and Rüschendorf [RR94]. These authors come back to the Cox, Ross and Rubinstein model (CRR) and with the help of a randomization method receive a new discrete trading model with large class of limit cases.

Following the Rachev and Rüschendorf [RR94] ideas we find the option price for the hyperbolic CRR model. Section 2 introduces the CRR model and its extensions. Section 3 contains basic facts about hyperbolic distributions. In Section 4 we find the option price formula for the hyperbolic CRR model. The last Section includes a numerical example giving a comparison of our option price formula with the earlier formula of Eberlein and Keller [EK94].

2   CRR model

In this section we recall basic ideas of the Cox, Ross and Rubinstein (CRR) model and its extension. We start with assuming that the stock price $S = (S_k)$ follows a multiplicative binomial process over discrete periods which divide a time interval $[0, t]$ into n parts of length h. Price changes are modeled by upward movements u with probability q and downward movements d with probability $1 - q$, thus $S_{k+1} = uS_k$, or dS_k with probabilities q and $1 - q$, respectively, and $u > d$ are real constants.

Let us define $U = \log u$ and $D = \log d$, and let $\epsilon_{n,i}$ be a sequence of independent random variables with $\mathbf{P}(\epsilon_{n,i} = 0) = q$ and $\mathbf{P}(\epsilon_{n,i} = 1) = 1 - q$. The cumulative return process can be defined as

$$\log \frac{S_k}{S_0} = \sum_{i=1}^{k} (\epsilon_{n,i} U + (1 - \epsilon_{n,i})D) = \sum_{i=1}^{k} X_{n,i}. \tag{1.1}$$

In order to find the price of a call with a striking price K, assume that the value of riskless bonds increases with the rate $\rho - 1$, i.e. $B_{k+1} = \rho B_k$, where $\rho > 1$ and $d \le \rho \le u$. In such a case Cox, Ross and Rubinstein determine the fair price of a call at the moment n (or equivalently t),

$$C_n = S_0 \Phi(a, n, p') - K\rho^{-n}\Phi(a, n, p), \tag{1.2}$$

where

$$p = \frac{\rho - d}{u - d}, \quad p' = \frac{u}{\rho}p, \quad a = 1 + \left[\log \frac{K}{S_0 d^n} \middle/ \log \frac{u}{d}\right]$$

$\Phi(a, n, p) = \mathbf{P}(\sum_{i=1}^{n} \epsilon_{n,i}^* \ge a)$ and $\epsilon_{n,i}^*$ is a sequence of independent r.v. with distribution $\mathbf{P}(\epsilon_{n,i}^* = 0) = p$ and $\mathbf{P}(\epsilon_{n,i}^* = 1) = 1 - p$. Here $[x]$ denotes an integer part of x. See Cox, Ross and Rubinstein [CRR79] or Shiryaev et al. [Shi94] for details.

Further, one assumes that the number of changes of the stock price in the interval $[0, t]$ increases to infinity and the parameters of the random walk $\log S_n/S_0$ depend on the number of jumps n and $U = U(n), D = D(n), q = q(n)$ and $\rho = \rho(n)$. Rachev and Rüschendorf [RR94] found necessary and sufficient conditions for convergence of sums $\sum_{k=1}^{n} X_{n,k}$ to the Gaussian law. Basing on their result we assume that

$$\sum_{k=1}^{n} X_{n,k} \xrightarrow{d} N(m, \sigma^2)$$

Moreover, we demand that we obtain Gaussian limits with parameters α, σ^2 and α', σ'^2 for probability measures for which the random walk $\log S_n/S_0$ exhibits upward movements with probabilities p and p', respectively. To omit complicated notations we always use ϵ to denote ϵ and ϵ^*.

In order to receive the Black–Scholes formula as a limit of (1.2) one also assumes that $\lim \rho^n(n) = r_0^t$. Then

$$C_n \to C_t = S_0 \Phi(x') - K r_0^t \Phi(x), \tag{1.3}$$

where $x = (\log S_0/K + \alpha)/\sigma$, $x' = (\log S_0/K + \alpha')/\sigma'$ and $\Phi(x) = \mathbf{P}(N(0,1) \geq x)$.

A natural method to extend a model is to randomize one of its parameters. Rachev and Rüschendorf [RR94] propose a model with a random number components. Introducing N_n, a positive integer valued random variable independent of the sequence $(X_{n,k})$, they define the stock price that exhibits the random number of jumps $X_{n,k}$ in the interval $[0, t]$ and then

$$\log \frac{S_t}{S_0} = \log \frac{S_{N_n}}{S_0} = \sum_{k=1}^{N_n} X_{n,k} \tag{1.4}$$

The interval $[0, t]$ is divided into a random number of subintervals of length $h = t/N_n$. One usually demands that intensity of the number of the jumps increases with n, for example $\mathbf{E}N_n = n$.

Similarly, as in the classical CRR model, we are interested in the limit distribution of $\log S_{N_n}/S_0$ when n tends to infinity. The following general theorem can be applied in our case.

Theorem 1 (Rachev–Rüschendorf) *Let $(X_{n,k})_{k\geq 1}$ be sequences of independent and identically distributed random variables (in each series), $n \geq 1$, and let (N_n) be a sequence of positive integer valued random variables independent of $(X_{n,k})$. If*

$$\sum_{k=1}^{n} X_{n,k} \xrightarrow{d} N(\alpha, \sigma^2), \quad \frac{N_n}{n} \xrightarrow{w} Y$$

then $\sum_{k=1}^{N_n} X_{n,k} \xrightarrow{d} Z$ where the characteristic function of Z is

$$\phi_Z(u) = \int_0^\infty \exp\left\{i\alpha z u - \frac{\sigma^2 z}{2} u^2\right\} dF_Y(z) \tag{1.5}$$

The last formula (1.5) suggests that all normal variance–mean mixtures can be obtained as a limit of sums of independent binomial random variables with a random number of components.

3 Hyperbolic distribution

The name *hyperbolic distribution* comes from the fact that its log density forms a hyperbola in contrast to the normal distribution, whose log density forms a parabola. It was pointed out by Barndorff-Nielsen [BN77], that this law can be represented as a normal variance–mean mixture where the mixing distribution is one of the generalized inverse Gaussian laws. Eberlein and Keller [EK94] find that the hyperbolic distributions provides an excellent fit to the distributions of daily returns, measured on the log scale, of stocks from a number of leading German enterprises. Using the ideas of Rachev and Rüschendorf [RR94] we will try to find the fair price of the call option.

We first introduce the generalized inverse Gaussian and the hyperbolic laws. A random variable Y has the generalized inverse Gaussian distribution, *i.e.* $Y \sim \mathrm{GIG}(\lambda, \chi, \psi)$ if its density has the following form

$$f_Y(x) = \frac{(\psi/\chi)^{\lambda/2}}{2K_\lambda(\sqrt{\chi\psi})} \, x^{\lambda-1} \exp\left\{-\frac{1}{2}[\chi\, x^{-1} + \psi\, x]\right\}, \quad x \geq 0$$

where the normalizing constant $K_\lambda(\cdot)$ is a modified Bessel function of the third kind with index λ. The generalized inverse Gaussian law is often characterized by its Laplace transform

$$\mathbf{E}\exp(-\theta\, Y) = \frac{\psi^{\lambda/2}}{(\psi + 2\theta)^{\lambda/2}} \frac{K_\lambda\left(\sqrt{\chi(\psi + 2\theta)}\right)}{K_\lambda\left(\sqrt{\chi\psi}\right)} \tag{1.6}$$

Now, define the hyperbolic distribution as a normal variance mean mixture with the generalized inverse Gaussian mixing distribution $\mathrm{GIG}(1, \chi, \psi)$. A random variable Z has the hyperbolic distribution if

$$(Z|Y) \sim \mathrm{N}\left(\mu + \alpha Y, Y\right)$$

where $Y \sim \mathrm{GIG}(1, \chi, \psi)$. It means that $Z \sim \mathrm{Hyp}(\chi, \psi, \alpha, \mu)$ can be represented in the form $Z = \mu + \alpha Y + \sqrt{Y} N(0,1)$ and its characteristic function is given by

$$\phi_Z(u) = \exp(iu\mu) \int_0^\infty \exp\left\{i\alpha zu - \frac{1}{2}zu^2\right\} dF_Y(z). \tag{1.7}$$

We can write its density function $f(x) = f(x; \chi, \psi, \alpha, \mu)$ in the form

$$f(x) = \frac{(\psi/\chi)^{1/2}}{2K_1(\sqrt{\chi\psi})}(\psi + \alpha^2)^{-\frac{1}{2}} \exp(-\sqrt{(\psi + \alpha^2)(\chi + (x - \mu)^2)} + \alpha(x - \mu))$$

Comparison of (1.5) and (1.7) implies that if $\mu = 0$ then the hyperbolic distribution is a limit law of sums with a random number of components.

It is sufficient to find N_n and the sequence $(X_{n,k})$. Recall that we need N_n such that

$$\frac{N_n}{n} \xrightarrow{d} Y \sim \mathrm{GIG}(1, \chi, \psi)$$

when $n \to \infty$. A good candidate for N_n is

$$\mathbf{P}(N_n = k) = \mathbf{P}(k - 1 < nY \leq k)$$

Further, we need to find $(X_{n,k})$ such that $\sum_{k=1}^{n} X_{n,k} \to X$ with $X \sim N(\alpha, \sigma^2)$. Then $\log S_{N_n}/S_0 = \sum_{k=1}^{N_n} X_{n,k} \xrightarrow{d} Z$ where the characteristic function of Z is given by

$$
\begin{aligned}
\phi_Z(u) &= \int_0^\infty \exp\left\{ i\alpha z u - \frac{1}{2} z \sigma^2 u^2 \right\} dF_Y(z) \\
&= \int_0^\infty \exp\left\{ i\frac{\alpha}{\sigma^2} z u - \frac{1}{2} z u^2 \right\} dF_{\sigma^2 Y}(z)
\end{aligned}
$$

To find the parameters of Z we notice that if $Y \sim \mathrm{GIG}(1, \chi, \psi)$ then $aY \sim \mathrm{GIG}(1, a\chi, \psi/a)$ and therefore $Z \sim \mathrm{Hyp}(\sigma^2\chi, \psi/\sigma^2, \alpha/\sigma^2, 0)$. Consequently we have

Corollary 1 *Under the assumptions of Theorem 1 and $N_n/n \xrightarrow{d} \mathrm{GIG}(1, \chi, \psi)$ the series $\sum_{k=1}^{N_n} X_{n,k}$ converges weakly to*

$$Z \sim Hyp\left(\sigma^2 \chi, \frac{\psi}{\sigma^2}, \frac{\alpha}{\sigma^2}, 0 \right).$$

4 Option pricing

Formula (1.2) gives the fair price for a call with the expiration date t and a fixed number of jumps in the interval $[0, t]$. We want to find the fair price for a call when the number of jumps in the interval is directed by the random variable N_n. Let $C(k)$ denote the fair price of a call option for the stock with k movements $X_{n,i}$ until time t, and $C(N_n)$ denote the fair price of a call option for the stock with a random number of jumps. In such a case it is natural to expect that $C(N_n)$ should be the mean value of the option prices $C(k)$,

$$C(N_n) = \sum_{k=1}^{\infty} C(k)\mathbf{P}(N_n = k) \tag{1.8}$$

From the above equation we receive the counterpart of equation (1.2)

$$
\begin{aligned}
C(N_n) &= \sum_{k=1}^{\infty} \left(S_0 \Phi(a_k, k, p') - K\rho^{-k}\Phi(a_k, k, p)\mathbf{P}(N_n = k) \right) \\
&= S_0 \mathbf{E}\Phi(a_{N_n}, N_n, p') - K\mathbf{E}\rho^{-N_n}\Phi(a_{N_n}, N_n, p) \tag{1.9}
\end{aligned}
$$

where expectation $\mathbf{E}$ is taken with respect to N_n and $p = p(n)$, $p' = p'(n)$

$$a_k = 1 + \left[\log \frac{K}{S_0 d(n)^k} \Big/ \log \frac{u(n)}{d(n)} \right] \qquad (1.10)$$

$$\Phi(a_k, k, p) = \mathbf{P}(\sum_{i=1}^{k} \epsilon_{n,i} \geq a_k) \quad \text{with } \mathbf{P}(\epsilon_{n,i} = 0) = p, \ \mathbf{P}(\epsilon_{n,i} = 1) = 1 - p.$$

Formula (1.9) is complicated from the computational point of view. For this reason it is natural to look for its limit case.

The theorem below applies a method of Rachev and Rüschendorf [RR94] to find the option price when the log price of a stock follows the hyperbolic distribution.

Theorem 2 *Let $\alpha, \alpha', \sigma, \sigma'$ be the parameters of the Gaussian laws obtained as limit distributions of $\log S_n/S_0$ for the different probability measures defined in Section 2, and let N_n be a positive integer valued random variable independent of the sequence $\log S_n/S_0$ such that*

$$\frac{N_n}{n} \xrightarrow{d} Y, \quad and \quad \rho^n \to r_0^t,$$

where Y has the generalized inverse Gaussian distribution $GIG(1, \chi, \psi)$. Then, the fair price of the call is given by

$$\begin{aligned}
C &= \lim_{n \to \infty} C(N_n) = S_0 \int_{\log K/S_0}^{\infty} f\left(x; \sigma'^2 \chi, \frac{\psi}{\sigma'^2}, \frac{\alpha'}{\sigma'^2}, 0\right) dx \\
&- K \, \mathbf{E} r_0^{-tY} \int_{\log K/S_0}^{\infty} f\left(x; \sigma^2 \chi, \frac{\psi + 2t \log r_0}{\sigma^2}, \frac{\alpha}{\sigma^2}, 0\right) dx, (1.11)
\end{aligned}$$

where $f(\cdot)$ denotes the density of the hyperbolic distribution with given parameters.

Proof: We first remark that to prove the theorem we have to find limits of the expectations in (1.9). From the definition of $\Phi(a_{N_n}, N_n, p')$ and $X_{n,k}$ we have

$$\begin{aligned}
\mathbf{E}\Phi(a_{N_n}, N_n, p') &= \sum_{k=1}^{\infty} \mathbf{P}\left(\sum_{i=1}^{k} \epsilon_{n,i} \geq a_k\right) \mathbf{P}(N_n = k) \\
&= \sum_{k=1}^{\infty} \mathbf{P}\left(\sum_{i=1}^{k} X_{n,i} - a_k R - kD \geq 0\right) \mathbf{P}(N_n = k) \\
&= \mathbf{P}\left(\sum_{i=1}^{N_n} X_{n,i} - a_{N_n} R - N_n D \geq 0\right)
\end{aligned}$$

Denote for simplicity $\tilde{X}(k) = \sum_{i=1}^{k} X_{n,i} - a_k R - kD$. We will find the characteristic function of $\tilde{X}(N_n)$ and its limit when n tends to infinity.

$$
\begin{aligned}
\mathbf{E}\exp\left(i\, u \tilde{X}(N_n)\right) &= \sum_{k=1}^{\infty} e^{(i\, u(-a_k R - kD))} \mathbf{E}\exp\left(i\, u \log \frac{S_k}{S_0}\right) \mathbf{P}(N_n = k) \\
&= \int_0^{\infty} e^{i\, u(-a_x R - xD)} \left(\phi_{X_{n,1}}(u)\right)^x d\mathbf{P}^{N_n}(x) \\
&= \int_0^{\infty} e^{i\, u(-a_{nx} R - nxD)} \left(\phi_{X_{n,1}}^n(u)\right)^x d\mathbf{P}^{\frac{N_n}{n}}(x)
\end{aligned}
$$

Since $\phi_{X_{n,1}}^n(u) \to \phi_{N(\alpha',\sigma'^2)}(u)$, $R = R(n) = U(n) - D(n) \to 0$ and consequently $a_{nx} R + nxD \to \log(K/S_0)$ (see (1.10)), we have

$$
\mathbf{E}\exp\left(iu\tilde{X}(N_n)\right) \to \int_0^{\infty} e^{i\, u \log(K/S_0)} \left(\phi_{N(\alpha',\sigma'^2)}(u)\right)^x d\mathbf{P}^Y(x).
$$

Now it is clear that

$$
\tilde{X}(N_n) \xrightarrow{d} Z' - \log \frac{K}{S_0}
$$

where $Z' \sim \text{Hyp}(\sigma'^2 \chi, \psi/\sigma'^2, \alpha'/\sigma'^2, 0)$. Finally we deduce that

$$
\begin{aligned}
\mathbf{E}\Phi(a_{N_n}, N_n, p') &\to \mathbf{P}\left(Z' \geq \log \frac{K}{S_0}\right) \qquad\qquad (1.12) \\
&= \int_{\log K/S_0}^{\infty} f\left(x; \sigma'^2 \chi, \frac{\psi}{\sigma'^2}, \frac{\alpha'}{\sigma'^2}, 0\right) dx
\end{aligned}
$$

Similarly we compute the second integral

$$
\begin{aligned}
\mathbf{E}\rho^{-N_n} \Phi(a_{N_n}, N_n, p) &= \sum_{k=1}^{\infty} \mathbf{P}\left(\sum_{i=1}^{k} \epsilon_{n,i} \geq a_k\right) \rho^{-k} \mathbf{P}(N_n = k) \\
&= \mathbf{E}\rho^{-N_n} \sum_{k=1}^{\infty} \mathbf{P}\left(\tilde{X}(k) \geq 0\right) \frac{\rho^{-k} \mathbf{P}(N_n = k)}{\mathbf{E}\rho^{-N_n}} \\
&= \mathbf{E}\rho^{-N_n} \mathbf{P}\left(\tilde{X}(N_n^\star) \geq 0\right)
\end{aligned}
$$

where $N_n^\star$ is a random variable with a distribution such that $\mathbf{P}(N_n^\star = k) = \rho^{-k} \mathbf{P}(N_n = k)/\mathbf{E}\rho^{-N_n}$. Since $\lim \rho^n = r_0^t$ we find that

$$
\mathbf{E}\rho^{-N_n} \to \mathbf{E}r_0^{-tY}
$$

where $Y \sim \text{GIG}(1, \chi, \psi)$. Further, we get that

$$
\tilde{X}(N_n^\star) \xrightarrow{d} Z^\star - \log \frac{K}{S_0},
$$

where $\mathrm{Law}(Z^\star)$ is the limit distribution of

$$\sum_{k=1}^{N_n^*} X_{n,k}$$

with $N_n^*/n \xrightarrow{d} Y^*$. To recognize the mixing distribution $\mathrm{Law}(Y^\star)$ we compute its Laplace transform

$$
\begin{aligned}
\mathbf{E}e^{-\theta \frac{N_n^*}{n}} &= \sum_{k=1}^{\infty} \exp\left\{-\theta\frac{k}{n}\right\} \frac{\rho^{-k}\mathbf{P}(N_n = k)}{\mathbf{E}\rho^{-N_n}} \\
&= \frac{1}{\mathbf{E}\rho^{-N_n}} \sum_{k=1}^{\infty} \mathbf{E}\exp\left(-\theta\frac{k}{n} - k\log\rho\right)\mathbf{P}(N_n = k) \\
&= \frac{1}{\mathbf{E}\rho^{-N_n}} \sum_{k=1}^{\infty} \mathbf{E}\exp\left(-\frac{k}{n}(\theta + t\log r_0 + \delta_n)\right)\mathbf{P}(N_n = k) \\
&= \frac{\mathbf{E}\exp\left(-\frac{N_n}{n}(\theta + t\log r_0 + \delta_n)\right)}{\mathbf{E}\rho^{-N_n}}
\end{aligned}
$$

with $\delta_n \to 0$. When n tends to infinity we obtain the Laplace transform of $Y^\star$

$$\mathbf{E}e^{-\theta\frac{N_n^*}{n}} \to \frac{\mathbf{E}e^{-Y(\theta + t\log r_0)}}{\mathbf{E}r_0^{-Yt}} \tag{1.13}$$

The above limit is a quotient of two Laplace transforms of the generalized inverse Gaussian distribution at the points $\theta + t\log r_0$ and $t\log r_0$. From (1.6) we have

$$\frac{\mathbf{E}e^{-Y(\theta + t\log r_0)}}{\mathbf{E}r_0^{-Yt}} = \frac{(\psi + 2t\log r_0)^{\frac{1}{2}}}{(\psi + 2\theta + 2t\log r_0)^{\frac{1}{2}}} \frac{K_1\left(\sqrt{\chi(\psi + 2\theta + 2t\log r_0)}\right)}{K_1\left(\sqrt{\chi(\psi + 2t\log r_0)}\right)}$$

Hence $Y^\star$ has the generalized inverse Gaussian law $\mathrm{GIG}(1, \chi, \psi + 2t\log r_0)$ and therefore

$$Z^\star \sim \mathrm{Hyp}\left(\sigma^2\chi, \frac{\psi + 2t\log r_0}{\sigma^2}, \frac{\alpha}{\sigma^2}, 0\right).$$

Moreover

$$\mathbf{E}\rho^{-N_n}\Phi(a_{N_n}, N_n, p) \to \mathbf{E}r_0^{-tY}\mathbf{P}\left(Z^\star \geq \log\frac{K}{S_0}\right) \tag{1.14}$$

$$= \mathbf{E}r_0^{-tY} \int_{\log K/S_0}^{\infty} f\left(x; \sigma^2\chi, \frac{\psi + 2t\log r_0}{\sigma^2}, \frac{\alpha}{\sigma^2}, 0\right) dx$$

Collecting the results (1.9), (1.12), and (1.14) we receive the statement, what completes the proof. $\square$

5   Example

Consider the following random walk $\log S_n/S_0$ on the interval $[0,t]$ with

$$U = \tilde{\sigma}\sqrt{\frac{t}{n}}, \quad D = -U = -\tilde{\sigma}\sqrt{\frac{t}{n}}, \quad \rho^n = r_0^t, \quad q = \frac{1}{2} + \frac{1}{2}\frac{m}{\sigma}\sqrt{\frac{t}{n}}.$$

Such a process converges to a Brownian motion with the drift and then $\log S_t/S_0 \stackrel{d}{=} N(\alpha, \sigma^2)$ where $\alpha = \mathbf{E}\log S_t/S_0 = t(m - \tilde{\sigma}^2/2)$, and $\sigma = \mathbf{Var}\log S_t/S_0 = t\tilde{\sigma}^2$.

However, we consider the random walk with a stochastic number of components driven by N_n, such that $N_n/n \stackrel{d}{\to} Y \sim \mathrm{GIG}(1,\chi,\psi)$. Tending with n to infinity we have a new limit distribution

$$\begin{aligned} Z &= \log S_t/S_0 = Y\,\alpha + \sqrt{Y}\sigma N(0,1) & (1.15) \\ &= \mathrm{Hyp}\left(\hat{\chi}, \hat{\psi}, \hat{m}, \hat{\mu},\right) = \mathrm{Hyp}\left(\sigma^2\chi, \frac{\psi}{\sigma^2}, \frac{m}{\sigma^2}, \hat{\mu}\right) \end{aligned}$$

Eberlein and Keller [EK94] assume that daily returns of the stock S_t have a hyperbolic distribution and therefore can be written in the form

$$\log S_1/S_0 = \mu + \hat{m}\hat{Y} + \sqrt{\hat{Y}}N(0,1)$$

with $\hat{Y} \sim \mathrm{GIG}(1,\hat{\chi},\hat{\psi})$. Together with the hypothesis of independent increments they obtain the Lévy motion process whose daily increments follow the hyperbolic law. In the hyperbolic CRR model, see (1.15), all cumulative returns are hyperbolic in contrast to the model of Eberlein and Keller.

To compute the option price, see Theorem 2, we find the densities of Z' and $Z^\star$

$$Z' \sim \mathrm{Hyp}\left(\sigma^2\chi, \frac{\psi}{\sigma^2}, \frac{\alpha+\sigma^2}{\sigma^2}, 0\right) \tag{1.16}$$

$$Z^\star \sim \mathrm{Hyp}\left(\sigma^2\chi, \frac{\psi+2t\log r_0}{\sigma^2}, \frac{\alpha}{\sigma^2}, 0\right) \tag{1.17}$$

where $\alpha = t(\log r_0 - \tilde{\sigma}^2/2)$.

Eberlein and Keller approximate the parameters of the hyperbolic law of daily returns of stocks in the German stock market and for the Deutsche Bank they receive

$$\hat{\chi} = 9\,10^{-6}, \quad \hat{\psi} = 11794.12, \quad \hat{m} = 0.3972, \quad \hat{\mu} = 0.$$

They establish also $t = 10$, $S_0 = 700\mathrm{DEM}$ and the annual interest rate $r = 0.08$. Notice that in (1.15)-(1.17) we have an additional parameter σ which can not be computed from the other parameters and therefore we calculate the option price for different σ's and the parameters given above

	$\sigma = 2.0$	$\sigma = 1.0$	$\sigma = 0.5$	$\sigma = 0.1$	$\sigma = 0.05$
$K = 680$	24.0118	24.0118	24.0119	24.0308	24.0553
$K = 700$	10.7714	10.7718	10.7717	10.7734	10.8259
$K = 720$	4.2417	4.2418	4.2419	4.2475	4.2641

Since in our case the parameter σ doesn't have a large influence on the price of the call, we set $\sigma = 1$. The table below presents prices of the call for different models: the Black-Scholes model (BS), the hyperbolic model of Eberlein and Keller, and the CRR hyperbolic model

strike K	BS price	hyperbolic EK price	hyperbolic CRR price
680	25.462	25.444	24.012
700	12.666	12.579	10.772
720	5.015	4.995	4.242

As a remarkable point let us mention that the hyperbolic CRR prices are different from the hyperbolic EK prices. In the classical Black-Scholes model for option pricing the fair price for an European call in continuous time coincides with the limit of the binomial option prices as the number of periods to the terminal time goes to infinity. As soon as we move away from the normal assumption for the asset returns, it is not clear that both approaches (continuous and discrete) will lead to the same fair price. Indeed, as it follows from the Eberlein and Keller [EK94], and Küchler et al. [KNSS94] the hyperbolic distributions provide much better fit to stock return data than the normal distribution for the leading German enterprises. Thus it is natural to consider a Black-Scholes type formula for asset returns following a hyperbolic law. The above example demonstrates clearly the sharp difference in prices obtained.

In our opinion, the observed difference between the Black-Scholes and the hyperbolic CRR prices is an expected feature as we move away from the normal assumption. Therefore, we believe that the discrete model is likely to give us more accurate option value.

Acknowledgments: The final version of this paper was prepared during the second author's stay at the Department of Statistics and Applied Probability, University of California Santa Barbara and was supported in part by a Fulbright research grant. He would like to thank Ole E. Barndorff-Nielsen and Zari T.Rachev for helpful comments.

6   References

[BN77]   Barndorff-Nielsen O.E. Exponentially decreasing distributions for the logarithm of particle size. *Proceedings of the Royal Society London A, 353, 401-419,* 1977.

[CRR79]   Cox J.C., Ross S.A. and Rubinstein M. Option pricing: a simplified approach. *J. Financial Economics 7, 229-263,* 1979.

[EK94]   Eberlein E. and Keller U. Hyperbolic distributions in finance. *Preprint,* 1994.

[KNSS94] Küchler, U., Neumann, K., Sorensen, M., and Streller, A., Stock returns and hyperbolic distributions, *Research report,* Humboldt Univ. Berlin 1994.

[Man63]   Mandelbrot B.B. New methods in statistical economics. *J. Political Economy 71, 421-440,* 1963.

[RR94]   Rachev S.T. and Rüschendorf L. Models for option prices. *Teor. Veroyatnost. i Primenen. 39, 150-199,* 1994.

[Shi94]   Shiryaev A.N., Kabanov Yu.M. Kramkov D.O. and Melnikov A.V. Towards the theory of pricing of options of both European and American types. I. Discrete time. *Teor. Veroyatnost. i Primenen. 39, 23-79,* 1994.

A class of shot noise models for financial applications
*†‡

Gennady Samorodnitsky

Cornell University

December 11, 1995

Abstract

We describe a class of non-Markov shot noise processes that can be used as models for rates of return on securities, exchange rate processes and other processes in finance. These are continuous time processes that can exhibit heavy tails that become lighter when sampling interval increases, clustering and long memory.

1 Introduction

A "typical" stochastic process $\{Y(t),\, t \geq 0\}$ can be the log-price process of a particular security S (a stock in particular), the log-exchange rate process between major currencies, etc. Understanding of the structure of the random process $\{Y(t),\, t \geq 0\}$ is of obvious importance, and many researchers in academia as well as those associated with the banking industry have taken a hard look on the data that accumulated throughout the years. Both the properties of the marginal distributions of the process and its dependence structure (in particular, correlations) have been thoroughly discussed, simply because those are the factors that affect the risk associated with the trading, and help in rational pricing of derivatives on securities (or on exchange rates). We begin with a discussion of some of the more important properties of the financial processes, which we formulate in terms of the *return process* (or increment process) $\{X(t) = Y(t+1) - Y(t),\, t \geq 0\}$. The latter process is usually regarded as being stationary (at

---

*This research was supported by the NSF grant DMS-94-00535, the NSA grant MDA904-95-H-1036 and United States - Israel Binational Science Foundation (BSF) Grant 92-00074/1 as well as by Forschungsinstitut für Mathematik and Institut für Operations Research of ETH, Zürich.

†*AMS 1991 subject classification.* Primary 60E07, 60G55, 62P05

‡*Keywords and phrases:* shot noise processes, non-Markov models, financial models, heavy tails, long-range dependence, regular variation, infinitely divisible processes, cluster Poisson point processes, Lévy measure, tail behavior of the distribution.

least, within reasonable time scale). There is not universal consensus that financial data always exhibit all of these features. There is a growing empirical evidence, nevertheless, supporting the findings below.

1. **The marginal distribution of $X(t)$ is** *heavy tailed.*

   This is always taken to mean that the tail probability $P(X(t) > \lambda)$ (and its counterpart on the negative side) is heavier than that of a normal distribution, and most often it is understood that this tail probability is regularly varying at infinity:

   $$P(X(t) > \lambda) \sim \lambda^{-\alpha} L(\lambda) \qquad (1.1)$$

   for some $\alpha > 0$, where L is a slowly varying function (that is, $L(xy)/L(x) \to 1$ as $x \to \infty$ for every $y > 0$). Again, the same conclusion is reached for the left tail, $P(X(t) < -\lambda)$, as $\lambda \to \infty$ (Mandelbrot [Man63], Fama [Fam65], Blattberg and Gonedes [BG74], Mittnik and Rachev [MR93]). There is some evidence that the tail exponent α lies, in many cases, between 2 and 4 (Akgiray and Booth [AB88], Loretan and Phillipps [LP94]), even though there is still some debate whether the estimated tail exponent exceeding two truly rules out the use of stable distributions in financial modeling (McCulloch [McC94]).

2. **The tails become less heavy as the sampling interval increases.**

   That is, daily returns appear to have heavier tails than weekly returns which, in turn, appear to have heavier tails than monthly returns, etc. In other words, the tail exponent α in (1.1) increases when we add a certain number of consecutive observations. This was first reported in Akgiray and Booth [AB88]; see also Guillaume *et al.* [GDD*94] who report similar phenomena for high frequency data.

3. **Clustering in the data.**

   This means that there are periods of high activity (and, thus, significant changes, or high volatility) and periods of low activity (low volatility). To say it differently, big changes come in clusters (Mandelbrot [Man63], Engle [Eng82]).

4. **Long-range dependence in the data.**

   This means that $X(t)$ and $X(t + s)$ are highly dependent even for very large time lag s. This is most often observed through the sample autocorrelation function of the volatility (absolute changes) in the prices (Bollerslev, Chou and Kroner [BCK92] and also Guillaume *et al.* [GDD*94] in the high frequency context).

The following plots represent the daily returns from the S&P composite index for the period (July 2, 1962 - December 31, 1991), Hill estimator plots for the daily returns and 4-day returns, and the empirical autocorrelation function for the absolute returns.

The plot of the data set (Fig. 1.1) demonstrates the clustering phenomenon. The "catastrophe" at observations numbered around 6300-6400 corresponds to the market crash of October 1987.

The Hill estimators of the tail exponents for the daily and 4-day returns are shown on Fig. 1.2. The reliability of this non-parametric technique for tail estimation is widely, though not universally, accepted.

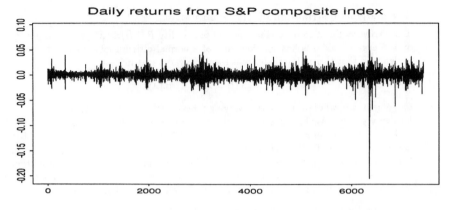

Figure 1.1

Hill estimator requires one to select the number of order statistics used in computing the estimator, and this is most often a difficult problem, typically solved in either rather heuristic way, or or by introducing certain assumptions on the actual marginal distribution of the daily returns.

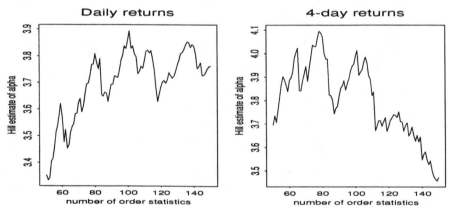

Figure 1.2

334

The fact that the data under consideration is heavy tailed seems to be well established, and it is also possible to observe it by plotting the the tail of the empirical distribution function against its argument in the log-log coordinates. Presently we have selected a Hill estimator based on a relatively short sample (this tends to increase the variance but reduce the bias of the estimator). It seems to indicate an increase in the tail exponent from about $\alpha = 3.75$ for daily returns to about $\alpha = 3.95$ for 4-day returns.

The empirical autocorrelation function depicted in Fig. 1.3 shows a very slow decay in correlations of absolute returns. It appears to be significant even after as much as 150-200 days.

Modeling of financial processes has developed, naturally, in two ways. The first is concerned with the dynamical structure of the processes, i.e. understanding how they develop in time. Here much work has been done along the lines of processes with stationary and independent increments ("the random walk hypothesis"), especially Brownian motion (Samuelson [Sam65], see also Föllmer and Schweizer [FS93] for a recent discussion), but also jump processes, like Lévy stable motion (Mandelbrot [Man63]) and others. Another body of work is devoted to ARCH and GARCH models, which are time series models that can account, for example, for the heavy tails and the clustering property of the financial processes.

Series : abs(nyse)

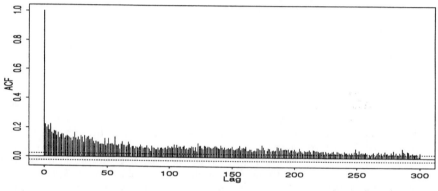

Figure 1.3

Secondly, it is important to find a good model for marginal distribution of the return $X(t)$, and there is a vast literature on the subject, beginning with the Gaussian and stable distributions mentioned above, up to very sophisticated families of distributions, showing an excellent fit to the data (alternative stable distributions of Mittnik and Rachev [MR93], hyperbolic distributions of Eberlein and Keller [EK94] and Küchler *et al.* [KNrS94], Student's t distribution of Spanos [Spa93] and inverse Gaussian distribution of Barndorff-Nielsen [Bar94], to mention just a few).

Of course, the above properties of financial processes show that the latter require careful modeling of both marginal distribution of $X(t)$ and the non-trivial dependence structure in the process $\{X(t),\ t \geq 0\}$. However, it is also important to have an *intuitive* model, one that seems to be in correspondence with the activity underlying the financial processes. For example, a process with stationary and independent increments has an obvious intuitive appeal by expressing the idea of "total randomness" of returns, even though the idea may not be universally accepted. See Taylor [Tay86] for a discussion. Stability (in a wide sense) property of financial models is another intuitive property, as emphasized by Mandelbrot [Man82] and Mittnik and Rachev [MR93].

The purpose of the present paper is to suggest a class of continuous time models for financial processes, the shot noise models, that arise naturally and intuitively if one thinks of the more significant activity in a market as caused by reaction of the market to various events (interest rate changes, mergers, announcements, expectations, rumors, political events, etc). Specifically, assume that the events affecting the market arrive at (random) times T_i, $i = 1, 2, \ldots$, the (random) initial effect of the event i on the market is Z_i, and the this effect changes with time according to a deterministic function $f(t)$, $t \geq 0$. The random variables $Z_1, Z_2, \ldots$ are assumed to be i.i.d. Since the effect of the event number i begins at the time of its arrival, T_i, we can think of the return at time t as the total of the effects caused up to this time, which gives

$$X(t) = \sum_{T_i \leq t} Z_i f(t - T_i).$$

Such a model is a simplification of the general shot noise model described usually as

$$X(t) = \sum_{T_i \leq t} W_i(t - T_i),$$

with i.i.d. stochastic processes $\{W_i(t),\ t \geq 0\}$. The deterministic nature of the dynamic effect of each event on the market leads to a simpler model (which by itself is very important), and we will argue in the sequel that even such a choice leads to a very rich class of models that can account for the properties of financial processes described above. One should mention that such processes are non-Markov moving averages, and that the many statistical problems associated with such processes are still being studied, and many others should yet be studied.

Shot noise processes are not a new phenomenon in probabilistic modeling. They are used in risk theory (Klüppelberg and Mikosch [KM93a] and [KM93b]), in modeling earthquake aftershocks (Vere-Jones [Ver70]) and computer failure times (Lewis [Lew64]), among others. We purport to show that these processes can be useful in financial modeling as well.

This paper is organized as follows. In the next section we define our model and discuss its basic properties. Section 3 contains a specific case of our model that exhibits many of the features of financial processes described in the present section, and Section 4 presents some simulation results and conclusions.

2   Shot noise processes

We consider a stochastic process of the form

$$X(t) = \sum_{i=1}^{\infty} Z_i f(t - T_i),\ t \in R, \tag{2.1}$$

where $f : R \to R$ is a measurable function vanishing on the negative half-line, $Z_1, Z_2, \ldots$ are i.i.d. random variables with a common distribution function F, and $T_1, T_2, \ldots$ constitute a *cluster Poisson process* independent of the sequence $Z_1, Z_2, \ldots$. Such a process is defined by three parameters: cluster arrival rate ρ, cluster size probabilities p_k, $k \geq 1$ and displacement distribution H. That is, cluster "centers" C_j, $j \geq 1$ arrive according to a time homogeneous Poisson process with rate ρ. The j'th cluster is of size $K_j = k$ with probability p_k, $k \geq 1$, and the points in this cluster are located at $C_j + D_{ij}$, $i = 1, \ldots, K_j$. The array $\{D_{ij}, i \geq 1, j \geq 1\}$ is an array of i.i.d. random variables with common distribution H, independent of the two independent sequences: that of Poisson arrivals $\{C_j\, j \geq 1\}$ and that of i.i.d. discrete random variables $\{K_j, j \geq 1\}$ with probabilities given by p_k, $k \geq 1$.

We refer to the stochastic process $\{X(t), t \in R\}$ as a cluster Poisson shot noise process. In the particular case when all clusters are of size 1 with probability 1 ($p_1 = 1$), the underlying point process $T_1, T_2, \ldots$ is just a time homogeneous Poisson process with rate ρ, leading to a simpler, Poisson shot noise model. The cluster nature of the more general model can, however, be useful for modeling the clustering property of financial processes.

One can see immediately from the definition of the cluster Poisson shot noise process that it suffers from an unidentifiability problem. Indeed, two "whole distributions" must be specified. In future work we plan to restrict the modeling to particular parametric classes of distributions, which can eliminate the unidentifiability problem. In certain cases one may want to decide on the form of the cluster size and displacement distributions from other considerations, based on the nature (or theories) of shocks affecting the market. It is this possibility of concentrating on the nature of the financial process that adds appeal to the proposed class of models.

We would like to emphasize from the beginning that, in spite of the deterministic nature of the function f in (2.1), the shot noise process $\{X(t), t \in R\}$ is not predictable between the jumps T_i, $i = 1, 2, \ldots$, because the shots of the noise overlap. Therefore, observing the sample path of the process until a time s, say, it is not generally possible to infer the locations T_i's and the magnitudes Z_i's of the jumps up to that time, and so it is impossible to predict the behavior of the process until the next jump. It is this feature of shot noise processes that makes them valuable as stochastic models.

The Poisson character of a cluster Poisson process implies immediately that the cluster Poisson shot noise process (2.1) is infinitely divisible, whenever it is well defined (that is, whenever it is defined by a convergent sum). We would like to understand its properties, for which purpose it is more convenient to regard this process as a stochastic integral

$$X(t) = \int_R f(t - s) M(ds),\ t \in R \tag{2.2}$$

337

with respect to an cluster compound Poisson random measure M defined by

$$M(A) = \sum_{j=1}^{\infty} \sum_{i=1}^{K_j} Z_{ij} \mathbf{1}\big((C_j + D_{ij}) \in A\big). \tag{2.3}$$

Here we have transformed, for obvious convenience, the i.i.d. sequence $Z_1, Z_2, \ldots$ into an i.i.d. array Z_{ij}, $i, j \geq 1$, with the same common distribution F. This form of representation of the cluster Poisson shot noise model shows that it is also a *moving average* with respect to the random measure M. The result of the following lemma is, undoubtedly, known. Nonetheless, it does not seem to be readily available, and we include it here for completeness. Notice that the lemma does not require the function f to vanish on the negative half-line.

LEMMA 2.1 *Suppose that*

$$\int_R \sum_{k=1}^{\infty} p_k E\Big(1 \wedge \Big|\big(\sum_{i=1}^{k} Z_{i1} f(x + D_{i1})\big)\Big|\Big) \, dx < \infty. \tag{2.4}$$

Then the stochastic integral $I(f) = \int_R f(s) M(ds)$ *defined by*

$$I(f) = \lim_{R \to \infty} \int_R f(s) M_R(ds) := \lim_{R \to \infty} I_R(f),$$

where

$$M_R(A) = \sum_{|C_j| \leq R} \sum_{i=1}^{K_j} Z_{ij} \mathbf{1}\big((C_j + D_{ij}) \in A\big),$$

exists as an almost sure limit, and is an infinitely divisible random variable with characteristic function

$$E e^{i\theta I(f)} = \exp\Big\{-\rho \int_R \big[1 - \psi_K\big(E e^{i\theta Z_{11} f(x + D_{11})}\big)\big] dx\Big\}, \tag{2.5}$$

where $\psi_K(s) = \sum_{k=1}^{\infty} p_k s^k$, $|s| \leq 1$ *is the generating function of the cluster size* K. *Moreover, the Lévy measure of* $I(f)$ *is given by*

$$\mu = \rho \sum_{k=1}^{\infty} p_k \mu_k, \tag{2.6}$$

where for $k \geq 1$ *the measure* μ_k *is defined by*

$$\mu_k = \big(Leb \otimes P\big) \circ T_k^{-1}.$$

Here T_k *is a map from* $R \times \Omega$ *to* R *defined by*

$$T_k(x, \omega) = \sum_{i=1}^{k} Z_{i1}(\omega) f(x + D_{i1}(\omega)), \quad x \in R, \ \omega \in \Omega. \tag{2.7}$$

338

PROOF: The required computation can be done in a number of ways, the most "logical" of which is to "mark" the Poisson process by the clusters. Computationally, nonetheless, the easiest way seems to be the following direct argument.

Recalling the definition of the random measure M, we condition on the number of Poisson points in $(-R, R)$ whose locations are then uniformly distributed in $(-R, R)$ and conclude that

$$E e^{i\theta I_R(f)} = \sum_{n=0}^{\infty} e^{-2\rho R} \frac{(2\rho R)^n}{n!} \left(E e^{i\theta S(f)} \right)^n$$

$$= \exp\left\{ -2\rho R \left(1 - E e^{i\theta S(f)} \right) \right\},$$

where

$$S(f) = \sum_{i=1}^{K_1} Z_{i1} f(RU + D_{i1}),$$

and U is a uniform random variable in $(-1, 1)$ independent of the rest of the random variables above. It follows that

$$E e^{i\theta S(f)} = \frac{1}{2R} \int_{-R}^{R} \sum_{k=1}^{\infty} p_k \left(E e^{i\theta Z_{11} f(x+D_{11})} \right)^k dx,$$

and therefore,

$$E e^{i\theta I_R(f)} = \exp\left\{ -\rho \int_{-R}^{R} \left[1 - \psi_K \left(E e^{i\theta Z_{11} f(x+D_{11})} \right) \right] dx \right\}. \tag{2.8}$$

From here we observe that the Lévy measure μ_R of $I_R(f)$ is given by (2.6), but the domains of the maps T_k's in (2.7) are restricted to $(-R, R) \times \Omega$. As $R \to \infty$ the measures μ_R increase to the measure μ given by (2.6). It follows easily from (2.4) that the characteristic function of $I_R(f)$ converges, as $R \to \infty$, to the expression in the right hand side of (2.5). Therefore, $I(f)$ exists as a limit in distribution, and its Lévy measure is given by (2.6). Finally, existence of $I(f)$ as an almost sure limit follows from the fact that, for a series of independent random variables, convergence in distribution is equivalent to the a.s. convergence. ∎

Remark Note that when the compounding distribution F is symmetric, the same argument shows that the integrability condition (2.4) may be replaced with a weaker condition

$$\int_R \sum_{k=1}^{\infty} p_k E \left(1 \wedge \left| \left(\sum_{i=1}^{k} Z_{i1} f(x + D_{i1}) \right) \right|^2 \right) dx < \infty. \tag{2.9}$$

The assumption of homogeneous arrivals for the underlying Poisson process clearly implies that, once the function f satisfies (2.4), the well defined stochastic process $\{X(t), t \geq 0\}$ given by (2.1) or (2.2) is a stationary process. Its second-order properties are described by the following lemma.

LEMMA 2.2 Let (2.4) hold, and assume that $EK_1^2 < \infty$, $EZ_{11}^2 < \infty$ and

$$\int_R f(y)^2 dy < \infty.$$

339

Then the stochastic process $\{X(t), t \geq 0\}$ has a finite second moment, and its covariance function is given by

$$R(t) = \rho E K_1 E Z_{11}^2 \int_R f(y) f(y+t) dy + \rho(E K_1^2 - E K_1)(E Z_{11})^2 \delta(t), \qquad (2.10)$$

where

$$\delta(t) = \int_R f(y+t) \Big(\int_R \int_R f(y + x_1 - x_2) H(dx_1) H(dx_2) \Big) dy. \qquad (2.11)$$

PROOF: The fact that the process has a finite second moment follows from (2.6) and the fact that $E I(f)^2 < \infty$ if and only if $\int_R x^2 \mu(dx) < \infty$, and (2.10) follows directly from (2.1) by using repeatedly the conditioning argument

$$\operatorname{cov}(X, Y) = E\Big(\operatorname{cov}(X, Y|\mathcal{F}) \Big) + \operatorname{cov}\Big(E(X|\mathcal{F}), E(Y|\mathcal{F}) \Big).$$

∎

3   The process $X_{\gamma,d,m}$

In this section we show that one can account for the properties exhibited by the financial processes and described in Section 1 by simple processes of the the shot noise type described in the previous section. We would like to emphasize that what we present here is only an example that exhibits these properties, and that many other cluster shot noise processes will share the latter. Let us take the compounding distribution F to be the Laplace distribution. That is, F is absolutely continuous with respect to Lebesgue measure, with the density

$$F'(x) = .5 e^{-|x|}, \ x \in R.$$

The choice of the Laplace density, though fairly arbitrary, leads to simpler computations, and permits one to view probability tails through Laplace transform-type of lenses (c.f. (3.9)) and to utilize Tauberian-type arguments.

For a $\gamma > 0$, $d > 0$ and an integer $m \geq 1$ let

$$f_{\gamma,d,m}(s) = j^\gamma \cos \pi \frac{j}{m} \text{ if } s \in \Big(j - (j+1)^{-(d+1)}, j + (j+1)^{-(d+1)} \Big), \ j \geq 1, \qquad (3.1)$$

and $f_{\gamma,d,m}(s) = 0$ is s does not belong to any of the above intervals. Define

$$X_{\gamma,d,m}(t) = \int_R f_{\gamma,d,m}(t-s) M(ds), \ t \geq 0, \qquad (3.2)$$

where M is a cluster compound Poisson random measure. The following theorem describes the properties of the process $X_{\gamma,d,m}$.

THEOREM 3.1 *(i) Suppose that $EK_1 < \infty$. Then $X_{\gamma,d,m}$ is a well defined stationary stochastic process.*

(ii) Assume that

$$EK_1^2 < \infty \tag{3.3}$$

and that the displacement distribution H is absolutely continuous with respect to the Lebesgue measure. Then

$$P(X_{\gamma,d,m}(t) > \lambda) \sim c_{d,\gamma,m}^{(1)} \rho EK_1 \lambda^{-d/\gamma}, \ \lambda \to \infty, \tag{3.4}$$

where

$$c_{d,\gamma,m}^{(1)} = \frac{1}{2\gamma m} \Gamma(d/\gamma) \sum_{i=0}^{2m-1} |\cos \pi \frac{i}{m}|^{d/\gamma}.$$

(iii) Under the assumptions of (ii), if $\gamma > 1$, then

$$P\left(\sum_{k=0}^{2m-1} X_{\gamma,d,m}(t+k) > \lambda\right) \sim \begin{cases} c_{d,\gamma,m}^{(2)} \rho EK_1 \lambda^{-(d+1)/\gamma} & \text{if } \frac{d+1}{\gamma} < \frac{d}{\gamma-1}, \\ c_{d,\gamma,m}^{(3)} \rho EK_1 \lambda^{-d/(\gamma-1)} & \text{if } \frac{d+1}{\gamma} > \frac{d}{\gamma-1}, \\ (c_{d,\gamma,m}^{(2)} + c_{d,\gamma}^{(3)}) \rho EK_1 \lambda^{-(d+1)/\gamma} & \text{if } \frac{d+1}{\gamma} = \frac{d}{\gamma-1}, \end{cases} \tag{3.5}$$

and if $\gamma \leq 1$ then

$$P\left(\sum_{k=0}^{2m-1} X_{\gamma,d,m}(t+k) > \lambda\right) \sim c_{d,\gamma,m}^{(2)} \rho EK_1 \lambda^{-(d+1)/\gamma} \tag{3.6}$$

as $\lambda \to \infty$. Here

$$c_{d,\gamma,m}^{(2)} = \frac{1}{m} \Gamma\left((d+\gamma+1)/\gamma\right) \sum_{i=1}^{2m-1} \sum_{j=1}^{2m} \left|\sum_{k=0}^{i-1} \cos \pi(j+k)/2m\right|^{(d+1)/\gamma} > 0$$

and

$$c_{d,\gamma,m}^{(3)} = \frac{1}{\gamma-1} (\gamma m)^{d/(\gamma-1)} \Gamma\left(d/(\gamma-1)\right).$$

(iv) Assume that (3.3) holds, and that

$$d > 2\gamma. \tag{3.7}$$

Then the process $X_{\gamma,d,m}$ has a finite variance, and its covariance function $R(n)$ satisfies, as $n \to \infty$,

$$R(n) \sim \kappa_{d,\gamma,m} \rho EK_1 n^{2\gamma-d} \sum_{i=1}^{2m} \cos \pi i/m \cos \pi(i+n)/m \tag{3.8}$$

in the sense that the product $n^{d-2\gamma} R(n)$ has at most $2m$ subsequential limits given by (3.8), all of which are finite, and some of which are positive. Here

$$\kappa_{d,\gamma,m} = 2m^{-1} \int_0^\infty x^\gamma (1+x)^{\gamma-d-1} ds.$$

Remark Note that the the probability tails of the process $X_{\gamma,d,m}$ are heavy, and that the tail of $\sum_{k=0}^{2m-1} X_{\gamma,d,m}(k)$ is lighter than that $X_{\gamma,d,m}(0)$, which corresponds to the observed properties of financial processes. Secondly, the covariance function of $X_{\gamma,d,m}$ decays like a power, and not as exponential, function, and, if $2\gamma < d \leq 2\gamma + 1$, then

$$\sum_{n=1}^{\infty} |R(n)| = \infty,$$

that is, $X_{\gamma,d,m}$ exhibits long range dependence in its usual meaning (Brockwell and Davis [BD91]).

PROOF OF THEOREM 3.1: Throughout the proof we will suppose, for simplicity of notation, that $\rho = 1$.

(i) We need to check (2.9). Observe that, for a fixed k and fixed displacements $D_{11}, \ldots, D_{k1}$

$$\{x \in R : \sum_{i=1}^{k} Z_{i1} f(x + D_{i1}) \neq 0\} \subset \bigcup_{i=1}^{k} (\mathcal{D} - D_{i1}),$$

where $\mathcal{D}$ is the domain of the function f:

$$\mathcal{D} = \bigcup_{j=1}^{\infty} \left(j - (j+1)^{-(d+1)}, j + (j+1)^{-(d+1)}\right).$$

Using the translation invariance of the Lebesgue measure we conclude that for any $k \geq 1$,

$$\int_R E\left(1 \wedge \left|\left(\sum_{i=1}^{k} Z_{i1} f(x + D_{i1})\right)\right|^2\right) dx \leq 2k \sum_{j=1}^{\infty} (j+1)^{-(d+1)},$$

and so (2.9) follows from $EK_1 < \infty$ and $d > 0$.

(ii) Suppose first that $K_1 = 1$ with probability 1. Then M is just a compound Poisson random measure, and we may assume that $D_{11} = 0$ with probability 1.

It follows from Lemma 2.1 that for any $\lambda > 0$, the Lévy measure μ of $X_{\gamma,d,m}(t)$ satisfies

$$\mu\{(\lambda, \infty)\} = \frac{1}{2} \int_{-\infty}^{\infty} \int_0^{\infty} 1(xf(s) > \lambda) e^{-x} \, dx \, ds + \frac{1}{2} \int_{-\infty}^{\infty} \int_0^{\infty} 1(xf(s) < -\lambda) e^{-x} \, dx \, ds$$

$$= \frac{1}{2} \int_0^{\infty} \int_0^{\infty} 1(x|f(s)| > \lambda) e^{-x} \, dx \, ds \tag{3.9}$$

$$= \frac{1}{2} \int_0^{\infty} \exp(-\lambda/|f(s)|) \, ds.$$

Substituting (3.1) into (3.9) we conclude that

$$\mu\{(\lambda, \infty)\} = \sum_{j=1}^{\infty} (j+1)^{-(d+1)} \exp(-\lambda j^{-\gamma}/|\cos \pi \frac{j}{m}|)$$

342

$$= \sum_{i=0}^{2m-1} \sum_{j=1}^{\infty} (2jm+i+1)^{-(d+1)} \exp(-\lambda(2jm+i)^{-\gamma}/|\cos\pi\frac{i}{m}|) \tag{3.10}$$

$$:= \sum_{i=0}^{2m-1} T_i(\lambda).$$

As $\lambda \to \infty$,

$$T_i(\lambda) \sim \frac{1}{2\gamma m}\Gamma(d/\gamma)|\cos\pi\frac{i}{m}|^{d/\gamma}\lambda^{-d/\gamma},$$

thus proving that

$$\mu\{(\lambda,\infty)\} \sim c_{d,\gamma,m}^{(1)}\lambda^{-d/\gamma}, \ \lambda \to \infty. \tag{3.11}$$

We remark that in the case $K_1 = 1$ a.s. (3.4) already follows from (3.11) and the equivalence of the tail of an infinitely divisible random variable with that of its Lévy measure when the latter is subexponential (Theorem 1 of Embrechts, Goldie and Veraverbeke [EGV79]).

In the general case we use (2.6) and appeal to Theorem 3.2 with $\Phi(y,x) = yf(x)$ to conclude that

$$\mu\{(\lambda,\infty)\} \sim EK_1\mu_1\{(\lambda,\infty)\}$$

$$\sim c_{d,\gamma,m}^{(1)}EK_1\lambda^{-d/\gamma}, \ \lambda \to \infty,$$

which establishes (3.4) in the same way as above.

(iii) We already know from the proof of part (ii) that it is enough to prove our statement in the case $K_1 = 1$ and $D_{11} = 0$ a.s.

The argument of (3.9) shows that for any $\lambda > 0$ the Lévy measure μ of $\sum_{k=0}^{2m-1} X_{\gamma,d,m}(t+k)$ is given by

$$\mu\{(\lambda,\infty)\} = \frac{1}{2}\int_R \exp(-\lambda/|\sum_{k=0}^{2m-1} f(s+k)|)\,ds. \tag{3.12}$$

For $j \geq 1$ and $i = 1, 2, \ldots, 2m$ we denote

$$I_i^{(j)} = \left(j - (j+i)^{-(d+1)}, j + (j+i)^{-(d+1)}\right).$$

Then

$$\int_R \exp(-\lambda/|\sum_{k=0}^{2m-1} f(s+k)|)\,ds = \int_{-(2m-1)}^{1-2^{-(d+1)}} \exp(-\lambda/|\sum_{k=0}^{2m-1} f(s+k)|)\,ds$$

$$+ \sum_{j=1}^{\infty} \sum_{i=1}^{2m-1} \int_{I_i^{(j)}-I_{i+1}^{(j)}} \exp(-\lambda/|\sum_{k=0}^{2m-1} f(s+k)|)\,ds \tag{3.13}$$

$$+ \sum_{j=1}^{\infty} \int_{I_{2m}^{(j)}} \exp(-\lambda/|\sum_{k=0}^{2m-1} f(s+k)|)\,ds$$

$$:= S_0(\lambda) + \sum_{i=1}^{2m-1} S_i(\lambda) + S_{2m}(\lambda).$$

343

Observe first that f is bounded from above on intervals bounded away from infinity. Therefore,

$$S_0(\lambda) = o(e^{-c\lambda}), \ \lambda \to \infty, \ c > 0. \tag{3.14}$$

Now,

$$S_{2m}(\lambda) = \sum_{j=1}^{\infty} 2(j + 2m)^{-(d+1)} \exp\left(-\lambda/|\sum_{k=0}^{2m-1} (j+k)^{\gamma} \cos \pi \frac{j+k}{m}|\right). \tag{3.15}$$

Let

$$a_j = |\sum_{k=0}^{2m-1} (j+k)^{\gamma} \cos \pi \frac{j+k}{m}|, \ j = 1, 2, \ldots.$$

It is a straightforward calculation that

$$a_j \sim m\gamma j^{\gamma-1}, \ j \to \infty. \tag{3.16}$$

Now there are two possibilities. If $\gamma \leq 1$ then it follows from (3.15) and (3.16) that

$$S_{2m}(\lambda) = o(e^{-c\lambda}), \ \lambda \to \infty, \ c > 0, \tag{3.17}$$

whereas if $\gamma > 1$ then

$$S_{2m}(\lambda) \sim \sum_{j=1}^{\infty} 2(j + 2m)^{-(d+1)} \exp\left(-\lambda/m\gamma j^{\gamma-1}\right) \tag{3.18}$$

$$\sim 2c_{d,\gamma,m}^{(3)} \lambda^{-d/(\gamma-1)}, \ \lambda \to \infty.$$

It remains to consider the terms $S_i(\lambda)$, $i = 1, \ldots, 2m - 1$ in (3.13). For each i as above we have

$$S_i(\lambda) = 2\sum_{j=1}^{\infty} \left((j+i)^{-(d+1)} - (j+i+1)^{-(d+1)}\right) \exp\left(-\lambda/|\sum_{k=0}^{i-1}(j+k)^{\gamma} \cos \pi(j+k)/m|\right). \tag{3.19}$$

Denote

$$b_j^{(i)} = |\sum_{k=0}^{i-1}(j+k)^{\gamma} \cos \pi(j+k)/m|, \ i = 1, \ldots, 2m - 1, \ j = 1, 2, \ldots.$$

Now, for each $i = 1, \ldots, 2m - 1$ and $r = 1, 2, \ldots, 2m$

$$\frac{b_j^{(i)}}{j^{\gamma}} \to |\sum_{k=0}^{i-1} \cos \pi \frac{r+k}{m}| := L(i,m,r)$$

as $j \to \infty$ along the subsequence $j = r + 2mJ$, $J = 1, 2, \ldots$. Clearly, each $L(i, m, r)$ is a finite nonnegative number, and it is straightforward to check that for every fixed $i = 1, \ldots, 2m - 1$

344

there is an $r = 1, 2, \ldots, 2m$ such that $L(i, m, r) > 0$. This means that if we let $\mathcal{L}^{(i)} = \{r = 1, 2, \ldots, 2m : L(i, m, r) > 0\}$, then $\mathcal{L}^{(i)} \neq \emptyset$. Write for $i = 1, \ldots, 2m - 1$

$$S_i(\lambda) = \sum_{r=1}^{2m-1} S_{i,r}(\lambda),$$

where

$$S_{i,r}(\lambda)$$
$$= 2 \sum_{J=0}^{\infty} \left((2mJ + r + i)^{-(d+1)} - (2mJ + r + i + 1)^{-(d+1)} \right) \exp\left(-\lambda / |\sum_{k=0}^{i-1} (2mJ + r + k)^{\gamma} \cos \pi \frac{r+k}{m}| \right).$$

For every $r \in \mathcal{L}^{(i)}$ we have, as $\lambda \to \infty$,

$$S_{i,r}(\lambda) \sim \frac{d+1}{2\gamma m} L(i, m, r)^{(d+1)/\gamma} \Gamma((d+1)/\gamma) \lambda^{-(d+1)/\gamma},$$

while for every $r \notin \mathcal{L}^{(i)}$ we have, as $\lambda \to \infty$,

$$S_{i,r}(\lambda) = o(\lambda^{-(d+1)/\gamma}).$$

Therefore, for every $i = 1, \ldots, 2m - 1$,

$$S_i(\lambda) \sim 2c_{d,\gamma,m}^{(2)} \lambda^{-(d+1)/\gamma} \quad \lambda \to \infty. \tag{3.20}$$

Now the conclusion of part (iii) of the theorem follows from (3.12), (3.13), (3.14), (3.17), (3.18), (3.20) and Theorem 1 of Embrechts, Goldie and Veraverbeke [EGV79].

(iv) Since

$$\int_R f(y)^2 \, dy = 2 \sum_{j=1}^{\infty} j^{2\gamma} (j+1)^{-(d+1)} \cos^2 \pi \frac{j}{m} < \infty$$

under (3.7), we conclude from Lemma 2.2 that the process $X_{\gamma,d,m}$ has a finite variance. Moreover, it follows from (2.10) that for every $n \geq 0$,

$$R(n) = 2EK_1 \int_R f(y) f(y+n) dy$$

$$= 4EK_1 \sum_{j=1}^{\infty} j^{\gamma} (j+n)^{\gamma} \cos \pi \frac{j}{m} \cos \pi \frac{j+n}{m} (j+n+1)^{-(d+1)}$$

$$= 4EK_1 \sum_{i=1}^{2m} \cos \pi \frac{i}{m} \cos \pi \frac{i+n}{m} \sum_{j=0}^{\infty} (i+2mj)^{\gamma} (i+2mj+n)^{\gamma} (i+2mj+n+1)^{-(d+1)}.$$

Now, for each $i = 1, \ldots, 2m$ we have

$$\sum_{j=0}^{\infty} (i+2mj)^{\gamma} (i+2mj+n)^{\gamma} (i+2mj+n+1)^{-(d+1)} \sim n^{2\gamma-d} (2m)^{-1} \int_0^{\infty} x^{\gamma} (1+x)^{\gamma-d-1} \, dx$$

345

as $n \to \infty$, and (3.8) follows. This completes the proof of the theorem. ∎

The remainder of this section is devoted to the proof of an ingredient in the argument of Theorem 3.1. Moreover, it concerns a question which is interesting in its own right. Given a sequence of i.i.d. random variables $X_1, X_2, \ldots$ with a common *subexponential* distribution F (that is, such that $\lim_{\lambda \to \infty} \frac{\overline{F*F}(\lambda)}{\overline{F}(\lambda)} = 2$), it turns out that

$$\lim_{\lambda \to \infty} \frac{\overline{F^{*m}}(\lambda)}{\overline{F}(\lambda)} = m$$

for any $m \geq 1$ and, moreover, if N is an independent of the i.i.d. sequence integer valued nonnegative random variable, then under quite general conditions we have

$$\lim_{\lambda \to \infty} \frac{P(\sum_{i=1}^{N} X_i > \lambda)}{P(X_1 > \lambda)} = EN. \tag{3.21}$$

see Chover, Ney and Wainger [CNW73], Embrechts, Goldie and Veraverbeke [EGV79] and Cline [Cli87]. In short, taking convolution powers of subexponential distributions preserves their tails.

Suppose now that taking convolution powers is performed *conditionally*. That is, let $(A, \mathcal{A}, \eta)$ be a σ-finite measure space, and let $(F(x, \cdot), \, x \in A)$ be a measurable family of distributions on R such that

$$\overline{G}(\lambda) := \int_A F(x, (\lambda, \infty)) \eta(dx) < \infty \tag{3.22}$$

for all λ large enough, and such that the distribution function $G = 1 - \min(\overline{G}, 1)$ belongs to the subexponential class. For each $x \in A$ let X_i^x, $i \geq 1$ be a sequence of i.i.d. random variables with common distribution $F(x, \cdot)$. Under what conditions is it true that

$$\lim_{\lambda \to \infty} \frac{\int_A P\left(\sum_{i=1}^{m} X_i^x > \lambda\right) \eta(dx)}{\int_A P(X_1^x > \lambda) \eta(dx)} = m \tag{3.23}$$

for every $m \geq 1$, and, more generally, under what conditions is the counterpart of (3.21) true:

$$\lim_{\lambda \to \infty} \frac{\int_A P\left(\sum_{i=1}^{N} X_i^x > \lambda\right) \eta(dx)}{\int_A P(X_1^x > \lambda) \eta(dx)} = EN \tag{3.24}$$

when N is independent of the sequence X_i^x, $i \geq 1$?

In this paper we consider only the case of a regularly varying $\overline{G}$, and the family $(F(x, \cdot), \, x \in A)$ generated in a rather special way. However, one can see even here that new phenomena arise, those not present when one takes unconditional convolution powers.

THEOREM 3.2 *Let* $(A, \mathcal{A}, \eta) = (R, \mathcal{B}, \mathrm{Leb})$, *and for an* $x \in R$ *let* $F(x, \cdot)$ *be the distribution of* $\Phi(Y_1, D_1 + x)$, *where* $\Phi : R^2 \to R$ *is a measurable function, and* $(Y_i, \, i \geq 1)$ *and* $(D_i, \, i \geq 1)$ *are independent sequences of i.i.d. random variables.*

Assume that the function $\overline{G}$ belongs to the class of regularly varying (at infinity) functions, with parameter $-p$ for some $p \geq 0$. Assume further that the distribution of D_1 is absolutely continuous with respect to the Lebesgue measure. Then (3.23) holds. More generally, if N is an independent of the two i.i.d. sequences integer valued nonnegative random variable, with $EN^2 < \infty$, then (3.24) holds as well.

PROOF: The crucial step in the proof is the following statement.

$$\lim_{\lambda \to \infty} \frac{\int_R \left[P\left(\Phi(Y_1, D_1 + x) > \lambda\right)\right]^2 dx}{\int_R P\left(\Phi(Y_1, D_1 + x) > \lambda\right) dx} = 0. \tag{3.25}$$

Suppose (3.25) has been proved. We have, for every $m \geq 2$, $\delta > 0$ and $\lambda > 0$ so big that (3.22) holds,

$$\int_R P\left(\sum_{i=1}^m \Phi(Y_i, D_i + x) > \lambda\right) dx$$

$$\geq m \int_R P\left(\Phi(Y_1, D_1 + x) > \lambda(1 + \delta)\right) \left[P\left(\Phi(Y_1, D_1 + x) \leq \lambda\delta/(m-1)\right)\right]^{m-1} dx$$

$$= m \int_R P\left(\Phi(Y_1, D_1 + x) > \lambda(1 + \delta)\right) dx$$

$$-m \sum_{i=1}^{m-1} \binom{m-1}{i} (-1)^{i-1} \int_R P\left(\Phi(Y_1, D_1 + x) > \lambda(1 + \delta)\right) \left[P\left(\Phi(Y_1, D_1 + x) > \lambda\delta/(m-1)\right)\right]^i dx$$

$$:= m\overline{G}(\lambda(1 + \delta)) - R(\lambda).$$

Now, it follows from (3.25) that

$$|R(\lambda)| \leq m(2^{m-1} - 1) \int_R \left[P\left(\Phi(Y_1, D_1 + x) > \lambda\delta/(m-1)\right)\right]^2 dx = o(\overline{G}(\lambda))$$

as $\lambda \to \infty$. We conclude that

$$\underline{\lim}_{\lambda \to \infty} \frac{\int_R P\left(\sum_{i=1}^m \Phi(Y_i, D_i + x) > \lambda\right) dx}{\overline{G}(\lambda)} \geq m(1 + \delta)^{-p},$$

and letting $\delta \to 0$ we obtain

$$\underline{\lim}_{\lambda \to \infty} \frac{\int_R P\left(\sum_{i=1}^m \Phi(Y_i, D_i + x) > \lambda\right) dx}{\overline{G}(\lambda)} \geq m \tag{3.26}$$

for every $m \geq 1$. Furthermore, it follows immediately from (3.26) and Fatou's lemma that

$$\underline{\lim}_{\lambda \to \infty} \frac{\int_R P\left(\sum_{i=1}^N \Phi(Y_i, D_i + x) > \lambda\right) dx}{\overline{G}(\lambda)} \geq EN \tag{3.27}$$

whether or not $EN^2 < \infty$.

For the upper bound, take an m and λ as before, and take also a $\delta \in (0,1)$. We have

$$\int_R P\Big(\sum_{i=1}^m \Phi(Y_i, D_i + x) > \lambda\Big) dx$$

$$\leq m \int_R P\Big(\Phi(Y_1, D_1 + x) > \lambda(1-\delta)\Big) dx + \frac{m(m-1)}{2} \int_R \big[P\big(\Phi(Y_1, D_1 + x) > \lambda\delta\big)\big]^2 dx.$$

From here, (3.25) and regular variation we get a matching upper bound to (3.26), thus proving (3.23). Furthermore,

$$\int_R P\Big(\sum_{i=1}^N \Phi(Y_i, D_i + x) > \lambda\Big) dx$$

$$\leq EN\overline{G}(\lambda(1-\delta)) + E\frac{N(N-1)}{2} \int_R \big[P\big(\Phi(Y_1, D_1 + x) > \lambda\delta\big)\big]^2 dx,$$

and using once again (3.25) and regular variation we obtain

$$\overline{\lim}_{\lambda\to\infty} \frac{\int_R P\big(\sum_{i=1}^N \Phi(Y_i, D_i + x) > \lambda\big) dx}{\overline{G}(\lambda)} \leq EN(1-\delta)^{-P},$$

from which (3.24) follows by letting $\delta \to 0$ and using (3.27).

It remains, therefore, to prove (3.25).

Assume first that the density h of the distribution of D_1 with respect to the Lebesgue measure is bounded. That is, there is a $C < \infty$ such that $h(x) \leq C$ for every $x \in R$. We have then

$$\int_R \big[P\big(\Phi(Y_1, D_1 + x) > \lambda\big)\big]^2 dx$$

$$= \int_R \Big[P\Big(\int_R \Phi(Y_1, x + y) > \lambda\Big) h(y) dy \int_R \Phi(Y_1, x + z) > \lambda\Big) h(z) dz\Big] dx$$

$$= \int_R \Phi(Y_1, y) > \lambda\Big) dy \int_R \Phi(Y_1, z) > \lambda\Big) dz \int_R h(y - x) h(z - x) dx$$

$$\leq C \Big[\int_R P\big(\Phi(Y_1, x) > \lambda\big) dx\Big]^2$$

$$= C \Big[\int_R P\big(\Phi(Y_1, D_1 + x) > \lambda\big) dx\Big]^2,$$

which proves (3.25) in that case.

We now remove the assumption of the boundedness of the density of D_1. For any $\epsilon > 0$ one can find two random variables, D_1' and D_1'', such that D_1'' has a bounded density, and for every Borel set A,

$$P(D_1 \in A) = \epsilon P(D_1' \in A) + (1 - \epsilon)P(D_1'' \in A). \qquad (3.28)$$

We have by (3.28)

$$\int_R \big[P\big(\Phi(Y_1, D_1 + x) > \lambda\big)\big]^2 dx$$

$$= \int_R \left[\epsilon P\Big(\Phi(Y_1, D_1' + x) > \lambda\Big) + (1 - \epsilon) P\Big(\Phi(Y_1, D_1'' + x) > \lambda\Big) \right]^2 dx$$

$$\leq (\epsilon^2 + 2\epsilon) \int_R P\Big(\Phi(Y_1, D_1' + x) > \lambda\Big) dx + (1 - \epsilon)^2 \int_R \left[P\Big(\Phi(Y_1, D_1'' + x) > \lambda\Big) \right]^2 dx.$$

Since (3.25) has already been proved for random variables D_1 that have a bounded density, we conclude that

$$\overline{\lim}_{\lambda \to \infty} \frac{\int_R \left[P\Big(\Phi(Y_1, D_1 + x) > \lambda\Big) \right]^2 dx}{\int_R P\Big(\Phi(Y_1, D_1 + x) > \lambda\Big) dx} \leq (\epsilon^2 + 2\epsilon),$$

from which (3.25) follows by letting $\epsilon \to 0$. ∎

Remarks

1. The surprising fact is that the statement of Theorem 3.2 is false without some assumption of regularity of the distribution of the random variable D_1. It is false, for example, if the latter has atoms. To see this, consider the case $D_1 = 0$ a.s, Y_1 is exponential random variable with mean 1, and $\Phi(y, x) = y|f(x)|$, with f given by (3.1). Then

$$\eta_1(\lambda) := \int_R P\Big(\Phi(Y_1, D_1 + x) > \lambda\Big) dx = \int_0^\infty \exp(-\lambda/|f(x)|) dx$$

is regularly varying at infinity with parameter $-p = -d/\gamma$. Furthermore,

$$\eta_2(\lambda) := \int_R P\Big(\Phi(Y_1, D_1 + x) + \Phi(Y_2, D_2 + x) > \lambda\Big) dx = \int_0^\infty \exp(-\lambda/|f(x)|)(1 + \lambda/|f(x)|) dx$$

$$= \eta_1(\lambda) + \lambda \eta_1'(\lambda).$$

It follows now from Proposition 0.7 of Resnick [Res87] that

$$\lim_{\lambda \to \infty} \frac{\int_R P\Big(\Phi(Y_1, D_1 + x) + \Phi(Y_2, D_2 + x) > \lambda\Big) dx}{\int_R P\Big(\Phi(Y_1, D_1 + x) > \lambda\Big) dx} = 1 + p,$$

which is, in general, different from 2.

2. We do not know whether the statement of Theorem 3.2 remains true if the distribution of D_1 is atomless, but not absolutely continuous with respect to the Lebesgue measure.

4 Simulation of the process $X_{\gamma, d, m}$ and conclusions

One can simulate the process $X_{\gamma, d, m}$ using directly the integral (or series) definition (2.2) and (2.3), truncating the sum over clusters by a certain finite number. The result of such a simulation is presented below. We emphasize that the whole problem of parameter estimation for a shot noise model is still under investigation, and the purpose of presenting a simulation result here is to give the reader an idea how a cluster shot noise process can behave.

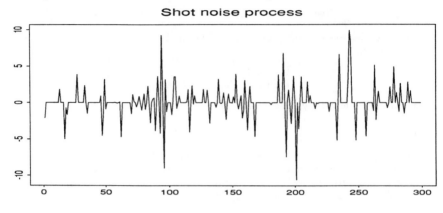

Figure 4.1

For the purpose of simulation we have selected the cluster size distribution to be geometric: $p_k = p(1-p)^{k-1}$, $k = 1, 2, \ldots$, with $p = .1$ (i.e. clusters have the average size of 10), the displacement distribution H is uniform in the interval $(0, .3)$ and the cluster arrival rate is $\rho = 1$. We have selected the parameters of the kernel $f_{\gamma,d,m}$ to be $m = 2, d = .2$ and $\gamma = .5$, thus corresponding to the probability tails of $P(X_{\gamma,d,m}(0) > \lambda) \sim c\lambda^{-4}$ and $P(X_{\gamma,d,m}(0) + X_{\gamma,d,m}(1) + X_{\gamma,d,m}(2) + X_{\gamma,d,m}(3) > \lambda) \sim c\lambda^{-6}$ as $\lambda \to \infty$. For the purpose of simulation we have taken 600 Poisson clusters.

The result of a simulation is presented on Fig. 4.1. Observe that the process $X_{\gamma,d,m}$ has a finite Lévy measure, so it will unavoidably take some zero values (this property can be actually useful in the context of high frequency data). The frequency with which zero values appear can be reduced by increasing the intensity of the point process ρ and the cluster size K_1.

A possible point of view of a model having the above properties is to regard it as modeling the *significant events* taking place on the market, with the rest of activity being regarded as, say, a Brownian motion, or another simple noise. We have added to the above shot noise process an independent standard white noise, with the result appearing on Fig. 4.2. Heavy tails and clustering phenomena are highly visible in the two plots.

Conclusions

We have described a class of cluster Poisson shot noise models that can naturally exhibit heavy tails that get lighter under summation of consecutive observations, long memory and clustering. Those processes are non-Markov moving averages with respect to certain random measures. Problems of parameter estimation, prediction and derivative pricing have to be studied for such models, but it seems clear that their intuitive structure, and the ability to explain, in a parsimonious way, many characteristics of financial processes, make them quite an attractive class of models for the latter.

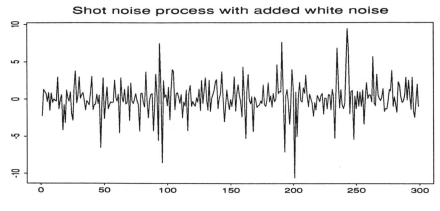

Shot noise process with added white noise

Figure 4.2

References

[AB88]   V. Akgirav and G.G. Booth. The stable-law model of stock returns. *J. Bus. Econ. Stat.*, 6:51–57, 1988.

[Bar94]   O.E. Barndorff-Nielsen. Gaussian\\Inverse Gaussian processes and the modelling of stock returns. 1994. Technical report, Aarhus University.

[BCK92]   T. Bollerslev, R.Y. Chow, and K.F. Kroner. ARCH modeling in Finance: a review of the theory and empirical evidence. *Journal of Econometrics*, 52:5–59, 1992.

[BD91]   P.J. Brockwell and R.A. Davis. *Time Series: Theory and Methods*. Springer-Verlag, New York, second edition, 1991.

[BG74]   R.C. Blattberg and N.J. Gonedes. A comparison of the stable and Student distributions as statistical models for stock prices. *Journal of Business*, 47:244–280, 1974.

[Cli87]   D.B.H. Cline. Convolutions of distributions with exponential and subexponential tails. *J. Austral. Math. Soc.*, 43:347–365, 1987. (Series A).

[CNW73]   J. Chover, P. Ney, and S. Wainger. Functions of probability measures. *J. Analyse Math.*, 26:255–302, 1973.

[EGV79]   P. Embrechts, C.M. Goldie, and N. Veraverbeke. Subexponentiality and infinite divisibility. *Z. Wahr. verw. Geb.*, 49:335–347, 1979.

[EK94]   E. Eberlein and U. Keller. Hyperbolic distributions in Finance. 1994. Technical report.

[Eng82]   R.F. Engle. Autoregressive conditional heteroskedasticity with estimates of the variance of U.K. inflation. *Econometrica*, 50:987–1008, 1982.

[Fam65]   E. Fama. The behavior of stock market prices. *Journal of Business*, 38:34–105, 1965.

[FS93]   H. Föllmer and M. Schweizer. A microeconomic approach to diffusion models for stock prices. *Math. Finance*, 3:1–23, 1993.

[GDD*94] D.M. Guillaume, M.M. Dagorodna, R.R. Davé, U.A. Müller, R.B. Olsen, and O.V. Pictet. From the bird's eye to the microscope: a survey of new stylized facts of the intra-daily foreign exchange markets. 1994. A discussion paper by the O& A Research Group.

[KM93a]   C. Klüppelberg and T. Mikosch. Explosive Poisson shot noise processes with applications to risk reserves. 1993. *Bernoulli*, to appear.

[KM93b]   C. Klüppelberg and T. Mikosch. Modelling delay in claim settlement. 1993. *Scand. Actuar. Journal*, to appear.

[KNrS94] U. Küchler, K. Neumann, M. Sørensen, and A. Streller. Stock returns and hyperbolic distributions. 1994. Technical report, Humboldt Universität zu Berlin.

[Lew64]   P.A. Lewis. A branching Poisson process model for the analysis of computer failure pattern. *Journal Royal Stat. Soc., Ser. B*, 26:398–456, 1964.

[LP94]   M. Loretan and P.C.B. Phillips. Testing the covariance stationarity of heavy-tailed time series. *J. Empirical Finance*, 1:211–248, 1994.

[Man63]   B.B. Mandelbrot. The variation of certain speculative prices. *Journal of Business*, 26:394–419, 1963.

[Man82]   B.B. Mandelbrot. *The Fractal Geometry of Nature*. W.H. Freeman and Co., San Francisco, 1982.

[McC94]   J.H. McCulloch. Measuring tail thickness in order to estimate the stable index α: a critique. 1994. Preprint.

[MR93]   S. Mittnik and S.T. Rachev. Modeling asset returns with alternative stable distributions. *Economics Reviews*, 1993.

[Res87]   S.I. Resnick. *Extreme Values, Regular Variation and Point Processes.* Springer-Verlag, New York, 1987.

[Sam65]   P. Samuelson. Rational theory of warrant pricing. *Industrial Management Review*, 6:13–32, 1965.

[Spa93]   A. Spanos. On modeling speculative prices: Student's t autoregressive model with dynamic heteroskedasticity. 1993. Technical report, University of Cyprus.

[Tay86]   S.J. Taylor. *Modelling Financial Time Series.* Wiley, New York, 1986.

[Ver70]   D. Vere-Jones. Stochastic models for earthquake occurencies. *Journal Royal Stat. Soc., Ser. B*, 532:1–42, 1970.

Gennady Samorodnitsky
School of Operations Research and Industrial Engineering
Cornell University
Ithaca, NY 14853
U.S.A.

Why discount? The Rationale of Discounting in Optimisation Problems

P. Whittle
Statistical Laboratory, University of Cambridge
16 Mill Lane, Cambridge CB2 1SB, UK

ABSTRACT The use of a discounted cost function seems to produce inconsistencies if the utility function is nonlinear, in that the optimal policy is then non-stationary, even for a time-invariant model. The reason for this is that, if discounting is motivated by the fact that capital can grow by compound interest, then there is an implication that one has the alternatives of operating the enterprise which is being controlled (and of accepting consequent costs or gains) or of letting one's capital grow at a constant interest rate (in the bank, say). The complete state variable for this model is then the state variable of the enterprise *plus* the amount held in the bank. If the utility function is linear or the enterprise deterministic then optimisation reduces to the minimisation of expected discounted enterprise costs. In other cases one has a non-degenerate problem of optimal allocation between enterprise and bank account.
Keywords: Discounting, risk-sensitivity, allocation.
Classification numbers: 90 A 43, 90 C 46, 93 E 20.

1   Discounting, Stationarity and Utility

Suppose that one is optimising the course of some process or enterprise in discrete time, and that r_τ is the reward yielded by this operation at time τ. Let us suppose that operation starts at time $\tau = 0$, that the reward function and the stochastic dynamics of the enterprise are time-homogeneous, and that one envisages indefinite operation. Then a criterion which is deeply rooted in the literature is that one chooses the policy (i.e. the rule by which the necessary operation decisions are taken) to maximise $E(\mathbb{R})$, where $\mathbb{R} = \sum_{\tau=0}^{\infty} \beta^\tau r_\tau$ is the *discounted total future reward* at time 0. Here β is the *discount factor*, a positive scalar which is strictly less than unity if discounting is *strict*.

There seem to be three motivations or justifications for the introduction of discounting. These are (i) the formal one of mathematical convenience, (ii) the accounting one, of the growth of capital by compound interest and (iii) the possibility of random termination, by mortality or technical change. Of these, the second is usually regarded as the most plausible and compelling. For point (i), discounting is indeed a convenient way of finding a measure

$\mathbb{R}$ of infinite-horizon reward which is (under appropriate assumptions and in the appropriate sense) finite. This measure is mathematically attractive to handle and reminiscent of the concept of Abel summation. However, convenience is not a reason; any regularising reformulation must have an operational justification.

Motivation (ii) is generally regarded as supplying the most natural such justification. If money in the bank attracts interest at a rate of $i\%$ per unit time then it grows by a factor of $\beta^{-1} = (1 + i/100)$ over unit time, so that an initial unit amount is worth β^{-s} after s time units. Viewed in reverse: a unit amount earned a time s in the future has a *present value* of β^s, whence the occurrence of the coefficient β^τ in the expression for $\mathbb{R}$.

Justification (iii) corresponds to the conviction that nothing goes on for ever. Suppose that the optimiser (who is to enjoy the fruits of the enterprise), having survived to time τ, has probability β of surviving to time $\tau + 1$, independently of τ and of system history. After his death he enjoys no further benefit. The probability that he reaches time σ and then dies is then $(1 - \beta)\beta^\sigma$, so that his reward if one takes expectations over σ alone is

$$E_\sigma \left(\sum_{\tau=0}^{\sigma} r_\tau \right) = \sum_{\tau=0}^{\infty} \beta^\tau r_\tau = \mathbb{R}.$$

An expectation over time of death thus has exactly the effect of discounting. A constant conditional probability of, say, technological change would have an analogous effect, with a modified (but generally non-zero) reward-rate after the change-point.

The distinctive feature of optimisation of a discounted criterion is that the optimisation from a given time can always be expressed as the maximisation of expected *present value*. So, if one has reached time t (≥ 0) then current and future decisions will be chosen so as to maximise

$$E(\mathbb{R} \mid W_t) = \sum_{\tau=0}^{t-1} r_\tau + \beta^t E(\mathbb{R}_t \mid W_t) \tag{1.1}$$

where W_t is the information available at time t and

$$\mathbb{R}_t = \sum_{\tau=0}^{\infty} \beta^\tau r_{t+\tau} \tag{1.2}$$

is the present value of current and future rewards *at time t*. The sum in (1) covers past rewards; we have assumed these are known at time t. Even if they were not (when one would have to take an expectation conditional on W_t) they would in any case be unaffected by policy from time t onwards. We see from the reward-decomposition (1) that the choice of policy from time t to maximise $E(\mathbb{R}) = E(\mathbb{R}_0 \mid W_0)$ (for given control actions before time t) is equivalent to that which maximises $E(\mathbb{R}_t \mid W_t)$, as asserted.

If one assumes that the observation structure of the process is also time-homogeneous then the passage of time will lead to a simple translation of the problem in time, and one will expect a similar time-invariance of the optimal policy. That is, that the optimal policy is *stationary*, in that optimal actions are determined from current observables in a time-invariant fashion.

However, suppose that instead of maximising $E(\mathbb{R})$ one maximises $E[U(\mathbb{R})]$, where U is a *utility function*, so that $U(\mathbb{R})$ represents the actual benefit derived from a reward $\mathbb{R}$. Then a surprising conclusion, noted early in [1], is that, if U is nonlinear, then the optimal policy is non-stationary. One sees this very clearly if one considers the exponential utility function

$$U(\mathbb{R}) = e^{\theta \mathbb{R}}. \tag{1.3}$$

Since this is increasing or decreasing in $\mathbb{R}$ according as the parameter θ is positive or negative, then one must maximise or minimise the expectation respectively. However, if we transform the criterion back to the original reward scale by inverting the U-transformation after having taken the expectation,

$$\rho = U^{-1}\left\{E[U(\mathbb{R})]\right\} = \theta^{-1}\log[E(e^{\theta \mathbb{R}})], \tag{1.4}$$

then ρ must be maximised in all cases. An application of Jensen's inequality to criterion (3) shows that the optimiser finds variability in $\mathbb{R}$ welcome or unwelcome according as θ is positive or negative. He is then regarded as *risk-seeking* or *risk-averse* respectively. In the transition case $\theta = 0$ expression (4) reduces to the previous criterion $E(\mathbb{R})$ and the optimiser is said to be *risk-neutral*. The parameter θ is referred to as the *risk-sensitivity parameter*. The exponential criterion (3) has been particularly investigated by Jacobson [3] and Whittle [5],[6] – see also references quoted in this last work – and has proved amenable under the assumptions of *LQG structure*: linear dynamics, quadratic reward function and Gaussian noise variables. Suppose that operation has moved forward to time t. Then the reward decomposition analogous to (1) is

$$E(e^{\theta \mathbb{R}} \mid W_t) = \exp\left(\theta \sum_{\tau=0}^{t-1} \beta^\tau r_\tau\right) E\left(e^{\theta \beta^t \mathbb{R}_t} \mid W_t\right), \tag{1.5}$$

and we see, as before, that the policy from time t onwards is to be chosen to extremise the final expectation in (5). But this depends explicitly on t, in that the risk-sensitivity parameter θ has been replaced by $\beta^t \theta$. This will induce a decreasing risk-sensitivity with time, and so a non-stationarity of the optimal policy.

This conclusion goes completely against one's intuition, in that one feels that the physical problem is still time-invariant, and should be optimised by a correspondingly time-invariant policy. The resolution of the matter, which is the principal point of this paper, will come in the next section. Note, however, that, if we assume the mortality mechanism rather than the accounting mechanism, then we are led to the expression

$$E\left[\exp\left(\theta \sum_{\tau=0}^{\sigma} r_\tau\right)\right] = E\left[\sum_{j=0}^{\infty}(1-\beta)\beta^j \exp\left(\theta \sum_{\tau=0}^{j} r_\tau\right)\right]$$

for the expected utility. One verifies easily that this criterion is invariant under shift to a later starting-point, implying stationarity of the optimal policy. (Or, rather, that if there are optimal policies then some of these are stationary.) However, there is nothing in this expression resembling reward-discounting, and LQG structure (if assumed for the process being controlled) is destroyed by the presence of the σ-expectation.

The resolution of the problem offered in the next section is based on the fact that the conventional model is in fact incomplete if growth of capital by compound interest is taken as the reason for discounting. An alternative resolution is that offered by [4]: these authors modify the utility function, so that it is generated by a nonlinear recursion.

2 The Necessity for Completion of the Discounting Model

One wonders how the correct view can be restored. The remedy is certainly not to renormalise the cost-function to present value at each stage; this would amount to working with a changing criterion and would be inconsistent with the principle that an optimisation from a given time can be embedded in an optimisation from an earlier time. It seems necessary to think through the concept of discounting afresh.

It is alternative (ii), the effect of compound interest, which is the usual justification for discounting. However, a complete model has to make the interest mechanism explicit. The introduction of interest implies the existence of two alternatives: either to use one's money to cover the costs of the enterprise (i.e. of the system being controlled) or to leave it in the bank to grow by compound interest. (More generally, one could consider the allocation of resources between several enterprises competing for support but the comparison of one uncertain enterprise with the certainty of an interest account is sufficient to make the point.) To the variable (x_t, say) which describes the state of the enterprise at time t we must now add another variable, y_t, say, which represents the capital of the plant-owner who is running the enterprise. This capital will obey the recursion

$$y_{t+1} = \beta^{-1}(y_t + r_t) \tag{1.6}$$

if capital in the bank grows by a factor β^{-1} at each stage. We are regarding r_t as the *net* reward which the enterprise yields to the owner/optimiser at time t; it is a random variable which may be positive or negative depending upon the balance between current returns and costs of the enterprise. Suppose for the moment that the plant-owner fixes upon a horizon-point $t = h$, when his capital will have grown to

$$y_h = \beta^{-h}\left(y_0 + \sum_{\tau=0}^{h-1} \beta^\tau r_\tau\right). \tag{1.7}$$

If he wishes to maximise terminal capital y_h then he will choose a policy which maximises $E[U(y_h)]$ for some utility function U. If U is linear or if there is no statistical variation then this will be equivalent to maximisation of the expected discounted net reward $E(\sum_{\tau=0}^{h-1} \beta^\tau r_\tau)$ from the enterprise. It will also be independent of y_0 (at least if a negative bank balance at any time in the period $0 \le t \le h$ is either permitted or improbable). In all other cases these assertions will not hold, and *the current capital y_t must figure as part of the state variable of the problem.*

The point, once made, is obvious, but it can be re-expressed in dynamic programming terms. Let $F(x, y, t)$ be the value function for the augmented process; the maximal expected final reward at time t conditional on current values x and y of enterprise-state and capital. Let u_t be the value of the control variable at time t and suppose that r_t has the explicit functional form $r(x_t, u_t)$. Then F obeys the dynamic programming equation

$$F(x, y, t) = \sup_u E\left[F\left(x', \beta^{-1}\left(y + r(x, u) \right), t+1 \right) \mid x, u \right], \qquad (t < h) \qquad (1.8)$$

where x' is the value of x_{t+1}, a random variable conditioned only by the values x and u of x_t and u_t. The value of u which yields the supremum in (8) (if there is one) is the optimal value of u_t.

Equation (8) has as many solutions as there are prescriptions of the terminal utility function. A linear terminal utility y_h corresponds to a solution of the form

$$F(x, y, t) = \beta^t [y + G(x, t)],$$

where G then necessarily obeys the equation

$$G(x, t) = \sup_u \left\{ r(x, u) + \beta E[G(x', t+1) \mid x, u] \right\}, \qquad (t < h). \qquad (1.9)$$

In (9) we recognise the dynamic programming equation for the enterprise alone, the interaction with external economics coming only from the presence of the β-coefficient, corresponding to a discounting.

For the more general case of a nonlinear utility function $U(y_h)$ one has to return to the more general equation (8), in which β does not separate out as a simple coefficient, but remains buried in the arguments of the value function. This corresponds to the fact that discounting has been lost as a simple concept. One now has in fact an allocation problem: that of optimally allocating capital between the certain investment represented by the bank account and the uncertain investment represented by the enterprise. This is precisely the moral of the story: that the problem is one of optimal resource allocation, and reduces to optimisation of the discounted enterprise in isolation only when the utility function is linear.

3   Some Simple Examples

The essential point has been made in the last section; once one is in the nonlinear-utility case then the concept of discounting scarcely survives in any clear sense. More specifically, it survives only in the sense that one of the investments one is considering (the bank account) has a fixed rate of return. There is interest, nevertheless, in seeing if any of these nonlinear cases can be solved.

Here we encounter a difficulty. The fact, evident from (7), that capital is growing exponentially (all being well, and in some stochastic sense) means that existence of an infinite-horizon limit and of a stationary optimal policy will be possible only for rather special choices of criterion. One such would be the minimisation of the probability of ruin: of the event that y will at some future time fall below zero. A related criterion, proposed by Bather in a reinsurance context and followed up by Dayananda [2], is the maximisation of the expected total return to shareholders before the moment of ruin.

Consider the following simple problem, phrased in continuous time. Suppose that capital y follows a diffusion process with drift coefficient $a(y, u)$ and diffusion coefficient $c(y, u)$. Here u labels an investment decision. The variable x has been omitted; this corresponds to the assumption that the alternative investment one considers is not so highly structured that one would need to define its state.

Suppose we wish to choose u for a given current y so as to minimise the probability of ultimate ruin. The minimal probability $f(y)$ will then obey the dynamic programming equation

$$\inf_u [af' + \frac{1}{2}cf''] = 0, \qquad (0 < y < \infty), \tag{1.10}$$

and the infimising value (if existing) will be optimal at y. This equation is subject to the boundary condition $f(0) = 1$, and also to the condition $\lim_{y \to \infty} f(y) = 0$ if the rates a and c are such that return from infinity has zero probability. We leave the reader to verify the very simple characterisation of optimality whch emerges: that u should be such as to maximise the ratio $(a(y, u)/c(y, u))$; of drift coefficient to diffusion coefficient. Suppose now that a and c are given the forms

$$a(y, u) = \alpha(y - u) + \lambda u - \mu, \tag{1.11}$$
$$c(y, u) = \nu u. \tag{1.12}$$

Here we interpret u as the amount of one's capital diverted to a risky investment and $y - u$ as the amount left in the bank, where it earns interest at a (continuous-time) rate α. Then λ represents the expected rate of return on the investment and μ the rate of expenditure needed to secure physical survival. We have chosen c proportional to u in (12). This corresponds to the assumption that the investment of an amount u buys u units of investment which behave as independent and identically distributed random variables. With choices (11) and (12) we find that the implied a/c criterion leads to the optimal policy

$$u = \begin{cases} y, & y < \mu/\alpha \\ 0, & y \geq \mu/\alpha. \end{cases} \tag{1.13}$$

The level μ/α of capital is the threshold value which assures an income of μ in perpetuity. If y is greater than this then there is no need to speculate. If y is less then speculation is the only means by which the threshold can be crossed – a crossing whose attempt cannot be postponed when slender resources are all the time being depleted by the unavoidable consumption at rate μ. These conclusions are independent of the relative magnitudes of λ and α; i.e. of whether the investment is advantageous or not in expectation. Suppose now that we modify assumption (12) to $c(y, u) = \nu u^2$. This would correspond to the assumption that one invests an amount u in a single risk, whose rate of return is then u times a random unit return. In the case $\lambda < \alpha$, when the investment is disadvantageous and its only merit is its statistical variability, we find that the optimal policy is again that of (13). However, in the case $\lambda > \alpha$, when it is advantageous, we find that the optimal policy makes u continuous as a function of y. For this policy one places all one's capital in the risky investment if $y < 2\mu/(\lambda + \alpha)$ and all one's capital in the bank account if $y > \mu/\alpha$. In the interval between u changes linearly and continuously, so that there is a mix of investments which shifts from one extreme to the other as y increases.

Comparing the two cases under the assumption $c \propto u^2$ we have the paradoxical result that an increase in the attractiveness of the risky investment leads one to use it less. This is because the mere fact that it is advantageous means that there is less need to use it; as y increases in the middle interval one increasingly values reliability of return relative to expected rate of return.

These examples do not of course take the discounting issue any further. That has been dealt with as far as it can be in Section 2; the point of the examples is simply to show that the allocation problem implied by a nonlinear utility can be solved in some cases.

4   References

[1] Chung, K.J. and Sobel, M.J. (1987) Discounted MDPs: distribution functions and exponential utility maximisation. *SIAM J. Control Optim.* **25**, 49–62.

[2] Dayananda, P.W.A. (1970) Optimal reinsurance. *J. Appl. Prob.* **7**, 134–156.

[3] Jacobson, D.H. (1973) Optimal stochastic linear systems with exponential performance criteria and their relation to deterministic differential games. *IEEE Trans. Automat. Control* AC–18, 124–131.

[4] Sargent, T. and Hansen, L. (1995) Discounted linear exponential quadratic control. *IEEE Trans. Automat. Control* AC–40, 968–971.

[5] Whittle, P. (1981) Risk-sensitive linear/quadratic/Gaussian control. *Adv. Appl. Prob.* **13**, 764–777.

[6] Whittle, P. (1990) *Risk-sensitive Optimal Control*. Wiley, Chichester and New York.

On Periodic Pollaczek Waiting Time Processes

J.W. Cohen

CWI

P.O. Box 94079, 1090 GB Amsterdam, The Netherlands

January, 1994

Abstract

Present-day modelling of traffic in broadband communication requires the use of rather sophisticated stochastic processes. Although a large class of suitable stochastic processes is known in the literature, their rather complicated structure limits their use because of the laborious numerical evaluation involved. The present study concerns the so-called periodic Pollaczek processes. The characteristics of the arrival process, such as service time τ_n and interarrival time σ_n, are here periodic functions of n, i.e. the vector (τ_n, σ_n) and $(\tau_{n+N}, \sigma_{n+N})$ have the same distribution, N being the period. The sequence $\mathbf{w}_n, n = 1, 2, \ldots$, defined by

$$\mathbf{w}_{n+1} = [\mathbf{w}_n - \tau_n - \sigma_n]^+$$

is investigated; as such the queue under consideration is a direct generalisation of the classical Pollaczek $GI/G/1$ queue. It appears that the model is a quite flexible one, and moreover very accessible for numerical evaluation if the distributions of all the service times, or of all the interarrival times, have rational Laplace-Stieltjes transforms.

AMS Subject Classification (1991): 90B22, 60K25
Keywords & Phrases: periodic arrival processes, actual waiting times, stationary distributions

1. INTRODUCTION

For the classical $GI/G/1$ waiting time model Pollaczek characterises the structure of the actual waiting time process $\mathbf{w}_n, n = 1, 2, \ldots$, by the relations

$$\begin{aligned} \mathbf{w}_{n+1} &= [\mathbf{w}_n + \rho_n]^+ \qquad n = 1, 2, \ldots, \\ \mathbf{w}_1 &= w_1 \geq 0, \end{aligned} \tag{1.1}$$

where $\mathbf{w}_n$ is the actual waiting time of the nth arriving customer, w_1 the initial waiting time and

$$\rho_n := \tau_n - \sigma_n; \tag{1.2}$$

with τ_n the service time of the nth arriving customer, σ_n the time between the nth and $(n + 1)st$ arrival. The $\tau_n, n = 1, 2, \ldots$, and similarly, the $\sigma_n, n = 1, 2, \ldots$, are i.i.d. nonnegative stochastic variables, and further $\{\tau_n, n = 1, 2, \ldots\}$ and $\{\sigma_n, n = 1, 2, \ldots\}$ are independent families.

A generalisation of this classical model may be formulated as follows. let

$$\mathbf{x}_n, \ n = 1, 2, \ldots, \tag{1.3}$$

be a discrete time Markov chain with a countable, irreducible state space $\mathcal{S}$ and with stationary transition probabilities. Here $(p_{ij}), i, j \in \mathcal{S}$, shall denote the one-step transition probabilities; $\mathbf{x}_1 = x_1$ shall be the initial state of the $\mathbf{x}_n$-process.

Further, let $\rho(j), j \in \mathcal{S}$, be a set of stochastic variables with distributions $r_j(\cdot), j \in \mathcal{S}$, and let

$$\rho_n \equiv \rho(\mathbf{x}_n), \quad n = 1, 2, \ldots, \tag{1.4}$$

361

be stochastic variables defined on the $\mathbf{x}_n$-process, such that for all $n = 1, 2, \ldots,$

$$\Pr\{\rho_{n+1} < \rho, \mathbf{x}_{n+1} = j | \mathbf{x}_m = i_m, \rho_m = \rho(i_m), m = 1, \ldots, n\} = r_{i_n}(\rho)p_{i_n j}, \tag{1.5}$$

for all $i_n \in S$.
Define the sequence $\mathbf{w}_n, n = 1, 2, \ldots,$ by

$$\mathbf{w}_{n+1} = [\mathbf{w}_n + \rho_n]^+, \quad n = 1, 2, \ldots, \tag{1.6}$$

for given initial $\mathbf{w}_1 = w_1 \geq 0$.

The process $(\mathbf{x}_n, \mathbf{w}_n), n = 1, 2, \ldots,$ as defined above will be called a Pollaczek waiting time process. Obviously if we take ρ_n to be the difference of two nonnegative stochastic variables $\tau(\mathbf{x}_n)$ and $\sigma(\mathbf{x}_n)$, so

$$\rho_n \equiv \tau(\mathbf{x}_n) - \sigma(\mathbf{x}_n), \tag{1.7}$$

then $\mathbf{w}_n$ may be interpreted as the actual waiting time of the nth arriving customer whose service time $\tau(\mathbf{x}_n)$ and next interarrival time $\sigma(\mathbf{x}_n)$ depend on the state of the $\mathbf{x}_n$-process.

The $(\mathbf{x}_n, \mathbf{w}_n)$-process has been introduced by ARJAS [2]. He presents a fairly detailed study, however, does not present results which are useful in actual performance studies. ASMUSSEN and THORISSON [3] study a similar model as Arjas, but with the $\mathbf{x}_n$-process replaced by a Markov process with a general state space. Their goal concerns the modelling of periodic queues. Indeed, the classical queueing models are not suited to model queueing situations with arrival processes, which lack a simple regenerative structure. The introduction of semi-Markov processes, see e.g. [7], [8], [17], has provided a greater flexibility in the modelling of the arrival, and also in that of the servicing process. The inherent numerical analysis is of course rather laborious, but generally within the limits of available computer facilities.

Queueing models with periodic arrival processes have been studied by several authors, see e.g. [3], [4], [5], [6]. The relevant studies mainly concentrate on the derivation of limit theorems and ergodicity conditions. Tangible results suited for numerical evaluation are hardly obtained. Actually, authors experience with the investigation of an $M/G/1$ model with a periodical arrival process with rate $\lambda(\cdot)$ and $\lambda(t + \mu) = \lambda(t)$ for all $t > 0, \mu$ being the period, is most disappointing indeed. Even for the very simple case with

$$\begin{aligned} \lambda(t) &= \lambda_1 \quad \text{for} \quad 0 \leq t \leq \frac{1}{2}\mu, \\ &= \lambda_2 \quad \text{,,} \quad \frac{1}{2}\mu < t < \mu; \end{aligned}$$

the analysis of the actual waiting time process is governed by a complicated integral equation, which seems hardly accessible for further analysis.

A more promising approach of periodical queueing processes is obtained whenever the characteristics of the queueing model are periodical functions of the number of arrivals. Such a model and actually a very general one is provided by the $(\mathbf{x}_n, \mathbf{w}_n)$-process introduced above, if we specify the state space S and the one-step transition probabilities as follows:

$$S = \{1, 2, \ldots, N\}, \tag{1.8}$$

$$\begin{aligned} p_{ij} &= 1 \text{ for } j = i + 1, \ i = 1, \ldots, \ N - 1, \text{ and for } i = N, \ j = 1, \\ &= 0 \text{ otherwise.} \end{aligned}$$

This special process will be called a periodic Pollaczek process. In section 2 the structure of this process is described and some notations are introduced. The ergodicity conditions are introduced, and it is assumed that they apply. These conditions follow from a general study of LOYNES [9].

In section 3 the functional equations for the Laplace-Stieltjes transforms of the stationary distributions of the actual waiting times $\mathbf{w}_1, \ldots, \mathbf{w}_N$ are formulated for the case that the service times and the interarrival times do not all have lattice distributions. In section 4 the functional equations are derived for the case that all these distributions are lattice variables with state space the set of positive integers.

In section 5 we briefly discuss the case $M = 1$, i.e. the classical Pollaczek model, as the well-known results given here are needed in the subsequent sections.

The functional equations derived in section 3 and 4 formulate a Hilbert Boundary Value Problem for a set of functions. Such boundary value problems have been extensively studied, see e.g. [10], [11] [12]. The somewhat special type of coefficients in our functional equations, which is due to the fact that they consist of the L.S.-transforms of the service time and interarrival time distributions, makes it possible to construct quite explicit solutions if the L.S.-transforms of all the service time distributions are rational functions without a need to specify those of the interarrival time distributions. This is also the case, if all the L.S.-transforms of the interarrival time distributions are rational.

In section 6 the periodical Pollaczek process is analysed for the case that all the service time distributions have a rational L.S.-transform. Section 7 discusses the case with the L.S.-tranforms of all the interarrival time distributions rational functions. Actually the results obtained in sections 6 and 7 are extensions of the $GI/K_n/1$ and the $K_m/G/1$ model, see [14].

The solutions obtained in sections 6 and 7 are quite explicit apart from the solution of a set of linear equations. The number of equations depends on the degrees of the polynomials in the denominators of the rational L.S.-transforms.

In section 8 some special variants of the periodic Pollaczek waiting process are shortly discussed; for a detailed analysis the reader is referred to [19]; the analysis concerns again the solution of standard Riemann Boundary Value Problems.

The analysis of the periodical Pollaczek waiting process shows that this process is quite flexible in modelling queueing processes with a rather complicated traffic structure, in particular they seem to be useful to describe models with bursty traffic. In the performance analysis of present day multiservice networks the models of classical queueing theory are frequently inadequate for modelling the actual queueing processes. This is due to the complicated character of the traffic to be processed by such systems. This complicated character stems from the demand to transport speach, data and video via the same network and to optimize such transport.

An intense world wide research activity is presently devoted to the modelling of traffic processing. The introduction of periodic Pollaczek waiting processes and its analysis is motivated by the need to incorporate bursty traffic in the modelling. Further algebraic and numerical analysis is needed to judge the usefulness of these period Pollaczek processes in actual performance analysis.

2. THE PERIODIC POLLACZEK $GI_N/G_N/1$ QUEUEING MODEL

With a slightly different notation we reformulate the $(\mathbf{x}_n, \mathbf{w}_n)$-process with the property (1.8) of the preceding section.

For a fixed integer $N \geq 1$ let

$$\boldsymbol{\rho}_n = (\rho_1^{(n)}, \ldots, \rho_N^{(n)}), \quad n = 1, 2, \ldots, \tag{2.1}$$

be a sequence of i.i.d. stochastic vectors, with the components of each vector independent stochastic variables.

The sequence of vectors $\mathbf{w}_n, n = 1, 2, \ldots,$

$$\mathbf{w}_n = (\mathbf{w}_1^{(n)}, \ldots, \mathbf{w}_N^{(n)}), \quad n = 1, 2, \ldots, \tag{2.2}$$

is recursively defined by: for $n = 1, 2, \ldots,$

$$\mathbf{w}_j^{(n)} = [\mathbf{w}_{j-1}^{(n)} + \boldsymbol{\rho}_{j-1}^{(n)}]^+ \text{ for } j = 2, \ldots, N, \tag{2.3}$$

$$\mathbf{w}_1^{(n+1)} = [\mathbf{w}_N^{(n)} + \boldsymbol{\rho}_N^{(n)}]^+,$$

with initial value

$$\mathbf{w}_1^{(1)} = w_1 \geq 0.$$

Concerning the $\boldsymbol{\rho}_j^{(n)}$ it will be assumed that it is the difference of two independent and positive variables,

$$\boldsymbol{\rho}_j^{(n)} = \boldsymbol{\tau}_j^{(n)} - \boldsymbol{\sigma}_j^{(n)}, \quad j = 1, \ldots, N; \quad n = 1, 2, \ldots. \tag{2.5}$$

The $\mathbf{w}_n$-process so defined will be called a *periodical* Pollaczek $GI_N/G_N/1$ queueing model with period N.

We introduce some further notations and assumptions.

By $\boldsymbol{\tau}_j, \boldsymbol{\sigma}_j$ and $\boldsymbol{\rho}_j, j = 1, \ldots, N$, we shall denote stochastic variables such that $\boldsymbol{\tau}_1, \boldsymbol{\sigma}_1, \ldots, \boldsymbol{\tau}_N, \boldsymbol{\sigma}_N$, are all independent and for $n = 1, 2, \ldots,$

$$\boldsymbol{\tau}_j \sim \boldsymbol{\tau}_j^{(n)}, \ \boldsymbol{\sigma}_j \sim \boldsymbol{\sigma}_j^{(n)}, \ \boldsymbol{\rho}_j \sim \boldsymbol{\rho}_j^{(n)}, \ j = 1, \ldots, N. \tag{2.6}$$

It will be always assumed that: for $j = 1, \ldots, N$,

i. $\beta_j := \mathrm{E}\{\boldsymbol{\tau}_j\} < \infty, \quad \alpha_j := \mathrm{E}\{\boldsymbol{\sigma}_j\} < \infty, \quad \gamma_j := \mathrm{E}\{\boldsymbol{\rho}_j\};$ (2.7)

ii. $\gamma = \beta - \alpha < 0;$

where

$$\beta := \sum_{j=1}^{N} \beta_j, \quad \alpha := \sum_{j=1}^{N} \alpha_j, \quad \gamma := \sum_{j=1}^{N} \gamma_j.$$

Further we define for $j = 1, \ldots, N$,

$$\begin{aligned}
\alpha_j(\rho) &:= \quad \mathrm{E}\{e^{-\rho\boldsymbol{\sigma}_j}\}, \quad \beta_j(\rho) := \mathrm{E}\{e^{-\rho\boldsymbol{\tau}_j}\}, \quad &&\mathrm{Re}\,\rho \geq 0, \\
\gamma_j(\rho) &:= \quad \alpha_j(-\rho)\beta_j(\rho), \quad &&\mathrm{Re}\,\rho = 0, \\
\alpha(\rho) &:= \quad \prod_{j=1}^{N} \alpha_j(\rho), \quad \beta(\rho) := \prod_{j=1}^{N} \beta_j(\rho), \quad &&\mathrm{Re}\,\rho \geq 0, \\
\gamma(\rho) &:= \quad \alpha(-\rho)\beta(\rho), \quad &&\mathrm{Re}\,\rho = 0.
\end{aligned} \tag{2.8}$$

Whenever the $\alpha_j(\rho)$ and/or the $\beta_j(\rho)$ are rational functions of ρ then we write: for $j = 1, \ldots, N$,

$$\alpha_j(\rho) = \frac{a_{1j}(\rho)}{a_{2j}(\rho)}, \quad \beta_j(\rho) = \frac{b_{1j}(\rho)}{b_{2j}(\rho)}, \quad \mathrm{Re}\,\rho \geq 0, \tag{2.9}$$

with

i. $a_{2j}(\rho)$ *a polynomial of degree* $m_j,$
      $b_{2j}(\rho)$,,   ,,   ,,   ,, $n_j;$ (2.10)

ii. $a_{1j}(\rho)$ *a polynomial of degree* $< m_j,$
      $b_{1j}(\rho)$,,   ,,   ,,   ,, $< n_j;$

iii. $a_{1j}(0) = a_{2j}(0) \neq 0,$
         $b_{1j}(0) = b_{2j}(0) \neq 0;$

iv. $a_k(\rho) = \prod_{j=1}^{N} a_{kj}(\rho) \quad b_k(\rho) = \prod_{j=1}^{N} b_{kj}(\rho), \quad k = 1, 2, \ \mathrm{Re}\,\rho \geq 0,$

$$\mu := \sum_{j=1}^{N} m_j, \qquad \nu := \sum_{j=1}^{n} n_j;$$

and it is assumed that the various rational functions in (2.9) are irreducible, so that

$$\begin{aligned} \alpha_j(\rho) \quad has \quad m_j \quad poles \quad in \quad \mathrm{Re}\,\rho < 0, \\ \beta_j(\rho) \quad\quad ,, \quad\quad n_j \quad\quad ,, \quad\quad ,, \quad \mathrm{Re}\,\rho < 0; \end{aligned} \tag{2.11}$$

these poles counted according to their multiplicities.

REMARK 2.1. Unless stated otherwise it will always be assumed that τ_j and σ_j, $j = 1, \ldots, N$, have absolutely continuous distributions, so that, cf. [16],

$$\begin{aligned} |\alpha_j(\rho_0)| = 1 \quad with \quad \mathrm{Re}\,\rho_0 = 0 \quad implies \quad \rho_0 = 0, \\ |\beta_j(\rho_0)| = 1 \quad\quad ,, \quad\quad ,, \quad\quad ,, \quad\quad ,, \ . \end{aligned} \tag{2.12}$$

This restriction is for the greater part of the following analysis rather inessential, but the case with all τ_j and all σ_j being lattice variables requires a slightly different approach. In such a case it is more convenient to work with the generating functions of the distributions of τ_j and σ_j, rather than using their Laplace-Stieltjes transforms, see section 4. $\square$

REMARK 2.2. The nth arriving customer with $k = n \bmod N$ will be called a type "k"-customer. $\square$

REMARK 2.3. Next to the vector sequence $\mathbf{w}_n$, $n = 1, 2, \ldots$, we introduce the vector sequence

$$\mathbf{i}_n = (\mathbf{i}_1^{(n)}, \ldots, \mathbf{i}_N^{(n)}), \quad n = 1, 2, \ldots,$$

with

$$\mathbf{i}_j^{(n)} := -[\mathbf{w}_j^{(n)} + \rho_j^{(n)}]^{-}, \quad j = 1, \ldots, N. \tag{2.13}$$

Obviously with probability one

$$\mathbf{i}_j^{(n)} \geq 0. \tag{$\square$ 2.14}$$

3. DERIVATION OF THE FUNCTIONAL EQUATIONS

From the definition of the $\mathbf{w}_n$-process in the previous secton it follows that the successive epochs of the sequence

$$\mathbf{w}_1^{(n)}, \ n = 1, 2, \ldots, \quad with \quad w_1 = 0, \tag{3.1}$$

at which $\mathbf{w}_1^{(n)} = 0$ are regeneration points of the $\mathbf{w}_n$-process. Note, however, that if $\tau_N > \sigma_N$ with probability one then the sequence does not have such epochs. But the assumption (2.7)ii implies that at least one $h \in \{1, \ldots N\}$ exists for which the sequence $\mathbf{w}_h^{(n+h)}$, $n = 1, 2, \ldots$, does have epochs at which $\mathbf{w}_h^{(m)} = 0$. Therefore the generality of the analysis is not restricted by assuming that $h = 1$, because the arrival epoch of every type h-customer may be taken as the starting point of the arrival process. So we may and do assume that (3.1) possesses epochs at which $\mathbf{w}_1^{(n)} = 0$.

Define

$$\mathbf{n}_1 := \min_{n=1,2,\dots} \{n : \mathbf{w}_1^{(n+1)} = 0 | \mathbf{w}_1 = 0\}, \tag{3.2}$$

so that $\mathbf{n}_1 N$ is the number of customers served in a type 1-busy period, such a period being defined as the time interval between the successsive arrival epochs of two type 1-customers with zero waiting time. Note: $\mathbf{n}_1$ is also the number of type j-customers served in a type 1-busy period.

Obviously, the $\mathbf{w}_1^{(n)}$-process is a random walk on $[0, \infty)$ with the zero state reflecting. It is readily seen that its drift is negative since $\gamma < 0$, cf. (2.7)ii, so the $\mathbf{w}_1^{(n)}$-process is positive recurrent, cf. [9]. Its state space is not a subset of $[0, \infty)$ if at least one of the distributions of τ_j or σ_j is absolutely continuous. It is readily seen that the same conclusions hold for each of the sequences

$$\mathbf{w}_j^{(n)}, \ n = 1, 2, \dots, \quad \text{with } j = 1, 2, \dots, N. \tag{3.3}$$

It follows that the $\mathbf{w}_j^{(n)}$-squences converge in distribution for $n \to \infty$ for every $j \in N$. Denote by $\mathbf{w}_j$ a stochastic variable with distribution the limiting distribution of the $\mathbf{w}_j^{(n)}$-sequence. It is readily shown that the sequence $\mathbf{i}_j^{(n)}, n = 1, 2, \dots$, cf. (2.13), also converges in distribution for $n \to \infty$; denote by $\mathbf{i}_j$ a stochastic variable with distribution this limiting distribution. It then follows from (2.3), (2.6) and (2.13): for $j = 2, \dots, N$,

$$\mathbf{w}_j \sim [\mathbf{w}_{j-1} + \rho_{j-1}]^+, \quad \mathbf{i}_{j-1} \sim -[\mathbf{w}_{j-1} + \rho_{j-1}]^-,$$
$$\mathbf{w}_1 \sim [\mathbf{w}_N + \rho_N]^+, \quad \mathbf{i}_1 \sim -[\mathbf{w}_N + \rho_N]^-. \tag{3.4}$$

From the identity: for real x

$$e^{-\rho x} + 1 = e^{-\rho [x]^+} + e^{-\rho [x]^-},$$

we obtain from (2.3) and (3.4)

$$e^{-\rho \mathbf{w}_j^{(n+1)}} = e^{-\rho (\mathbf{w}_{j-1}^{(n)} + \rho_{j-1}^{(n)})} + 1 - e^{\rho \mathbf{i}_{j-1}^{(n)}}, \quad j = 2, \dots, N, \tag{3.5}$$

$$e^{-\rho \mathbf{w}_1^{(n+1)}} = e^{-\rho (\mathbf{w}_N^{(n)} + \rho_N^{(n)})} + 1 - e^{\rho \mathbf{i}_N^{(n)}}.$$

Hence by taking expectations in (3.5) and by noting that $\mathbf{w}_j^{(n)}$ and $\rho_j^{(n)}$ are independent, cf. section 2, it follows, cf. (2.8): for $\mathrm{Re}\,\rho = 0$; $j = 2, \dots, N$,

i. $\quad \mathrm{E}\{e^{-\rho \mathbf{w}_j}\} - \gamma_{j-1}(\rho)\mathrm{E}\{e^{-\rho \mathbf{w}_{j-1}}\} = -\rho \mathrm{E}\{\mathbf{i}_{j-1}\} \dfrac{1 - \mathrm{E}\{e^{\rho \mathbf{i}_{j-1}}\}}{-\rho \mathrm{E}\{\mathbf{i}_{j-1}\}};$

$\qquad \qquad \qquad \qquad \qquad \qquad \qquad \qquad \qquad \qquad \qquad \qquad \qquad \qquad \qquad$ (3.6)

ii. $\quad \mathrm{E}\{e^{-\rho \mathbf{w}_1}\} - \gamma_N(\rho)\mathrm{E}\{e^{-\rho \mathbf{w}_N}\} = -\rho \mathrm{E}\{\mathbf{i}_N\} \dfrac{1 - \mathrm{E}\{e^{\rho \mathbf{i}_N}\}}{-\rho \mathrm{E}\{\mathbf{i}_N\}}.$

REMARK 3.1. Note that if at least one of the distributions of τ_j or σ_j, $j = 1, \dots, N$, is absolutely continuous then all $\mathbf{w}_j$ and $\mathbf{i}_j$ are nonlattice variables. Further it is well known that

$$\frac{1 - \mathrm{E}\{e^{\rho \mathbf{i}_j}\}}{-\rho \mathrm{E}\{\mathbf{i}_j\}}, \tag{3.7}$$

is the Laplace-Stieltjes transform of an absolutely continuous distribution with support $(-\infty, 0)$; the ergodicity of the $\mathbf{w}_j^{(n)}$-sequence implies that $\mathrm{E}\{\mathbf{i}_j\}$ is finite. $\qquad \square$

Put for $j = 1, \ldots, N$,

$$\Phi_j(\rho) \quad := \mathrm{E}\{e^{-\rho \mathbf{W}_j}\}, \qquad\qquad\qquad \mathrm{Re}\,\rho \geq 0,$$

$$\Psi_j(\rho) \quad := \frac{\mathrm{E}\{\mathbf{i}_j\}}{\alpha - \beta} \frac{1 - \mathrm{E}\{e^{\rho \mathbf{i}_j}\}}{-\rho \mathrm{E}\{\mathbf{i}_j\}}, \quad \text{if} \quad \mathrm{E}\{\mathbf{i}_j\} > 0, \quad \mathrm{Re}\,\rho \leq 0, \qquad (3.8)$$

$$\qquad\quad := 0 \qquad\qquad\qquad\qquad\qquad\quad = 0.$$

Next introduce the vectors and matrices

$$\Phi(\rho) := (\Phi_1(\rho), \ldots, \Phi_N(\rho)), \qquad\qquad\qquad\qquad\qquad\qquad (3.9)$$

$$\Psi(\rho) := (\Psi_1(\rho), \ldots, \Psi_N(\rho)),$$

$$\Gamma_{ij}(\rho) := \gamma_j(\rho)\delta_{ij}, \quad i, j \in \{1, \ldots, N\},$$

$$\delta_{ij} \text{ the Kronecker symbol,}$$

$$P = (p_{ij}) \text{ with } p_{ij} \text{ as in (2.8)},$$

I the identity matrix.

Hence the relations in (3.6) may be rewritten as: for $\mathrm{Re}\,\rho = 0$,

$$\Phi(\rho)[I - \Gamma(\rho)P] = (\beta - \alpha)\rho\,\Psi(\rho)P. \qquad\qquad\qquad\qquad (3.10)$$

It is readily verified that, cf. (2.8), for $\mathrm{Re}\,\rho = 0$,

$$D_N(\rho) := \|I - \Gamma(\rho)P\| = 1 - \gamma(\rho), \qquad\qquad\qquad\qquad\qquad (3.11)$$

$$D_N(\rho)P[I - \Gamma(\rho)P]^{-1} = \qquad\qquad\qquad\qquad\qquad\qquad\qquad\quad (3.12)$$

$$\begin{bmatrix} \hat{\gamma}_2\hat{\gamma}_3\ldots\hat{\gamma}_N & 1 & \hat{\gamma}_2 & \cdots & \hat{\gamma}_2\hat{\gamma}_3\ldots\hat{\gamma}_{N-1} \\ \hat{\gamma}_3\ldots\hat{\gamma}_N & \hat{\gamma}_1\hat{\gamma}_3\ldots\hat{\gamma}_N & 1 & & \hat{\gamma}_3\ldots\hat{\gamma}_{N-1} \\ & & \hat{\gamma}_1\hat{\gamma}_2\hat{\gamma}_4\ldots\hat{\gamma}_N & \cdots & \\ \cdots & \cdots & \cdots & & \cdots \\ \hat{\gamma}_N & \hat{\gamma}_1\hat{\gamma}_N & \cdots & & 1 \\ 1 & \hat{\gamma}_1 & \hat{\gamma}_1\hat{\gamma}_2 & \cdots & \hat{\gamma}_1\hat{\gamma}_2\ldots\hat{\gamma}_{N-1} \end{bmatrix},$$

where for brevity we have written

$$\hat{\gamma}_j \equiv \gamma_j(\rho).$$

It follows from (3.10) that : for $\mathrm{Re}\,\rho = 0$,

$$\Phi(\rho) = (\beta - \alpha)\rho\,\Psi(\rho)\,P[I - \Gamma(\rho)P]^{-1}, \qquad\qquad\qquad\qquad (3.13)$$

note that $D_N(\rho)$ has in $\rho = 0$ a zero with multiplicity one, because

$$\frac{d\gamma(\rho)}{d\rho}\Big|_{\rho=0} = \beta - \alpha < 0. \qquad\qquad\qquad\qquad\qquad\qquad (3.14)$$

From the definition of $\Phi_j(\rho)$ and $\Psi_j(\rho)$, $j = 1, \ldots, N$, it follows that:

i. $\Phi_j(\rho)$ is regular for $\operatorname{Re}\rho > 0$, continuous for $\operatorname{Re}\rho \geq 0$, (3.15)

$\Phi_j(\rho) = 1$ for $\rho = 0$,

$\Phi_j(\rho) = O(1)$ for $|\rho| \to \infty$, $|\arg\rho| \leq \dfrac{1}{2}\pi$;

ii. $\Psi_j(\rho)$ is regular for $\operatorname{Re}\rho < 0$, continuous for $\operatorname{Re}\rho \leq 0$;

$\Psi_j(\rho) = \dfrac{E\{i_j\}}{\alpha - \beta}$ for $\rho = 0$,

$\Psi_j(\rho) = O(\dfrac{1}{|\rho|})$ for $|\rho| \to \infty$, $\dfrac{1}{2}\pi \leq \arg\rho \leq 1\dfrac{1}{2}\pi$.

The set of equations (3.13) for the unknown vectors $\Phi(\rho)$ and $\Psi(\rho)$, and similarly the equivalent set (3.10), formulate together with the conditions (3.15) a homogeneous Hilbert Boundary Value Problem for the $2N$ unknown components of $\Phi(\rho)$ and $\Psi(\rho)$, with the imaginary axis $\operatorname{Re}\rho = 0$ the line of discontinuity, cf. [10]. p. 384, see also [11], [12]. Actually, the functional equation (3.13) differs slightly from that discussed in [10] because the line of discontinuity is not a closed contour bounding a finite domain.

The construction of explicit solutions of these types of boundary value problem is generally hardly possible except in the three cases:

i. all $\beta_j(\rho)$ are rational functions,
ii. ,, $\alpha_j(\rho)$,,   ,,   ,,
iii. for a subset of N all $\beta_j(\rho)$ are rational
and for the complementary subset all $\alpha_j(\rho)$
are rational.

In the context of Queueing Theory the Hilbert Boundary Value Problem formulated above have been studied in a slightly more general form by Miller [13]. In his approach it is shown that the matrix $I - \Gamma(\rho)P$ can be written as the product of two matrices, one with all its elements regular for $\operatorname{Re}\rho < 0$, the other with its elements regular for $\operatorname{Re}\rho > 0$. The existence of such a factorisation is also shown in [10], however, it leads to a rather formal solution of the problem. The cases, i and ii mentioned above will be analysed in the next section which quickly leads to results accessible for numerical evaluation.

REMARK 3.2. It is readily verified that for $\rho = 0$ the relation (3.10) is an identity; note that $\Gamma(0)$ is the identity matrix and $\Phi_j(0) = 1$, $j = 1, \ldots, N$. We next consider the relation (3.10) for $\operatorname{Re}\rho = 0$, $\rho \to 0$.

Because of (2.7) we have for $\operatorname{Re}\rho = 0, \rho \to 0$,

$$\begin{aligned} \Gamma_{ij}(\rho) &= 1 - \rho\gamma_j + o(\rho) \quad \text{for} \quad i = j, \\ &= 0 \qquad\qquad\qquad ,, \quad i \neq j, \end{aligned}$$

(3.16)

so

$$\Gamma(\rho) = I + \rho\dfrac{\mathrm{d}}{\mathrm{d}\rho}\Gamma(\rho) + o(\rho)I.$$

(3.17)

Since P has an inverse we have from (3.10) for $\operatorname{Re}\rho = 0$,

$$\Phi(\rho)[P^{-1} - \Gamma(\rho)] = (\beta - \alpha)\rho\,\Psi(\rho),$$

and so by using (3.17) for $\operatorname{Re}\rho = 0$, $\rho \to 0$,

$$\Phi(\rho)[P^{-1} - I - \rho\frac{\mathrm{d}}{\mathrm{d}\rho}\Gamma(\rho) + o(\rho)I] = (\beta - \alpha)\rho\,\Psi(\rho). \tag{3.18}$$

Multiply (3.18) on the right by 1^T, the transposed of the unit vector $(1,\ldots,1)$, then since

$$[P^{-1} - I]1^T = (0,\ldots,0),$$

we obtain for $\mathrm{Re}\,\rho = 0$, $\rho \to 0$,

$$[-\Phi(\rho)\frac{\mathrm{d}}{\mathrm{d}\rho}\Gamma(\rho)|_{\rho=0} + o(1)I]1^T = (\beta - \alpha)\,\Psi(\rho)1^T, \tag{3.19}$$

Since, cf. (3.8) and (3.15),

$$\Phi(0) = (1,\ldots,1), \quad \Psi(0) = \frac{1}{\alpha - \beta}(\mathrm{E}\{\mathbf{i}_1\},\ldots,\mathrm{E}\{\mathbf{i}_N\}),$$

we have from (3.16) and (3.19) by letting $\rho \to 0$ that

$$\alpha - \beta = \sum_{j=1}^{N}\gamma_j = \sum_{j=1}^{N}\mathrm{E}\{\mathbf{i}_j\} = \sum_{j=1}^{N}-\mathrm{E}\{[\mathbf{w}_j + \boldsymbol{\rho}_j]^-\}. \tag{3.20}$$

The relation (3.20) shows that the average total idle time during a 1-cycle, i.e. the time between successive arrivals of type 1-customers is equal to $\alpha - \beta$; it is readily seen that the same result applies for a j-cycle; a j-cycle being defined analogously as a 1-cycle. □

4. THE FUNCTIONAL EQUATIONS FOR THE LATTICE CASE.

In the derivations of the preceding section we have used the Laplace-Stieltjes transforms of the various distributions involved. When, however, the state space of all τ_j and σ_j, $j = 1,\ldots,N$, is the set of positive integers it is preferable to use generating functions instead of Laplace-Stieltjes transforms. In this section we shall derive the functional equation for the case of lattice variables. Exactly the same notation will be used but p shall stand for the variable of the various generating functions. So we have for $j = 1,\ldots,N$,

$$\alpha_j(p) := \mathrm{E}\{p^{\sigma_j}\}, \qquad \beta_j(p) := \mathrm{E}\{p^{\tau_j}\}, \quad |p| \le 1,$$

$$\gamma_j(p) := \alpha_j(p^{-1})\beta_j(p), \quad \gamma(p) := \prod_{j=1}^{N}\gamma_j(p), \quad |p| = 1, \tag{4.1}$$

and if $\alpha_j(p)$ or $\beta_j(p)$ is a rational function of p then

$$\alpha_j(p) = \frac{a_{1j}(p)}{a_{2j}(p)}, \quad \beta_j(p) = \frac{b_{1j}(p)}{b_{2j}(p)}, \quad |p| \le 1, \tag{4.2}$$

with

i. $a_{2j}(p)$ *a polynomial in p with degree* m_j,
     $b_{2j}(p)$,,   ,,   ,, p ,,   ,, n_j;

ii. $a_{1j}(p)$,,   ,,   ,, p ,,   ,, $< m_j$,   (4.3)
     $b_{1j}(p)$,,   ,,   ,, p ,,   ,, $< n_j$;

iii. $a_{1j}(0) = a_{2j}(0) \ne 0$,

$$b_{1j}(0) = b_{2j}(0) \neq 0;$$

iv. $a_{2j}(p)$ has m_j poles   in $|p| > 1$,

      $b_{2j}(p)$,, n_j ,,   ,, $|p| > 1$.

It will always be assumed that $a_{1j}(p)$ and $a_{2j}(p)$ have no common factors, similarly for $b_{1j}(p)$ and $b_{2j}(p)$ and that the τ_j and σ_j are aperiodic lattice variables, i.e.

$$|\alpha_j(p)| = 1 \quad for \quad a \quad p \quad with \quad |p| = 1 \quad implies \quad p = 1,$$

$$|\beta_j(p)| = 1 \quad ,, \quad ,, \quad p \quad ,, \quad |p| = 1 \quad ,, \quad p = 1. \tag{4.4}$$

Again it is assumed that the sequences $\mathbf{w}_j^{(n)}$, $n = 1, 2, \ldots$, converge in distribution for $n \to \infty$. The variables $\mathbf{w}_j$, $\mathbf{i}_j$, $j = 1, \ldots, N$, are introduced as in the previous section. We start from the identity: for real y,

$$p^{[y]^+} + p^{[y]^-} = p^y + 1.$$

Hence

$$p^{\mathbf{w}_j} = p^{\mathbf{w}_{j-1} + \boldsymbol{\rho}_{j-1}} + 1 - p^{-\mathbf{i}_{j-1}}, \quad j = 2, \ldots, N, \tag{4.5}$$

$$p^{\mathbf{w}_1} = p^{\mathbf{w}_N + \boldsymbol{\rho}_N} + 1 - p^{-\mathbf{i}_N}.$$

So for $|p| = 1$, $j = 2, \ldots, N$,

$$\mathrm{E}\{p^{\mathbf{w}_j}\} - \gamma_{j-1}(p)\,\mathrm{E}\{p^{\mathbf{w}_{j-1}}\} = -(1 - \frac{1}{p})\mathrm{E}\{\mathbf{i}_{j-1}\}\frac{1 - \mathrm{E}\{p^{-\mathbf{i}_{j-1}}\}}{-(1 - \frac{1}{p})\mathrm{E}\{\mathbf{i}_{j-1}\}}, \tag{4.6}$$

$$\mathrm{E}\{p^{\mathbf{w}_1}\} - \gamma_N(p)\,\mathrm{E}\{p^{\mathbf{w}_N}\} = -(1 - \frac{1}{p})\mathrm{E}\{\mathbf{i}_N\}\frac{1 - \mathrm{E}\{p^{-\mathbf{i}_N}\}}{-(1 - \frac{1}{p})\mathrm{E}\{\mathbf{i}_N\}}.$$

Put for $j = 1, \ldots, N$,

$$\Phi_j(p) := \mathrm{E}\{p^{\mathbf{w}_j}\}, \qquad\qquad |p| \leq 1,$$

$$\Psi_j(p) := \frac{\mathrm{E}\{\mathbf{i}_j\}}{\alpha - \beta}\frac{1 - \mathrm{E}\{p^{\mathbf{i}_j}\}}{-(1 - \frac{1}{p})\mathrm{E}\{\mathbf{i}_j\}}, \quad |p| \geq 1, \tag{4.7}$$

and introduce the vectors and matrices

$$\Phi(p) := (\Phi_1(p), \ldots, \Phi_N(p)),$$

$$\Psi(p) := (\Psi_1(p), \ldots, \Psi_N(p)), \tag{4.8}$$

$$\Gamma_{ij}(p) := \gamma_j(p)\delta_{ij}, \quad i, j \in \{1, \ldots, N\}.$$

The relations (4.6) may now be rewritten as: for $|p| = 1$,

$$\Phi(p)[I - \Gamma(p)P] = (\beta - \alpha)(1 - p^{-1})\,\Psi(p)P. \tag{4.9}$$

From (4.9) we obtain as before: for $|p| = 1$, $p \neq 1$,

$$\Phi(p) = (\beta - \alpha)(1 - \frac{1}{p})\,P[I - \Gamma(p)P]^{-1}. \tag{4.10}$$

From the definition of $\Phi_j(p)$ and $\Psi_j(p)$, $j = 1,\ldots,N$, it follows that

i. $\Phi_j(p)$ is regular for $|p| < 1$, continuous for $|p| \leq 1$,   (4.11)

   $\Phi_j(1) = 1$,

ii. $\Psi_j(p)$ is regular for $|p| > 1$, continuous for $|p| \geq 1$,

   $\Psi_j(1) = \dfrac{E\{i_j\}}{\alpha - \beta}$,

   $\Psi_j(p) = O(1)$ for $|p| \to \infty$.

The relation (4.9) for the vectors $\Phi(p)$ and $\Psi(p)$, and similarly the equivalent relation (4.10), formulates together with the condition (4.11) a homogeneous Hilbert Boundary Value Problem with the unit circle $|p| = 1$ the contour of discontinuity, cf. [10] p. 384.

5. SOME REMARKS ON THE CASE $N = 1$

For the case $N = 1$ the relation (3.10) becomes: for $\operatorname{Re} \rho = 0$,

$$E\{e^{-\rho w}\}\frac{1-\gamma(\rho)}{(\beta-\alpha)\rho} = \frac{E\{i\}}{\alpha-\beta}\frac{1-E\{e^{\rho i}\}}{-\rho E\{i\}},$$ (5.1)

where we have written

   $\mathbf{w}$ for $\mathbf{w}_1$ and $\mathbf{i}$ for $\mathbf{i}_1$.

By taking $\rho = 0$ in (5.1) it follows, cf. also (3.20),

   $E\{i\} = \alpha - \beta$.   (5.2)

Let $\mathbf{v}$ be a positive stochastic variable with distribution

$$\Pr\{\mathbf{v} < v\} = \frac{1}{E\{i\}}\int_0^v \Pr\{i \geq u\}du, \quad v > 0,$$
$$= 0, \qquad\qquad\qquad v \leq 0,$$ (5.3)

then

$$E\{e^{-\rho \mathbf{v}}\} = \frac{1 - E\{e^{-\rho i}\}}{\rho E\{i\}}, \quad \operatorname{Re} \rho \geq 0.$$ (5.4)

Hence, it follows from (5.1), (5.2) and (5.4), that: for $\operatorname{Re}\rho = 0$,

$$E\{e^{-\rho w}\}\frac{1-\gamma(\rho)}{(\beta-\alpha)\rho} = E\{e^{\rho \mathbf{v}}\}.$$ (5.5)

The present case is actually the classical $GI/G/1$ model with $\alpha(\rho)$ and $\beta(\rho)$ the Laplace-Stieltjes transforms of the interarrival time and of the service time distribution, with $\mathbf{w}$ a stochastic variable with distribution the stationary distribution of the actual waiting time process and $\mathbf{i}$ the idle time in a busy cycle, cf. [14], [15]. From the results of Queueing Theory of the $GI/G/1$ model, or from those of Fluctuation Theory, cf. [14], II.5 and I.6.6, it follows that there exists a unique pair of functions

$K_-(\rho)$ and $K_+(\rho)$ such that

i. $K_-(\rho)$ is regular for $\operatorname{Re}\rho > 0$, continuous for $\operatorname{Re}\rho \geq 0$,   (5.6)

$$K_-(0) = 1, \quad K_-(\rho) = O(1) \text{ for } |\rho| \to \infty, \quad |\arg \rho| \leq \frac{\pi}{2},$$

$K_-(\rho)$ has no zeros in $\operatorname{Re}\rho \geq 0$,

ii. $K_+(\rho)$ is regular for $\operatorname{Re}\rho < 0$, continuous for $\operatorname{Re}\rho \leq 0$,

$$K_+(0) = 1, \quad K_+(\rho) = O(-\log |\rho|) \text{ for } |\rho| \to \infty, \frac{1}{2}\pi \leq \arg \rho \leq \frac{3}{2}\pi,$$

iii. $e^{K_+(\rho) - K_-(\rho)} = \dfrac{1 - \gamma(\rho)}{(\beta - \alpha)\rho}, \quad \operatorname{Re}\rho = 0,$

provided $\beta - \alpha < 0$, and

$$\sum_{j=1}^{N} (\tau_j - \sigma_j)$$

is not a lattice variable and nonzero with positive probability. For integral expressions expressing $K_+(\cdot)$ and $K_-(\cdot)$ as functionals of $\gamma(\cdot)$, see [14], p. 143 and [15].

Actually, the relation (5.6)iii describes the factorisation of

$$\frac{1 - \gamma(\rho)}{(\beta - \alpha)\rho}, \quad \operatorname{Re}\rho = 0,$$

as the product of two functions with one the boundary value of a function regular in $\operatorname{Re}\rho < 0$, the other regular in $\operatorname{Re}\rho > 0$. A further result from $GI/G/1$ queueing theory is that:

$$\begin{aligned} E\{e^{-\rho\mathbf{W}}\} &= e^{K_-(\rho)}, \quad \operatorname{Re}\rho \geq 0, \\ E\{e^{\rho\mathbf{V}}\} &= e^{K_+(\rho)}, \quad \operatorname{Re}\rho \leq 0. \end{aligned} \quad (5.7)$$

6. THE CASE WITH ALL $\beta_j(\rho)$ RATIONAL

In this section we shall consider the case that all $\beta_j(\rho)$, $j = 1, \ldots, N$, are rational functions of ρ, and no assumptions are made concerning the $\alpha_j(\rho)$, $j = 1, \ldots, N$.

We start from the relation (3.13) written as: for $j = 1, 2, \ldots, N$, and $\operatorname{Re}\rho = 0$,

$$\Phi_j(\rho) = \frac{(\beta - \alpha)\rho}{1 - \gamma(\rho)}[\Psi(\rho) \, P[I - \Gamma(\rho) \, P]^{-1}]_j \, (1 - \gamma(\rho)), \tag{6.1}$$

$$\beta_j(\rho) = \frac{b_{1j}(\rho)}{b_{2j}(\rho)}, \; \gamma_j(\rho) = \frac{b_{1j}(\rho)}{b_{2j}(\rho)} \, \alpha(-\rho), \; b_2(\rho) = \prod_{j=1}^{N} b_{2j}(\rho). \tag{6.2}$$

By using (5.6)iii we rewrite this relation as: for $j = 1, \ldots, N$, $\operatorname{Re}\rho = 0$,

$$\frac{b_2(\rho)}{b_{2j}(\rho)} \, e^{-K_-(\rho)} \, \Phi_j(\rho) = e^{-K_+(\rho)} \frac{b_2(\rho)}{b_{2j}(\rho)}[\Psi(\rho) \, P[I - \Gamma(\rho)P]^{-1}]_j \, (1 - \gamma(\rho)). \tag{6.3}$$

From (3.10), (3.15) and (5.6)i it is seen that the function in the lefthand side of (6.2) is the boundary value of a regular function in $\mathrm{Re}\,\rho > 0$ which for $|\rho| \to \infty$, $|\arg \rho| < \frac{1}{2}\pi$, behaves as $|\rho|^{\nu - n_j}$. By noting that in the elements of the jth column of the matrix in (3.12) the factor $\gamma_j(\rho)$ does not occur it is readily seen from (3.15) and (5.6)ii that the righthand side of (6.2) is the boundary value of a function regular in $\mathrm{Re}\,\rho < 0$ and that for $|\rho| \to \infty$, $\frac{1}{2}\pi < \arg \rho < \frac{3}{2}\pi$, it behaves as $|\rho|^{\nu - n_j}$. Consequently, these functions for $\mathrm{Re}\,\rho > 0$ and $\mathrm{Re}\,\rho < 0$, respectively are each other's analytic continuation. Hence by applying Liouville's theorem it follows that these functions are both a polynomial say, $P_j(\rho)$, of degree $\nu - n_j$, i.e. for $j = 1, \ldots, N$,

$$\Phi_j(\rho) = \frac{b_{2j}(\rho)}{b_2(\rho)}\, e^{K_-(\rho)}\, P_j(\rho), \quad \mathrm{Re}\,\rho \geq 0, \tag{6.4}$$

$P_j(\rho)$ a polynomial in ρ of degree $\nu - n_j$, $\nu = \sum\limits_{j=1}^{N} n_j$ the degree of $b_2(\rho)$.

Substitution of (6.4) into (3.6) leads to: for $j = 2, \ldots, N$, $\mathrm{Re}\,\rho = 0$,

$$\frac{b_{2j}(\rho)}{b_2(\rho)} e^{K_-(\rho)} P_j(\rho) - \frac{b_{2j-1}(\rho)}{b_2(\rho)} \gamma_{j-1}(\rho) e^{K_-(\rho)} P_{j-1}(\rho) = -\rho\, \Psi_{j-1}(\rho),$$

or by using (5.6)iii: for $j = 2, \ldots, N$, $\mathrm{Re}\,\rho = 0$,

$$[b_{2j}(\rho)P_j(\rho) - b_{2j-1}(\rho)\gamma_{j-1}(\rho)P_{j-1}(\rho)]\, e^{K_+(\rho)} = \frac{1 - \gamma(\rho)}{(\beta - \alpha)\rho} b_2(\rho)\rho\, \Psi_{j-1}(\rho). \tag{6.5}$$

The function in the lefthand side of (6.5) is the boundary value of a function regular for $\mathrm{Re}\,\rho < 0$, cf. (5.6)ii and (6.4), that in the righthand side is also such a function, cf. (3.9) and note that $\{1 - \gamma(\rho)\}b_2(\rho)/\rho$ has no poles in $\mathrm{Re}\,\rho \leq 0$, since all $\beta_j(\rho)$ are rational. Hence by analytic continuation (6.5) holds for $\mathrm{Re}\,\rho \leq 0$. It follows from the continuity of $\Psi_{j-1}(\rho)$ in $\mathrm{Re}\,\rho \leq 0$ that the zeros of

$$\frac{1 - \gamma(\rho)}{(\beta - \alpha)\rho}\, b_2(\rho) \text{ in } \mathrm{Re}\,\rho \leq 0, \tag{6.6}$$

should be zeros of

$$b_{2j}(\rho)\, P_j(\rho) - b_{2j-1}(\rho)\gamma_{j-1}(\rho)\, P_{j-1}(\rho), \quad j = 2, \ldots, N; \tag{6.7}$$

note that $K_+(\rho)$ is continuous in $\mathrm{Re}\,\rho \leq 0$.

By applying Rouché's theorem it is readily shown, cf. [14], p. 323, that the function in (6.6) has for $\beta - \alpha < 0$ exactly ν zeros $\epsilon_1, \ldots \epsilon_\nu$, say, in $\mathrm{Re}\,\rho \leq 0$ and their real parts are all negative. Hence, if all these zeros have multiplicitly one; see remark 6.1 below, then: for $\rho = \epsilon_k$, $k = 1, \ldots, \nu$,

i. $\quad b_{2j}(\rho)P_j(\rho) - b_{2j-1}(\rho)\gamma_{j-1}(\rho)P_{j-1}(\rho) = 0, \quad j = 2, \ldots, N,$ $\hfill (6.8)$

ii. $\quad b_{21}(\rho)P_1(\rho) - b_{2N}(\rho)\gamma_N(\rho)P_N(\rho) = 0;$

the relation in (6.8)ii follows from the last realtion in (3.6) by using the same arguments as above. From (3.15)i, (5.6)i and (6.4) we have

$$\frac{b_{2j}(0)}{b_2(0)}\, P_j(0) = 1, \quad j = 1, \ldots, N. \tag{6.9}$$

Because $P_j(\rho)$ is a polynomial of degree $\nu - n_j$, cf. (6.4), and so has $\nu - n_j + 1$ coefficients the determination of these polynomials requires the determination of

$$\sum_{j=1}^{N}(\nu - n_j + 1) = N + (N-1)\nu, \tag{6.10}$$

unknowns. Because $1 - \gamma(\rho)$ is the main determinant of the set of equations (3.6) for the $\Phi_j(\rho)$, cf. (3.11), it follows that the relation (6.8)ii depends linearly on those in (6.8)i. So it is seen that the conditions (6.8)i and (6.9) represent $(N-1)\nu+N$ inhomogeneous linear equations for the $N+(N-1)\nu$, cf. (6.10), coefficients of the polynomials $P_j(\rho)$, $j = 1,\ldots,N$. Because the condition $\beta - \alpha < 0$ implies that the $\mathbf{w}_n$-process has a unique stationary distribution, it follows that there exists a unique $\Phi(\rho)$, satisfying (3.15)i, and consequently, the system of linear, inhomogeneous equations (6.8)i and (6.9) has a unique solution if all zeros $\rho = \epsilon_k$, $k = 1,\ldots,\nu$, have multiplicity one.

REMARK 6.1. The sum of the multiplicities of the zeros of the function in (6.9) is always equal to n. If such a zero, say ϵ_1, has a multiplicity larger than one, then ϵ_1 should be also a zero with the same multiplicity of (6.7), and it is readily seen that again a system of linear equations for the coefficients of the $P_j(\cdot)$ is obtained which determines these coefficients. □

From the above it is seen that the polynomials $P_j(\rho)$, $j = 1,\ldots,N$, may be considered to have been determined. Hence, for the ultimate determination of the vectors $\Phi(\rho)$ and $\Psi(\rho)$ it remains to determine $K_+(\rho)$ and $K_-(\rho)$. From the preceding section it follows, cf. (5.7), that $\exp K(\rho)$, $\mathrm{Re}\,\rho \geq 0$ is the Laplace-Stieltjes transform of the stationary distribution of the actual waiting time of a $GI/K_n/1$ queue, since $\beta(\rho) = \prod_{j=1}^{N}\beta_j(\rho)$ is a rational function of ρ. That transform has been determined in [14], section II.5.10. From the results obtained there it follows that, cf. (II.5.190): for $\mathrm{Re}\,\rho \geq 0$,

$$e^{K_-(\rho)} = \frac{b_2(\rho)}{b_2(0)}\prod_{k=1}^{\nu}\frac{\epsilon_k}{\epsilon_k - \rho}. \tag{6.11}$$

Hence from (6.4), (6.9) and (6.11) it follows that: for $j = 1,\ldots,N$, $\mathrm{Re}\,\rho \geq 0$,

$$\Phi_j(\rho) = \frac{b_{2j}(\rho)P_j(\rho)}{b_{2j}(0)P_j(0)}\prod_{k=1}^{\nu}\frac{\epsilon_k}{\epsilon_k - \rho}, \qquad \mathrm{Re}\,\rho \geq 0, \tag{6.12}$$

if it is assumed that all ϵ_k have multiplicity one. Note that $b_{2j}(\rho)P_j(\rho)$ is a polynomial of degree ν.

So we have reached the following result.

For $\beta - \alpha < 0$ the Laplace-Stieltjes transforms $\Phi_j(\rho)$, $j = 1,\ldots,N$; $\mathrm{Re}\,\rho \geq 0$, of the stationary distributions of the actual waiting times $\mathbf{w}_j$, $j = 1,\ldots,N$ (in an arrival-cycle) are given by

i. $\quad \Phi_j(\rho) = \dfrac{b_{2j}(\rho)P_j(\rho)}{b_{2j}(0)P_j(0)}\prod_{k=1}^{\nu}\dfrac{\epsilon_k}{\epsilon_k - \rho}, \qquad j = 1,\ldots,N; \quad \mathrm{Re}\,\rho \geq 0,$ \hfill (6.13)

with

ii. $\quad \epsilon_k$, $k = 1,\ldots,\nu$, the zeros of

$$1 - \prod_{j=1}^{N}\frac{b_{1j}(\rho)}{b_{2j}(\rho)}\alpha_j(-\rho) \text{ in } \mathrm{Re}\,\rho < 0,$$

iii. $\quad P_j(\rho)$, $j = 1,\ldots,N$, polynomials of degree $\nu - n_j$, determined by the conditions

$$b_{2j}(0)\,P_j(0) = b_2(0), \quad j = 1,\ldots,N,$$

and

for $\rho = \epsilon_k$, $k = 1, \ldots, \nu$,

$$b_{21}(\rho)P_1(\rho) - \gamma_N(\rho)b_{2N}(\rho) \, P_N(\rho) = 0,$$

$$b_{2j}(\rho)P_j(\rho) - \gamma_{j-1}(\rho)b_{2j-1}(\rho) \, P_{j-1}(\rho) = 0, \quad j = 2, \ldots, N.$$

7. THE CASE WITH ALL $\alpha_j(\rho)$ RATIONAL

In this section we consider the case that all $\alpha_j(\rho)$, $j = 1, \ldots, N$, are rational functions of ρ, and no assumptions will be made concerning the $\beta_j(\rho)$, $j = 1, \ldots, N$.

We start from the relations (3.6) and by using (2.9) we have, cf. (3.8): for $\mathrm{Re}\,\rho = 0$, $j = 2, \ldots, N$,

i) $\quad a_{2j-1}(-\rho)\Phi_j(\rho) - a_{2j-1}(-\rho)\gamma_{j-1}(\rho)\Phi_{j-1}(\rho) = (\beta - \alpha)\rho\Psi_{j-1}(\rho)a_{2j-1}(-\rho),$ (7.1)

ii) $\quad a_{2N}(-\rho)\Phi_1(\rho) - a_{2N}(-\rho)\gamma_N(\rho)\Phi_n(\rho) = (\beta - \alpha)\rho\Psi_N(\rho)a_{2N}(-\rho).$

From (2.10) and (3.15) it is seen that the lefthand side of (7.1)i is the boundary value of a function regular in $\mathrm{Re}\,\rho > 0$, continuous in $\mathrm{Re}\,\rho \geq 0$, which for $|\rho| \to \infty$, $|\arg\rho| \leq \frac{1}{2}\pi$ behaves as $|\rho|^{m_J-1}$. Similarly, it is seen that the righthand side in (7.1)i is the boundary value of a function regular in $\mathrm{Re}\,\rho < 0$, continuous in $\mathrm{Re}\,\rho \leq 0$ and which behaves as $|\rho|^{m_j-1}$ for $|\rho| \to \infty$, $\frac{1}{2}\pi \leq \arg\rho \leq \frac{3}{2}\pi$. Hence these functions for $\mathrm{Re}\,\rho < 0$ and $\mathrm{Re}\,\rho > 0$ are each other's analytic continuation, and so by applying Liouville's theorem it follows that : for $\mathrm{Re}\,\rho \leq 0$, $j = 1, \ldots, N$,

$$(\beta - \alpha)\rho a_{2j}(-\rho)\Psi_j(\rho) = Q_j(\rho),$$ (7.2)

$Q_j(\rho)$ *a polynomial in ρ of degree m_j, $Q_j(0) = 0$, cf. (3.8.)*

For the present case we have, cf. (3.10), (3.11), (5.6)iii: for $\mathrm{Re}\,\rho = 0$,

$$\frac{D_N(\rho)}{(\beta - \alpha)\rho} = \frac{1 - \gamma(\rho)}{(\beta - \alpha)\rho} = \frac{a_2(-\rho) - a_2(-\rho)\gamma(\rho)}{(\beta - \alpha)\rho a_2(-\rho)} = e^{K_+(\rho) - K_-(\rho)}.$$ (7.3)

Because $\beta - \alpha < 0$ it follows readily, cf. [14], p. 330, that for the present case $D_N(\rho)$ has μ zeros δ_k, $k = 1, \ldots, \mu$, in $\mathrm{Re}\,\rho \geq 0$, and $\delta_1 = 0$, $\mathrm{Re}\,\delta_k > 0$, $k = 2, \ldots, \mu$, a result easily obtained by applying Rouché's theorem.

REMARK 7.1. For the sake of simplicity it is again assumed that these zeros all have multiplicity one; see also remark 6.1.
□

From [14] p. 330, and from section 5 we obtain:

$$e^{K_+(\rho)} = \frac{a_2(0)}{a_2(-\rho)} \prod_{k=2}^{\mu} \frac{\delta_k - \rho}{\delta_k}, \qquad\qquad \mathrm{Re}\,\rho \leq 0,$$ (7.4)

$$e^{K_-(\rho)} = \frac{-a_2(0)(\alpha - \beta)\rho}{a_2(-\rho) - a_1(-\rho)b(\rho)} \prod_{k=2}^{\mu} \frac{\delta_k - \rho}{\delta_k}, \qquad\qquad \mathrm{Re}\,\rho \geq 0.$$

For $\mathrm{Re}\,\rho = 0$, we have, cf. (3.13),

$$\Phi(\rho) = \frac{(\beta - \alpha)\rho}{1 - \gamma(\rho)}\Psi(\rho)D_n(\rho)P[I - \Gamma(\rho)P]^{-1},$$

and so from (7.3): for $\mathrm{Re}\,\rho = 0$,

$$\Phi(\rho)e^{-K_-(\rho)} = \frac{e^{-K_+(\rho)}}{a_2(-\rho)}\Psi(\rho)a_2(-\rho)D_N(\rho)P[I - \Gamma(\rho)P]^{-1}. \tag{7.5}$$

Hence by using (7.2) and (7.3) we have: for $\mathrm{Re}\,\rho = 0$, $j = 1, \ldots, N$,

$$\Phi_j(\rho)e^{-K_-(\rho)} = \frac{\prod\limits_{k=2}^{m}\delta_k}{a_2(0)} \frac{a_2(\rho)}{\prod\limits_{k=2}^{m}(\delta_k - \rho)} \sum_{i=1}^{N} \frac{Q_i(\rho)}{\alpha_{2i}(\rho)} \frac{D_N(\rho)}{(\beta - \alpha)\rho}[P[I - \Gamma(\rho)P]^{-1}]_{ij}. \tag{7.6}$$

From the structure of the matrix in (3.12) and from the fact that the factor $a_2(-\rho)/a_{2i}(-\rho)$ occurs in the righthand side it is seen that this righthand side has a meromorphic continuation in $\mathrm{Re}\,\rho > 0$, since $Q_i(\rho)$ is a polynomial. The lefthand side of (7.6) has an analytic continuation in $\mathrm{Re}\,\rho > 0$, see (3.15)i and (5.6)i. Because these continuations are unique it follows that: for $j = 1, \ldots, N$,

$$\sum_{i=1}^{N} \frac{Q_i(\rho)}{\alpha_{2i}(-\rho)} \frac{D_n(\rho)}{(\beta - \alpha)\rho}[P[I - \Gamma(\rho)P]^{-1}]_{ij} = 0 \text{ for } \rho = \delta_k, \qquad k = 2, \ldots, \mu, \tag{7.7}$$

note that these δ_k, $k = 2, \ldots, \mu$, are the only poles of the mentioned meromorphic continuation.

Next we multiply (7.1) by $e^{-K_-(\rho)}$. Then by using (7.2), (7.3) and (7.4) we obtain: for $\mathrm{Re}\,\rho = 0$, $j = 2, \ldots, N$,

i.
$$\Phi_j(\rho)e^{-K_-(\rho)} = \frac{1}{\alpha_{2j-1}(-\rho)}[\alpha_{2j-1}(-\rho)\gamma_{j-1}(\rho)\Phi_{j-1}(\rho)e^{-K_-(\rho)}+ \tag{7.8}$$

$$Q_{j-1}(\rho)\frac{a_2(-\rho)}{a_2(0)}D_N(\rho)\prod_{k=2}^{\mu}\frac{\delta_k}{\delta_k - \rho}],$$

ii.
$$\Phi_1(\rho)e^{-K_-(\rho)} = \frac{1}{\alpha_{2N}(-\rho)}[\alpha_{2N}(-\rho)\gamma_N(\rho)\Phi_N(\rho)e^{-K_-(\rho)} + Q_N(\rho)\frac{a_2(-\rho)}{a_2(0)}D_N(\rho)\prod_{k=2}^{\mu}\frac{\delta_k}{\delta_k - \rho}];$$

note that for $\rho = 0$ the relations (7.8) imply that

$$\Phi_1(0) = \Phi_2(0) = \ldots = \Phi_N(0). \tag{7.9}$$

To determine the polynomials $Q_j(\rho)$, $j = 1, \ldots, N$, note that $Q_j(\rho)$ contains m_j coefficients because $Q_j(0) = 0$, cf. (7.2); so in total $\mu = m_1 + \ldots + m_N$ coefficients have to be determined. Now consider (7.7) for $j = 1$, then these conditions lead to $\mu - 1$ equations for the μ unknown coefficients in the polynomials Q_j, $j = 1, \ldots, N$. Hence together with the norming condition $\Phi_1(0) = 1$, i.e. by using (7.6) for $j = 1$ and $\rho = 0$, we obtain a system of μ linear inhomogeneous equations for the μ unknowns. Assume for the present that this system possesses a unique solution. Then $\Phi_1(\rho)$, $\mathrm{Re}\,\rho \le 0$, is determined by (7.6). By taking successively $j = 2, \ldots, N$ in (7.8)i explicit expressions for $\Phi_j(\rho)$, $\mathrm{Re}\,\rho = 0$, are obtained. It also follows from (7.8)i, or equivalently from (7.1)i and (7.2), that for the expressions so obtained $\alpha_{2j-1}(-\rho)\Phi_j(\rho)$, $\mathrm{Re}\,\rho > 0$, $j = 2, \ldots, N$, has a unique analytic continuation in $\mathrm{Re}\,\rho > 0$. Consequently, it follows from (7.6) that the relations (7.8)i for $j = 2, \ldots, N$, are identically satisfied. Hence, by using (7.6) for $j = 2, \ldots, N$, it follows that the expressions for $\Phi_j(\rho)$, $j = 2, \ldots, N$, as constructed above, are regular for $\mathrm{Re}\,\rho > 0$ (note the structure of the matrix in (3.12)), which implies that the zeros of $\alpha_{2j-1}(-\rho)$ in $\mathrm{Re}\,\rho > 0$ are zeros of the sum between the square brackets in (7.8)i for every $j = 2, \ldots, N$.

REMARK 7.2. The analysis above is based, apart from the assumption introduced above, on the relations (7.8)i, for $j = 2, \ldots, N$, and on the relation (7.7) for $j = 1$. It is readily verified that this set of relations is equivalent with the system (7.1), cf. also (3.10), and also equivalent with (3.13), Hence the solution constructed above satisfies (7.1)ii, and similarly (7.8)ii. Note further, cf. also (7.9), that this solution leads to a unique determination of $\Psi_j(\rho)$, $\mathrm{Re}\,\rho \le 0$, $j = 1, \ldots, N$, and that this solution possesses the properties (3.15). The verification of this is not difficult but requires some algebras. □

To complete our analysis it remains to justify the assumption introduced above concerning the uniqueness of the solution of the linear system of inhomogeneous equations. This justification is obtained as in the previous section. Viz. the condition $\beta - \alpha < 0$ implies that the w_n-process has a unique stationary distribution and so there exits a unique $\Phi(\rho)$ satisfying (3.15)i, and consequently the system of equations (7.1)i for $j = 2, \ldots, N$, together with (7.6) for $j = 1$, should have a unique solution satisfying (3.15)i, and this is only possible if the system (7.7) for $j = 1$, together with $\Phi_1(0) = 1$ determines the μ coefficients of the $Q_j(\rho)$, $j = 1, \ldots, N$, uniquely.

8. SOME SPECIAL CASES

The discussions in the two preceding sections show that an effective analysis of the periodic Pollaczek waiting processes is possible for quite general cases. In this section we comment briefly on a number of special cases of this Pollaczek model.

A first variant is the mixed case, i.e.: the $\alpha_j(\rho), j \in A$, and the $\beta_j(\rho), j \in B$ are rational, here A and B are disjoint set and their union is the set $\{1, 2, \ldots, N\}$. The analysis for general N, A and B is possible, but quite intricate for $N > 2$. For $N = 2$ this case has been discussed in detail in [19].

A second variant concerns the case where the $\tau_j = t_j = 2, \ldots, N$ and the $\sigma_j, j = 1, \ldots, N - 1$, are constants, $\tau_j = t_j$ and $\sigma_j = s_j$, so only τ_1 and σ_N are true stochastic variables. Again an explicit solution can be constructed, cf.[19]. Actually, the functional equation to be solved here is a standard Riemann Boundary Value Problem. A special case of this variant is the periodic GI/G/1 queueing model where every first customer of the cycle generates the arrival of a fixed number i.e. $N - 1$, of subsequent customers with constant service - and interarrival times, each service time being equal to the next interarrival time. This variant may serve as a simple model for bursty traffic.

A third variant of interest concerns the case with all σ_j constant, $\sigma_j = s_j$, say, $j = 1, \ldots, N$. Put $s = s_j + \ldots + s_N$. Consider for this process the embedded process $w_1^{(n)}$, $n = 1, 2, \ldots$, and compare it with the single server model D/G/1 with interarrival times equal to s and service time distribution that of the sum of $\tau_1, \ldots, \tau_N$. For this D/G/1 process the supply to the workload of the server occurs at the equally spaced arrival moments, where as in the $w_1^{(n)}$-process the supply to the workload is distributed, (discretely and unevenly) over each of the intervals of length s. Hence the supply of the workload for the $w_1^{(n)}$-process is a smoother process than that in the D/G/1 system. Consequently, the workload process in this periodic Pollaczek process is smoother, i.e. its jumps are less pronounced than those of the workload process of the D/G/1 system.

ACKNOWLEDGEMENT
The author is grateful to S.C. Borst, O.J. Boxma, M.B. Combé and G. Koole for their helpful comments in preparing this manuscript.

REFERENCES
1. POLLACZEK, F., Problémes Stochastique Posés par la Phénomène de Formation d'une Queue d'Attente à un Guichet et par des Phénomènes Apparantés, Gauthier Villars, Paris, 1957.

2. ARJAS, E., On the use of a fundamental identity in the theory of semi-Markov chains, Adv. Appl. Prob. 4 (1972) 271-284.

3. ASMUSSEN, S., THORISSON, H., A Markov chain approach to periodic queues, J. Appl. Prob. 24 (1987) 215-225.

4. HARRISON, J.M., LEMOINE, A.J., Limit theorems for periodic queues, J. Appl. Prob. **14** (1977) 566-576.

5. LEMOINE, A.J., Waiting time and workload in queues with periodic Poisson input, J. Appl. Prob. **26** (1989) 390-397.

6. BAMBOS, N., WALRAND, J., On queues with periodic input, J. Appl. Prob. **26** (1989) 382-389.

7. ÇINLAR, E., Queues with semi-Markovian arrivals, J. Appl. Prob. **4** (1967) 365-379.

8. NEUTS, Matrix-Geometric Solutions in Stochastic Models; an algorithmic approach, John Hopkins University Press. Baltimore, 1981.

9. LOYNES, R.M., The stability of a queue with nonindependent inter-arrival and service times, Proc. Camb. Phil. Soc. **58** (1962) 499-520.

10. MUSKHELISHVILI, N.I., Singular Integral equations, Noordhoff, Groningen, 1953.

11. MICHLIN, S.G., PRÖSZDORF, S., Singuläre Integral-Operatoren, Akademie Verlag, Berlin, 1980.

12. ZABREYKO, P.P., E.O., Integral equations, a reference text, Noordhoff, Intern. Publishers, Groningen, 1975.

13. MILLER, H.D., A matrix factorization problem in the theory of random variables defined on a finite Markov chain. Proc. Camb. Phil. Soc. **58** (1962) 268-285.

14. COHEN, J.W., The Single Server Queue, North-Holland Publ. Comp., Amsterdam, 1982, 2nd edition

15. COHEN, J.W., Complex functions in Queueing Theory, Pollaczek Memorial Volume, Arch. Elektr. Uebertragung, **47** (1993) 300-310.

16. LUKACS, E., Characteristic Functions, Griffin, London, 1960.

17. LUCANTONI, D.M., The BMAP/G/1 queue: a tutorial, Performance Evaluation of Computer and Communication Systems, ed. L. DONATIELLO & R. NELSON, Lect. Notes Comp. Sc. Springer-Verlag, Berlin, 1993.

18. COHEN, J.W., BOXMA, O.J., Boundary Value Problems in Queueing System Analysis, North Holland Publ. Co, Amsterdam, 1983.

19. COHEN, J.W., On periodic Pollaczek waiting processes, report BS-R9407, Dept. Op. Res., Stat., Syst. Theory, Inst. Math., Comp. Sc., C.W.I., Amsterdam, Januari, 1994.

Random Walk Approach to Relaxation in Disordered Systems

Marcin Kotulski[1]
Karina Weron

ABSTRACT A detailed study of the limiting probability distributions of $R(t)$ — the location at time t — for one-dimensional random walks with waiting-time distributions having long tails, is presented. In the framework of the random walk approach the nonexponential character of relaxation as a function of time, given by the tail of the Mittag-Leffler distribution, has been obtained. As a consequence, the frequency-domain response takes the well-known form of the empirical Cole-Cole function.

1 Introduction

Relaxation — the time-dependent change of macroscopic properties of physical systems — discovered as early as in the 19th century, is one of the oldest unsolved problems in physics. There is a considerable progress in understanding the phenomenology of relaxation processes in a wide variety of physical systems, including amorphous semiconductors, insulators, polymers, and many glassy materials, but still the formal rules governing dynamics in such complex physical systems are puzzling. One of the most characteristic features is the nonexponential (non-Debye) character of relaxation as a function of time [Jon83, RR87]. It is a well-known fact, that despite the variety of materials used and of experimental techniques employed, the relaxation behavior is very similar and can be represented in terms of simple empirical fitting functions. Good fits to the frequency-domain observations have been obtained with the Havriliak-Negami [HH94] function

$$\chi(\omega) = \frac{1}{[1 + (i\omega\tau)^\alpha]^\beta} ,\qquad (1.1)$$

where τ is a characteristic relaxation time and α and β are parameters ranging between 0 and 1. Since almost all experimental data can be fitted with the above empirical law, it is generally expected that there must be some common nature of microscopic relaxation mechanisms. This common

---

[1]kotulski@im.pwr.wroc.pl; Papers: http://www.im.pwr.wroc.pl/~kotulski

physical nature, if it exists, should have a common mathematical structure [KS86, DH87, Wer91, WJ93] underlying the "universal" relaxation
law (1.1).

In the present paper, we study the properties of the time-domain relaxation function defined as follows [GY95]

$$\Phi_n(k,t) = \mathrm{E}\exp\{ikt^{\alpha/2}W_{\alpha/2}\}\,, \tag{1.2}$$

for the nonbiased (i.e., without drift) random walk process. The random
variable $W_{\alpha/2}$ is defined by the limiting distribution of the distance $R(t)$
reached at time t by a walking particle starting at the origin (see Theorem 2
below) and $k > 0$ is the wave number. For the biased process we deal with
a random variable H_α, defined by the limiting distribution of the distance
$R(t)$, supported on the non-negative half-line (see Theorem 1 below) and
hence the physical sense [Kam87] of the relaxation function is given by

$$\Phi_b(k,t) = \mathrm{E}\exp\{-kt^\alpha H_\alpha\}\,. \tag{1.3}$$

In practice, experimental data are represented in terms of the frequency-
domain response function $\chi(k,\omega)$:

$$\chi(k,\omega) = -\int_0^\infty e^{-i\omega t}\frac{d\Phi(k,t)}{dt}dt\,. \tag{1.4}$$

In contrast to the recent investigations of the random walk approach to
dynamical systems [WWH89, KZ94, GY95], we are able to derive the densities of the limiting distribution of the location $R(t)$ in explicit forms. This
result follows directly [Kot95] from the limiting properties of $R(t)$ being a
random sum of independent and identically distributed random variables.
In the present work Theorem 1 and Theorem 2 (Section 2) summarize the
results for biased and nonbiased walks, respectively. In Section 3 we show
that the relaxation functions (1.2) and (1.3) are given by the tail of the
Mittag-Leffler distribution. Consequently, the Fourier transform (1.4) takes
the well-known form of the empirical Cole-Cole function to which reduces
the Havriliak-Negami function (1.1) for $\beta = 1$, with a different characteristic relaxation time τ in the biased and nonbiased cases.

2   Limiting behavior of the Continuous-Time Random Walk $R(t)$

We consider the process $R(t)$, see Fig. 1, which is a random sum of N_t
random variables R_i,

$$R(t) = \sum_{i=0}^{N_t} R_i\,, \quad \text{where} \quad N_t = \max\{k : \sum_{i=0}^{k} T_i \leq t\}\,, \tag{1.5}$$

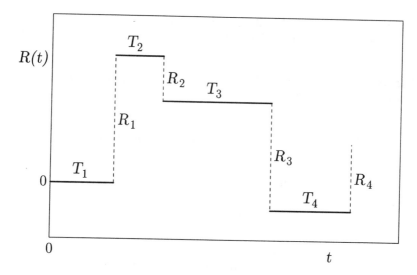

FIGURE 1. Single realization of a process $R(t)$.

$\{R_i,\ i \geq 1\}$ is a sequence of independent and identically distributed random variables representing jumps and $\{T_i,\ i \geq 1\}$ is a sequence of independent and identically distributed *positive* random variables representing waiting-times; $R_0 = 0$, $T_0 = 0$. The distribution of the pair (R_i, T_i) is the same for each $i \geq 1$. In general, R_i and T_i are dependent. The independence between R_i and T_i is denoted in the physical literature as *"decoupled memory"*.

The process $R(t)$ in stochastic modeling is known as the cumulative process [Smi55]. It was independently introduced by Montroll and Weiss [MW65] into statistical physics under the name Continuous-Time Random Walk (CTRW).

In the following we assume the finite variance of jumps R_i,

$$\mathrm{Var}\,R_i < \infty , \tag{1.6}$$

and a long tail distribution of waiting-times T_i between successive jumps R_i, that is

$$\Pr(T_i > t) \sim At^{-\alpha}, \quad \text{where} \quad 0 < \alpha < 1, \tag{1.7}$$

and A is a positive constant. We call the process $R(t)$ *biased* if $\mathrm{E}R_i \neq 0$, and *nonbiased* if $\mathrm{E}R_i = 0$. Note, that the assumption (1.7) leads to $\mathrm{E}T_i = \infty$.

In order to derive the limiting distributions of $R(t)$, under the conditions (1.6) and (1.7), recall the basic fact about the density function of a positive stable random variable. The density $s_\alpha(y)$ of a positive stable random variable S_α is defined by its characteristic function

$$\int_0^\infty e^{i\omega y} s_\alpha(y)\,dy = \exp\{-|\omega|^\alpha \exp[-i(\pi/2)\alpha\,\mathrm{sign}(\omega)]\} , \tag{1.8}$$

where α is the index of stability. We use here the form (B) of the stable characteristic function [Zol86] because of its very convenient analytical properties.

Theorem 1 *Assuming (1.6) and (1.7), for the biased process $R(t)$ and any dependence between R_i and T_i, we have*

$$\Pr\left(\frac{R(t)/\mu}{ct^\alpha} \leq x\right) \to H_\alpha(x) = \int_0^x \rho_\alpha(r)dr,\qquad(1.9)$$

as $t \to \infty$, where $\mu =ER_i$ and

$$c = \frac{1}{A\Gamma(1 - \alpha)}\qquad(1.10)$$

is a positive constant expressed in terms of parameters α and A defined in (1.7). The function $\rho_\alpha(x)$ is the density of the positive random variable $\{1/S_\alpha\}^\alpha$, such that

$$\rho_\alpha(x) = \begin{cases} (1/\alpha)x^{-1/\alpha-1}s_\alpha(x^{-1/\alpha}) & , \quad x > 0 \\ 0 & , \quad x \leq 0 \end{cases},\qquad(1.11)$$

where $s_\alpha(y)$ is defined in (1.8).

Plots of probability densities $\rho_\alpha(x)$ corresponding to selected α's are given in Fig. 2. Note that the density function $\rho_\alpha(x)$ obtains for $\alpha = 1/2$ the cutoff Gaussian form, $\rho_{1/2}(x) = (2/\sqrt{4\pi})\exp(-x^2/4)$, $x > 0$, while (see [Zol86] 2.9) $\rho_\alpha(x) \to \delta_1(x)$ as $\alpha \uparrow 1$ and $\rho_\alpha(x) \to e^{-x}$, $x > 0$, as $\alpha \downarrow 0$.

Proof: We need the limiting behavior of N_t. As $t \to \infty$

$$\Pr\left(\frac{N_t}{ct^\alpha} \leq x\right) \to H_\alpha(x), \quad 0 < \alpha < 1\qquad(1.12)$$

where $H_\alpha(x)$ is defined in (1.9) and the constant c in (1.10). For derivation of (1.12), but without the explicit form (1.10) of the constant c, see also [Fel66] XI.5.5. However, the explicit form of the constant c is important for computer simulations in statistical physics.

Since $N_t \to \infty$ with probability 1, using the strong law of large numbers, we obtain

$$\lim_{t\to\infty} \frac{R(t)/\mu}{N_t} = 1\qquad(1.13)$$

with probability 1. Hence, by combining Eqs. (1.12) and (1.13) we obtain formula (1.9) which ends the proof.

In the proof of Theorem 2 we need the following known result about products of certain powers of positive stable r.v.'s. (See for instance, Example (h) of Chap. VI of [Fel66]; Problem 13 of Chap. XIII of [Fel66] is also pertinent. We thank Professor R. Pyke for this reference.)

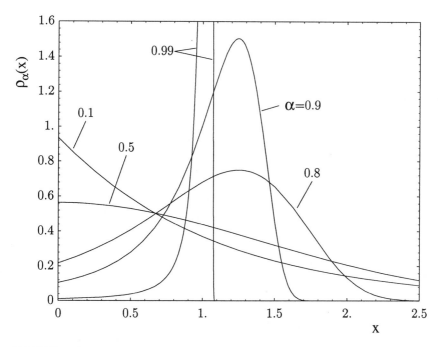

FIGURE 2. Plots of probability densities $\rho_\alpha(x)$ corresponding to selected α's; $\alpha = 0.1, 0.5, 0.8, 0.9,$ and 0.99. Note that $\rho_\alpha(x) = 0$ for $x \leq 0$.

Lemma 1 *Let a positive random variable X have the cutoff Gaussian density*

$$f(x) = \begin{cases} (2/\sqrt{4\pi})\exp(-x^2/4) & , \quad x > 0 \\ 0 & , \quad x \leq 0 \end{cases} , \qquad (1.14)$$

and a positive stable random variable S_α have density defined in (1.8). If X is independent of S_α, then the random variable $X\{S_\alpha\}^{-\alpha/2}$ has the same distribution as the random variable $\{S_{\alpha/2}\}^{-\alpha/2}$, or in terms of densities

$$\int_0^\infty r^{\alpha/2} f\left(x r^{\alpha/2}\right) s_\alpha(r)dr = \rho_{\alpha/2}(x) .$$

Proof: It may be of interest to show how this known result may be proved using Mellin transforms [Zol86] which are particularly suited for products of independent r.v.'s. Note that the Mellin transform of a product of independent positive r.v.'s is the product of the Mellin transforms of these random variables. We denote by $M(s, \alpha)$ the Mellin transform of S_α,

$$M(s, \alpha) = \mathrm{E}\{S_\alpha\}^s = \frac{\Gamma(1 - s/\alpha)}{\Gamma(1 - s)}, \quad -1 < s < \alpha, \qquad (1.15)$$

and by $M(s, 2)$ the Mellin transform of the random variable X,

$$M(s, 2) = \mathrm{E}X^s = \frac{2\sin(\pi s/2)}{\sin(\pi s)} \frac{\Gamma(1 - s/2)}{\Gamma(1 - s)}, \quad -1 < s < 2. \qquad (1.16)$$

Equations (1.15) and (1.16) are proved in [Zol86].

The Mellin transform of $X\{S_\alpha\}^{-\alpha/2}$ is

$$E[(X\{S_\alpha\}^{-\alpha/2})^s] = M(s,2)M(-\alpha s/2, \alpha) . \qquad (1.17)$$

Substituting Eqs. (1.15) and (1.16) into (1.17) we get

$$
\begin{aligned}
E[(X\{S_\alpha\}^{-\alpha/2})^s] &= \frac{2\sin(\pi s/2)}{\sin(\pi s)} \frac{\Gamma(1-s/2)}{\Gamma(1-s)} \frac{\Gamma(1+(\alpha s)/(2\alpha))}{\Gamma(1+\alpha s/2)} \\
&= \frac{\pi s}{\sin(\pi s)} \frac{1}{\Gamma(1-s)} \frac{1}{\Gamma(1+\alpha s/2)} \\
&= \frac{\Gamma(1+s)}{\Gamma(1+\alpha s/2)} = M\left(-\frac{\alpha}{2}s, \frac{\alpha}{2}\right) . \qquad (1.18)
\end{aligned}
$$

The last term $M(-\alpha s/2, \alpha/2)$ in formula (1.18) is the Mellin transform of $\{S_{\alpha/2}\}^{-\alpha/2}$. Because probability distributions are uniquely determined by their Mellin transforms, Lemma 1 is proved.

To see that this result is a consequence of the known result referred to above, simply note that by taking appropriate powers, Lemma 1 is equivalent to saying that $(X^{-2})^{1/\alpha}S_\alpha$ is equal in law to $S_{\alpha/2}$. However, it is known that random variable X^{-2} is stable with index of stability $1/2$ (see [Fel66], Chap. XIII.3).

It is interesting to mention, that Lemma 1 can be also justified by physical arguments [Tun75].

Theorem 2 *Assuming (1.6) and (1.7) for the nonbiased (i.e., without drift) process $R(t)$ and R_i independent of T_i, we have*

$$\Pr\left(\frac{R(t)}{Ct^{\alpha/2}} \le x\right) \rightarrow W_{\alpha/2}(x) = \int_{-\infty}^{x} \varphi_{\alpha/2}(r)dr , \qquad (1.19)$$

as $t \to \infty$, where

$$C = \sqrt{\frac{\mathrm{Var}R_i}{2A\Gamma(1-\alpha)}} \qquad (1.20)$$

is a positive constant expressed in terms of parameters α and A defined in (1.7). The density function

$$\varphi_{\alpha/2}(x) = \frac{1}{2}\,\rho_{\alpha/2}(|x|) \qquad (1.21)$$

is the symmetrization of density $\rho_{\alpha/2}(x)$ defined in (1.11).

Plots of probability densities $\varphi_{\alpha/2}(x)$ corresponding to selected α's are given in Fig. 3.

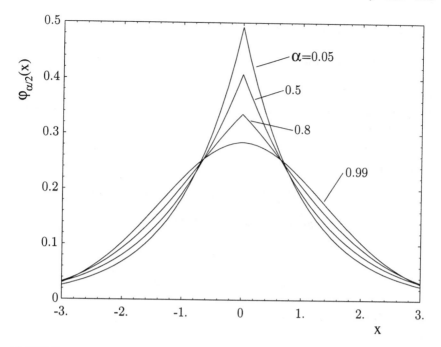

FIGURE 3. Plots of probability densities $\varphi_{\alpha/2}(x)$ corresponding to selected α's; $\alpha = 0.05$, 0.5, 0.8, and 0.99.

Proof: Using Theorem 1 of Dobrushin [Dob55] and (1.9) we obtain the integral equation for $W_{\alpha/2}(x)$,

$$W_{\alpha/2}(x) = \int_0^\infty \Psi\left(xr^{-1/2}\right) \rho_\alpha(r)dr,\qquad (1.22)$$

and also the constant $C = ((\operatorname{Var}R_i)/2)^{1/2}c^{1/2}$, where c is defined in (1.10) and $\Psi(\cdot)$ is the Gaussian distribution with mean 0 and variance 2. Note that formula (1.20) is important for applications in statistical physics.

Equation (1.22) by substituting (1.11) yields

$$W_{\alpha/2}(x) = \int_0^\infty \Psi\left(xr^{\alpha/2}\right) s_\alpha(r)dr = \Pr\left(Y\{S_\alpha\}^{-\alpha/2} \le x\right),\qquad (1.23)$$

where Y is a random variable having the Gaussian distribution $\Psi(\cdot)$ with mean 0 and variance 2, and S_α is a random variable having the stable density $s_\alpha(x)$ defined in (1.8); variables Y and S_α are independent. Due to the fact that $X\{S_\alpha\}^{-\alpha/2}$ has density $\rho_{\alpha/2}(x)$ (see Lemma 1) and the symmetry of Y, the random variable $Y\{S_\alpha\}^{-\alpha/2}$ has density $\varphi_{\alpha/2}(x)$.

Let us remark that Theorem 2 can not be applied in the case of dependent R_i and T_i, (see [Kot95]). Theorems 1 and 2 were derived in [WWH89] using the steepest descends method. Although they did not obtain the explicit form of limiting densities $\rho_\alpha(x)$, $\varphi_{\alpha/2}(x)$ and relation (1.21) between them.

3   The Cole-Cole response function

First let us consider the *nonbiased* one-dimensional process $R(t)$, defined in Section 2, Eq. (1.5). In this case, the limiting distribution has been derived under the condition that a random jump R_i and its waiting-time T_i are independent (see Theorem 2). The explicit form of the relaxation function $\Phi_n(k,t)$, Eq. (1.2), is given by the following theorem.

Theorem 3 *The relaxation function $\Phi_n(k,t)$ in the nonbiased case, that is when $\mathrm{E}R_i = 0$, equals*

$$\Phi_n(k,t) = \mathrm{E}\exp\{ikt^{\alpha/2}W_{\alpha/2}\} = 1 - G_\alpha(k^{2/\alpha}t),$$

where $G_\alpha(k^{2/\alpha}t)$ is the Mittag-Leffler distribution and $k > 0$.

Proof: Observe that for any complex z and $\rho_\alpha(x)$ defined in (1.11):

$$E_\alpha(-z) = \int_0^\infty \exp(-zx)\rho_\alpha(x)dx , \qquad (1.24)$$

(see [Zol86] (2.10.8) and [Fel66] XIII.8.3), where $E_\alpha(z)$ is the Mittag-Leffler special function, which has the series representation

$$E_\alpha(z) = \sum_{n=0}^\infty z^n/\Gamma(1 + n\alpha) .$$

By means of Eq. (1.24) the relaxation function (1.2) can be expressed as

$$\begin{aligned}
\Phi_n(k,t) &= \mathrm{E}\exp(ikt^{\alpha/2}W_{\alpha/2}) \\
&= \frac{1}{2}\int_{-\infty}^\infty \exp(ikt^{\alpha/2}x)\rho_{\alpha/2}(|x|)dx \\
&= \frac{1}{2}[E_{\alpha/2}(-ikt^{\alpha/2}) + E_{\alpha/2}(ikt^{\alpha/2})] \\
&= E_\alpha(-k^2t^\alpha) = 1 - G_\alpha(k^{2/\alpha}t),
\end{aligned}$$

where $G_\alpha(k^{2/\alpha}t)$ is the Mittag-Leffler distribution [Pil90] defined as

$$G_\gamma(y) = \begin{cases} 1 - E_\gamma(-y^\gamma) &, \quad y > 0 \\ 0 &, \quad y \le 0 \end{cases}, \quad \text{where } 0 < \gamma \le 1, \qquad (1.25)$$

This ends the proof. Note that $G_1(y) = 1 - e^{-y}$ is the exponential distribution.

Properties of the Mittag-Leffler distribution [Pil90, Fel66] imply that the frequency-domain response $\chi(k,\omega)$, Eq. (1.4), takes the form of the well known empirical Cole-Cole function

$$\chi(k,\omega) = \int_0^\infty e^{-i\omega t}d(G_\alpha(t/\tau)) = \frac{1}{1 + (i\omega\tau)^\alpha} , \qquad (1.26)$$

with $0 < \alpha < 1$ and the characteristic relaxation time $\tau = k^{-2/\alpha}$.

Now let us consider the *biased* case of the one-dimensional process $R(t)$. In this case, the derivation of the limiting distribution does not require the independence between a random jump R_i and its waiting-time T_i (see Theorem 1).

Theorem 4 *The relaxation function* $\Phi_b(k,t)$ *in the biased case, i.e. when* $\mathrm{E}R_i \neq 0$, *equals*

$$\Phi_b(k,t) = \mathrm{E}\exp\{-kt^\alpha H_\alpha\} = 1 - G_\alpha(k^{1/\alpha}t),$$

where $G_\alpha(k^{1/\alpha}t)$ *is the Mittag-Leffler distribution defined in (1.25).*

Proof: In this case, by means of Eqs. (1.24) and (1.25), we obtain

$$\begin{aligned}
\Phi_b(k,t) &= \mathrm{E}\exp(-kt^\alpha H_\alpha) \\
&= \int_0^\infty \exp(-kt^\alpha x)\rho_\alpha(x)dx \\
&= E_\alpha(-kt^\alpha) = 1 - G_\alpha(k^{1/\alpha}t).
\end{aligned}$$

As in the nonbiased case the frequency-domain response $\chi(k,\omega)$ takes the Cole-Cole form (1.26) with $\tau = k^{-1/\alpha}$, $k > 0$.

Acknowledgments: This work was supported by NSF grant no INT 92-20285. We thank Prof. Ron Pyke and the referee for their detailed reading and comments on the first version of this paper.

4 References

[DH87] Dissado L. A. and Hill R. M. Self-similarity as a fundamental feature of the regression of fluctuations. *Chem. Phys. 111, 193–207,* 1987.

[Dob55] Dobrushin R. L. Lemma on the limit of composed random function. *Uspekhi Mat. Nauk 10(64), 157–159,* 1955. (cf. *Math. Rev. 17, 48,* 1956).

[Fel66] Feller W. *An Introduction to Probability Theory and its Applications, Vol. 2.* Wiley, New York, 1966.

[GY95] Gomi S. and Yonezawa F. Anomalous relaxation in the fractal time random walk model. *Phys. Rev. Lett. 74, 4125–4128,* 1995.

[HH94] Havriliak S. and Havriliak S. J. Results from an unbiased analysis of nearly 1000 sets of relaxation data. *J. Non-Cryst. Solids 172–174, 297–310,* 1994.

[Jon83] Jonscher A. K. *Dielectric Relaxation in Solids.* Chelsea Dielectric Press, London, 1983.

[Kam87] van Kampen N. G. *Stochastic Processes in Physics and Chemistry.* North-Holland, Amsterdam, 1987.

[Kot95] Kotulski M. Asymptotic distributions of continuous-time random walks: a probabilistic approach. *J. Stat. Phys. 81, 777–792,* 1995.

[KS86] Klafter J. and Shlesinger M. F. On the relationship among three theories of relaxation in disordered systems. *Proc. Natl. Acad. Sci. USA 83, 848–851,* 1986.

[KZ94] Klafter J. and Zumofen G. Probability distributions for continuous-time random walks with long tails. *J. Phys. Chem. 98, 7366–7370,* 1994.

[MW65] Montroll E. W. and Weiss G. H. Random walks on lattices. II. *J. Math. Phys. 6, 167–181,* 1965.

[Pil90] Pillai R. N. On Mittag-Leffler functions and related distributions. *Ann. Inst. Statist. Math. 42, 157–161,* 1990.

[RR87] Ramakrishnan T. V. and Raj Lakshmi M., eds. *Non-Debye Relaxation in Condensed Matter.* World Scientific, Singapore, 1987.

[Smi55] Smith W. L. Regenerative stochastic processes. *Proc. Roy. Soc. A, 232, 5–31,* 1955.

[Tun75] Tunaley J. K. E. Some properties of the solutions of the Montroll-Weiss equation. *J. Stat. Phys. 12, 1–10,* 1975.

[WWH89] Weissman H., Weiss G. H., and Havlin S. Transport properties of the continuous-time random walk with a long-tailed waiting-time density. *J. Stat. Phys. 57, 301–316,* 1989.

[Wer91] Weron K. A probabilistic mechanism hidden behind the universal power law for dielectric relaxation: general relaxation equation. *J. Phys.: Condens. Matter 3, 9151–9162,* 1991.

[WJ93] Weron K. and Jurlewicz A. Two forms of self-similarity as a fundamental feature of the power-law dielectric response. *J. Phys. A: Math. Gen. 26, 395–410,* 1993.

[Zol86] Zolotarev V. M. *One-dimensional Stable Distributions.* American Mathematical Society, Providence, Rhode Island, 1986.

Inference for a Class of Causal Spatial Models

I. V. Basawa
University of Georgia
Athens, Georgia 30602 U.S.A.

ABSTRACT A new class of lattice models, whose joint densities are products of unilateral conditional densities, is introduced. Maximum likelihood estimation, for causal conditional exponential families on a two dimensional lattice, is discussed. The technique proposed enables one to construct a rich class of lattice models with a parsimoneous parameterization.
Key Words: Latice processes; Markov random fields; Unilateral (causal) models; Maximum likelihood estimation; Martingales.

1 Introduction

Inference problems for lattice random fields have been studied by several authors including Bartlett (1978), Besag (1974), Cliff and Ord (1975), Haining (1979), Pickard (1987), Tjostheim (1978, 1983), and Whittle (1954), among others. See Cressie (1991) for a review of the literature. Broadly speaking, lattice models can be divided into two groups: (a) simultaneously specified models, and (b) conditionally specified models. In many cases, conditionally specified models have proved to be more convenient to work with, especially for non-Gaussian fields. Since spatial dependence is typically multi-lateral (in all directions) rather than unilateral (in one direction, such as time), the conditional density of a Markov random field typically depends on a conditioning set (nearest neighbor set) consisting of observations in all directions of a given observation. Unlike in the unilateral case, the joint density of a given set of observations from a lattice model, is not the same as the product of all conditional densities corresponding to observations in that set. If the conditional densities belong to an exponential family with pair-wise only dependence, the joint density can be written down in a compact way. However, this latter joint density involves a factor (called the partition function) which depends on the model parameters, and it is typically not available in closed form. Consequently, in special cases, various alternatives to the likelihood such as pseudolikelihood (Besag (1974)) are used. More generally, Monte Carlo Markov Chain methods are used to get a simulation-approximation for the likelihood. See, for instance, Geyer (1992).

In this paper, we propose a method of building joint densities via a product of conditional densities constructed in a unilateral direction. In the case of linear spatial models, such a unilateral scheme was used successfully by Tjostheim (1978, 1983), Basawa et al. (1992), and Basu and Reinsel (1993). See, also Pickard (1980). We propose here, a general method of unilateral model building for lattice processes. It must be noted that the joint densities obtained in this way, typically, capture multi-lateral dependence, in the sense that the conditional density of a specified observation, given all other observations in a set, depends on nearest neighbors in all directions.

The proposed technique provides a rich class of models for which parameter estimation can be carried out via the martingale framework. Causal (unilateral) lattice models are introduced in Section 2. Section 3 is concerned with maximum likelihood parameter estimation for causal exponential families. Some generalizations are presented in Section 4.

2   Causal Lattice Models

Let $\{X_{i,j}\}, i = 1, ..., m$ and $j = 1, ..., n$, denote observations from a process defined on a two dimensional regular lattice. Label the data in a linear order, $X_{i,j} \equiv X_t$, with $t = j + n(i - 1)$, as follows:

$$
\begin{pmatrix} X_{m,1} & X_{m,2} & \cdots & X_{m,n} \\ \vdots & \vdots & & \vdots \\ \cdots & \cdots & \cdots & \cdots \\ X_{2,1} & X_{2,2} & \cdots & X_{2,n} \\ X_{1,1} & X_{1,2} & \cdots & X_{1,n} \end{pmatrix} \rightarrow \begin{pmatrix} X_{(m-1)n+1} & X_{(m-1)n+2} & \cdots & X_{mn} \\ \vdots & \vdots & & \vdots \\ X_{n+1} & X_{n+2} & \cdots & X_{2n} \\ X_1 & X_2 & \cdots & X_n \end{pmatrix}
$$

Thus, the vector of observations is

$$
\begin{aligned}
X(N) &= (X_{1,1}, X_{1,2}, ...X_{1,n}, X_{2,1}, X_{2,2}, ...X_{2,n}, ...X_{m,n}) \\
&= (X_1, X_2, ..., X_n, X_{n+1}, X_{n+2}, ...X_N),
\end{aligned}
$$

where $N = mn$, the total number of observations. Denote $X_{(i,j)} = \{X_{kl} : 1 \le k < i, 1 \le l \le n\} \cup \{X_{il} : 1 \le l < j\}$, the observations in the lower half-plane, and

$$
X^{(1)}_{(i,j)} = \begin{cases} (X_{i-1,j}, X_{i,j-1}, X_{i-1,j-1}), & i \ge 2, j \ge 2 \\ X_{i-1,1}, & i \ge 2, j = 1 \\ X_{1,j-1}, & i = 1, j \ge 2, \end{cases}
$$

the set of observations in the lower first quadrant. In building up the joint distribution for the $\{X_{i,j}\}, i = 1, ..., m$ and $j = 1, ..., n$, we shall view $X_{(i,j)}$

as the "past" observations and $X^{(1)}_{(i,j)}$ as the first-order nearest-neighbor "past" of the observation $X_{i,j}$. Consider the Markov assumption:

$$P(X_{i,j} \leq x_{i,j} | X_{(i,j)}) = P(X_{i,j} \leq x_{i,j} | X^{(1)}_{(i,j)}). \tag{2.1}$$

Markov dependence on lower quadrants of order higher than the first order can be defined in a similar fashion. Eq. (2.1) can, alternatively, be written as

$$P(X_t \leq x_t | X_{t-1}, ..., X_1) = \begin{cases} P(X_t \leq x_t | X_{t_1}, X_{t_2}, X_{t_3}), & i \geq 2 \text{ and } j \geq 2 \\ P(X_t \leq x_t | X_{t_1}), & i \geq 2 \text{ and } j = 1 \\ P(X_t \leq x_t | X_{t_2}), & i = 1 \text{ and } j \geq 2 \end{cases}$$

where $t = j + n(i-1), t_1 = j + n(i-2), t_2 = j - 1 + n(i-1)$, and $t_3 = j - 1 + n(i-2)$. In order to simplify the notation, denote

$$X_{(t)} = (X_{t-1}, ..., X_1),$$

$$X^{(1)}_{(t)} = \begin{cases} (X_{t_1}, X_{t_2}, X_{t_3}), & \text{for } i \geq 2, \text{ and } j \geq 2 \\ X_{t_1}, & \text{for } i \geq 2, \text{ and } j = 1 \\ X_{t_2}, & \text{for } i = 1, \text{ and } j \geq 2 \end{cases}.$$

We thus have the assumption

$$P(X_t \leq x_t | X_{(t)}) = P(X_t \leq x_t | X^{(1)}_{(t)}). \tag{2.2}$$

Let $f(x_t | x^{(1)}_{(t)})$ denote the conditional transition density corresponding to $P(X_t \leq x_t | X^{(1)}_{(t)})$. The joint density of $X(N) = (X_1, ..., X_N)$ given $X_1 = x_1$, is then given by

$$p(x(N)) = \prod_{t=2}^{N} f(x_t | x^{(1)}_{(t)}). \tag{2.3}$$

It may be noted that, even though the joint density $p(x(N))$ in (2.3) was constructed via a particular sequence of conditional densities, this should not be misinterpreted to imply that the conditional density of $X_{i,j}$ given all $X_{k,l}, (k, l) \neq (i, j)$, depends only on the first-order lower-quadrant. In fact, the latter conditional density computed via (2.3) will, typically depend on nearest neighbors on all sides. See Ex 2.2 below.

Ex. 2.1. Consider the set of observations

$$\begin{pmatrix} x_{3,1} & x_{3,2} & x_{3,3} & x_{3,4} \\ x_{2,1} & x_{2,2} & x_{2,3} & x_{2,4} \\ x_{1,1} & x_{1,2} & x_{1,3} & x_{1,4} \end{pmatrix} \rightarrow \begin{pmatrix} x_9 & x_{10} & x_{11} & x_{12} \\ x_5 & x_6 & x_7 & x_8 \\ x_1 & x_2 & x_3 & x_4 \end{pmatrix}.$$

Under the assumption (2.1) (or equivalently, (2.2)), the joint density (conditional on x_1) is given by

$$p(x(12)) = \prod_{t=2}^{12} f(x_t | x^{(1)}_{(t)})$$

$$= [f(x_2|x_1)f(x_3|x_2)f(x_4|x_3)]$$
$$\times \quad [f(x_5|x_1)f(x_6|x_1,x_2,x_5)f(x_7|x_2,x_3,x_6))f(x_8|x_3,x_4,x_7)]$$
$$\times \quad [f(x_9|x_5)f(x_{10}|x_5,x_6,x_9)f(x_{11}|x_6,x_7,x_{10})f(x_{12}|x_7,x_8,x_{11})].$$

This example is presented here merely to illustrate the simplicity of the construction of the joint densities from certain unilateral conditional densities.

Ex. 2.2. Causal Autoregressive Process

Consider the model

$$X_{i,j} = \theta_1 X_{i-1,j} + \theta_2 X_{i,j-1} + \theta_3 X_{i-1,j-1} + \epsilon_{i,j},$$

where $\{\epsilon_{i,j}\}$ are independent $N(0,\sigma^2)$ random variables. Here $f(x_{i,j}|x_{(i,j)}^{(1)})$ is a normal density with mean $m_{i,j}(\theta) = \theta_1 x_{i-1,j} + \theta_2 x_{i,j-1} + \theta_3 x_{i-1,j-1}$, and variance σ^2. The joint density (given x_1) is

$$p(x(N)) = \prod_{\substack{i,j \\ (i,j)\neq(1,1)}}^{m,n} f(x_{i,j}|x_{(i,j)}^{(1)}).$$

From the above joint density, it is not hard to verify (see, for instance, Basu and Reinsel (1993)) that the conditional density of $X_{i,j}$ given all $\{X_{k,l}\}, (k,l) \neq (i,j)$, is normal with mean

$$m_{ij}^*(\theta) = (1 + \theta_1^2 + \theta_2^2 + \theta_3^2)^{-1}[(\theta_1 - \theta_2\theta_3)(x_{i-1,j} + x_{i+1,j})$$
$$+(\theta_2 - \theta_1\theta_3)(x_{i,j-1} + x_{i,j+1}) + \theta_3(x_{i-1,j-1} + x_{i+1,j+1})$$
$$-\theta_1\theta_2(x_{i-1,j+1} + x_{i+1,j-1})],$$

and variance

$$\sigma^{2*} = (1 + \theta_1^2 + \theta_2^2 + \theta_3^2)^{-1}\sigma^2.$$

Note that m_{ij}^* depends on the eight nearest neighbors on all sides of (i,j), not just on the lower quadrant. Also, note the symmetry in $m_{ij}^*(\theta)$ with respect to its dependence on the nearest neighbors.

3   Causal Conditional Exponential Families

Consider the conditional density

$$f_\theta(x_{i,j}|x_{(i,j)}^{(1)}) \propto \exp[x_{i,j}h(x_{(i,j)}^{(1)},\theta) + g(h(x_{(i,j)}^{(1)},\theta))], \qquad (3.1)$$

where

$$h(x_{(i,j)}^{(1)},\theta) = \begin{cases} \theta_0 + \theta_1 x_{i-1,j} + \theta_2 x_{i,j-1} + \theta_3 x_{i-1,j-1}, & i \geq 2, \text{and } j \geq 2 \\ \theta_0 + \theta_1 x_{i-1,1}, & i \geq 2, \text{and } j = 1 \\ \theta_0 + \theta_2 x_{1,j-1}, & i = 1, \text{and } j \geq 2, \end{cases}$$

and the function $g(.)$ is determined such that $f_\theta(x_{ij}|x_{i,j}^{(1)})$ is a probability density.

The likelihood function, conditional on x_{11}, is given by

$$p_\theta(x(N)) \propto \exp[\sum_{i,j}\{x_{i,j}h(x_{(i,j)}^{(1)},\theta) + g(h(x_{(i,j)}^{(1)},\theta))\}], \qquad (3.2)$$

and the likelihood score function is

$$S_N(\theta) = \frac{\partial \log p_\theta(X(n))}{\partial \theta} = \sum_{i,j}(X_{i,j} - m_{i,j}(\theta))\frac{\partial h(X_{(i,j)}^{(1)},\theta)}{\partial \theta}, \qquad (3.3)$$

where

$$m_{i,j}(\theta) = \frac{-\partial g}{\partial h} = E(X_{i,j}|X_{(i,j)}^{(1)}). \qquad (3.4)$$

Using the notation of Section 2, we can write

$$S_N(\theta) = \sum_{t=2}^{N}(X_t - m_{(t)}^{(1)}(\theta))\frac{\partial h(X_{(t)}^{(1)},\theta)}{\partial \theta}, \qquad (3.5)$$

where

$$m_{(t)}^{(1)}(\theta) = E(X_t|X_{(t)}^{(1)}).$$

It is easily verified that $\{S_N(\theta)\}$, $N \geq 2$, is a zero mean martingale with respect to the σ-field generated by $\{X_N, X_{N-1}, ...\}$.

Define

$$C_N(\theta) = \sum_{t=1}^{N}\sigma_t^2(\theta)(\frac{\partial h}{\partial \theta})(\frac{\partial h}{\partial \theta})^T, \qquad (3.6)$$

where

$$\sigma_t^2(\theta) = \frac{-\partial^2 g}{\partial h^2} = Var(X_t|X_{(t)}^{(1)}). \qquad (3.7)$$

Under appropriate regularity conditions, see, for instance, Basawa and Prakasa Rao (1980), Basawa and Scott (1983), and Hall and Heyde (1980), we can establish the following theorem.

Theorem 3.1. As $N \to \infty$,

(a) $C_N^{-1/2}(\theta)S_N(\theta) \overset{d}{\to} N(0, I)$.

(b) There exists a consistent solution $\hat{\theta}_N$ of the equation $S_N(\theta) = 0$, with probability tending to 1.

(c) If $\hat{\theta}_N$ is any consistent solution of $S_N(\theta) = 0$, then $C_N^{1/2}(\theta)(\hat{\theta}_N - \theta) \overset{d}{\to} N(0, I)$.

We now present some examples of the causal conditional exponential family.

Ex. 3.1. Causal Autoregressive Process
Refer to Ex 2.2. The conditional density is seen to be

$$f_\theta(x_{i,j}|x_{(i,j)}^{(1)}) \propto \exp[x_{i,j}m_{i,j}(\theta)\sigma^{-2} - \frac{m_{i,j}^2(\theta)}{2\sigma^2}]$$

with

$$m_{i,j}(\theta) = \theta_1 X_{i-1,j} + \theta_2 X_{i,j-1} + \theta_3 X_{i-1,j-1}, i \geq 2 \text{ and } j \geq 2.$$

Here

$$h(x_{(i,j)}^{(1)}, \theta) = m_{i,j}(\theta)\sigma^{-2}, \text{and } g = -\frac{1}{2}h^2\sigma^2.$$

Also, note that

$$Var(X_{i,j}|X_{(i,j)}^{(1)}) = \frac{-\partial^2 g}{\partial h^2} = \sigma^2.$$

Ex. 3.2. Causal Autologistic Model
This model is specified by

$$f_\theta(x_{i,j}|x_{(i,j)}^{(1)}) = \exp[x_{i,j}h(x_{(i,j)}^{(1)}, \theta) + g(h(x_{(i,j)}^{(1)}, \theta))]$$

with

$$\begin{aligned} h(x_{(i,j)}^{(1)}, \theta) &= \log(\frac{p(x_{(i,j)}^{(1)}, \theta)}{1 - p(x_{(i,j)}^{(1)}, \theta)}) \\ &= \theta_0 + \theta_1 x_{i-1,j} + \theta_2 x_{i,j-1} + \theta_3 x_{i-1,j-1}, i \geq 2, j \geq 2, \end{aligned}$$

and

$$g = \log(1 - p(x_{(i,j)}^{(1)}, \theta)).$$

Also, note that

$$E(X_{i,j}|X_{(i,j)}^{(1)}) = \frac{\exp(h)}{1 + \exp(h)} = p(X_{(i,j)}^{(1)}, \theta),$$

and

$$Var(X_{i,j}|X_{(i,j)}^{(1)}) = p(X_{(i,j)}^{(1)}, \theta)(1 - p(X_{(i,j)}^{(1)}, \theta)).$$

Ex. 3.3. Causal Auto-Poisson Model
Consider the density

$$\begin{aligned} f_\theta(x_{i,j}|x_{(i,j)}^{(1)}) &= (x_{i,j}!)^{-1}(m_{i,j}(\theta))^{x_{i,j}} \exp(-m_{i,j}(\theta)) \\ &\propto \exp[x_{i,j}\log(m_{i,j}(\theta)) - m_{i,j}(\theta)], \end{aligned} \quad (3.8)$$

with

$$\log(m_{i,j}(\theta)) = \theta_0 + \theta_1 x_{i-1,j} + \theta_2 x_{i,j-1} + \theta_3 x_{i-1,j-1}.$$

Here

$$E(X_{i,j}|X_{(i,j)}^{(1)}) = Var(X_{i,j}|X_{(i,j)}^{(1)}) = m_{i,j}(\theta).$$

4   Comments on Extensions

In the previous sections (Sections 2 and 3), it is assumed that the effective conditioning set $X_{(i,j)}^{(1)}$ consists only of observations in the first order lower quadrant. This assumption can be relaxed in several ways. Some examples are given below. We shall use $X_{(i,j)}^*$ as a generic conditioning set.

(a) <u>First order lower half-plane</u>

$$X_{(i,j)}^* = \{X_{i,j-1}, X_{i-1,j-1}, X_{i-1,j}, X_{i-1,j+1}\}.$$

(b) <u>Second order lower quadrant</u>

$$X_{(i,j)}^* = \{X_{i,j-1}, X_{i,j-2}, X_{i-1,j-2}, X_{i-1,j-1}, X_{i-1,j}, X_{i-2,j-2}, X_{i-2,j-1}, X_{i-2,j}\}.$$

(c) <u>Second order lower half-plane</u>

$$X_{(i,j)}^* = \{X_{i,j-1}, X_{i,j-2}, X_{i-1,j-2}, X_{i-1,j-1}, X_{i-1,j},$$

$$X_{i-1,j+1}, X_{i-1,j+2}, X_{i-2,j-2}, X_{i-2,j-1}, X_{i-2,j}, X_{i-2,j+1}, X_{i-2,j+2}\}.$$

In general, $X_{(i,j)}^*$ may consist of any finite set of nearest neighbors in the lower half-plane $X_{(i,j)}$. Recall that

$$X_{(i,j)} = \{X_{k,l} : 1 \le k < i, 1 \le l \le n\} \cup \{X_{i,l} : 1 \le l < j\}.$$

The most general set-up within this framework is obtained when $X_{(i,j)}^* = X_{(i,j)}$. Often, a much smaller subset of $X_{(i,j)}$ such as $X_{(i,j)}^{(1)}$ or the examples in (a) - (c) above would give an adequate conditioning set to generate a rich class of models.

More general conditional exponential families than those in Section 2 are obtained by replacing $X_{(i,j)}^{(1)}$ in Section 2 by $X_{(i,j)}^*$ discussed above.

5   Concluding Remarks

An interesting question that arises is as to whether it is always possible to find a unilateral (or causal) representation of $X_{i,j}$ in terms of the half plane $X_{(i,j)}$ such that a given set of autocorrelations is generated by the causal model. This question has been answered affirmatively by Whittle (1954), and Tjøstheim (1978, 1983), under mild regularity conditions, for the linear processes. For the more general case considered in this paper, one may ask whether, given a conditionally specified multilateral model (such as those in Besag (1974)), one can find a unilateral conditional model of the type discussed in this paper, such that the two models have the same likelihood function, or at least the same conditional second-order moment properties. This question appears worth pursuing at least for the models belonging to the conditional exponential families. However, our goal in this paper has been to explore new models obtained from unilateral conditional densities, rather than to give unilateral representations for the existing multilateral conditional models.

It is worth noting that the conditional exponential family assumption is not crucial for the validity of Theorem 3.1. For instance, starting with the likelihood in (2.3), one may impose regularity conditions directly on the conditional densities (see, for instance, Basawa, Feigin and Heyde (1976)), and establish the result in Theorem 3.1.

Finally, we refer the reader to the recent monograph by Guyon (1995) for a good discussion of various kinds of competing random field models.

ACKNOWLEDGEMENTS

This research is partially supported by grants from the Office of Naval Research and the National Science Foundation. Thanks are due to a referee whose comments led to the discussion in Section 5.

References

Bartlett, M. S. (1978). Nearest neighbour models in the analysis of field experiments. *J. Roy. Statist. Soc.* B 40, 147-158.

Basawa, I. V., Brockwell, P. J. and Mandrekar, V. (1992). Inference for spatial time series. *Interface-90 Proceedings* 301-302, Springer-Verlag, New York.

Basawa, I. V., Feigin, P. D. and Heyde, C. C. (1976). Asymptotic properties of maximum likelihood estimators for stochastic processes. *Sankhya* A, 38, 259-270.

Basawa, I. V. and Prakasa Rao, B. L. S. (1980). *Statistical Inference for Stochastic Processes.* Academic Press, London.

Basawa, I. V. and Scott, D. J. (1993). *Asymptotic Optimal Inference for Non-ergodic Models.* Lecture Notes in Statistics, Vol 17, Springer-Verlag, New York.

Basu, S. and Reinsel, G. C. (1993). Properties of the spatial unilateral first order ARMA model. *Adv. Appl. Prob.* 25, 631-648.

Besag, J. E. (1974). Spatial interaction and the statistical analysis of lattice systems. *J. Roy. Statist. Soc.* B 36, 192-225.

Cliff, A. D. and Ord, J. K. (1975). Model building and analysis of spatial pattern in human geography. *J. Roy. Statist. Soc.* B 37, 297-328.

Cressie, N. A. C. (1991). *Statistics for Spatial Data*, Wiley, New York.

Greyer, C. J. (1992). Practical Markov chain Monte Carlo (with discussion). *Statist. Sci.* 7, 473-511.

Guyon, X. (1995). *Random Field on a Network: Modeling, Statistics and Applications*, Springer-Verlag, New York.

Haining, R. P. (1979). Statistical tests and process generators for random field models. *Geographical Analysis* 11, 45-64.

Hall, P. and Heyde, C. C. (1980). *Martingale Limit Theory and Its Application*. Academic Press, New York.

Pickard, D. K. (1980). Unilateral Markov fields. *Adv. Appl. Prob.* 12, 655-671.

Pickard, D. K. (1987). Inference for discrete Markov fields: The simplest nontrivial case. *J. Amer. Statist. Assn.* 82, 90-96.

Tjostheim, D. (1978). Statistical spatial series modeling. *Adv. Appl. Prob.* 10, 130-154.

Tjostheim, D. (1983). Statistical spatial series modelling II: Some further results on unilateral lattice processes. *Adv. Appl. Prob.* 15, 562-584.

Whittle, P. (1954). On stationary processes in the plane. *Biometrika* 41, 434-449.

Printed: December 4, 1995

PROBLEMS IN THE MODELLING AND
STATISTICAL ANALYSIS OF EARTHQUAKES

Y. Y. KAGAN[1] AND D. VERE-JONES[2]

[1] Institute of Geophysics and Planetary Physics, University of California, Los Angeles, California 90095-1567, USA. (E-mail: kagan@cyclop.ess.ucla.edu)

[2] Institute of Statistics and Operations Research, Victoria University of Wellington, P.O. Box 600, Wellington, New Zealand. (E-mail: dvj@isor.vuw.ac.nz)

1. INTRODUCTION

The purpose of this paper is to set out some of the statistical and probabilistic problems that arise from the work of the first author over the past two decades. Nearly all of this work has been published in geophysical journals, where the major emphasis is on the physical interpretation of the model structure and subsequent statistical analysis. One consequence is that many of the mathematical issues inherent in the model structures and conceptions have never been fully explored (see also Vere-Jones, 1994). In fact they raise many issues for discussion, from the adequacy of existing stochastic models for processes exhibiting self-similar or fractal behaviour, to specific questions concerning the testing of statistical models for random 3-dimensional rotations. Our hope is that by reviewing these papers in the context of the present conference, we may encourage attempts to resolve some of the more mathematical issues which arise, or relate them to relevant recent work.

The work under review can be roughly divided into three phases. The first phase, including papers K1-K4 (see reference section), dates to the early 1980's and was concerned with the moment structure of earthquake spatial locations. The thrust of the work was to expose the basic self-similarity of the earthquake process. There is a broad indication that for $k = 2$, 3 and 4, the probability density for the distribution of k points selected at random from the catalogue is inversely proportional to the 'volume' of the minimal convex set spanned by the k points in question. The underlying question here is, what sort of models, if any, exhibit these general features? This first phase is reviewed in section 2.

The second phase relates to Kagan's attempts, outlined in K5 and K6 in particular, to develop simulation models which would comprehensively mirror the above and related features of the observed catalogue data. Here the viewpoint taken is that the crust is in a process of continual elementary fracturing, and it is the eye of the beholder, rather than any intrinsic feature of the process, which selects out large clusters of elementary events and identifies them as earthquakes. The resulting simulation model is very successful in this aim, but it is necessarily complex, even though it depends on only a small number of fitted parameters. In particular it incorporates three-dimensional rotational effects which give the model a special geometrical character of its own. As a result it is difficult to relate the statistical features of the simulations to theoretical results on the many different types of behaviour possible with spatial branching processes. The model, and some preliminary attempts to identify its behaviour in theoretical terms, are described in section 3.

At best the simulation model provides a purely kinematic picture of earthquake occurrence. A first attempt to introduce dynamical aspects comes with attempts to model statistical aspects of the stress field in which earthquakes take place, or the modifications to the existing stress field caused by the occurrence of a new event. Here further geometric questions arise. These are closely related to the description of the fault motion in terms of a "double couple" mechanism (Aki and Richards, 1980), essentially equivalent to a 3-dimensional rotation with additional symmetries. A key role is played by rotational stable distributions which are the analogues here for the stable distributions which dominate the earlier discussion of spatial locations. These and related issues are taken up in section 4, covering the papers K7-K11 in particular.

The paper concludes with a brief listing of the main problems noted in the earlier sections, together with a number of more immediate statistical questions.

2. THE SPATIAL MOMENTS OF EARTHQUAKE CATALOGUES

2.1. Empirical Results

We are concerned here with the catalogue as a list of epicentres (points on the surface of the earth), or hypocentres (points within the earth's crust) of earthquake events identified and located within a particular geographical region over some stated observational period. The earthquake size is characterized by the scalar seismic moment (Aki and Richards, 1980), as well as by several 'magnitudes' (M), the most important of which is the moment magnitude (M_w), which is based on measurement of the seismic moment. For historical reasons these magnitudes are adjusted to be close to the Richter magnitude. Typical examples of earthquake epicenter maps are shown in Figs. 1(a), 1(b). Catalogues considered range from catalogues of major global events with magnitudes $M > 6$, regional catalogues from New Zealand or Japan of events with magnitudes $M > 4$, local catalogues for selected parts of California with magnitudes $M > 1.5$. Problems of completeness, accuracy, etc., are discussed in detail in papers K1-K4 and the references therein, and need to be carefully taken into account in the interpretation of any statistical analysis. However we shall leave such aspects temporarily aside and just consider the key results, verified for catalogues on a range of scales, reported in these four papers.

The first three papers are devoted to studies of the empirical 2-point, 3-point, and 4-point moments of the sets of epicentres and hypocentres (see Fig. 2). The principal quantities studied are the proportions of k-tuples ($k = 2, 3, 4$) of points from the catalogue with the property that the maximum distance between any two points in the k-tuple does not exceed r, as a function of r, and the joint density function of the coordinates of the points forming such a k-tuple.

Write

$$q_k(r) = N_k(r)/N, \tag{2.1}$$

where $N_k(r)$ is the number of k-tuples with the stated property, and N_k is the total number of k-tuples from the catalogue. The quantities $q_k(r)$ are computed first for the epicentres, as points in R^2, and then for the hypocentres, as points in R^3. As with Ripley's k-function (Ripley, 1977), they can be interpreted as the average number of k-tuples within a distance r of an "average" point of the catalogue.

In order to overcome the biases in such estimates which come from boundary effects (see e.g. Ripley, 1977; Stoyan et al., 1987; Stoyan and Stoyan, 1994 for details), the above ratios were then

400

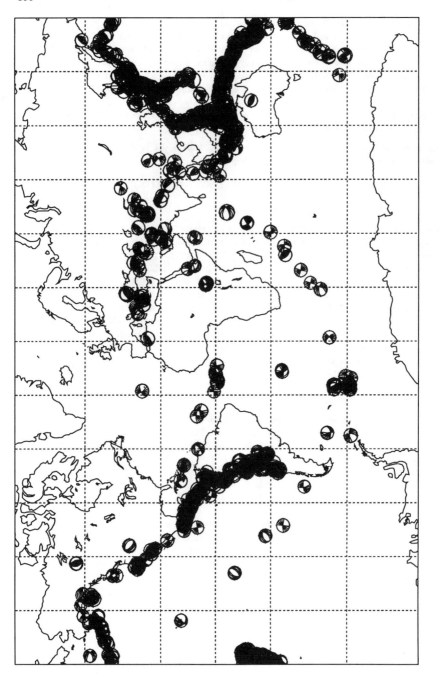

Fig. 1. (a) Focal mechanisms of earthquakes for global data 1977-1994 with moment magnitude $M_w \geq 6.5$ and (b) Epicentral map for California data 1932-1994 with local magnitude $M_L \geq 3.0$.

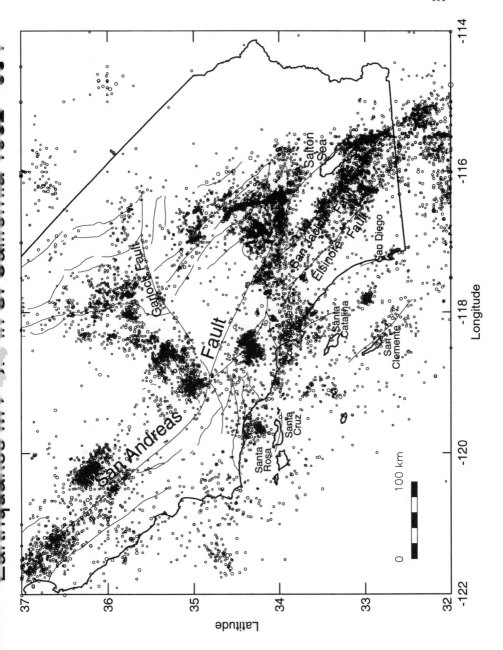

Longitude

Latitude

San Andreas Fault

Garlock Fault

San Jacinto F...

Elsinore Fault

Salton Sea

San Diego

Santa Catalina

San Clemente

Santa Cruz

Santa Rosa

100 km

0

compared to the corresponding values for a simulated Poisson catalogue, of the same size and extent as the original catalogue, but with the epicentral coordinates uniformly distributed over the region, and the depth distribution matched to that of the actual catalogue.

The resulting ratios

$$Q_k(r) = q_k(r)/\tilde{q}_k(r), \tag{2.2}$$

where the tilde refers to the simulated catalogue, were then tabulated and graphed in various ways. The general result, from all three papers, and for epicentres as well as hypocentres, is that the ratios decay approximately proportionately to $1/r$. This corresponds to a dimensional deficit (see also comments below) of about 1 in all cases, i.e., the epicentres behave roughly as if concentrated along a line and the hypocentres as if concentrated on a plane.

More precisely, the graphs of the ratios $Q_k(r)$ against r typically display three ranges: an initial range, in which the ratio is very high and approximately constant; a middle range, often extended, over which the $1/r$ behaviour is observed and a final range, in which the ratio approaches 1, as r approaches the diameter of the observation region. The first range is interpreted by Kagan as being dominated by measurement errors; the very high values illustrate the strength of local clustering. The second range is interpreted as being illustrative of self-similar behaviour, and the third as dominated by boundary effects.

Further aspects of the 3- and 4-point distributions were then considered, essentially by studying the numbers of events in appropriate "cells" defined by different radial and angular regions, and compared with the expected results from a range of hypothetical models. The most interesting results were those referred to in the introduction: to a first order of approximation, the results matched those which would be expected if the moments had densities inversely proportional to the distance D between pairs of points in the case $k = 2$, to the area S of the triangle formed by triplets of points in the case $k = 3$, and the volume V of the tetrahedron formed by quadruplets of points in the case $k = 4$ (see again Fig. 2).

Thus the key results can be summarized as follows:

(A) The growth rates of the moment functions are consistent with a dimensional deficit of approximately 1. (In fact these are not exact: later studies reported in K12, see also Ogata and Katsura (1991), suggest fractal dimensions of closer to 1.5 for epicentral maps and 2.2 for hypocentral maps.)

(B) To within an order of magnitude over different radial and angular combinations,
• i) the distribution of a pair of points selected at random from the catalogue is consistent with density inversely proportional to the distance $1/D$;
• ii) the distribution of a triplet of points selected at random from the catalogue is consistent with a density inversely proportional to the area $1/S$;
• iii) the distribution of a quadruplet of points selected at random from the catalogue is (for the hypocentres only) consistent with a density inversely proportional to the volume $1/V$.

The basic problem posed by these results, is to find a stochastic model which exhibits these features.

This question is not addressed as such in Kagan's papers, although he stresses the point that the results are all indicative of self-similar behaviour, and that (iii) rules out the possibility that hypocentres are concentrated on a single fault plane.

Moment
Function

Distribution
Density

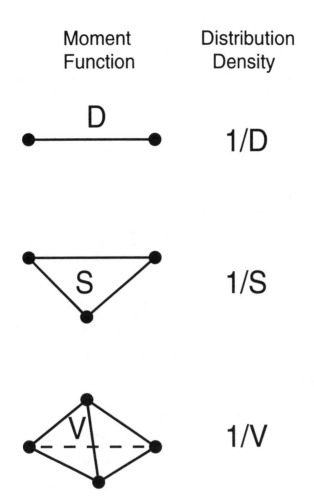

1/D

1/S

1/V

Fig. 2. Schematic representation of 2-, 3-, and 4-point moment functions and their suggested densities.

2.2. Modelling Considerations

To expose some of the issues involved in trying to find an answer to the above problems, consider first the expected value of the quantity $N_2(r)$ arising in the definition of $q_2(r)$. If we adopt a standard point process notation (see Daley and Vere-Jones, 1988) and assume that the first and second moment measures exist, then we can write (ignoring boundary effects)

$$E[N_2(r)] = \int_W M_1(dx) M_1(S_r(x)|x), \qquad (2.3)$$

where $M_1(dx) = E[N(dx)]$ is the first moment measure, $M_1(dy|x) = E[N(dy)|N(dx) = 1]$ is the first order Palm moment measure, W is the observation region, and $S_r(x)$ is the "sphere" radius r, centre x.

We see that the behaviour of $E[N_2(r)]$ depends both on the character of $M_1(dx)$ and on the local clustering structure described by $M_1(dy|x)$.

In the case of a homogeneous process, with mean density m, $M_1(dx) = m\,dx$ and $M_1(dy|x)$ approaches $m\,dy$ for large values of the distance $y - x$. In this case $E[N_2(r)]$ will ultimately grow as r^d, where d is the dimension of the space under study. This means that the ratios $Q_2(r)$ will be asymptotically constant for large r, and will not therefore exhibit the inverse power law behaviour that is claimed.

A number of ways can be suggested to get around this apparent difficulty: the power law behaviour may relate to the correlations rather than the moments; the point process may be "Palm-homogeneous" (see definition below) rather than strictly homogeneous; or the observed behaviour may reflect purely the properties of $M_1(dx)$, as in a Poisson process with singular parameter measure. We explore each of these possibilities briefly below. In none of these, however, does an easy answer to the basic question seem available. We remain uncertain, therefore, whether or not a point process model can be defined which exhibits the basic features A, B claimed by Kagan for the empirical earthquake data.

(a) Power law correlations with a homogenous model

The covariance measure for a point process model can be defined by

$$C_2(dx \times dy) = E[dN(x)dN(y)] - E[dN(x)]E[dN(y)]. \qquad (2.4)$$

In the homogenous case, and assuming derivatives exist, this can be simplified to

$$C_2(y - x) = m[m_1(y|x) - m], \qquad (2.5)$$

where the Palm intensity $m_1(y|x)$ is also a function of $y - x$ only. In the earthquake applications, $m_1(y|x) >> m$ for $y - x$ small, and it is a matter of indifference, and probably impossible to determine from a finite data set, whether the power decay relates to $m_1(y|x)$ directly, or to the difference $m_1(y|x) - m$. By taking the latter point of view, as in the studies by Vere-Jones (1978) and Chong (1983), one can remain formally within the context of spatially homogeneous processes, yet still retain approximate power-law behaviour for the second moment measure, and an appearance of self-similarity.

An approach of this kind has been taken recently by Stoyan (1994), in part to emphasize the care that needs to be taken in distinguishing point process models that have non-standard fractal dimension from those which exhibit self-similarity. Roughly speaking, the correlation dimension of a point process model can be defined as the limit of the ratio of $\log E\big(N_2(r)/\log(r)\big)$ as r approaches zero. Non-standard dimensions can be achieved even for such a simple example as the Neyman-Scott model by choosing a sufficiently irregular form for $m_1(y|r)$. However, strict self-similarity is not possible for a point process model as usually defined, since it requires infinite accumulations of points in bounded regions; still less can it be achieved for a homogeneous model, which implies the existence of a mean distance between points, and hence a distinguished distance scale. The situation if the process is non-homogeneous is not so clear. The assumption of finite moment measures, required to ensure the existence of the quantities appearing in the definition of the fractal dimension, does not of itself rule out self-similarity, at least in an approximate sense. However, if appropriate examples exist, they are elusive.

Unfortunately it appears to be self-similarity, rather than an anomalous fractal dimension as such, which characterises the earthquake process. Moreover it becomes increasingly difficult to match the behaviour of any model to the apparent behaviour of the higher moments of the earthquake catalogues. To take the 3-point moment for the epicenters as an example, one has to find a moment measure which behaves as if it had a density in $R^2 \times R^2$ proportional to $1/R_1 R_2 \sin(\theta)$, where R_1 and R_2 are distances between x and y and x and z in a triple (x, y, z), and θ is the angle between the lines (x, y) and (x, z). The radial factors integrate out, but the term in $\sin \theta$ produces a singularity which cannot be integrated out: there is too great a concentration on long skinny triangles. The situation is even worse for the four-point moments. For the hypocentres, the function $(1/R_1 R_2 \sin \theta)$, considered now as a function in $R^3 \times R^3$, is integrable at zero, and is therefore a conceivable candidate for a density. On the other hand the $1/V$ form for the 4-point moment density again gives rise to a singularity. Thus the inverse area and inverse volume properties force one either to reject the model for sufficiently small angles in the configuration, or to seek a model outside the framework of point processes with absolutely continuous factorial moment or cumulant measures.

From a physical point of view it is of course true that the observed behaviour ultimately breaks down as one reaches the scale of individual rock grains, or certainly individual molecules or atoms. If one accepts this limitation, the problem is to find a cluster structure with the right form of power law behaviour over at least a very extended range. We pick up this problem below in a slightly wider context.

(b) Palm-stationary point processes

Kagan refers at several points to the suggestion of Mandelbrot (1983) that point process models for self-similar behaviour should be sought not within the class of homogeneous processes but within the class of processes for which the behaviour relative to a given point of the process is independent of the point selected as origin.

One interpretation of this requirement is that the Palm distributions of the process should be the same for all points of the process. This would imply in particular that the moment measures of the Palm distributions were functions of the differences between the arguments only, for example, that $M_1(dy/x)$ was again a function of $y - x$ only, even though the process was not itself homogeneous. We shall refer to such processes as Palm-stationary point processes. In the 1-dimensional case they correspond to processes which are interval-stationary but not necessarily stationary.

Examples of such Palm-stationary point processes include the random walk (or "random flight") examples of Mandelbrot (1975). Mandelbrot shows that by choosing the step-length to have a Pareto distribution, processes can be found for which

$$m_1(S_r(x)|x) \propto r, \tag{2.6}$$

Note that Mandelbrot's distributions are truncated near $r = 0$, so the power law behaviour does not persist for indefinitely small values of r. For larger values of r, however, they exhibit many properties of self-similarity.

Unfortunately, according to Kagan (K3), they do not exhibit the higher moment properties of the earthquake process, in particular the $1/V$ property of the four-point moment function. It is possible, however, that more complex examples could do so. Indeed, the branching process models considered in section 3 could well include examples of this type.

What hampers us in pursuing this discussion, is that we are not aware of a well established theory for such Palm-stationary point processes. For example, given a candidate family of Palm distributions, when does there exist a well-defined point process with these distributions? And is that point process unique? Any references which deal with these or related questions would be appreciated.

Further examples of Palm-stationary processes suggested by Mandelbrot, such as the Levy dust model or the zeros of Brownian motion, have a very complex point set structure, including finite accumulation points, and cannot therefore be modelled within the standard point process framework. Mandelbrot hints that more general stochastic frameworks could be developed for this purpose, but again we do not know whether any systematic attempts have been made to build up such a framework, nor, if they have, whether they include processes which can match the $1/S$ and $1/V$ behaviour observed by Kagan. Again references would be appreciated.

(c) Singular Poisson processes

Here we consider the opposite type of explanation for the observed behaviour of the moment functions to that suggested in (a). For a pure Poisson process there is no cluster structure, so that $M_1(dy/x) = M_1(dy) = \mu(dy)$ where μ is the parameter measure of the process. In this case, therefore

$$E[N_2(r)] = \int_W \mu[S_r(x)]\mu(dx). \tag{2.7}$$

This shows immediately that the correlation dimension for the point process, interpreted as the limit of the ratio of the logarithm of (2.7) to $\log r$ as r approaches zero, coincides with the correlation dimension of the parameter measure μ. The same thing happens also with the higher moments $E\big(N_k(r)\big)$; after appropriate rescaling they yield in the limit the k-th order Renyi moment dimension of μ, defined by

$$Q_k = \lim_{r \searrow 0}\Big[\frac{1}{k-1}\log\int \mu(S_r)^{k-1}\mu|dx\Big]/\log r, \tag{2.8}$$

whenever the limit exists (see, for example, Cutler, 1991; Geilikhman et al., 1990; or Vere-Jones et al., 1995 for further discussions of these dimensions).

Although there are many examples of measures which have fractional moment dimensions, it is rather less easy to match the self-similar behaviour of the process to corresponding properties of μ. Well-known examples such as the Cantor measure exhibit scale invariance only with regard to a fixed sequence of scales (in that case powers of 3). Much of the Russian discussion of earthquake processes is couched in the terminology of "hierarchical block structures", where, again, self-similar behaviour is observed in a sequence of increasing or decreasing scales, but not on all scales.

Once again we are lacking information on whether there has been any attempt to develop a theory of point process models based on this weaker concept of "intermittent self-similarity". Kagan's results do not suggest any such restrictions, but if the point process is to be self-similar at all scales, the underlying measure has to be fully scale invariant, and we already know that for R^d, the only such measures are a product of angular and radial components about a fixed origin, where the radial component has density proportional to $1/r$. (See the discussion on p. 325 of Daley and Vere-Jones, 1988). This type of behaviour seems inappropriate as a model for earthquakes, because of its dependence on a fixed origin. Once again the required model seems elusive.

3. THE KAGAN-KNOPOFF BRANCHING MODEL

Branching processes have been used extensively to describe different aspects of earthquake occurrence. Early applications include those of Hawkes and Adamopoulos (1973) and Kagan (1973); the closely related ideas of Otsuka (1972) should also be mentioned. The branching structure of Hawkes' model was elucidated in Hawkes and Oakes (1974). Kagan (1973) used an explicit branching model to model magnitudes, times, and locations, with magnitude as the controlling variable. Vere-Jones (1976; 1978) following earlier suggestions by Otsuka (1972), Saito et al. (1973), discussed the development of a crack during the fracture process, and showed that a schematic representation as a critical branching process provided a plausible interpretation of the frequency-magnitude distribution. In the critical or near-critical case, the tail of the frequency magnitude distribution is controlled by the total size of the branching episode, which in the critical case has a universal asymptotic form $P(N > n) \propto C n^{-1/2}$, and implies a power law form for the total energy or seismic moment release. This form persists even if each "offspring" contributes a random increment to the total energy release, provided the tail of the energy increment does not itself decay more slowly than $E^{-1/2}$; in the latter case it is the tail of the increment distribution which controls the tail of the total energy release.

The model set out by Kagan and Knopoff in K5, K6 embraces both the crack propagation aspect and the time/location aspect. The process is highly clustered, but they contend that it is the process of measurement which tends to isolate the clusters and to give them a separate identity as individual earthquake events. In reality the process is a continuum, in time as well as space, effectively self-similar in all its aspects, down to the scale of elementary dislocations.

To capture these effects Kagan and Knopoff (K5, K6) develop a simulation model, in fact a multi-dimensional branching process, for the elementary dislocations, and pursue this through procedures for identifying larger-scale events from accumulations of elementary events. The model incorporates space, time, direction, orientation and magnitude effects, and thus provides a comprehensive model for the earthquake processes occurring on a regional scale. Although of necessity somewhat complex it involves remarkably few free parameters: effectively only the parameter governing the degree of dispersion in the orientation of the dislocation, and a criticality parameter. Other parameters are built into the model, and have values suggested on physical grounds and confirmed by further simulation studies.

Despite the interest of the model, and the many questions which it raises, it has not (to our knowledge) been the subject of any significant analytical investigations. In particular it is not clear how key aspects, such as its evident self-similarity, or the clustering of events subsequently identified as earthquakes, can be deduced from its defining elements. The main purposes of this section are to briefly summarize the model structure, and to indicate some possible directions for pursuing these questions.

3.1. Description of the Simulation Model

The individuals in the branching process are elementary dislocations, i.e., discs, with a direction of motion (slip vector) across each disc, and are characterized by the following coordinates (see Fig. 3):

A(i) three spatial coordinates to determine the location of the centre of the disc;

A(ii) two angular coordinates to describe its orientation;

A(iii) one angular coordinate to determine the direction of slip across the disc;

A(iv) one time coordinate for when the slip took place.

The dislocations are taken to be fixed in size and shape, i.e. discs with constant radius, and the stress or moment release associated with each elementary event is also taken to be constant (although aligned along the selected direction of motion). Any such dislocation can be identified with a "double couple" which is the traditional method of describing the focal mechanism of an earthquake.

Each 'parent' dislocation produces 'offspring' according to the following rules:

• B(i) The number of offspring is governed by a Poisson distribution with mean $1 - k$ where k is the criticality parameter;

• B(ii) The centres of the offspring discs are uniformly and independently distributed around the circumference of the parent disc;

• B(iii) The orientation of each offspring disc is independently selected according to a 2-dimensional angular Cauchy distribution (see below), with concentration parameter c, centred on the orientation of the parent disc;

• B(iv) The direction of motion across each offspring disc is selected according to a 1-dimensional angular Cauchy distribution with concentration parameter d centred on the direction across the parent disc.

• B(v) The time delay between the occurrence of a parent dislocation and one of its offspring is distributed according to a non-negative stable distribution with Laplace transform $\exp(-s^{1/2})$, i.e. with density

$$\frac{1}{\sqrt{2\pi t^3}} \exp\left(-\frac{1}{2t}\right).$$

The angular Cauchy distributions referred to above may be defined as follows. Let the position of a point X on the surface of the (hyper-) sphere in $(k + 1)$ dimensions be denoted by the pair (q, y), where q is the colatitude (angle with a fixed reference axis) and y is a point on the sphere in k dimensions. (Note when $k = 1$, the sphere reduces to the point pair $(-1, +1)$). Now suppose X is uniformly distributed over the "surface" of its hypersphere, and the point $Y = (q', y')$ is defined

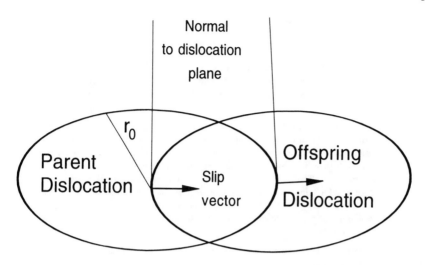

Fig. 3. Representation of elementary dislocations ('individuals') in Kagan's branching process model.

on the same hypersphere by the equations

$$\tan q' = c \tan q; \quad y' = y. \tag{3.1}$$

Then Y is said to have a k-dimensional isotropic angular Cauchy distribution with concentration parameter c. When c is very small, the distribution of Y is highly concentrated around the N. Pole; when $c = 1$, Y again has a uniform distribution; and when c becomes large Y is concentrated around the 'equator'.

Note that in B(iii) the N. Pole is taken to be the orientation of the parent dislocation, and that we identify q with $\pi - q$. An alternative point representation of B(ii) and B(iii) can be obtained by treating the pair (orientation; direction of motion) as a point on the 3-dimensional surface S^3 of a 4-dimensional ball; then both aspects of the offspring distribution are controlled by a single concentration parameter r. The representation of such a 'focal mechanism' as a point on the 2-dimensional surface S^2 of a 3-dimensional ball is familiar in seismology, where the focal mechanism is depicted as a (2-dimensional picture of) a sphere divided into quadrants by a pole (point on S^2) and a set of axes about that pole. Typical examples are shown in Fig 1(a); see also Fig 6 and the further discussion in section 4.1.

The simulation proceeds in two stages. In the first stage the branching family trees are started from a number of initial ancestors, and the spatial coordinates (location of disc centre, orientation, and direction) are recorded. From the results of this stage, it is possible to obtain a visual picture of the resulting "fractures" by plotting the intersection of the elementary discs with a fixed plane (see Fig. 4). It is partly from such pictures that the angular Cauchy distribution, rather than some other angular analogue of the stable distributions, has been chosen. Experimentally the plots obtained reproduce visually, and in the case of some quantitative characteristics, such as the fractal dimensions, also quantitatively, the same sorts of features that are observed from geological mappings of fault traces.

The second stage of simulation involves adding the time delays between the appearance of parent and offspring. With this information available, a cumulative plot of the number of elementary events against time can be obtained. (In seismological terms, each elementary event is supposed to contribute a fixed amount to a scalar moment release, so that cumulative plots can be interpreted as analogues to the cumulative moment-release plots used in discussing real earthquake catalogues). The intense clustering of the near critical process results in this cumulative plot taking on a self-similar, step-function appearance. By convoluting the derivative of this cumulative function with a suitably shaped template, i.e. by examining a function of the form

$$X(t) = \int_{t-\tau}^{t} H(t-u) dN(u),$$

a time-series record can be obtained which may be compared with the trace of a seismograph, or its envelope, in the real situation. By applying similar criteria to those used in reality to identify particular events, Kagan is then able to produce from the time series record a list of simulated 'events', each with its own 'magnitude' corresponding to the size of a local maximum in the plot of the time series (see Fig. 5). Moreover, by averaging the locations of the elementary dislocations resulting in such an 'event', an approximate location for the 'hypocentre' of the event may be determined, representing roughly the centre of gravity (centroid) of the locations of the elementary

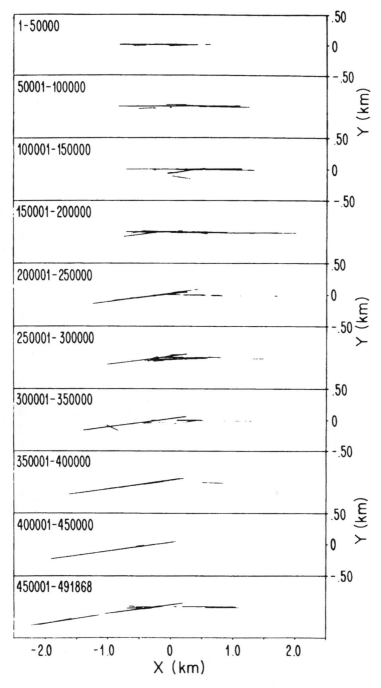

Fig. 4. Stages in the evolution of an episode in Kagan's branching
simulation model: intersection with a fixed plane of the
dislocation discs.

412

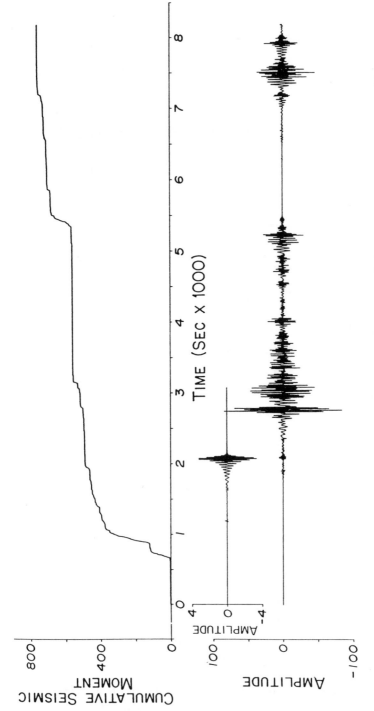

Fig. 5. Cumulative event curve (interpreted here as "cumulative seismic moment" for a realisation of Kagan-Knopoff's branching process model with (below) an illustration of the filtered signal (using the theoretical seismogram in the middle) from which events and their "seismic moments" can be determined.

events contributing to the cluster. In this way a synthetic seismic catalogue can be produced, in which events are listed in time sequence and associated with a hypocentre, magnitude and focal mechanism.

Such a catalogue can then be treated in essentially the same way as the real catalogues. In particular, it is possible to compute the empirical the 2-, 3- and 4-point moment functions for the synthetic catalogues, and see whether they exhibit features similar to those described in the previous sections. According to K6, similar properties are in fact observed.

The question now arises as to how far the properties observed for the simulated catalogues can be related to known properties of spatial branching processes. Some preliminary suggestions concerning this question are set out below.

3.2. Some Theoretical Considerations

Since many different limit relations exist for different types of branching processes, a desirable first step is to identify more precisely the contexts in which answers to the above question might be sought. The process lives in a space of 7 dimensions, so a natural first step is to consider its projections onto time and space axes only. Because of the assumed independence of the different components, this at first sight would seem to lead to a branching process in the lower dimensional space, with the other components simply ignored. However one important approximation has to be made in so doing, which in the case of distributions with long tails could prove to be of crucial importance. The caveat concerns the effect of simulating the process only within a bounded region. It is known, for example, that in spaces of dimension 2 or higher, stable regimes can exist even for critical branching processes which would "explode" in a space of dimension 1. In such situations what happens at the boundaries is all-important, so that integrating out the unwanted variables may give a very misleading impression of the limit behaviour to be expected. The angular variables are not so likely to cause trouble in this respect, since they take their values in a compact space; it is the space-time interaction which is likely to introduce problems.

Bearing this possible difficulty in mind, let us nevertheless consider first the purely temporal aspects of the process, ignoring spatial and angular coordinates of the points in the process. We have then a branching process along the time axis, in which the "gestation period" – the time lag between the appearance of an ancestor and its offspring – is distributed according to a non-negative stable distribution. In the model the value of the shape parameter is set at $1/2$, but we may consider the more general stable distribution with parameter a, $0 < a < 1$. The choice $a = 1/2$ is motivated by consideration of the first passage time to a fixed level (interpreted as a critical stress level) of a Brownian motion. Since stress is a tensor, and moreover the ambient distributions are more likely to be stable than Gaussian, a first question is how robust is the first passage time distribution to perturbations of the underlying Gaussian model?

Another justification of the choice $a = 1/2$ relates to the ability of the model with this choice of parameter to reproduce important features of the observed process, in particular the so-called Omori Law for the decay of aftershock sequences. This law asserts that the frequency of aftershocks diminishes with time after the main shock at a rate proportional to $(c + t)^{-1-h}$, where h is small and often taken as zero. Within the model, the decay of an aftershock sequence will correspond roughly to the decay of the expected number of offspring from an initial large number of ancestors, i.e. it will be proportional to the first order intensity of the process started from a single ancestor.

This obeys the renewal equation

$$h(t) = \delta_0(t) + m \int_0^t f(x)h(t-x)dx, \qquad (3.2)$$

(see, for example, Jagers, 1989; or Daley and Vere-Jones, 1988), the solution of which has Laplace transform

$$h^*(s) = [1 - mf^*(s)]^{-1}.$$

This solution has some rather special features when the inter-event distribution f has an infinite mean and is of stable form. If the mean number of offspring $m < 1$, then $h(x)$ decays asymptotically as x^{-1-a}, i.e at the same rate as $f(x)$. If $m = 1$, then instead of approaching a constant (as it would if the mean time delay were finite) $h(x)$ decays as x^{-1+a}. This suggests that if the process is imbedded into a process with immigration, then values of a close to zero, rather than $a = 1/2$, would produce the better agreement with Omori's Law. However, in addition to the caveat mentioned above, there is also the fact that the events which are claimed to follow Omori's law, are not the infinitesimal elements of the original branching process, but groupings of such events selected because they stand out above some minimal background level. This selection procedure alone could produce substantial changes to the tail behaviour. It seems to us that further study, both of the model and of the simulation results, is needed to clarify the relationship to Omori's Law, and to further aspects the observed process such as the occurrence of "foreshocks" as well as "aftershocks" associated with a major event.

It may be worth pointing out that this projection of the K-K model onto the time axis is in fact an example of the Hawkes' self-exciting model with an "infectivity function" (see Daley and Vere-Jones, 1988, p. 367) proportional to the density of a stable distribution. Thus the K-K model is really a fore-runner, in a space of higher dimension, of the space-time versions of the Hawkes process considered recently by authors such as Musmeci and Vere-Jones (1992).

Consider next the purely spatial aspects of the process. A simpler, 2-dimensional, version of the model can be developed by replacing the dislocation loops by linear steps in the plane, each step having a centre (x,y) and direction q. Offspring are located with equal probability at either end of the step, with the offspring orientation following a 1-dimensional angular Cauchy distribution, with concentration parameter r, centred on the orientation of the parent step. The directions of slip may be taken as always parallel to the step, with a high probability of following the sense of the parent step.

The result is a rather more easily visualized branching process of lines in the plane. Since each step can be represented as a point in $R^2 \times S^1$, the process could also be represented as a branching random walk along the surface of a 3-dimensional cylinder. If the step length and the time interval were rescaled suitably, it could no doubt be represented in the limit by an infinitely divisible process for which the first two coordinates had a Brownian motion-like behaviour and the third was a stable (Cauchy) process. What can be said about the fractal dimensions of the trajectory of such a process, or its projection back on to the plane as a process of lines?

Let us now add in the time component, and consider the full process in space-time. Two principal questions which need to be considered here are whether there exists a theoretical basis for the clustering properties which lead to the extraction of individual earthquake 'events', and whether the catalogue of events so defined can be shown to have the self-similarity and moment properties suggested by Kagan's empirical studies.

Here the context is that of limit behaviour for critical or just sub-critical space-time branching processes. Circumstances are certainly known in which such processes can exhibit limiting cluster properties (e.g., Dawson and Fleischmann, 1988) but it is not immediately clear whether a limit theorem could be formulated to embrace the properties described by Kagan's simulations. Our initial impression is that such limit theorems, if indeed they exist at all, would require extensions to the existing theory.

Finally we raise some questions concerning the use of the angular Cauchy distribution, rather than angular variants of other stable distributions, to describe the deviations of the focal mechanisms (orientation and direction of motion) of the elementary faults. We do not know if there is a physical interpretation, analogous to the first passage time distribution referred to in connection with the time-delay distribution, for the angular Cauchy distribution. Its use was suggested by comparisons of the visual appearance and other characteristics of the simulation model and the observed process.

This leads to the problem of estimating parameters within the model. As mentioned earlier, the model belongs to the general family of Hawke's processes used successfully by many authors in conjunction with a maximum likelihood estimation scheme. However, because the model relates only to elementary events, whereas the available data relate to clusters of elementary events, it cannot be directly fitted to the observations. The model therefore has to be fitted on the extent to which it reproduces broader, macroscopic features of the observed process. The problem then is how to formulate effective procedures for estimating and testing particular parameter values, such as those for the stable laws used to model the angular distributions.

4. DISTRIBUTIONS FOR THE STRESS AND FOCAL MECHANISMS.

It is only in the last decade or so that the focal mechanisms of earthquakes have been published with sufficient consistency and accuracy to make possible statistical analyses of the focal mechanisms similar to those described in section 2 for the spatial coordinates. Since 1977 the Harvard catalogue (Dziewonski et al., 1994) now routinely publishes focal mechanisms for events on a world-wide basis with moment magnitude $M_w > 5.8$, and more restricted catalogues are available for smaller events in some regions. Starting from the mid-eighties, the first author has initiated a series of studies making use of this material, opening up in doing so many new questions and avenues of study. A good account of progress to date, written mainly for physicists but reviewing many of the key statistical ideas, is in K12. Here we look only at a few selected problems, based mainly on K9-K11.

4.1. The 2-point Correlation Function for Focal Mechanisms.

Fig. 6 displays the equal-area projection (Aki and Richards, 1980, p. 110) for an example of an earthquake focal mechanism. This representation of a mechanism is standard in seismological literature. Focal mechanisms in Fig. 1a can be considered as 3-dimensional rotations of the Fig. 6 diagram.

The "distance" between two focal mechanisms will be interpreted as the rotation required to bring one point on the "surface" of the three-dimensional normalized sphere into coincidence with the second point. Following Moran (1975), Kagan has suggested representing each such point in

416

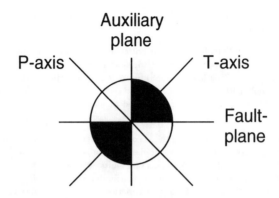

Fig. 6. Schematic diagram of earthquake focal mechanism, representing the equal-area projection of quadrupole radiation patterns. The null (N) axis is orthogonal to the T- and P-axes, or is located on the intersection of fault and auxiliary planes.

quaternion notation

$$\mathbf{q} = q_0 + q_1 * \mathbf{i} + q_2 * \mathbf{j} + q_3 * \mathbf{k},$$

where **ijk** are mutually orthogonal unit vectors satisfying $\mathbf{i} * \mathbf{i} = \mathbf{j} * \mathbf{j} = \mathbf{k} * \mathbf{k} = -1$ and $\mathbf{i} * \mathbf{j} = \mathbf{k}$, $\mathbf{j} * \mathbf{k} = \mathbf{i}$, $\mathbf{k} * \mathbf{i} = \mathbf{j}$, and the scalars q_i satisfy $q_1{}^2 + q_2{}^2 + q_3{}^2 + q_4{}^2 = 1$. If we write $q_0 = \cos(\phi/2)$, $q_1 = \sin(\phi/2)\cos(\theta)$, $q_2 = \sin(\phi/2)\sin(\theta)\cos(\psi)$, $q_3 = \sin(\phi/2)\sin(\theta)\sin(\psi)$, then $\mathbf{q}$ can be interpreted as a rotation from one event to the second through an angle ϕ about an axis with spherical polar coordinates θ, ψ.

Purely random (uniform) rotations correspond to selecting a direction at random on the 3-dimensional surface of 4-dimensional ball (Moran, 1975); in this case the rotation angle ϕ has a cumulative distribution of the form

$$F(\phi) = (1/\pi)[\phi - \cos(\phi)]. \qquad (4.1)$$

More generally we can generate an isotropic angular Cauchy distribution following the prescription given in section 2, in which case ϕ has the marginal distribution

$$F(\phi) = (1/\pi)[\phi^* - \cos(\phi^*)]. \qquad (4.2)$$

where $\tan(\phi) = K\tan(\phi^*)$. (In fact the same marginal will be obtained whenever the rotation angle is independent of the orientation of the axis)

To prepare the data for an analysis of the 2-point moment functions of the fault mechanisms, the key step is to develop an inversion procedure for determining the rotation q required to take a mechanism $\mathbf{q}_1$ into a mechanism $\mathbf{q}_2$. Details of the procedure are outlined in K9.

The situation is much complicated by the existence of symmetries, which introduce a form of aliasing, or confounding, into the description of the distribution of ϕ. For example, the representation of a stress tensor in terms of quaternions is invariant under rotations of π about any of the axes, corresponding to multiplication of the quaternion by any of the unit vectors $\mathbf{i}, \mathbf{j}, \mathbf{k}$. In relation to fault mechanisms, such symmetries arise because in the routine determination of focal mechanism from the seismic records, it is not possible to distinguish between the fault plane and the plane orthogonal to it and the direction of motion, nor to distinguish the sense of the fault motion. The result is that there are in general four rotations that will take $\mathbf{q}_1$ into one or other of the equivalent representations of $\mathbf{q}_2$. From these the one with the smallest value of ϕ may be selected. This has the effect of folding back some of the original ϕ-distribution onto itself. In particular it turns out that there is always at least one rotation with a ϕ of $2\pi/3$ or less, so the modified distribution of ϕ is restricted to the range $(0, 2\pi/3)$. Again we refer to K9 for details.

Since it is impossible to measure the total stress at seismogenic depths where earthquakes originate, Kagan (K9) measured the angles of focal mechanism rotation for any pair of earthquakes in the Harvard catalog. If the stress is a smooth spatial function, the angles of neighbouring events would be small, thus these angles indicate the degree of the stress irregularity. In Fig. 7 we show an example of the angle distribution for earthquakes separated by various distances and situated along a fault plane (around the N-axis, see Fig. 6) or outside the plane (around the T- and P-axes). Two theoretical curves show the rotational Cauchy and the uniform random rotation (see more in K9). For small distances between the earthquake centroids the angle distributions are approximated by the Cauchy curve, whereas for large distances only the events aligned along a fault plane are Cauchy distributed; for other earthquakes the rotation angle is uniformly random.

418

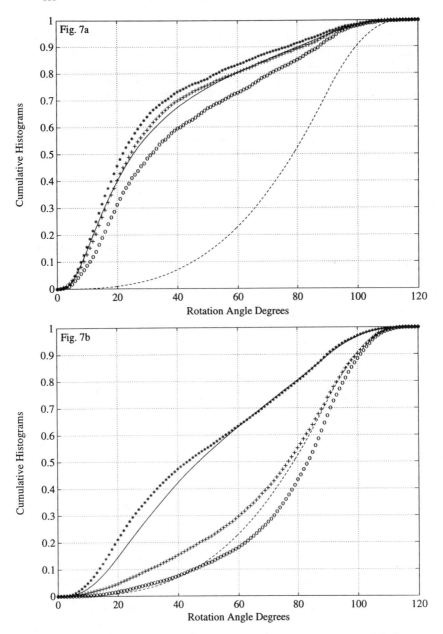

Fig. 7a,b. Distributions of rotation angles for pairs of focal mechanisms of shallow earthquakes from the Harvard catalog; hypocenters are separated by distances a) 0-50 km; b) 400-500 km; circles – hypocenters in 30° cones around the T-axis; plusses – hypocenters in 30° cones around the P-axis; stars – hypocenters in 30° cones around the N-axis. Solid line is for the Cauchy rotation with a) $\kappa = 0.1$; b) $\kappa = 0.2$; dashed line is for the random rotation.

The empirical results are set out in K9 in more detail, and may be briefly summarised as follows.

- (i) For most distances, angular separations and time intervals from the initial event in the pair, the distribution of the rotation angle ϕ appears to be reasonably well approximated by an angular Cauchy distribution of the general form (4.2), but not well approximated by other angular distributions such as the Fisher distribution.

- (ii) The concentration parameter k, on the other hand, varies appreciably with these factors. At short distance and time intervals, the distribution is most highly concentrated along the direction of fault motion, but becomes less concentrated as the distances, time intervals and angular separations are increased.

- (iii) the distribution of the parameters θ and ψ, representing the shift in orientation of the fault plane, does not appear to be uniformly distributed over the unit sphere; it shows some tendency to concentrate on smaller values, but the high dimensionality of the study makes it difficult to ascertain any clear pattern in the variation of the distribution as a function of distance, angular separation or time.

4.2. Distributions of Stress Fields.

The branching model described in section 3 is purely kinematic; forces do not appear. In a continuous medium, forces are traditionally described in terms of a stress field, with the stress at any point represented by a 3×3 stress tensor, or matrix. Each element $\sigma(i,j)$ of the matrix represents the component of force acting in direction j on an element of surface perpendicular to direction i, where i, j, k correspond to three orthogonal reference directions in the medium. It is not entirely clear that such a description is appropriate for a highly heterogeneous, fractured medium such as the earth's crust, but nevertheless stress calculations play a highly useful role in rock mechanics and in seismology, and they seem the natural starting point in trying to develop a dynamic model.

We need to recall a few further terms and properties of the stress tensor. The matrix under wide conditions is symmetric, and therefore has three real eigenvalues associated with three real eigenvectors. The latter determine the axes of principal stress. The forces acting across an element of area orthogonal to one of these axes are purely normal to the surface. For all other directions they contain both a normal and a tangential or shear component. The latter is a maximum for directions which bisect the angles between the principal axes. The coefficients in the characteristic polynomial $\det(\lambda * I - \Sigma)$ of the stress matrix Σ form the three stress invariants, the first representing just the sum of the principal stresses. The stress invariants are independent of the choice of the reference axes, and determine the size and shape of the stress ellipsoid or other conic.

Any discontinuity, such as a dislocation, in the medium will cause a local perturbation to the stress field. It will be important in the sequel that the effects of such perturbations decay as r^{-3}, where r is the distance from the perturbation, at least for r outside the immediate vicinity of the perturbation. Zolotarev and Strunin (1971) (see also Zolotarev, 1986) have shown that if the perturbations are distributed uniformly at random (i.e., following a Poisson process) through space, the components of the stress tensor at an arbitrary point of the space follow a Cauchy distribution. In K8 and K9, using partly analytic and partly simulation methods, some special cases are examined in detail, and used to justify the use of Cauchy distributions, both spatial and angular, in the branching model. As the assumptions underlying this argument are important, we briefly reproduce it below.

Suppose the sources are located at points $x_k(u_k, \theta_k, \phi_k)$, where to take advantage of the symmetry we have taken a signed distance u $(-\infty < u < \infty)$, with restriction of the latitude θ $(0 < \theta < \pi/2)$, in place or the usual r, θ, ψ of spherical polar coordinates. Now consider the net contribution from all sources to a given component of the stress tensor at the origin. This takes the form of the sum

$$S = \sum f(\theta_k, \psi_k)/|(u_k)|^3 = \int\limits_{-\infty}^{\infty} \int\limits_{0}^{\pi/2} \int\limits_{0}^{2\pi} \operatorname{sign}(u) f(\theta, \psi)/|(u)|^3 dN(u, \theta, \psi),$$

where N denotes counting measure and $f(\theta_k, \phi_k)$ describes the resolution of a unit vector, directed towards x_k, along the chosen direction of the stress field. We have then to find the characteristic function of the distribution of S, $c(t) = E[\exp(itS)]$. Examination of this expectation shows that it is nothing other than the characteristic functional of the point process N, evaluated for the special choice of carrier function $\xi \exp\{itf()/|(u)|^3\}$. The ch. fl. of the homogeneous Poisson process $\Phi(\xi)$ is given by

$$\Phi(\xi) = \exp\left\{\int [\exp(i.\xi(x)) - 1]dx\right\}.$$

Substituting for ξ and changing variable from u to $y = f()/u^3$ we can rewrite the characteristic function $c(t)$ in the form

$$c(t) = \int\limits_{-\infty}^{\infty} \{[\exp(ity) - 1]/|y|^{(1+\alpha)}\}dy \int \{f(\theta, \psi)^\alpha d\theta d\psi\}, \qquad (4.3)$$

where $\alpha = 1$ in the present case. The y-integral is recognisable as the characteristic function of the symmetric stable distribution with parameter α, i.e., the Cauchy distribution in the present instance. Exactly similar argument can be used to derive the Holtzmark distribution $(\alpha = 3/2)$ for the gravitational field due to Poisson distributed stars. It should be noted that both these fields have long-tailed distributions, distinguishing them from examples such as fractional Brownian motion and their relatives, in which the field variables have short-tailed distributions.

There are a number of questions and comments which arise in relation to the above derivation.

• (i) The argument as given applies to a single scalar component of the stress field. It would seem possible in principle to extend the argument to the joint distribution of the stress components. Is this indeed possible, and does it lead to the Cauchy angular distribution for the direction of principal stress?

• (ii) Is the conclusion affected by introducing distributions for the sizes and orientations of the random perturbations?

• (iii) Can the argument be extended to situations where the sources have a self-similar distribution, resulting in a change of index to the derived stable distribution.

The last of these questions is of particular interest. Kagan has several times suggested that this may be the case, (see, e.g., comments in K10-K12), and has found empirically that distributions which appear to be of stable type with index different from one arise in simulations of self-similar patterns, and for the data from the Harvard catalogue, using standard formulae for the stress field produced at distance for the double-couple model of earthquake fault mechanism. Justifying this result analytically, however, runs into the same sort of difficulty as that described in section 2c: exactly what are we to assume for the distribution of sources? The argument leading to (4.3) depends rather crucially on the Poisson character of the underlying point process. This would seem to restrict us to the processes discussed in section 2.2c (Poisson with singular parameter measure), and hence to the difficulties noted in that section.

A more practical question related to this discussion is

• (iv) Find robust methods of inference for the index of a symmetric stable distribution. The robustness is needed because the empirical distribution will usually be contaminated by measurement errors or boundary effects at the ends of the distribution, so that only some part of the distribution is properly available for inference.

To conclude this section we note the work in K11 and K12 in which similar ideas are used to compute the increments to the local stress field caused by recent earthquakes in the California region. Clearly the results are very tentative, and in themselves take no account of any pre-existing stress field, or increases due to plate motion. Nevertheless one can begin to see from the patterns produced how the stress regime over a region may evolve, and in its turn serve to initiate future earthquakes, thus exhibiting on a large and complex scale the same branching, self-organising behaviour studied on a small scale and in idealised fashion with the simulation model.

5. SUMMARY AND FURTHER REMARKS.

We summarize below the main points to arise from the preceding discussion.

(i) There seems to be some general difficulty in knowing how best to approach models for point process data that exhibit self similarity and fractal behaviour. For example, we do not know if any legitimate models exist for processes which exhibit the $1/V$ dependence for the 4-point moment function claimed by Kagan for earthquake catalogues. An underlying question is whether the standard point process framework is adequate to discuss problems of this kind, and if not, how it should be extended.

(ii) There are a number of related questions on which we lack current information. How far has the idea of "Palm stationarity" been explored? What can be said about processes exhibiting "intermittent" self-similarity? What are the most appropriate extensions of fractal and multifractal dimension to point process models? References and suggestions would be welcome.

(iii) The analytical properties of the K-K branching model remain almost completely unexplored. Two immediate questions which arise concern the fractal dimension of realizations of the process and the existence of cluster limit theorems which might provide a theoretical underpinning to Kagan's empirical results on the self-similarity and other properties of the derived process of simulated "earthquakes".

(iv) Many statistical and modelling questions arise in the representation of focal mechanisms and stress tensors as points on a three-dimensional space. Kagan's rotational Cauchy distributions appear to play a key role here, but again they lack a proper theoretical underpinning. Can they be derived from an extension of Zolotarev's derivation of the ordinary Cauchy distribution for a scalar component of the stress field? Do further extensions of Zolotarev's theory exist for the stress field in a non-homogeneous but still self-similar field of randomly distributed defects?

(v) A number of more immediate statistical questions arise out of Kagan's work, which space has precluded us from treating in detail, but which we mention here by way of conclusion. (a) What are efficient estimation and testing procedures for the 3-D rotational Cauchy distributions? (b) Given that branching process models tend to suggest distributions of the form

$$\text{Prob}(X > x) = x^{-\beta} \exp[-\gamma(x - 1)],$$

what are the best ways of testing for constancy of the parameters in the spatial domain? (c) What are robust ways of testing the value of the parameter α in a symmetric stable distribution (robust here meaning that data from the centre and the extremes of the distribution may be suspect)?

Acknowledgments.

This research was supported in part by Grant VIC406 of New Zealand Foundation for Research, Science & Technology. We are grateful for useful discussions with R. B. Davies. Publication 4290, Institute of Geophysics and Planetary Physics, University of California, Los Angeles.

References

Aki, K., and P. Richards, 1980. *Quantitative Seismology*, W. H. Freeman, San Francisco, 2 Vols, 557 and 373 pp.

Chong, F. S., 1983. Time-space-magnitude interdependence of upper crustal earthquakes in the main seismic region of New Zealand, *N. Z. J. Geol. Geophys.*, **26**, 7-24.

Cutler, C. D., 1991. Some results on the behaviour and estimation of the fractal dimensions of distributions on attractors, *J. Statist. Phys.*, **62**, 651-708.

Daley, D. J., and D. Vere-Jones, 1988. *An Introduction to the Theory of Point Processes*, New York, Springer-Verlag, pp. 702.

Dawson, D. A., and K. Fleischmann, 1988. Strong clumping of critical branching models in subcritical dimensions, *Stochastic Process. Appl.*, **30**, 193-208.

Dziewonski, A. M., G. Ekstrom, and M. P. Salganik, 1994. Centroid-moment tensor solutions for January-March, 1994, *Phys. Earth Planet. Inter.*, **86**, 253-261.

Geilikhman, M. B., T. V. Golubeva, and V. F. Pisarenko, 1990. Multifractal patterns of seismicity, *Earth and Plan. Sci. Letters*, **99**, 127-132.

Hawkes, A. G., and L. Adamopoulos, 1973. Cluster models for earthquakes - Regional comparisons, *Bull. Int. Statist. Inst.*, **45(3)**, 454-461.

Hawkes, A. G., and D. Oakes, 1974. A cluster representation of a self-exiting process, *J. Appl. Prob.*, **17**, 493-503.

Jagers, P., 1989. General branching processes as a random field, *Stochastic Process. Appl.*, **32**, 183-212.

Kagan, Y. Y., 1973. Statistical methods in the study of the seismic process, *Bull. Int. Statist. Inst.*, **45(3)**, 437-453.

Kagan, Y. Y., 1981a. Spatial distribution of earthquakes: the three-point moment function, *Geophys. J. Roy. astr. Soc.*, **67**, 697-717 (K2).

Kagan, Y. Y., 1981b. Spatial distribution of earthquakes: the four-point moment function, *Geophys. J. Roy. astr. Soc.*, **67**, 719-733 (K3).

Kagan, Y. Y., 1982. Stochastic model of earthquake fault geometry, *Geophys. J. R. astr. Soc.*, **71**, 659-691, (K6).

Kagan, Y. Y., 1990. Random stress and earthquake statistics: spatial dependence, *Geophys. J. Int.*, **102**, 573-583, (K8).

Kagan, Y. Y., 1991. Fractal dimension of brittle fracture, *J. Nonlinear Sci.*, **1**, 1-16, (K4).

Kagan, Y. Y., 1992. Correlations of earthquake focal mechanisms, *Geophys. J. Int.*, **110**, 305-320, (K9).

Kagan, Y. Y., 1994a. Incremental stress and earthquakes, *Geophys. J. Int.*, **117**, 345-364, (K10).

Kagan, Y. Y., 1994b. Distribution of incremental static stress caused by earthquakes, *Nonlinear Processes in Geophysics*, **1**, 172-181, (K11).

Kagan, Y. Y., 1994c. Observational evidence for earthquakes as a nonlinear dynamic process, *Physica D*, **77**, 160-192, (K12).

Kagan, Y. Y. and L. Knopoff, 1980. Spatial distribution of earthquakes: the two-point correlation function, *Geophys. J. Roy. astr. Soc.*, **62**, 303-320, (K1).

Kagan, Y. Y. and L. Knopoff, 1981. Stochastic synthesis of earthquake catalogs, *J. Geophys. Res.*, **86**, 2853-2862, (K5).

Kagan, Y. Y. and L. Knopoff, 1987. Random stress and earthquake statistics: time dependence, *Geophys. J. R. astr. Soc.*, **88**, 723-731, (K7).

Mandelbrot, B. B., 1975. Sur un modele decomposable d'Universe hierarchise: deduction des correlations galactiques sur la sphere celeste. *C. R. Acad. Sc. Paris*, **280**, Ser. A, 1551-1554.

Mandelbrot, B. B., 1983. *The Fractal Geometry of Nature*, W. H. Freeman, San Francisco, Calif., 2nd edition, pp. 468.

Moran, P. A. P., 1975. Quaternions, Haar measure and estimation of paleomagnetic rotation, in: *Perspectives in Probability and Statistics*, ed. J. Gani, Acad. Press, 295-301.

Musmeci, F., and D. Vere-Jones, 1992. A space-time clustering model for historical earthquakes, *Ann. Inst. Statist. Math.*, **44**, 1-11.

Ogata, Y., and K. Katsura, 1991. Maximum likelihood estimates of the fractal dimension for random point patterns, *Biometrika*, **78**, 463-474.

Otsuka, M., 1972. A chain-reaction-type source model as a tool to interpret the magnitude-frequency relation of earthquakes, *J. Phys. Earth*, **20**, 35-45.

Ripley, B. D., 1977. Modelling spatial patterns, *J. Roy. Stat. Soc.*, **B39**, 172-212.

Saito, M., M. Kikuchi, and M. Kudo, 1973. Analytical solution of "Go-game model of earthquakes", *Zisin*, **26**(2), 19-25.

Stoyan, D., 1994. Caution with fractal point patterns. *Statistics*, **25**, 267-270.

Stoyan, D., W. S. Kendall, and J. Mecke, 1987. *Stochastic Geometry and its Applications*, New York, Wiley, 345 pp.

Stoyan, D., and H. Stoyan, 1994. *Fractals, Random Shapes, and Point Fields: Methods of Geometrical Statistics*, New York, Wiley, 389 pp.

Vere-Jones, D., 1976. A branching model for crack propagation, *Pure Appl. Geophys.*, **114**, 711-725.

Vere-Jones, D., 1978. Space time correlations for microearthquakes – a pilot study, *Adv. Appl. Prob.* (Suppl. – *Spatial Patterns and Processes*), **10**, 73-87.

Vere-Jones, D., 1994. Statistical models for earthquake occurrence: clusters, cycles and characteristic earthquakes, in *Proc. First US/Japan Conf. on Frontiers of Statistical Modeling: An Informational Approach*, ed. by H. Bozdogan, pp. 105-136, Kluwer Publ., Netherlands.

Vere-Jones, D., and D. Harte, 1995. Dimension estimates of earthquake epicentres and hypocentres, (VUW preprint: in preparation).

Vere-Jones, D., R. B. Davies, D. Harte, T. Mikosch and Q. Wang, 1995. Problems and examples in the estimation of fractal dimension from meteorological and earthquake data, Ed. T. Subba Rao, *Proc. Int. Conf. on Application of Time Series in Physics, Astronomy and Meteorology*, Padua, (to appear).

Zolotarev, V. M., and B. M. Strunin, 1971. Internal-stress distribution for a random distribution of point defects, *Soviet Phys. Solid State*, **13**, 481-482 (English translation).

Zolotarev, V. M., 1986. *One-Dimensional Stable Distributions*, Amer. Math. Soc., Providence, R.I., pp. 284.

ON THE EXISTENCE OF UMVU ESTIMATORS FOR BERNOULLI EXPERIMENTS IN THE NON-IDENTICALLY DISTRIBUTED CASE WITH APPLICATIONS TO THE RANDOMIZED RESPONSE METHOD AND THE UNRELATED QUESTION MODEL

A. Dannwerth and D. Plachky

Institute of Mathematical Statistics
University of Münster
Einsteinstr. 62
D-48149 Münster

Abstract

It is shown that there does not exist any UMVU estimator of the type $d^*(\sum_{j=1}^{n} X_j)$, based on independent random variables $X_1, \ldots, X_n$, where X_j has a Bernoulli distribution with corresponding probability $a_j p + b_j \in [0,1]$ for $p \in [0,1]$, $a_j \neq 0$, $j = 1, \ldots, n$, and where $d^* : \{0, \ldots, n\} \to \mathbb{R}$ is one-to-one, if the condition $a_j = a_k$ and $b_j = b_k$ or $a_j = -a_k$ and $b_j = 1 - b_k$ is not valid for all $(j, k) \in \{1, \ldots, n\}^2$. In particular, it is proved that the model parameter a_j, b_j, $j = 1, \ldots, n$, in connection with the unrelated question model must be kept fixed, if there should exist some UMVU estimator for p of this type. Furthermore, it is shown that the existence of some non-constant UMVU estimator for p according to the randomized response method implies that the randomization parameter $p_j \in [0,1] \backslash \{\frac{1}{2}\}$, $j = 1, \ldots, n$, must satisfy the condition $p_j = p_k$ or $p_j + p_k = 1$ for all $(j, k) \in \{1, \ldots, n\}^2$.

1   Introduction and Auxiliary Results

Let $\mathcal{L}_2(\Omega, \mathcal{A}, P)$ denote the set consisting of all square-integrable functions in connection with the probability space $(\Omega, \mathcal{A}, P)$ and let $\mathcal{P}$ stand for some non-empty set of probability measures P defined on the σ-algebra $\mathcal{A}$ of subsets of the set Ω. Furthermore, some $d^* \in \bigcap_{P \in \mathcal{P}} \mathcal{L}_2(\Omega, \mathcal{A}, P)$ is called uniformly minimum variance unbiased estimator (UMVU estimator), if and only if $\mathrm{Var}_P(d^*) = \inf\{\mathrm{Var}_P(d) : d \in \bigcap_{P \in \mathcal{P}} \mathcal{L}_2(\Omega, \mathcal{A}, P), E_P(d) = E_P(d^*), P \in \mathcal{P}\}$ holds true for all $P \in \mathcal{P}$. It is usual to call d^* to be a UMVU estimator for $\delta : \mathcal{P} \to \mathbb{R}$ defined by $\delta(P) = E_P(d^*)$, $P \in \mathcal{P}$. According to the covariance method (cf. Lehmann (1983), p. 77) the property of $d^* \in \bigcap_{P \in \mathcal{P}} \mathcal{L}_2(\Omega, \mathcal{A}, P)$ to be a UMVU estimator is equivalent to $\mathrm{Cov}_P(d^*, d_0) = 0$, $P \in \mathcal{P}$, for all unbiased estimators $d_0 \in \bigcap_{P \in \mathcal{P}} \mathcal{L}_2(\Omega, \mathcal{A}, P)$ of zero, i.e. $E_P(d_0) = 0$, $P \in \mathcal{P}$, is valid. This characterization of UMVU estimators implies the following well-known properties:

426

(i) *Uniqueness:* d_j^*, $j = 1, 2$, UMVU estimators with $E_P(d_1^*) = E_P(d_2^*)$, $P \in \mathcal{P}$, implies $d_1^* = d_2^*$ P-a.e., $P \in \mathcal{P}$.

(ii) *Linearity:* d_j^*, $j = 1, 2$, UMVU estimators, $a_j \in \mathbb{R}$, $j = 1, 2$, implies that $a_1 d_1^* + a_2 d_2^*$ is a UMVU estimator.

(iii) *Multiplicativity:* d_j^*, $j = 1, 2$, UMVU estimators, d_1^* or d_2^* bounded, implies that $d_1^* d_2^*$ is a UMVU estimator.

(iv) *Closedness:* d_n^*, $n \in \mathbb{N}$, UMVU estimators satisfying $\lim_{n\to\infty} E_P((d_n^* - d)^2) = 0$, $P \in \mathcal{P}$, for some $d \in \bigcap_{P \in \mathcal{P}} \mathcal{L}_2(\Omega, \mathcal{A}, P)$, implies that d is a UMVU estimator, too.

These properties imply that some bounded, UMVU estimator d^* has the property that $|d^*|$ is also a UMVU estimator because of $|\frac{d^*}{2M}| = ((\frac{d^*}{2M})^2 - 1 + 1)^{1/2} = \sum_{k=0}^\infty \binom{1/2}{k}((\frac{d^*}{2M})^2 - 1)^k$ with $M = \sup\{|d^*(\omega)| : \omega \in \Omega\}$ and that the indicator function $I_{\{d^* > a\}}$ of $\{d^* > a\} \in \mathcal{A}$, $a \in \mathbb{R}$, is also a UMVU estimator because of $\lim_{n\to\infty} \min\{|nd^*|, 1\} = I_{\{d^* \neq 0\}}$, which implies $\lim_{n\to\infty} \min\{n(d^* - a)^+, 1\} = I_{\{d^* > a\}}$, $a \in \mathbb{R}$, where $(d^* - a)^+$ stands for the positive part of $d^* - a$, i.e. $(d^* - a)^+ = \max\{d^* - a, 0\}$.

Example. *(Representation of all UMVU estimators for Bernoulli experiments in the i.i.d. case)*
Let Ω stand for $\{0, 1\}^n$, $\mathcal{A}$ for the set $\mathcal{P}(\Omega)$ consisting of all subsets of Ω, P for the Bernoulli distribution with $P(\{1\}) = p$, $P(\{0\}) = 1 - p$, where $p \in [0, 1]$ is unknown, and $\mathcal{P}$ for the set consisting of the n-fold direct products $\bigotimes_{i=1}^n P_j$, $P_j = P$, $i = 1, \ldots, n$. Now, $d_k^* : \{0, 1\}^n \to \mathbb{R}$ defined by $d_k^*(\omega_1, \ldots, \omega_n) = \left(\begin{smallmatrix} \sum_{j=1}^n \omega_j \\ k \end{smallmatrix}\right) / \binom{n}{k}$, $(\omega_1, \ldots, \omega_n) \in \{0, 1\}^n$, is the UMVU estimator for $\delta : \mathcal{P} \to \mathbb{R}$ defined by $\delta(P) = p^k$, $P \in \mathcal{P}$, $k \in \{0, \ldots, n\}$, which follows from a well-known argument based on sufficiency and completeness of the mapping $\{0, 1\}^n \to \mathbb{R}$ defined by $(\omega_1, \ldots, \omega_n) \to \sum_{i=1}^n \omega_j$, $(\omega_1, \ldots, \omega_n) \in \{0, 1\}^n$. A direct approach might be based on the covariance method by starting from the equation $\sum_{\substack{\omega_j \in \{0,1\} \\ j=1,\ldots,n}} d_0(\omega_1, \ldots, \omega_n) \binom{n}{k} p^{\sum_{j=1}^n \omega_j} (1 - p)^{n - \sum_{i=1}^n \omega_j} = 0$, $p \in [0, 1]$, for some unbiased estimator $d_0 : \{0, 1\}^n \to \mathbb{R}$ of zero, which leads to $\sum_{\substack{\omega_j \in \{0,1\} \\ j=1,\ldots,n}} d_0(\omega_1, \ldots, \omega_n) \binom{n}{k} x^{\sum_{j=1}^n \omega_j} = 0$ for all $x > 0$, where x stands for $\frac{p}{1-p}$, $p \in [0, 1)$. If one differentiates the last equation k-times with respect to x, one arrives at $\text{Cov}_p(d_k^*, d_0) = 0$ for all $p \in [0, 1]$. Therefore, the class of all UMVU estimators coincides with the set consisting of all estimators $d^* : \{0, 1\}^n \to \mathbb{R}$ of the type $d^*(\omega_1, \ldots, \omega_n) = \sum_{k=0}^n a_k \left(\begin{smallmatrix} \sum_{j=1}^n \omega_j \\ k \end{smallmatrix}\right)$, $a_k \in \mathbb{R}$, $k = 0, \ldots, n$, $(\omega_1, \ldots, \omega_n) \in \{0, 1\}^n$, since $E_p(d) = \sum_{k=0}^n a_k p^k$, $a_k \in \mathbb{R}$, $k = 0, \ldots, n$, $p \in [0, 1]$, holds true for $d \in \bigcap_{P \in \mathcal{P}} \mathcal{L}_2(\Omega, \mathcal{A}, P)$.

2 Main Result and Applications

The starting point is the following modification of the preceding example: $\Omega = \{0, 1\}^n$, $\mathcal{A} = \mathcal{P}(\Omega)$, and $\mathcal{P}$ coincide with the family of all direct products $\bigotimes_{j=1}^n P_j$, where P_j is the Bernoulli distribution with corresponding probability of the type

$a_j p + b_j \in [0,1]$, $p \in [0,1]$, $a_j \neq 0$, $j = 1, \ldots, n$. In connection with the randomized response method· $a_j = 2p_j - 1$, $b_j = 1 - p_j$ is valid, where p_j $(\neq \frac{1}{2})$ denotes the probability associated with the corresponding randomization procedure depending on the j-th experiment, $j = 1, \ldots, n$. The model parameter a_j, b_j concerning the unrelated question method are $a_j = p_j$, $p_j \neq 0$, $b_j = (1 - p_j)\pi_j$, where p_j, π_j are the probabilities associated with the corresponding randomization procedure depending on the j-th experiment, $j = 1, \ldots, n$ (cf. Chaudhuri and Mukerjee (1988)). It will be shown that the model parameter must be independent of $j \in \{1, \ldots, n\}$ up to the effect resulting in the interchanging of p_j and $1 - p_j$ in connection with the randomized response model resp. must be strictly independent of $j \in \{1, \ldots, n\}$ concerning the unrelated question method, if there should exist a UMVU estimator of the type $(\omega_1, \ldots, \omega_n) \to d^*(\omega + \ldots + \omega_n)$, $(\omega_1, \ldots, \omega_n) \in \{0,1\}^n$, with d^* : $\{0, \ldots, n\} \to \mathbb{R}$ one-to-one for $\delta : \mathcal{P} \to \mathbb{R}$ defined by $\delta(\bigotimes_{j=1}^n P_j) = p$, $p \in [0,1]$. For this purpose the following main result will be proved:

Theorem 1 *The existence of some UMVU estimator of the type* $(\omega_1, \ldots, \omega_n) \to d^*(\omega_1 + \ldots + \omega_n)$, $(\omega_1, \ldots, \omega_n) \in \{0,1\}^n$, *with* d^* : $\{0, \ldots, n\} \to \mathbb{R}$ *one-to-one, in connection with the Bernoulli experiment* $(\{0,1\}^n, \mathcal{P}(\{0,1\}^n), \{\bigotimes_{j=1}^n P_j : P_j$ *Bernoulli distribution with parameter* $a_j p + b_j \in [0,1]$, $p \in [0,1]$, $a_j \neq 0$, $j = 1, \ldots, n\})$, $n \geq 2$, *implies* $a_j = a_k$, $b_j = b_k$ *or* $a_j = -a_k$, $b_j = 1 - b_k$ *for all* $(j,k) \in \{1, \ldots, n\}^2$. *Under these assumptions on* a_j, b_j, $j = 1, \ldots, n$, *the estimator* $d^* : \{0,1\}^n \to \mathbb{R}$ *defined by* $d^*(\omega_1, \ldots, \omega_n) = \frac{1}{n} \sum_{j=1}^n \frac{\omega_j - b_j}{a_j}$, $(\omega_1, \ldots, \omega_n) \in \{0,1\}^n$, *is UMVU for* δ *introduced by* $\delta(\bigotimes_{j=1}^n P_j) = p$, $p \in [0,1]$.

Proof: In the case, where $a_j = a_k$, $b_j = b_k$ or $a_j = -a_k$, $b_j = 1 - b_k$ holds true for any $(j,k) \in \{1, \ldots, n\}^2$, the random variables $\frac{X_j - b_1}{a_1}$, $j = 1, \ldots, n$, are independent and identically distributed, where the common distribution is concentrated on $\{\frac{1-b_1}{a_1}, -\frac{b_1}{a_1}\}$ with probability p and $1 - p \in [0,1]$, and where $X_j : \{0,1\}^n \to \{0,1\}$ stands for the j-th projection, $j = 1, \ldots, n$. Therefore, based on a sufficiency and completeness argument the estimator occuring in the second part of Theorem 1 is UMVU for δ. For the proof of the first part of Theorem 1 one starts from the special case $n = 2$, where $(\omega_1, \omega_2) \to d^*(\omega_1, \omega_2)$, $(\omega_1, \omega_2) \in \{0,1\}^2$, $d^*\{0,1,2\} \to \mathbb{R}$ one-to-one is some UMVU estimator. Then one gets by the covariance method the equation $\sum_{\substack{\omega_i \in \{0,1\} \\ i=1,2}} d^*(\omega_1, \omega_2)[\frac{1}{a_1}(\omega_1 - b_1) - \frac{1}{a_2}(\omega_2 - b_2)] \prod_{i=1}^2 \alpha_i^{\omega_i}(1 - \alpha_i)^{1-\omega_i} = 0$ with $\alpha_i = a_i p_i + b_i$, $i = 1, 2$, from which $\sum_{\substack{\omega_i \in \{0,1\} \\ i=1,2}} d^*(\omega_1, \omega_2)[a_2\omega_1 - a_2 b_1 - (a_1\omega_2 - a_1 b_2)] \prod_{i=1}^n \alpha_i^{\omega_i}(1 - \alpha_i)^{1-\omega_i} = 0$ follows. Therefore,

$$
\begin{aligned}
&d^*(0,0)[-a_2 b_1 + a_1 b_2](1 - \alpha_1)(1 - \alpha_2) + \\
&d^*(0,1)[-a_2 b_1 + a_1 b_2 - a_1](1 - \alpha_1)\alpha_2 + \\
&d^*(1,0)[-a_2 b_1 + a_1 b_2 + a_2]\alpha_1(1 - \alpha_2) \\
&d^*(1,1)[-a_2 b_1 + a_1 b_2 - a_1 + a_2]\alpha_1\alpha_2 = 0
\end{aligned}
\tag{1}
$$

holds true. Now the condition for d^* that $\operatorname{card}(d^{*-1}(\{M\})) = 1$, $\operatorname{card}(d^{*-1}(\{m\})) = 3$ or $\operatorname{card}(d^{*-1}(\{M\})) = 3$, $\operatorname{card}(d^{*-1}(\{m\})) = 1$ is valid with $M = \sup\{d^*(\omega_1, \omega_2) : (\omega_1, \omega_2) \in \{0,1\}^2\}$ and $m = \inf\{d^*(\omega_1, \omega_2) : (\omega_1, \omega_2) \in \{0,1\}^2\}$ is not possible according to the assumption on the structure of d^*. For the discussion of the

428

other cases $\operatorname{card}(d^{*-1}(\{M\})) = \operatorname{card}(d^{*-1}(\{m\})) = 1$ and $\operatorname{card}(d^{*-1}(\{M\})) = 2$ or $\operatorname{card}(d^{*-1}(\{m\})) = 2$ one derives from (1) the equation

$$d^*(0,0)[-a_2 b_1 + a_1 b_2] - d^*(0,1)[-a_2 b_1 + a_1 b_2 - a_1] - \qquad (2)$$
$$d^*(1,0)[-a_2 b_1 + a_1 b_2 + a_2] + d^*(1,1)[-a_2 b_1 + a_1 b_2 - a_1 + a_2] = 0$$

by differentiation of (1) with respect to p, if one takes $\frac{d}{dp}\alpha_i^{\omega_i}(1-\alpha_i)^{1-\omega_i} = (-1)^{\omega_i+1}a_i$, $\omega_i \in \{0,1\}$, $i = 1,2$, into consideration. Now the case $\operatorname{card}(d^{*-1}(\{M\})) = \operatorname{card}(d^{*-1}(\{m\})) = 1$, i.e. $\operatorname{card}(d^{*-1}(\mathbb{R}\backslash\{m,M\})) = 2$ holds true for $m \neq M$, leads together with (2) to the equations

$$-a_2 b_1 + a_1 b_2 = -a_2 b_1 + a_1 b_2 - a_1 + a_2 = 0 \quad \text{or} \qquad (3)$$
$$-a_2 b_1 + a_1 b_2 - a_1 = -a_2 b_1 + a_1 b_2 + a_2 = 0,$$

since according to the preceding first section one might replace d^* in (2) by the indicator function I_A of A defined by $d^{*-1}(\mathbb{R}\backslash\{m,M\})$ and since $a_1 = 0$ or $a_2 = 0$ is not possible by assumption. By the same reason one arrives at (3) in connection with the case $\operatorname{card}(d^{*-1}(\{M\})) = 2$ or $\operatorname{card}(d^{*-1}(\{m\})) = 2$ and $m \neq M$, if one replaces d^* in (2) by the indicator function of A defined by $d^{*-1}(\{M\})$ (or $d^{*-1}(\{m\})$). Finally, (3) is equivalent to $a_1 = a_2$, $b_1 = b_2$ or $a_1 = -a_2$, $b_1 = 1 - b_2$. The proof will now be finished by observing the following effect for any UMVU estimator $d^* : \{0,1\}^n \to \mathbb{R}$, $d_2^* : \{0,1\}^2 \to \mathbb{R}$ defined by $d_2^*(\omega_n, \omega_{n+1}) = d^*(\omega_1,\dots,\omega_{n-1},\omega_n,\omega_{n+1})$, $(\omega_n,\omega_{n+1}) \in \{0,1\}^2$, $(\omega_1,\dots,\omega_{n-1}) \in \{0,1\}^{n-1}$ fixed, is some UMVU estimator, too. This follows by the covariance method, since $\tilde{d}_0 : \{0,1\}^{n+1} \to \mathbb{R}$ defined by $\tilde{d}_0(\omega_1,\dots,\omega_{n+1}) = I_A(\omega_1,\dots,\omega_{n-1}) \cdot d_0(\omega_n,\omega_{n+1})$, $(\omega_1,\dots,\omega_{n+1}) \in \{0,1\}^{n+1}$, is for any subset A of $\{0,1\}^{n-1}$ and all unbiased estimators $d_0 : \{0,1\}^2 \to \mathbb{R}$ of zero (for the sample size 2), also some unbiased estimator of zero (for the sample size $n + 1$). □

Example. *(Randomized Response Method, Unrelated Question Model)*
The underlying model for the randomized response method is the Bernoulli experiment $(\{0,1\}^n, \mathcal{P}(\{0,1\}^n), \{\bigotimes_{j=1}^n P_j : P_j \text{ Bernoulli distribution with parameter } a_j p + b_j, a_j = 2p_j - 1, b_j = 1 - p_j, p_j \in [0,1]\backslash\{\frac{1}{2}\}, j = 1,\dots,n, p \in [0,1]\}), n \geq 2$. Therefore, there exists according to the preceding theorem a UMVU estimator for δ with $\delta(\bigotimes_{j=1}^n P_j) = p$, $p \in [0,1]$ of the type described by Theorem 1, if and only if $p_j = p_k$ or $p_j + p_k = 1$ is valid for all $(j,k) \in \{1,\dots,n\}^2$ and $d^* : \{0,1\}^n \to \mathbb{R}$ defined by $d^*(\omega_1,\dots,\omega_n) = \frac{1}{n}\sum_{j=1}^n \frac{\omega_j - 1 + p_j}{2p_j - 1}$, $(\omega_1,\dots,\omega_n) \in \{0,1\}^n$, is a UMVU estimator for δ, if the corresponding assumptions on p_j, $j = 1,\dots,n$, is fulfilled. The unrelated question model might be described by the Bernoulli experiment $(\{0,1\}^n, \mathcal{P}(\{0,1\}^n), \{\bigotimes_{j=1}^n P_j : P_j \text{ Bernoulli distribution with parameter } p_j p + (1 - p_j)\pi_j, p_j \in (0,1], \pi_j \in [0,1], j = 1,\dots,n, p \in [0,1]\}), n \geq 2$. Therefore, there exists by the preceding theorem some UMVU estimator of the type described there for δ with $\delta(\bigotimes_{j=1}^n P_j) = p$, $p \in [0,1]$, if and only if $p_j = p_0$, $\pi_j = \pi_0$, $j = 1,\dots,n$, for some $p_0 \in (0,1]$, $\pi_0 \in [0,1]$, holds true and $d^* : \{0,1\}^n \to \mathbb{R}$ defined by $d^*(\omega_1,\dots,\omega_n) = \frac{1}{n}\sum_{j=1}^n \frac{\omega_j - (1-p_0)\pi_0}{p_0}$, $(\omega_1,\dots,\omega_n) \in \{0,1\}^n$, is a UMVU estimator for δ.

It will now be shown that the assumption on the special structure of some UMVU estimator in the first part of the last example in connection with the randomized response method might be replaced by the much simpler condition to be not equal to some constant.

Theorem 2 *There exists some UMVU estimator for p based on independent random variables $X_1, \ldots, X_n$, $n \geq 2$, where X_j has a Bernoulli distribution with probability $pp_j + (1-p)(1-p_j)$, $p_j \neq \frac{1}{2}$, $j = 1, \ldots, n$, $p \in [0,1]$, if and only if $p_j = p_k$ or $p_j + p_k = 1$, $j, k \in \{1, \ldots, n\}$, $j \neq k$, holds true. In this case $\frac{1}{n} \sum_{i=1}^{n} \frac{X_i - 1 + p_i}{2p_i - 1}$ is some UMVU estimator of p, where $p_i \in \{p_0, 1 - p_0\}$ for some $p_0 \in [0,1] \backslash \{\frac{1}{2}\}$, is valid.*

Proof: Let $\{0,1\}^n$ denote the underlying sample space. Then any estimator d^* : $\{0,1\}^n \to \mathbb{R}$ is a UMVU estimator for the real-valued parameter function defined by the expectation $E_p(d(X_1, \ldots, X_n))$, $p \in [0,1]$, if and only if according to the covariance method $\mathrm{Cov}_p(d^*(X_1, \ldots, X_n), d_0(X_1, \ldots, X_n)) = 0$ holds true for all $p \in [0,1]$ and for any unbiased estimator $d_0 : \{0,1\}^n \to \mathbb{R}$ of zero, i.e. $E_p(d_0(X_1, \ldots, X_n)) = 0$, $p \in [0,1]$, is valid. In particular, $d^* : \{0,1\}^n \to \mathbb{R}$ defined by $d^*(x_1, \ldots, x_n) = \frac{1}{n} \sum_{j=1}^{n} \frac{x_j - 1 + p_j}{2p_j - 1}$, $(x_1, \ldots, x_n) \in \{0,1\}^n$, is some UMVU estimator of p, if $p_j = p_k$ or $p_j + p_k = 1$, $j, k \in \{1, \ldots, n\}$, $j \neq k$, is valid. In this case $X_1, \ldots, X_n$ might be replaced by i.i.d. random variables with underlying Bernoulli distribution with probability $pp_0 + (1-p)(1-p_0)$, $p_0 \neq \frac{1}{2}$, $p \in [0,1]$, since $p_j = p_0$ or $p_j = 1 - p_0$, $j = 1, \ldots, n$, must hold true for some $p_0 \neq \frac{1}{2}$, where X_j is replaced by $1 - X_j$, if $p_j = 1 - p_0$ is valid. Now the property of $d^* : \{0,1\}^n \to \mathbb{R}$ defined by $d^*(x_1, \ldots, x_n) = \frac{1}{n} \sum_{j=1}^{n} \frac{x_j - 1 + p_0}{2p_0 - 1}$, $(x_1, \ldots, x_n) \in \{0,1\}^n$ to be some UMVU estimator for p follows from the equation $E_p(d_0(X_1, \ldots, X_n)) = 0$ for unbiased estimators $d_0 : \{0,1\}^n \to \mathbb{R}$ of zero. Differentiation of $E_p(d_0(X_1, \ldots, X_n)) = \sum_{\substack{x_i \in \{0,1\} \\ j=1,\ldots,n}} d_0(x_1, \ldots, x_n)(pp_0 + (1 - p)(1-p_0))^{\sum_{i=1}^{n} x_i} \cdot (1 - pp_0 - (1-p)(1-p_0))^{n - \sum_{j=1}^{n} x_i}$ with respect to p yields the equation $(2p_0 - 1)(1 - pp_0 - (1-p)(1-p_0))E_p(d_0(X_1, \ldots, X_n) \cdot \sum_{i=1}^{n} X_i) + (1 - 2p_0)(pp_0 + (1 - p)(1 - p_0))E_p(d_0(X_1, \ldots, X_n) \cdot (n - \sum_{i=1}^{n} X_i)) = 0$, $p \in [0,1]$, from which $E_p(d_0(X_1, \ldots, X_n) \sum_{i=1}^{n} X_i) = 0$, i.e. $E_p(d_0 \circ (X_1, \ldots, X_n) \cdot \frac{1}{n} \sum_{i=1}^{n} \frac{X_i - 1 + p_0}{2p_0 - 1}) = 0$, $p \in [0,1]$, follows. Furthermore, $E_p(d^* \circ (X_1, \ldots, X_n)) = p$, $p \in [0,1]$, holds true, which implies that $d^*(x_1, \ldots, x_n) = \frac{1}{n} \sum_{j=1}^{n} \frac{x_j - 1 + p_j}{2p_j - 1}$, $(x_1, \ldots, x_n) \in \{0,1\}^n$ defines some UMVU estimator for p, if $p_j = p_k$ or $p_j + p_k = 1$ is valid for $j, k \in \{1, \ldots, n\}$, $j \neq k$. Conversely, it will now be shown that the existence of some UMVU estimator $d^* : \{0,1\}^n \to \mathbb{R}$, $n \geq 2$, for p implies that $p_j = p_k$ or $p_j + p_k = 1$ holds true for $j, k \in \{1, \ldots, n\}$, $j \neq k$. For this purpose it will be proved that any UMVU estimator $d^* : \{0,1\}^n \to \mathbb{R}$ is equal to some $c \in \mathbb{R}$, if $p_j = p_k$ or $p_j + p_k = 1$ is not valid for $j, k \in \{1, \ldots, n\}$. At first it follows from the proof of Theorem 1 that the existence of some non-constant UMVU estimator $d_2^* : \{0,1\}^2 \to \mathbb{R}$ in connection with the Bernoulli-experiment $(\{0,1\}^2, \mathcal{P}(\{0,1\}), \{P_1 \otimes P_2 : P_j$ Bernoulli distribution with parameter $a_j p + b_j \in [0,1]$, $p \in [0,1]$, $a_j \neq 0$, $b_j \notin \{0,1\}$, $j = 1,2\})$ implies that one of the following equations $\frac{a_1}{b_1} = \frac{a_2}{b_2}$, $\frac{a_1}{1 - b_1} = \frac{a_2}{1 - b_2}$, $\frac{a_1}{b_1} = -\frac{a_2}{1 - b_2}$, $\frac{a_2}{b_2} = -\frac{a_1}{1 - b_1}$ must hold true, from which in the special case $a_j = 2p_j - 1$, $b_j = 1 - p_j$, $j = 1, 2$, the equations $p_1 = p_2$ or $p_1 + p_2 = 1$ follows.

It will now be shown by induction on n that any UMVU estimator $d^* : \{0,1\}^n \to \mathbb{R}$ is equal to some constant $c \in \mathbb{R}$, if $p_j = p_k$ or $p_j + p_k = 1$ is not valid for $j, k \in \{1, \ldots, n\}$. For this purpose one takes into consideration that for any UMVU estimator $d^* : \{0,1\}^{n+1} \to \mathbb{R}$ the estimator $d_2^* : \{0,1\}^2 \to \mathbb{R}$ defined by $d_2^*(x_n, x_{n+1}) =$

430

$d^*(x_1, \ldots, x_{n-1}, x_n, x_{n+1})$, $(x_n, x_{n+1}) \in \{0,1\}^2$ and keeping $(x_1, \ldots, x_{n-1}) \in \{0,1\}^{n-1}$ fixed is some UMVU estimator for the sample size $n = 2$. This follows from the covariance method, since $\tilde{d}_0 : \{0,1\}^{n+1} \to \mathbb{R}$ defined according to $\tilde{d}_0(x_1, \ldots, x_{n-1}, x_n,$ $x_{n+1}) = I_A(x_1, \ldots, x_{n-1}) d_0(x_n, x_{n+1})$, $(x_1, \ldots, x_{n+1}) \in \{0,1\}^{n+1}$, is some unbiased estimator of zero for the sample size $n+1$, where A is any subset of $\{0,1\}^{n-1}$ and d_0 is some unbiased estimator of zero for the sample size $n = 2$. If now in addition neither $p_j = p_k$ nor $p_j + p_k = 1$ is valid for $j, k \in \{1, \ldots, n+1\}$, one might assume without loss of generality by choosing otherwise some suitable permutation of $\{1, \ldots, n+1\}$ that $p_n = p_{n+1}$ or $p_n + p_{n+1} = 1$ does not hold true. Therefore, $d_2^* : \{0,1\}^2 \to \mathbb{R}$ defined above must be equal to some constant $c \in \mathbb{R}$ according to the considerations for the sample size $n = 2$, i.e. $d^*(x_1, \ldots, x_{n+1})$, $(x_1, \ldots, x_{n+1}) \in \{0,1\}^{n+1}$, does not depend on $(x_n, x_{n+1}) \in \{0,1\}^2$. By induction on n there is also independence of $(x_1, \ldots, x_{n-1}) \in \{0,1\}^{n-1}$, if $p_j = p_k$ or $p_j + p_k = 1$ is not valid for $j, k \in \{1, \ldots, n-1\}$. Finally, in the case, where $p_j = p_k$ or $p_j + p_k = 1$ holds true for $j, k \in \{1, \ldots, n-1\}$, the estimator $d^* : \{0,1\}^{n+1} \to \mathbb{R}$, depending only on $(x_1, \ldots, x_{n-1}) \in \{0,1\}^{n-1}$, is some UMVU estimator for the sample size $n + 1$ and also for the sample size $n - 1$, where $p_j = p_0$ or $p_j = 1 - p_0$, $p = 1, \ldots, n - 1$, is valid for some $p_0 \in [0,1] \backslash \{\frac{1}{2}\}$. Defining now $Y_j = X_j$ if $p_j = p_0$ resp. $Y_j = 1 - X_j$, if $p_j = 1 - p_0$, holds true, $j = 1, \ldots, n - 1$, yields i.i.d. random variables with a Bernoulli distribution with parameter $p p_0 + (1 - p)(1 - p_0)$, $p \in [0,1]$. Now any UMVU estimator d^* based on $Y_1, \ldots, Y_{n-1}$ has the representation $d^*(y_1, \ldots, y_{n-1}) = \sum_{k=0}^{n-1} a_k \binom{\sum_{j=1}^{n-1} y_j}{k}$, $(y_1, \ldots, y_{n-1}) \in \{0,1\}^{n-1}$ (cf the example concluding the first section). Finally, the covariance method together with the property that d^* is also some UMVU for the sample size $n + 1$ yields the equation $E_p(\sum_{k=0}^{n-1} a_k \binom{\sum_{j=1}^{n-1} Y_j}{k}(\frac{Y_{n-1} + p_0 - 1}{2p_0 - 1} - \frac{X_n + p_n - 1}{2p_n - 1})) = 0$, $p \in [0,1]$, where $Y_1, \ldots, Y_{n-1}$, X_n are independent and X_n has a Bernoulli distribution with parameter $p p_n + (1 - p)(1 - p_n)$, $p_n \in [0,1] \backslash \{\frac{1}{2}\}$ fixed, $p \in [0,1]$. Hence, $\sum_{k=0}^{n-1} \frac{a_k}{2p_0 - 1} E_p(Y_{n-1} \binom{\sum_{j=1}^{n-1} Y_j}{k}) + \sum_{k=0}^{n-1} a_k \frac{p_0 - 1}{2p_0 - 1} E_p(\binom{\sum_{j=1}^{n-1} Y_j}{k}) - \sum_{k=0}^{n-1} a_k \binom{n-1}{k} \tilde{p}^k p = 0$ is valid for $p \in [0,1]$, where $\tilde{p}$ stands for $p_0 p + (1 - p_0)(1 - p)$, i.e. $p = \frac{\tilde{p} - 1 + p_0}{2p_0 - 1}$. Furthermore, $E_p(Y_{n-1} \binom{\sum_{j=1}^{n-1} Y_j}{k}) = \tilde{p} E_p(\binom{\sum_{j=1}^{n-2} Y_j + 1}{k}) = \tilde{p}(E_p(\binom{\sum_{j=1}^{n-2} Y_j}{k-1}) + E_p(\binom{\sum_{j=1}^{n-2} Y_j}{k})) = \tilde{p}(\binom{n-2}{k} \tilde{p}^k + \binom{n-2}{k-1} \tilde{p}^{k-1})$ for $k \in \{1, \ldots, n-2\}$ resp. $\tilde{p}$ for $k = 0$, and $\tilde{p}^{n-1}$ for $k = n - 1$ holds true. This implies

$0 = \frac{a_0}{2p_0 - 1} \tilde{p} + \frac{a_{n-1}}{2p_0 - 1} \tilde{p}^{n-1} + \sum_{k=1}^{n-2} \frac{a_k}{2p_0 - 1}(\binom{n-2}{k} \tilde{p}^{k+1} + \binom{n-2}{k-1} \tilde{p}^k) + \sum_{k=0}^{n-1} a_k \frac{p_0 - 1}{2p_0 - 1} \binom{n-1}{k} \tilde{p}^k - \sum_{k=0}^{n-1} a_k \binom{n-1}{k} \tilde{p}^k \frac{\tilde{p} + p_0 - 1}{2p_0 - 1}$

$= \frac{a_0}{2p_0 - 1} \tilde{p} + \frac{a_{n-1}}{2p_0 - 1} \tilde{p}^{n-1} - \sum_{k=1}^{n-2} \frac{a_k}{2p_0 - 1}(\binom{n-1}{k} - \binom{n-2}{k}) \tilde{p}^{k+1} - \frac{a_0}{2p_0 - 1} \tilde{p} - \frac{a_{n-1}}{2p_0 - 1} \tilde{p}^n + \sum_{k=1}^{n-2} \frac{a_k}{2p_0 - 1} \binom{n-2}{k-1} \tilde{p}^k = \frac{a_{n-1}}{2p_0 - 1} \tilde{p}^{n-1} - \frac{a_{n-1}}{2p_0 - 1} \tilde{p}^n$

$- \frac{a_{n-1}}{2p_0 - 1} \tilde{p}^n - \sum_{k=1}^{n-2} \frac{a_k}{2p_0 - 1} \binom{n-2}{k-1} \tilde{p}^{k+1} + \sum_{k=1}^{n-2} \frac{a_k}{2p_0 - 1} \binom{n-2}{k-1} \tilde{p}^k = \frac{a_{n-1}}{2p_0 - 1} \tilde{p}^{n-1} - \frac{a_{n-1}}{2p_0 - 1} \tilde{p}^n$

$- \sum_{k=2}^{n-1} \frac{a_{k-1}}{2p_0 - 1} \binom{n-2}{k-2} \tilde{p}^k + \sum_{k=1}^{n-2} \frac{a_k}{2p_0 - 1} \binom{n-2}{k-1} \tilde{p}^k = \frac{a_{n-1}}{2p_0 - 1} \tilde{p}^{n-1} - \frac{a_{n-1}}{2p_0 - 1} \tilde{p}^n - \frac{a_{n-2}}{2p_0 - 1} \binom{n-2}{n-3} \tilde{p}^{n-1} +$

$\frac{a_1}{2p_0 - 1} \tilde{p} + \sum_{k=2}^{n-2} \frac{1}{2p_0 - 1}(a_k \binom{n-2}{k-1} - a_{k-1} \binom{n-2}{k-2}) \tilde{p}^k = \frac{a_1}{2p_0 - 1} \tilde{p} - \frac{a_{n-1}}{2p_0 - 1} \tilde{p}^n + \sum_{k=2}^{n-2} \frac{1}{2p_0 - 1}(a_k \binom{n-2}{k-1} -$

$a_{k-1} \binom{n-2}{k-2}) \tilde{p}^k$, $\tilde{p} \in [\min\{p_0, 1 - p_0\}, \max\{p_0, 1 - p_0\}]$, from which $a_k = 0$, $k \in \{1, \ldots, n - 1\}$, follows, i.e. $d^* = a_0$ is valid, which has to be shown. $\quad \square$

The method of proof for Theorem 2 concerning the randomized response method yields the following corollary about the unrelated-question model (cf. Chaudhuri and

Mukerjee (1988)).

Corollary 3 *There exists some UMVU estimator for p based on independent random variables $X_1, \ldots, X_n$, $n \geq 2$, where X_j has a Bernoulli distribution with probability $pp_j + f(p_j) \in [0,1]$, $p \in [0,1]$, $p_j \in (0,1)$, $j = 1, \ldots, n$, with $f : \{p_1, \ldots, p_n\} \to (0,1)$ such that $j \to \frac{f(p_j)}{p_j}$ and $j \to \frac{f(p_j)}{1-p_j}$ are one-to-one for $j \in \{1, \ldots, n\}$, if and only if $p_j = p_k$, $j, k \in \{1, \ldots, n\}$, holds true.*

Bibliography

Chaudhuri, A. and R. Mukerjee (1988): Randomized Response. M. Dekker, New York

Lehmann, E. L. (1983): Theory of Point Estimation. Wiley, New York

ON A THREE-SAMPLE TEST

LAJOS TAKÁCS
Case Western Reserve University,
Cleveland, Ohio

SUMMARY

In 1958 H.T. David defined a statistic for testing the hypothesis that three samples of equal size n are drawn from the same continuous distribution. In this paper explicit formulas are given for the distribution and the asymptotic distribution of this statistic.

AMS 1991 subject classifications. Primary 62G10, 62G30, 62E15.
Key words and phrases. Three samples, hypothesis testing, order statistics, empirical distributions, limit distributions.

1. INTRODUCTION

Let $\xi_1, \xi_2, \ldots, \xi_{rn}$ be independent random variables each having the same continuous distribution function $F(x)$. Denote by $F_{n,1}(x), F_{n,2}(x), \ldots, F_{n,r}(x)$ the empirical distribution functions of the samples $(\xi_1, \xi_2, \ldots, \xi_n)$, $(\xi_{n+1}, \xi_{n+2}, \ldots, \xi_{2n}), \ldots, (\xi_{(r-1)n+1}, \xi_{(r-1)n+2}, \ldots, \xi_{rn})$ respectively. The distribution function $F_{n,i}(x)$ is defined as the number of variables $\xi_{(i-1)n+1}, \xi_{(i-1)n+2}, \ldots, \xi_{in}$ less than or equal to x divided by n. Define

$$\delta_{i,j}(n) = \sup_{-\infty < x < \infty} [F_{n,i}(x) - F_{n,j}(x)] \tag{1}$$

for $i, j = 1, 2, \ldots, r$. Since $F(x)$ is a continuous distribution function, the joint distribution of the random variables $\delta_{i,j}(n) (i, j = 1, 2, \ldots, r)$ does not depend on $F(x)$, that is, the random variables $\delta_{i,j}(n) (i, j = 1, 2, \ldots, r)$ are distribution-free statistics. We are interested in studying the distribution and the asymptotic distribution of the random variable

$$\delta_r(n) = \max[\delta_{1,2}(n), \delta_{2,3}(n), \ldots, \delta_{r-1,r}(n), \delta_{r,1}(n)] \tag{2}$$

for $r = 2, 3, \ldots$. The statistic (2) can be used to test the hypothesis that the r samples are drawn from the same continuous distribution. For other related statistics we refer to the papers of V. Ozols [15], M. Fisz [6], [7], [8], L.C. Chang and M. Fisz [1], [2], I.I. Gikhman [9] and J. Kiefer [12].

The asymptotic distribution of $\delta_2(n)$ was found by N.V. Smirnov [16] in 1939 and the distribution of $\delta_2(n)$ by B.V. Gnedenko and V.S. Korolyuk [10] in 1951. The distribution and the asymptotic distribution of $\delta_3(n)$ were determined in 1958 by H.T. David [3], but his formulas are involved and not convenient for numerical calculations. In this paper we determine the distribution of $\delta_3(n)$ and give explicit formulas for the limit distribution function

$$\lim_{n \to \infty} \mathbf{P}\{\sqrt{n}\delta_3(n) \leq x\} = H(x) \tag{3}$$

and its moments

$$M_r = \int_0^\infty x^r \, dH(x) \tag{4}$$

for $r = 0, 1, 2, \ldots$.

2. THE DISTRIBUTION OF $\delta_2(n)$

In 1951 by B.V. Gnedenko and V.S. Korolyuk [10] proved that

$$\binom{2n}{n} \mathbf{P}\{n\delta_2(n) < k\} = \sum_{|j| \leq n/k} (-1)^j \binom{2n}{n + jk} \tag{5}$$

for $k = 1, 2, \ldots$. In 1939 N.V. Smirnov [16] proved that

$$\lim_{n \to \infty} \mathbf{P}\{\sqrt{\frac{n}{2}}\delta_2(n) \leq x\} = K(x) \tag{6}$$

where $K(x)$ is the Kolmogorov distribution function. If $x > 0$, we have

$$K(x) = 1 - 2 \sum_{j=1}^{\infty} (-1)^{j-1} e^{-2j^2 x^2} = \frac{\sqrt{2\pi}}{x} \sum_{j=0}^{\infty} e^{-(2j+1)^2 \pi^2 / (8x^2)}, \tag{7}$$

and if $x \leq 0$, then $K(x) = 0$. See A.N. Kolmogorov [13].

434

3. THE DISTRIBUTION OF $\delta_3(n)$

Let us combine the three samples $(\xi_1, \xi_2, \ldots, \xi_n)$, $(\xi_{n+1}, \xi_{n+2}, \ldots, \xi_{2n})$ and $(\xi_{2n+1}, \xi_{2n+2}, \ldots, \xi_{3n})$ into a single sample in such a way that the $3n$ elements are arranged in nondecreasing order of magnitude. In the combined sample denote by $\alpha_i, \beta_i, \gamma_i$ the numbers of elements among the first i elements belonging to the first, second and third sample respectively. Then $\alpha_i + \beta_i + \gamma_i = i$ for $1 \le i \le 3n$ and $\alpha_{3n} = \beta_{3n} = \gamma_{3n} = n$. By using these notations we can write that

$$\mathbf{P}\{n\delta_3(n) < k\} = \mathbf{P}\{\alpha_i - \beta_i < k,\ \beta_i - \gamma_i < k,\ \gamma_i - \alpha_i < k \text{ for } 1 \le i \le 3n\} \tag{8}$$

if $k = 1, 2, \ldots$.

We shall associate a random walk with the variables $(\alpha_i, \beta_i, \gamma_i), 1 \le i \le 3n$. Let us suppose that a particle performs a random walk on the $x + y + z = 0$ plane in the three-dimensional Euclidean space. The particle starts at the origin $(0, 0, 0)$ and takes $3n$ steps. In the ith step $(i = 1, 2, \ldots, 3n)$ it moves a distance $\sqrt{2}$ in one of three specified directions which are determined by the vectors $(1, 0, -1), (-1, 1, 0)$ and $(0, -1, 1)$ according to whether the ith element of the combined sample belongs to the first or second or third sample respectively. The successive displacements are supposed to be independent and in each step the three vectors $(1, 0, -1), (-1, 1, 0)$ and $(0, -1, 1)$ are equally probable. At the end of the ith step the position of the particle is described by the point

$$Q(\alpha_i, \beta_i, \gamma_i) = (\alpha_i - \beta_i, \beta_i - \gamma_i, \gamma_i - \alpha_i) \tag{9}$$

where the coordinates are the (x, y, z) coordinates of this point in the plane $x + y + z = 0$. Obviously, $\alpha_i + \beta_i + \gamma_i = i$.

To find the probability that $n\delta_3(n) < k$ we should determine the probability that during the $3n$ steps the particle never touches the lines $x = k$, $y = k$ and $z = k$ in the plane $x + y + z = 0$ given that at the end of the $3n$ steps the position of the particle is $Q(n, n, n) = (0, 0, 0)$, that is, given that the particle returns to the initial position in $3n$ steps. We can imagine that there are three absorbing barriers (lines) in the plane $x + y + z = 0$, namely, $x = k$, $y = k$ and $z = k$, and during the $3n$ steps the particle remains inside the equilateral triangle determined by the three lines.

Let us define

$$p(a, b, c) = \frac{(a + b + c)!}{a! b! c!} \left(\frac{1}{3}\right)^{a+b+c} \tag{10}$$

for $a \ge 0, b \ge 0, c \ge 0$ and $p(a, b, c) = 0$ otherwise. Furthermore, let

$$\Phi_k(a, b, c) = \sum_i \sum_j p(a + ik, b + jk, c - (i + j)k) \tag{11}$$

where $p(a, b, c)$ is given by (10) and the sum is extended over all $i, j = 0, \pm 1, \pm 2, \ldots$. We observe that $\Phi_k(a, b, c)$ is invariant under the six permutations of (a, b, c). Moreover,

$$\Phi_k(a + k, b - k, c) = \Phi_k(a, b + k, c - k) = \Phi_k(a + k, b, c - k) = \Phi_k(a, b, c). \tag{12}$$

435

THEOREM 1. *If* $k = 1, 2, \ldots$, *we have*

$$p(n, n, n)\mathbf{P}\{n\delta_3(n) < k\} = \Phi_{3k}(n, n, n) - 3\Phi_{3k}(n + k, n - k, n)+$$

$$\tag{13}$$

$$\Phi_{3k}(n + k, n + k, n - 2k) + \Phi_{3k}(n - k, n - k, n + 2k)$$

where the terms on the right-hand side are defined by (11).

PROOF. The right-hand side of (13) is the probability that in $3n$ steps the particle returns to the initial position $(0, 0, 0)$ and in the meantime never touches any of the three absorbing barriers, $x = k$, $y = k$ and $z = k$. To obtain (13) we form the repeated images of the point $Q(n, n, n) = (0, 0, 0)$ in the three straight lines $x = k$, $y = k$ and $z = k$ in the plane $x + y + z = 0$. The respective images of any point $Q(\alpha, \beta, \gamma) = (\alpha - \beta, \beta - \gamma, \gamma - \alpha)$ in the three lines are

$$Q(\beta + k, \alpha - k, \gamma) = (\beta - \alpha + 2k, \alpha - \gamma - k, \gamma - \beta - k), \tag{14}$$

$$Q(\alpha, \gamma + k, \beta - k) = (\alpha - \gamma - k, \gamma - \beta + 2k, \beta - \alpha - k), \tag{15}$$

and

$$Q(\gamma - k, \beta, \alpha + k) = (\gamma - \beta - k, \beta - \alpha - k, \alpha - \gamma + 2k). \tag{16}$$

We calculate $p(\alpha, \beta, \gamma)$ for each image point $Q(\alpha, \beta, \gamma)$ and form the sum of $\pm p(\alpha, \beta, \gamma)$ for all distinct image points of $Q(n, n, n)$. We take the plus sign if $Q(\alpha, \beta, \gamma)$ can be obtained from $Q(n, n, n)$ by an even number of reflections and take the minus sign if $Q(\alpha, \beta, \gamma)$ can be obtained from $Q(n, n, n)$ by an odd number of reflections. We add also $p(n, n, n)$ to this sum to obtain the right-hand side of (13).

We can also express (13) in the following simpler form.

THEOREM 2. *If* $k = 1, 2, \ldots$, *we have*

$$p(n, n, n)\mathbf{P}\{n\delta_3(n) < k\} = \Phi_k(n, n, n) - 9\Phi_{3k}(n + k, n - k, n) \tag{17}$$

where the right-hand side can be obtained by (11).

PROOF. Let us express $\Phi_k(n, n, n)$ by (11) where the sum extends over all $i, j = 0, \pm 1, \pm 2, \ldots$. Let us divide this sum into nine parts according as $i \equiv 0, 1, -1 \pmod 3$ and $j \equiv 0, 1, -1 \pmod 3$. Thus we obtain that

$$\Phi_k(n, n, n) = \Phi_{3k}(n, n, n) + 6\Phi_{3k}(n + k, n - k, n)+$$

$$\tag{18}$$

$$\Phi_{3k}(n + k, n + k, n - 2k) + \Phi_{3k}(n - k, n - k, n + 2k).$$

436

A comparison of (13) and (18) proves (17). Both terms on the right-hand side of (17) are finite sums. Formula (17) is convenient for numerical calculations.

By (17) we obtain the following limit theorem.

THEOREM 3. *The limit*

$$\lim_{n \to \infty} \mathbf{P}\{\sqrt{n}\delta_3(n) \le x\} = H(x) \tag{19}$$

exists and $H(x)$ is the distribution function of a positive random variable. If $x > 0$, then

$$H(x) = A(x) - 9B(x) \tag{20}$$

where

$$A(x) = \sum_i \sum_j e^{-x^2(i^2+ij+j^2)} \tag{21}$$

and

$$B(x) = \sum_i \sum_j e^{-x^2[(3i+1)^2+(3i+1)(3j-1)+(3j-1)^2]}. \tag{22}$$

If $x \le 0$, then $H(x) = 0$.

PROOF. If $k = [x\sqrt{n}]$ where $x > 0$, then by the multinomial limit theorem

$$\lim_{n \to \infty} \frac{p(n+ik, n+jk, n-(i+j)k)}{p(n,n,n)} = e^{-x^2(i^2+ij+j^2)} \tag{23}$$

for $i, j = 0, \pm 1, \pm 2, \ldots$ and

$$\lim_{n \to \infty} \frac{p(n+k+3ik, n-k+3jk, n-3(i+j)k)}{p(n,n,n)} = e^{-x^2[(3i+1)^2+(3i+1)(3j-1)+(3j-1)^2]} \tag{24}$$

for $i, j = 0, \pm 1, \pm 2, \ldots$. Now let us express the right-hand side of (17) by (11), write $k = [x\sqrt{n}]$ and divide each term by $p(n,n,n)$. Then divide the result into two sums. The first is extended over $|i+j| \le m$ and the second over $|i+j| > m$. If $n \to \infty$, then by (23) and (24) we obtain (19). In the first sum we can form the limit term by term, and for any fixed $x > 0$ the second sum can be made arbitrarily close to 0 if m is chosen sufficiently large.

As we shall see, formulas (21) and (22) can significantly be simplified and also can be determined explicitly. For this purpose we need some results from the theory of binary quadratic forms. The next section contains the necessary auxiliary theorems.

437

4. BINARY QUADRATIC FORMS

First, let us consider the quadratic form

$$Q(x, y) = x^2 + xy + y^2. \tag{25}$$

According to a result found in 1840 by P.G.Lejeune Dirichlet [5] p. 219, the number of representations of a positive integer m as $Q(i, j)$ where $i, j = 0, \pm 1, \pm 2, \ldots$ is $6\, \varepsilon(m)$ where

$$\varepsilon(m) = \sum_{d|m} \left(\frac{-3}{d}\right). \tag{26}$$

In (26)

$$\left(\frac{-3}{d}\right) = \begin{cases} 1 & \text{if } d \equiv 1 \pmod 3, \\ -1 & \text{if } d \equiv 2 \pmod 3, \\ 0 & \text{if } d \equiv 0 \pmod 3 \end{cases} \tag{27}$$

is the Kronecker symbol. See also E. Landau [14] p. 187 and L.E. Dickson [4] p. 78. Accordingly, $\varepsilon(m)$ is equal to the excess of the number of divisors $3r + 1$ of m over the number of divisors $3r + 2$. If $\Re(s) > 1$, then by (26)

$$\sum_{m=1}^{\infty} \frac{\varepsilon(m)}{m^s} = \zeta(s) L_{-3}(s) \tag{28}$$

where $\zeta(s)$ is the Riemann zeta function and $L_{-3}(s)$ is a Dirichlet L-function. If $\Re(s) > 1$, then

$$\zeta(s) = \sum_{m=1}^{\infty} \frac{1}{m^s} \tag{29}$$

and

$$L_{-3}(s) = \sum_{m=1}^{\infty} \left(\frac{-3}{m}\right) \frac{1}{m^s} = \sum_{r=0}^{\infty} \frac{1}{(3r+1)^s} - \sum_{r=0}^{\infty} \frac{1}{(3r+2)^s}. \tag{30}$$

We note that

$$L_{-3}(s) = [\zeta(s, 1/3) - \zeta(s, 2/3)]/3^s \tag{31}$$

for $\Re(s) > 1$ where

$$\zeta(s, a) = \sum_{r=0}^{\infty} \frac{1}{(r+a)^s} \tag{32}$$

is the Hurwitz zeta function defined for $\Re(s) > 1$ and $a > 0$.

Table 1 contains $\varepsilon(m)$ for $m \leq 236$. If m is not included in the table, $\varepsilon(m) = 0$.

Table 1: The function $\varepsilon(m)$

m	1	3	4	7	9	12	13	16	19	21	25	27	28	31	36
$\varepsilon(m)$	1	1	1	2	1	1	2	1	2	2	1	1	2	2	1
m	37	39	43	48	49	52	57	61	63	64	67	73	75	76	79
$\varepsilon(m)$	2	2	2	1	3	2	2	2	2	1	2	2	1	2	2
m	81	84	91	93	97	100	103	108	109	111	112	117	121	124	127
$\varepsilon(m)$	1	2	4	2	2	1	2	1	2	2	2	2	1	2	2
m	129	133	139	144	147	148	151	156	157	163	169	171	172	175	181
$\varepsilon(m)$	2	4	2	1	3	2	2	2	2	2	3	2	2	2	2
m	183	189	192	193	196	199	201	208	211	217	219	223	225	228	229
$\varepsilon(m)$	2	2	1	2	3	2	2	2	2	4	2	2	1	2	2

We can easily calculate $\varepsilon(m)$ by the following observations: If $(m,n) = 1$, then $\varepsilon(mn) = \varepsilon(m)\varepsilon(n)$. If $\alpha \geq 1$, then $\varepsilon(3^\alpha) = 1$. If $p = 3r + 1$ is prime and $\alpha \geq 1$, then $\varepsilon(p^\alpha) = \alpha + 1$. If $p = 3r+2$ is prime, then $\varepsilon(p^\alpha) = 1$ for $\alpha =$ even, and $\varepsilon(p^\alpha) = 0$ for $\alpha =$ odd. We have $\varepsilon(3m) = \varepsilon(m)$ and $\varepsilon(4m) = \varepsilon(m)$ for all $m \geq 1$, and $\varepsilon(2m) = 0$ if $m =$ odd.

If $m = i^2 + ij + j^2$, then $4m = (i - j)^2 + 3(i + j)^2$. Consequently, we can interpret $6\varepsilon(m)$ also as the number of representations of $4m$ as $u^2 + 3v^2$ where $u, v = 0, \pm 1, \pm 2, \ldots$.

Denote by $\gamma(m)$ the number of representations of $3m$ as $(3i+1)^2+(3i+1)(3j+1)+(3j+1)^2$ where $i, j = 0, \pm 1, \pm 2, \ldots$. If $3m = (3i+1)^2+(3i+1)(3j+1)+(3j+1)^2$, then $4m = (3i+3j+2)^2+3(i-j)^2$. Consequently, we can interpret $\gamma(m)$ also as the number of representations of $4m$ as $u^2 + 3v^2$ where $u, v = 0 \pm 1, \pm 2, \ldots$ and $u \equiv 2 \pmod 3$. If $u \equiv -2 \pmod 3$, then the number of solutions of $4m = u^2 + 3v^2$ $(u, v = 0, \pm 1, \pm 2, \ldots)$ is again $\gamma(m)$. If $m \equiv 0 \pmod 3$, then obviously, $\gamma(m) = 0$. If $(m,3) = 1$, then $2\gamma(m) = 6\varepsilon(m)$. Accordingly,

$$\gamma(m) = \begin{cases} 3\varepsilon(m) & \text{if } (m,3) = 1, \\ 0 & \text{if } m \equiv 0 \pmod 3. \end{cases} \tag{33}$$

By (33) we obtain that if $\Re(s) > 1$, then

$$\sum_{m=1}^{\infty} \frac{\gamma(m)}{m^s} = 3\left(1 - \frac{1}{3^s}\right)\zeta(s)L_{-3}(s). \tag{34}$$

Here we used that $\varepsilon(3m) = \varepsilon(m)$ for all $m = 1, 2, \ldots$.

Let us introduce here the following elliptic theta functions which we shall need in what follows.

439

If $|q| < 1$, define

$$\Theta_2(q) = \sum_{m=-\infty}^{\infty} q^{(m-\frac{1}{2})^2} = 2q^{\frac{1}{4}} \sum_{m=0}^{\infty} q^{m(m+1)}, \tag{35}$$

$$\Theta_3(q) = \sum_{m=-\infty}^{\infty} q^{m^2} = 1 + 2 \sum_{m=1}^{\infty} q^{m^2}, \tag{36}$$

and

$$\Theta_4(q) = \sum_{m=-\infty}^{\infty} (-1)^m q^{m^2} = 1 + 2 \sum_{m=1}^{\infty} (-1)^m q^{m^2}. \tag{37}$$

5. THE DISTRIBUTION FUNCTION $H(x)$

By (20) we have

$$H(x) = A(x) - 9B(x) \tag{38}$$

for $x > 0$ where $A(x)$ is defined by (21) and $B(x)$ by (22).

By (26) we obtain that

$$A(x) = 1 + 6 \sum_{m=1}^{\infty} \varepsilon(m) e^{-mx^2} \tag{39}$$

for $x > 0$.

Now we shall prove that

$$B(x) = [A(x) - A(x\sqrt{3})]/6. \tag{40}$$

Let us define

$$C(x) = \sum_i \sum_j e^{-x^2[(3i+1)^2 + (3j+1)^2 + (3i+1)(3j+1)]} \tag{41}$$

for $x > 0$. By (33) and (39) we obtain that

$$C(x) = \sum_{m=1}^{\infty} \gamma(m) e^{-3mx^2} = 3 \sum_{(m,3)=1} \varepsilon(m) e^{-3mx^2} = [A(x\sqrt{3}) - A(3x)]/2 \tag{42}$$

for $x > 0$. Here we used that $\varepsilon(3m) = \varepsilon(m)$ for $m = 1, 2, \ldots$.

If we divide the sum (21) into 9 parts according as $i, j \equiv 0, 1, -1 \pmod 3$, then we obtain that

$$A(x) = A(3x) + 6B(x) + 2C(x) \tag{43}$$

for $x > 0$. By (42) and (43) we obtain (40).

440

Accordingly, we have the following result.

THEOREM 4. *If $x > 0$, then*

$$H(x) = [3A(x\sqrt{3}) - A(x)]/2 \qquad (44)$$

where $A(x)$ is given by (21) or by (39).

PROOF. We obtain (44) by (38) and (40).

We shall see later that

$$A(x) = \frac{2\pi}{\sqrt{3}x^2} A\left(\frac{2\pi}{\sqrt{3}x}\right) \qquad (45)$$

for $x > 0$. If $x \geq (2\pi/\sqrt{3})^{1/2} = 1.904625614...$, then it is convenient to calculate $A(x)$ by (39). If $x < (2\pi/\sqrt{3})^{1/2}$, then it is convenient to use (45). The function $A(x)$ is decreasing for $0 < x < \infty$ and $\lim_{x \to \infty} A(x) = 1$. If $x \to 0$, then $x^2 A(x) \to 2\pi/\sqrt{3}$. We can write that

$$A(x) = \frac{2\pi}{\sqrt{3}x^2} + Q(x) \qquad (46)$$

where $Q(x)$ is the distribution function of a positive random variable. By (46)

$$H(x) = \frac{3Q(x\sqrt{3}) - Q(x)}{2} \qquad (47)$$

for $x > 0$.

We can express the distribution function $H(x)$ explicitly with the aid of the elliptic theta functions $\Theta_2(q)$, $\Theta_3(q)$ and $\Theta_4(q)$ defined by (35), (36) and (37).

Let us introduce the notations

$$K(x) = \Theta_4(e^{-2x^2}) = \frac{\sqrt{2\pi}}{2x}\Theta_2(e^{-\pi^2/(2x^2)}) \qquad (48)$$

and

$$V(x) = \Theta_3(e^{-2x^2}) = \frac{\sqrt{2\pi}}{2x}\Theta_3(e^{-\pi^2/(2x^2)}) \qquad (49)$$

for $x > 0$. Here $K(x)$ is the Kolmogorov distribution function defined by (7) and

$$V(x) = L(x) + \frac{\sqrt{2\pi}}{2x} \qquad (50)$$

for $x > 0$. In (50) $L(x)$ is the distribution function of a positive random variable and is given by

$$L(x) = \frac{\sqrt{2\pi}}{x} \sum_{j=1}^{\infty} e^{-j^2\pi^2/(2x^2)} = 1 - \frac{\sqrt{2\pi}}{2x} + 2\sum_{j=1}^{\infty} e^{-2j^2x^2} \qquad (51)$$

441

for $x > 0$.

We note that

$$K(x) = 2V(2x) - V(x) \tag{52}$$

and

$$V(x) = K(x) + \frac{\sqrt{2\pi}}{2x} K(\frac{\pi}{4x}) \tag{53}$$

if $x > 0$. By (49) we have also

$$V(x) = \frac{\sqrt{2\pi}}{2x} V(\frac{\pi}{2x}) \tag{54}$$

for $x > 0$. The function $V(x)$ is a monotone decreasing function of x in the interval $(0, \infty)$. We have $\lim_{x \to \infty} V(x) = 1$ and $\lim_{x \to 0} V(x)x = \sqrt{\pi/2}$.

THEOREM 5. *We have*

$$A(x) = \frac{1}{2} \left[K(\frac{x}{\sqrt{8}}) K(\frac{x\sqrt{3}}{\sqrt{8}}) + V(\frac{x}{\sqrt{8}}) V(\frac{x\sqrt{3}}{\sqrt{8}}) \right] \tag{55}$$

for $x > 0$ where $K(x)$ is given by (48) and $V(x)$ by (49).

PROOF. Denote by $\psi(m)$ the number of solutions of $m = u^2 + 3v^2$ ($u, v = 0, \pm 1, \pm 2, \ldots$). By (36) we obtain that

$$1 + \sum_{m=1}^{\infty} \psi(m) q^m = \Theta_3(q) \Theta_3(q^3) \tag{56}$$

for $|q| < 1$. We have already seen that $\psi(4m) = 6\varepsilon(m)$ for $m = 1, 2, \ldots$. Evidently, $\psi(2m) = 0$ if $m =$ odd. ($\psi(m) = 2\varepsilon(m)$ if $m \not\equiv 0 \pmod 4$.) Since $\Theta_3(-q) = \Theta_4(q)$, by (56) we obtain that

$$1 + \sum_{m=1}^{\infty} \psi(2m) q^{2m} = \frac{1}{2}[\Theta_3(q)\Theta_3(q^3) + \Theta_4(q)\Theta_4(q^3)] \tag{57}$$

or

$$1 + \sum_{m=1}^{\infty} \psi(4m) q^{4m} = \frac{1}{2}[\Theta_3(q)\Theta_3(q^3) + \Theta_4(q)\Theta_4(q^3)]. \tag{58}$$

If we put $q = e^{-x^2/4}$ in (58), we obtain (55).

To calculate $A(x)$ numerically, we can conveniently use the program MATHEMATICA by Wolfram Research [17]. In this program the elliptic theta functions are built-in functions.

Tables 2 and 3 contain $H(x)$ and $H'(x)$ for $0 < x < 3$ and Figures 1 and 2 depict the graphs of $H(x)$ and $H'(x)$.

If in (55) we express $K(x)$ by (52) and apply (54), then we obtain (45).

442

Table 2: The distribution function $H(x)$

x	$H(x)$	x	$H(x)$	x	$H(x)$	x	$H(x)$
0.05	0.000000	0.80	0.017945	1.55	0.732770	2.30	0.984875
0.10	0.000000	0.85	0.034768	1.60	0.770750	2.35	0.988013
0.15	0.000000	0.90	0.059754	1.65	0.804516	2.40	0.990547
0.20	0.000000	0.95	0.093424	1.70	0.834273	2.45	0.992582
0.25	0.000000	1.00	0.135429	1.75	0.860288	2.50	0.994209
0.30	0.000000	1.05	0.184692	1.80	0.882862	2.55	0.995501
0.35	0.000000	1.10	0.239640	1.85	0.902313	2.60	0.996522
0.40	0.000000	1.15	0.298463	1.90	0.918962	2.65	0.997325
0.45	0.000000	1.20	0.359325	1.95	0.933121	2.70	0.997953
0.50	0.000001	1.25	0.420531	2.00	0.945090	2.75	0.998441
0.55	0.000018	1.30	0.480616	2.05	0.955146	2.80	0.998819
0.60	0.000154	1.35	0.538403	2.10	0.963545	2.85	0.999110
0.65	0.000798	1.40	0.593006	2.15	0.970521	2.90	0.999332
0.70	0.002876	1.45	0.643813	2.20	0.976282	2.95	0.999501
0.75	0.007941	1.50	0.690457	2.25	0.981012	3.00	0.999630

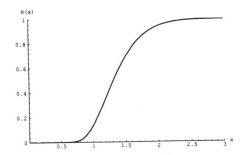

Figure 1: The distribution function $H(x)$

443

Table 3: The density function $H'(x)$

x	$H'(x)$	x	$H'(x)$	x	$H'(x)$	x	$H'(x)$
0.05	0.000000	0.80	0.262626	1.55	0.802723	2.30	0.069566
0.10	0.000000	0.85	0.414870	1.60	0.716889	2.35	0.056336
0.15	0.000000	0.90	0.586311	1.65	0.634417	2.40	0.045373
0.20	0.000000	0.95	0.759268	1.70	0.556761	2.45	0.036345
0.25	0.000000	1.00	0.917267	1.75	0.484847	2.50	0.028956
0.30	0.000000	1.05	1.047907	1.80	0.419179	2.55	0.022945
0.35	0.000000	1.10	1.143913	1.85	0.359938	2.60	0.018084
0.40	0.000000	1.15	1.202827	1.90	0.307061	2.65	0.014176
0.45	0.000002	1.20	1.225992	1.95	0.260316	2.70	0.011054
0.50	0.000069	1.25	1.217323	2.00	0.219351	2.75	0.008573
0.55	0.000890	1.30	1.182163	2.05	0.183741	2.80	0.006614
0.60	0.005757	1.35	1.126377	2.10	0.153021	2.85	0.005075
0.65	0.023035	1.40	1.055707	2.15	0.126711	2.90	0.003874
0.70	0.065334	1.45	0.975368	2.20	0.104334	2.95	0.002941
0.75	0.143967	1.50	0.889826	2.25	0.085431	3.00	0.002221

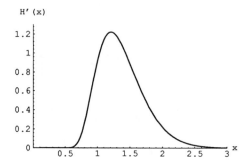

Figure 2: The density function $H'(x)$

444

Table 4: The moments M_r

r	M_r	r	M_r
1	1.366756642	6	17.002350808
2	1.992662271	7	33.719318653
3	3.099750204	8	70.540540006
4	5.140763822	9	155.129203720
5	9.073696213	10	357.418614734

The moments M_r, defined by (4), exist for every $r = 0, 1, 2, \ldots$. We have $M_0 = 1$ and

$$M_r = r \int_0^\infty x^{r-1}[1 - H(x)]\,dx \tag{59}$$

if $r \geq 1$. By (39) and (44) we obtain that

$$M_1 = -\frac{3\sqrt{\pi}}{2}(\sqrt{3} - 1)\zeta(\frac{1}{2})L_{-3}(\frac{1}{2}), \tag{60}$$

$$M_2 = \frac{\pi \log 3}{\sqrt{3}} \tag{61}$$

and

$$M_r = \frac{3r}{2}\left(1 - \frac{3}{3^{r/2}}\right)\Gamma(\frac{r}{2})\zeta(\frac{r}{2})L_{-3}(\frac{r}{2}) \tag{62}$$

for $r \geq 3$ where $\zeta(s)$ is the Riemann zeta function defined by (29) and $L_{-3}(s)$ is the Dirichlet L-function defined by (30). Table 4 contains M_r for $r \leq 10$.

6. JOINT DISTRIBUTIONS

In the same way as we proved Theorem 1 we can determine the joint distribution of the random variables $\delta_{1,2}(n)$, $\delta_{2,3}(n)$ and $\delta_{3,1}(n)$. If k, ℓ and m are positive integers, we have

$$p(n, n, n)\mathbf{P}\{n\delta_{1,2}(n) < k, n\delta_{2,3}(n) < \ell, n\delta_{3,1}(n) < m\} =$$

$$\Phi(n, n, n) - \Phi(n + k, n - k, n) - \Phi(n + \ell, n - \ell, n) - \tag{63}$$

$$\Phi(n + m, n - m, n) + \Phi(n + k, n + \ell, n - k - \ell) + \Phi(n - k, n - m, n + k + m)$$

445

where
$$\Phi(a, b, c) = \Phi_{k+\ell+m}(a, b, c) \tag{64}$$

is defined by (11). The limit of the joint distribution function of $\sqrt{n}\delta_{1,2}(n)$, $\sqrt{n}\delta_{2,3}(n)$ and $\sqrt{n}\delta_{3,1}(n)$ as $n \to \infty$ can be obtained from (63) by using the multinomial limit theorem.

If in (63) $m > n$, then we obtain the following marginal distribution

$$p(n, n, n)\mathbf{P}\{n\delta_{1,2}(n) < k \text{ and } n\delta_{2,3}(n) < \ell\} =$$

$$p(n, n, n) - p(n + k, n - k, n) - p(n + \ell, n - \ell, n) + \tag{65}$$

$$p(n + k, n + \ell, n - k - \ell) - p(n + k + \ell, n - k - \ell, n) + p(n - k, n - \ell, n + k + \ell).$$

If in (65) we put $k = [x\sqrt{n}]$, $x > 0$ and $\ell = [y\sqrt{n}]$, $y > 0$ and let $n \to \infty$, we obtain that

$$\lim_{n\to\infty} \mathbf{P}\{\sqrt{n}\delta_{1,2}(n) \le x \text{ and } \sqrt{n}\delta_{2,3}(n) \le y\} =$$

$$\tag{66}$$

$$1 - e^{-x^2} - e^{-(x+y)^2} - e^{-y^2} + 2e^{-(x^2+xy+y^2)}$$

This formula is in agreement with a result of V. Ozols [15] found in 1956.

For any $r > 3$ the joint distribution of $\delta_{1,2}(n), \delta_{2,3}(n), \ldots, \delta_{r-1,r}(n), \delta_{r,1}(n)$ can be determined in a way similar to the proof of Theorem 1. For $r > 3$ we consider a random walk on the r-dimensional hyperplane $x_1 + x_2 + \cdots + x_r = 0$. The particle starts at the origin $(x_1, x_2, \ldots, x_r) = (0, 0, \ldots, 0)$ and takes rn steps. In each step, independently of the others, it moves a distance $\sqrt{2}$ in one of r specified directions determined by r vectors which are the r cyclic permutations of $(1, 0, \ldots, 0, -1)$. The successive displacements are independent and in each step the r vectors are equally probable. To find the probability that $n\delta_{1,2}(n) < k_1, n\delta_{2,3}(n) < k_2, \ldots, n\delta_{r-1,r}(n) < k_{r-1}, n\delta_{r,1}(n) < k_r$ we should determine the probability that in the random walk the particle returns to the origin in rn steps and in the meantime it never touches the barriers $x_1 = k_1, x_2 = k_2, \ldots, x_r = k_r$ and divide this probability by the probability that starting from the origin the particle returns to the origin in rn steps. The probability that during the rn steps the particle remains inside the $(r - 1)$-dimensional simplex $x_1 = k_1, x_2 = k_2, \ldots, x_r = k_r$ can be calculated by using the reflection principle. The appropriate limit distribution function can be derived by using the multinomial central limit theorem.

In particular, for the marginal distribution of $\delta_r(n)$ we have the following limit theorem

$$\lim_{n\to\infty} \mathbf{P}\{\sqrt{n}\delta_{1,2}(n) \le x, \ldots, \sqrt{n}\delta_{r-1,r}(n) \le x\} = \prod_{i=1}^{r-1} \left(1 - e^{-ix^2}\right)^{r-i} \tag{67}$$

for $x > 0$. A direct proof of (67) can be accomplished by using a result of S. Karlin and J. McGregor [11].

446

In calculating the tables and printing the graphs of $H(x)$ and $H'(x)$ I used the program MATH-EMATICA of Wolfram Research [17].

REFERENCES

[1] L.C. Chang and M. Fisz: Asymptotically independent linear functions of empirical distribution functions. *Science Record* 1 (1957) 335-340.

[2] L.C. Chang and M. Fisz: Exact distributions of the maximal values of some functions of empirical distribution functions. *Science Record* 1 (1957) 341-346.

[3] H.T. David: A three-sample Kolmogorov-Smirnov test. *The Annals of Mathematical Statistics* 29 (1958) 842-851.

[4] L.E. Dickson: *Introduction to the Theory of Numbers.* University of Chicago Press, 1929. [Reprinted by Dover, New York, 1957.]

[5] P.G. Lejeune Dirichlet: *Vorlesungen über Zahlentheorie.* Fourth edition. Braunschweig, 1893. [Reprinted by Chelsea, New York, 1968.]

[6] M. Fisz: A limit theorem for empirical distribution functions. *Bulletin de l'Académie Polonaise des Sciences Cl. III.* 5 (1957) 695-698.

[7] M. Fisz: A limit theorem for empirical distribution functions. *Studia Mathematica* 17 (1958) 71-77.

[8] M. Fisz: Some non-parametric tests for the k-sample problem. *Colloquium Mathematicum* 7 (1960) 289-296.

[9] I.I. Gikhman: On a nonparametric homogeneity criterion for k samples. *Theory of Probability and its Applications* 2 (1957) 369-373.

[10] B.V. Gnedenko and V.S. Korolyuk: On the maximum discrepancy between two empirical distributions. (Russian) *Doklady Akademii Nauk SSSR* 80 (1951) 525-528. [English translation: *Selected Translations in Mathematical Statistics and Probability.* Vol. 1. American Mathematical Society, Providence, RI, 1961, pp. 13-16.]

[11] S. Karlin and J. McGregor: Coincidence probabilities. *Pacific Journal of Mathematics* 9 (1959) 1141-1164.

[12] J. Kiefer: K-sample analogues of the Kolmogorov-Smirnov and Cramér-v. Mises tests. *The Annals of Mathematical Statistics* 30 (1959) 420-447.

[13] A. Kolmogoroff: Sulla determinazione empirica di una legge di distribuzione. *Giornale dell' Istituto Italiano degli Attuari* 4 (1933) 83-91.

[14] E. Landau: *Elementary Number Theory.* Chelsea, New York, 1958.

[15] V. Ozols: Generalization of the theorem of Gnedenko-Korolyuk to three samples in the case of two one-sided boundaries. (Latvian. Russian summary) *Latvijas PSR Zinātņu Akadēmijas Vēstis.* No. 10 (111) (1956) 141-152.

447

[16] N.V. Smirnov: On the estimation of the discrepancy between empirical curves of distribution for two independent samples. *Bulletin Mathématique de l'Université de Moscou. Série Internationale.* 2 No. 2 (1939) 3-16.

[17] S. Wolfram: *Mathematica. A System for Doing Mathematics by Computer.* Second edition. Addison-Wesley, Redwood City, California, 1991.

Lajos Takács
2410 Newbury Drive
Cleveland Heights, Ohio 44118

Lecture Notes in Statistics

For information about Volumes 1 to 31
please contact Springer-Verlag

PM835-3-TN
38